Texts in Computing

Volume 19

Computational Logic

Volume 1:

Classical Deductive Computing with Classical Logic

Third Edition

Texts in Computing Series Editor
Ian Mackie   mackie@lix.polytechnique.fr

Computational Logic

Volume 1:

Classical Deductive Computing with Classical Logic

Luis M. Augusto

© Individual author and College Publications 2018. All rights reserved.
Second edition, 2020, Third edition, 2022.

ISBN 978-1-84890-280-0

College Publications
Scientific Director: Dov Gabbay
Managing Director: Jane Spurr

http://www.collegepublications.co.uk

Cover produced by Laraine Welch

Contents

Contents

List of Figures

List of Figures

List of Tables

List of Algorithms

Preface to the first edition

It is often the case that computer science is considered merely a branch of mathematics. This (still) often motivates the belief that logic is required for computer science just because it is required for mathematics, namely for proofs. However, logic in computing goes well beyond the context of mathematical proof, being present today in fields such as artificial intelligence and cognitive science, and having significant engineering and industrial applications. This impressive plethora of computational applications of *logic* could not be possible without a large variety of *logics*, which for our purposes can be elegantly–i.e. by means of the English connector *and*–segregated in two major classes: *classical logic(s)* and *non-classical logics*.

Yet another, but perhaps not so elegant, segregation must be contemplated when speaking of computing today: *classical computing* or *non-classical computing*. While in the latter kind one can include a large variety of computation models and computers (e.g., quantum computers, artificial neural networks, evolutionary computing), we shall consider *classical computing* to be the processing of information carried out by the von Neumann, or industrial-scale digital computer, which has as a major theoretical foundation the Turing computing paradigm. This paradigm, concretized in the Turing machine, sees computation as a spatial-temporal discrete business over symbols that can best be carried out in binary code. While this paradigm does not take into account the resources available for computation, the von Neumann computer is in fact constrained by physical–i.e. spatial and temporal–resources, which means that classical computing has more or less clearly established limitations.

When logic, whether classical or non-classical, is applied in computing, either classical or non-classical, we speak of *computational logic*. This is an important label in at least two senses. Firstly, it captures the fact that there is a subfield of formal logic that can be applied in a computational setting. This subfield might be obtained by imposing restrictions (for example, on the sets of operators), but also by extensions or just plain variations. Secondly, it helps us to distinguish clearly between computation carried out with a *logical language* from computation carried out with *other* formal languages. In effect, while the latter typically is concerned with preserving the legality of symbol strings (legal strings are processed into further legal strings), the former often aims at *truth-preservation*. Say that we have a theory and wish to know whether some assertion follows logically from it, i.e. belongs to it, or is true in it. The deduction theorem allows us to express this *logical following* in a single symbol string, known as a logical formula, and our question is notoriously best concretized in the *validity* and *satisfiability problems*, which ask whether a logical formula is always true, or is true in some interpretation, respectively. When these problems–in particular the latter–are posed in a computational context, we accordingly speak of *deductive computation*. When the computational solution is to be found by means of classical computing, we then speak of *classical deductive computing*.

In this book we elaborate on *classical deductive computing with classical logic*, and we do so without a specific regard to the field of application. Our foci are first and foremost two main subjects in which classical deductive computing with classical logic has a prominent role: *automated theorem proving* and *logic programming*.

This is thus a book on *applied logic*. Furthermore, this is a book on *applied mathematical logic*. We take here the label *mathematical logic* as synonymous with *formal logic*, and this in a very narrow sense: Formal logic is logic whose foundations lie in mathematical objects and structures. Although these mathematical foundations may be inconspicuous at the object-language level, at the metalanguage level they do become more conspicuous or even explicit. Interestingly enough–though not surprising anymore–, the mathematical structures and objects usually required in mathematical logic are precisely those needed for classical deductive computing; we talk here of lattices, graphs, trees, etc., all known as *discrete structures and objects*. This accounts for a whole chapter (Chapter 1) dedicated to the topics of discrete mathematics required for a satisfactory grasping of the material in this book. More specifically, we restrictively provide the mathematical notions that are foundational for both the theory of classical computing and classical deduction. Chapter 1 constitutes Part II of this volume, Part I being the Introduction.

Were this book on formal logic alone, there would be no need for a chapter on the *theory of computing*. Although logical languages are first and foremost formal languages, outside a computational context no issues of computability or complexity arise–certainly not in the usual treatment of logic for philosophy courses, but not even in pure mathematical logic textbooks. These issues arise when we need to compute with logical languages (e.g., Turing-completeness of programming languages). Because these issues arise here, we need to approach Turing machines, which, in turn, require the fundamentals of formal languages and models of computation, in order to be satisfactorily understood. We thus provide the basics of the *general* theory of classical computing, which includes the study of formal languages and grammars, models of computation, and computability theory. As a matter of fact, we provide more than the basics, doing so in the belief that such knowledge often comes in handy for anyone interested in computational logic. This material constitutes Chapter 2, which is Part III.

This book is one–the first–of two volumes addressing the topic of classical deductive computing. In it we focus on computing with classical logic. Although new technologies have opened a path that led to a proliferation of *new* logics, the so-called non-classical logics, classical logic remains as the standard logical system which the other, newer, systems extend or from which they diverge. This would be reason enough to justify this volume, but the fact is that, despite the many technological advances witnessed in the last decades, classical logic is still the logical system of choice for many technological applications requiring what in this book we call deductive computation.

Although the literature on classical logic is prolific, with many good introductions to the subject, with self-containment in view we provide a whole chapter (Chapter 4) on *classical logic*. This follows a comprehensive discussion on *formal logic, deduction,* and *deductive computation* carried out in Chapter 3, in which such fundamental notions as logical language, from the viewpoints of both form and meaning, and logical consequence, in relation to inference and deductive systems, as well as to computation, are thoroughly discussed.

The decision problem in computational logic is overwhelmingly tackled by checking for (un)satisfiability, namely by means of the so-called SAT testers or solvers. However, we thought that a working knowledge of *classical validity testing methods* is also required. These–the *classical calculi*–we present in Chapter 5, which is followed, in Chapter 6, by the different *semantics* that provide a foundation for *meaning in classical logic*.

Chapters 3 to 6, constituting Part IV of this book, comprise our discussion of classical deduction and classical logic.

In Part V, we begin by elaborating on the *(classical) satisfiability problem*, already introduced in Chapter 2, and by providing the means to computerize classical logic with a view to finding computational solutions to this problem. This *satisfiability testing* is extensively discussed in the remaining Sections of this Chapter 7. We then proceed with extensive treatments of the aforementioned main fields of computational logic, to wit, automated theorem proving (Chapter 8) and logic programming (Chapter 9). With respect to the former, we give an equal weight to resolution and analytic tableaux. This is uncommon, as the resolution calculus has all but obliterated the analytic tableaux calculus in the context of automated theorem proving, but we think this obliteration is not justified and hope to contribute to the reassessment of the pay-offs of further automating the analytic tableaux calculus. Precisely due to this imbalance our treatment of this calculus is not as comprehensive as our elaboration on resolution. As far as logic programming is concerned, we naturally focus on Prolog, as this is the major (family of) language(s) in this programming paradigm. It is our belief that by mastering the essential aspects of Prolog related to its deductive capabilities, as well as the general theory of logic programming, the reader will be well equipped to tackle most tasks involving this programming paradigm, as well as other (sub-)languages thereof, such as Datalog and Answer Set Programming.

We restrict our elaboration on classical computing to first-order predicate logic, which is known to be adequate (i.e. sound and complete) and as such provides us with a reliable means for classical deductive computation. This by no means entails that we disfavor higher-order logics, but we leave their inclusion in this text to possible future editions thereof.

As said above, this is the first of two volumes. Born in the late 1960s / early 1970s, computational logic has quickly grown to have many subfields or subjects; many, indeed (see Introduction). Clearly, this proliferation cannot be covered by a single volume, and we decided to divide the material we find essential in two volumes, the main segregation between both being that we dedicate this (first) volume to computing with classical logic, and we shall elaborate on computation with non-classical logics in a second volume. This segregation is justified not only by the fact that classical and non-classical logics have very different computational assumptions and applications, but also by the sheer quantity of topics that need to be addressed; a single book would certainly be too voluminous and readers may be interested in only one of these, classical or non-classical logics.

An advantage of this project over other works in the field is the breadth of its covering: The reader has in it far more content on computational logic than is usually the case in a single monograph or textbook. This, like any advantage, comes at a price, though: Depth had to be relinquished. This is, however, remediated by bibliographical references to works of a more limited breadth but with greater depth of treatment. Moreover, this work contains a large selection of exercises on all the approached topics. Having in mind both that most specialized monographs and handbooks lack any exercises and the large variety of topics here approached, this is indeed yet another advantage, at least for the reader of a more practical persuasion. In our selection of exercises we included novel material (e.g., theorems not given in the main text), so that the reader is expected also to approach problems in computational logic in a creative way. Exercises asking the reader to reflect on some statements or passages, as well as to engage in research, are also included. These latter exercises are meant to complement the main text with some topics that, while not being secondary, would require some extended discussion, making of this a much larger volume.

Some final remarks: Some of the material in this volume draws on two books of ours also published in College Publications, to wit, Augusto (2020a, b). This material either is as was first published, or has been submitted to some, often substantial, revisions and extensions. As was or revised/extended, it is mostly to be found in Chapters 1, 7, and 8, as well as in all Chapters of Part IV, though not in all Sections thereof. Chapters 2 and 9, as well as many Sections in Part IV (e.g., Sections 3.1-3), are completely novel, drawing only from folklore or from works by other authors. These are orthodoxly cited and indicated in the bibliographical references, but not always did we see it necessary to do so, especially with respect to material that has to some extent already acquired the character of mathematical or logical folklore.

Being a book on computational logic, this is, as said,–also–a book on mathematical logic. This explains the usual distinction in the main text of statements into definitions (abbreviated **Def.**), propositions (**Prop.**), and the odd undistinguished paragraph that for ends of internal reference is referred to as "§"; these are all given a number indicating the Section (two digits separated by a dot) and the order in the Section. For example, **2.1.3 (Def.)** indicates Definition 3 in Section 2.1. Theorems, as well as their companion lemmas and corollaries, are numbered in the same way but separately from the other numbered statements, and the same holds for examples. Exercises are numbered according to not only Section, but also Subsection.[1]

It is usual to provide the reader with a schematic guide for the reading of a book in the fields that are our foci. With this in mind, but not wishing to direct the reader more than the Table of Contents already is expected to do, we think that in order for the lay reader to have a *minimal* satisfactory grasping of classical deductive computing with classical logic the following topics are essential: The system of classical

---

[1]In the present 3rd edition this has been standardized as, for example, **Definition. 2.1.3.**, and the references by "§" were replaced by *Remarks*.

logic CL and the logic **CL** (Chapter 4), Herbrand semantics (concentrated in Sections 6.2 and 7.3.3), and Sections 7.1-2 for the satisfiability problem and for the necessary means to make logical formulae of CL amenable to computation. These are *sine-qua-non* requirements for a good understanding of automated theorem proving (Chapter 8) or logic programming (Chapter 9), or both. The novice reader wishing to gain a full grasp of our main topic cannot eschew the reading of the whole volume. It should be remarked, however, that some Chapters are self-standing in the sense that they can be used independently from the rest of the volume. This is particularly true of Chapter 2, which is largely conceived as a condensed treatment–with the usual selection of exercises–of the theory of classical computing, and thus can be of use for readers whose interest might fall exclusively on this topic.

For reasons to do with time, we do not include solutions to any of the exercises in this edition, but sooner or later they are expected to be provided, either online or in later editions. Readers wishing to contribute with original solutions to problems other than the most basic ones (e.g., proofs of theorems) are welcome to contact me for this end.

My thanks go to Dov M. Gabbay for including this work in this excellent series of College Publications, and to Jane Spurr for her usual impeccable assistance in the publication process.

Madrid, June 2018

Luis M. S. Augusto

Preface to the second edition

The first edition of the present work was rather hastily completed for many reasons. This hastiness contributed to addenda and errata lists longer than I feel comfortable with, as well as to the omission of some contents that I consider important in a comprehensive introduction to the large field of classical deductive computing with classical logic. Thus, this second edition improves on the first by both eliminating (hopefully most) addenda and errata, and including the mentioned contents. These are largely constituted by Datalog, on which I elaborate at length in a wholly new chapter (Chapter 9.3) for mainly two reasons: Firstly, Datalog has an intrinsic interest from the viewpoint of databases, thus expanding on the applications of logic programming; secondly, it provides an important illustration of the equation *Algorithm = Logic* in computational logic, to be contrasted with the case of Prolog, which concretizes the equation *Algorithm = Logic + Control*. On a more personal level, Datalog is a highy rewarding topic to research into; more specifically, how such a frugal logical language as Datalog can call for impressively complex formal semantics promises to keep researchers busy for a long time to come.

A few more exercises, in particular exercises aiming at connecting Part III and Parts IV-V, were added in this edition. Further minor improvements were made by redrawing some of the figures and by making minor changes to the main text.

Madrid, January 2020

Luis M. S. Augusto

Preface to the third edition

The present edition–the third–of this course book has an increased focus on algorithms, both in the theory of formal languages and automata and in computational logic. There are now fourteen clearly isolated algorithms in this book, all designed in pseudo-code for the sake of generality. This focus is also shown in the addition of a wholly new section (Section 1.4) on algorithms and programs.

Some new concepts were introduced, whether as definitions or in exercises, sometimes with additional examples, and in order to keep the book within a manageable size the contents on deterministic context-free languages and $LR(k)$ grammars were almost entirely removed, as they are not required in the rest of the book and are more relevant to compilation than to basic formal language theory.

Given the importance of the Turing machine for the field of computational logic the contents of Section 2.2.3 were largely revised and more examples of, or exercises with, Turing machines are now provided.

I thank College Publications, London, for the willingness to publish this new edition so shortly after the second one.

Madrid, January 2022

Luis M. S. Augusto

Part I.

Introduction

0.1. Symbolic computation and classical computing

By *computation* in this book we mean the process of, given some information (the *input*), obtaining some further information (the *output*) by means of successive operations and/or changes in a closed information-processing system. By *closed system* it is meant that the output has a strict affinity with the input. This closure is not a necessary requirement for a definition of computation. A biological organ(ism) is essentially a computing system, in the sense that it is an information-processing system, but it is often open to the environment, exchanging information of different kinds with it. For instance, the human eye first receives photons and processes them ultimately into visual images. These visual images can be stored in memory and recalled in a format that clearly is no longer a visual image, but rather a mental image thereof. The relation between photons and a visual image is a tricky psychological question, and that of a visual image and the mental image thereof is a knotty philosophical one, but by *strict affinity* we mean that the output is of the same kind as the input. In a broader sense, we mean that both the input and the output are, say, symbolic. In a stricter sense, we have for instance a string of symbols as input resulting in a string of symbols as output.

A traditional artificial neural network, for instance, is a computing system, but here we shall not be concerned with it, because our focus is on *symbolic computation*, i.e. computation as carried out on (strings of) symbols.[1] Having thus restricted the definition of computation as a symbolic affair, we can relax this definition to include computing systems that do not provide a concrete output, but merely operate on symbol strings. For instance, a finite automaton need not produce a string of symbols when accepting or rejecting an input string of symbols, but its working remains symbolic through and through. We can further relax this definition and consider computation in a partially closed system, as long as it carries out at least one processing of information that is closed in the senses above. This being so, we can include humans in our definition of computation. Humans are open information-processing systems, but a human solving an equation is a closed computing system, as s/he processes an input symbol string, say, "$2x = 4{+}2$" into the output symbol string "3" by applying successive operations over the input and the successively resulting strings ($2x = 6; x = 6 \div 2; x = 3$).

---

[1] By "traditional," we mean here an artificial neural network operating solely quantitatively. Hybrid, quantitative-symbolic, networks already exist.

Symbolic computation (henceforth often just computation) can be carried out by a plethora of different machines, both physical and abstract, called *computers*. The physical computing machines are commonly referred to as *hardware*, and the abstract ones are typically *(mathematical) models* of computation. Despite the large variety of computers, of which the human brain–the "wetware"–is one, in this book we shall be mostly concerned with the digital computer and with other, more abstract, discrete structures that can be seen as models of computation related to the digital computer. The models of computation tightly associated to the digital computer that we shall be concerned with here are the Turing machine, concretizing the logic-operational model of computation, and the Kleene-Church more mathematical model. As will be seen, these two models of computation are in fact equivalent.

We emphasize now their importance and briefly specify how these models are approached in this book. We begin with the mathematical bases of our definition of computation and move on then to the simpler abstract models up to the more powerful Turing machine.

Information in a von Neumann computer–the commonly marketed digital computer[2]–is stored and processed all-in-all in a discrete manner. This entails the well-known fact that the mathematics required for computing is essentially the branch called *discrete mathematics*. In particular, computer languages just are sets of strings, which means that special emphasis needs to be taken with respect to sets and their associated notions (e.g., relations and functions). Also, automata can be defined in terms of labeled digraphs, a particular kind of discrete structures. The fundamental computational results concerning the (recursive) enumerability of sets are based on the discreteness of the integers. Etc. We accordingly provide, in Chapter 1, the topics of discrete mathematics that are mandatory in this book.

These topics are mandatory not only because symbolic computation is essentially based on discrete mathematical objects and structures, but also because discrete mathematical objects and structures are at the very bases of the *metatheory of classical logic* (*mathematical logic*, in short), the language we are mostly concerned with for computation ends. Some examples of this "foundationalism," which at the same time is a binary or bivalent "foundationalism," are classical logical consequence defined as a complete lattice based on the power set of $\{0,1\}$, the characteristic function as mapping to the set $\{0,1\}$, classical proofs as labeled binary trees (in the tableaux and resolution calculi, for instance). Etc.

---

[2]Henceforth, if the word "computer" is used without further specification, the reader may assume we mean the digital computer.

The three core topics of formal languages, complexity, and computability constitute the large field of the theory of (symbolic) computation. When this is considered in the perspective of the Turing paradigm together with physical constraints, we then speak of *classical computation* or *computing*. It is well known that the syntactical structure of most programming languages is represented by the kind of formal grammar known as context-free. However, regular languages are also important for programming languages, and issues of complexity and computability are best treated at the level of recursively enumerable languages, associated to the famous Turing machines. Hence, we saw no alternative to discussing the whole Chomsky hierarchy in some detail–with a single exception to the context-sensitive languages that, as is usual in the field, are more briefly discussed.

A final note: We could be here concerned with the human brain, or just with human computers, as most logic books implicitly are, were it not for the fact that humans are unreliable computers for many reasons, not the least of which is error-proneness, especially so when confronted with high levels of symbolic complexity. And not because biological neurons are hopelessly slower than electronic logic gates; as we shall see, symbolic complexity, as considered in classical computing, can be such as to make even the most powerful computer useless in practical terms.

0.2. Logic: Formal, symbolic, deductive, and classical

Logic is the science of reasoning. As a science, it can be formal or informal, the main distinction being that formal logic is largely a field of mathematics and informal logic is more cognate with psychology. *Formal logic*, i.e. logic whose foundations are mostly–but not only–to be found in mathematics, describes rigorous principles and processes of reasoning that can be equally followed by either human or artificial agents. In particular, it is highly appropriate for the implementation of reasoning processes in the form of *algorithms*. It is not certain that humans can think in a non-algorithmic way, i.e. without following a series of well-defined successive steps that end in the attainment of a goal or in failure thereof, but *informal logic* concerns itself also with, say, heuristic and commonsense reasoning, types of reasoning that are perhaps not wholly amenable to our conception of algorithms.

In fact, this is not an exclusive distinction, as the contents of formal and informal logic often overlap. Heuristic and commonsense reasoning, a topic often seen as belonging to informal logic, are adequate for contexts characterized by uncertainty, inconsistencies and lacunae in knowledge, approximation goals, etc. Clearly, this characterizes most of human reasoning; after all, we are *satisficers* rather than *optimizers*, perhaps innately so but also because we are constrained by temporal and spatial limitations. But this type of reasoning can be, and has been, formalized, namely with computational implementations in view in contexts with the features above. In effect, in order to be computationally implemented reasoning has to be formalized, i.e. given an unequivocal, contentless form, such as symbol strings, on which mathematical operations can be carried out mechanically. Thus, rather that being interested in a vague distinction between formal and informal logic, we are more concerned with *symbolic logic*, i.e. logic as formalized by means of symbols such as constants, variables, predicates, logical operators, and quantifiers. But because a symbolic logical language is essentially a formal language, with a well-defined vocabulary and a precisely-ruled grammar, we wish to keep the label "formal logic" as a synonym for symbolic logic.

Our main reason to choose symbolic over some non-symbolic, or non-formalized (which, as seen, does not mean necessarily "informal") logic, is that systems of symbolic logic can be worked with as calculi, and their symbols are, or can be, readily accepted by computer devices for many and diverse implementations, from the calculation of basic arithmetical functions to, say, the diagnosis of clinical conditions presenting complex, often tentatively defined, symptoms.

We thus settled that for us symbolic logic and formal logic are synonyms. Having settled this, we have now at least two major further distinctions to consider: We need to address what makes a logic deductive vs. non-deductive, and we have to segregate classical logic from its rivals or extensions, the so-called non-classical logics. Luckily, this is an easier task, compared to the above, due to the less vague character of these distinctions.

We begin with *deduction*. Although this term is itself not unequivocal, we can say that *deductive logic* is *truth-preserving logic* in the sense that, given some argument, if the *premises* are true, then the *conclusion*, or *consequence*, is necessarily true, too. As the famous argument runs, if all humans are mortal and Socrates is a human (the premises), then–necessarily–Socrates is mortal (the conclusion). In fact, the necessary character of this conclusion holds also by this argument's form alone: A conclusion is necessarily deducible if there is a succession of (sub)proofs ending in it.[3]

---

[3]This is a less intuitive feature than truth, and we thus keep the latter for the

However, conclusions may follow logically from premises without doing so necessarily or preserving truth; this is the case of abductive and inductive logics, the two main *non-deductive logics*.

In effect, in *abductive logic*, a conclusion follows from a set of premises as an explanation. For instance, let us say that on waking up John sees that the lawn is wet. John is in possession of a few facts: it is the rainy season; the roofs of houses and the street pavements are also wet; there has been no flood or tsunami; no one has watered the lawn. From these facts, that work as premises for John, he can conclude that the lawn is wet because it rained–more correctly, it must have rained, reason why the lawn is wet. But this is just a (best) explanation, not guaranteeing at all that if the premises are true, and the lawn is indeed wet, then necessarily it rained; dew might be an equally plausible explanation, though no facts concerning this phenomenon feature in John's database.

In *inductive logic*, a conclusion follows from a set of premises not in terms of truth, but in terms of strength, i.e. an inductive argument is more or less strong. For instance, one has often seen black crows, and in fact only black crows, so that one may feel inclined to conclude that all crows are black. The strength of the argument lies, in this case, on how large the number of crows one has seen is. Other than strength, plausibility and probability can also be factors considered in inductive arguments.

We now move on to what is arguably the greatest distinction in modern logic. *Classical logic* is, above all, *bivalent*: in it, a statement is either true or false. This entails that there are only two truth values, to wit, truth and falsity, reason why we can also say that classical logic is *Boolean*. This bivalence is then expressed in the *law of excluded middle*, stated as, given a proposition or statement P, either P or not-P holds (or is true). Classical logic is also *truth-functional*, i.e. the truth value of a complex proposition is a function of the truth values of its subconstituents. Given a truth-testing method–say, a truth table–a complex formula may be valuated as true always, sometimes, or never, in which cases it is a tautology, a contingency, or a contradiction, respectively.

This essentially semantical characterization of classical logic comes in handy when we wish to distinguish it from the *non-classical logics*. In effect, these can be either *extensions* or *deviations* with respect to classical logic: When they are extensions, their set of tautologies is larger than that of classical logic; smaller, when they are deviations.[4] Examples

---

time being. Below, in Chapter 3, we shall discuss deduction from the syntactical viewpoint at length.

[4]We can also see mere *variations* in notation preserving the content as non-classical logics (e.g., many-sorted logics), but with the semantical characterization above

7

of the former are modal logic and the modal logics (e.g., epistemic logic, temporal logics, dynamic logics) and the intensional logics; examples of the latter are intuitionistic logic and the many-valued logics.[5]

It is possible to remain in classical terrain while working in abductive or inductive logic. In order to do so we only require a bivalent stand. However, we do not usually consider these logics as classical, because in fact they are *supra-classical*, i.e. their sets of consequences are not closed under classical logical consequence. Other ways to characterize supra-classicality are non-monotonicity and ampliation, features that can be obtained by the addition of background assumptions, by restrictions on the set of valuations, and/or by augmenting the set of inference rules.[6]

On the contrary, deductive logic is closed under classical consequence. In particular, it is *monotonic*, as well as *reflexive* and *transitive*: Respectively, adding more premises to an argument does not change its deductive character; every sentence or statement is a consequence of itself; and derived statements can be reused as premises. Moreover, deductive logic obeys the *deduction-detachment theorem* that states that a statement Q is a conclusion of a premise P iff the statement "If P, then Q" is a tautology. Below, we shall see what this in fact implies.

We thus settle that classical logic is (here) synonymous with deductive logic.[7]

0.3. Computational logic and its subfields

Symbolic logic, classical or non-classical, deductive or non-deductive, provides some of the many languages that can be used for computation, namely with the digital computer. This particular approach in the vast field of formal logic can be roughly labeled *computational logic*.

Although this label–"computational logic"–has been around since the early 1970s, it is not (always) easy to use it in an unequivocal way. Often, this label captures, for instance, the use of logic for intelligent agents and multi-agent systems, as well as logics for reasoning with uncertainty, fundamental components of contemporary computing. Some of its many subfields are deductive databases, description logics, equational logics, and logics for the semantic web, to name but a few. Tellingly, an editorial note by J. H. Siekmann and Dov M. Gabbay for volume 9–*Computa-*

---

it is unwarranted to see such logics as non-classical.

[5]In the second volume of this work, we shall discuss the topic of non-classicality at length. See also Augusto (2020a).

[6]See Augusto (2020b) for an elaboration on non-deductive logics from the viewpoint of supra-classical consequence.

[7]Note, however, that there are non-classical deductive logics. See Augusto (2020b).

tional logic; Siekmann (2014)–of the prestigious *Handbook of the history of logic* informs the reader that not all topics of computational logic could be covered by chapters in this volume, for space (as well as time) considerations.

This proliferation is accounted for by the many and diverse large fields in which computational logic is today central: linguistics (e.g., natural language processing), mathematics (e.g., theorem proving), artificial intelligence (e.g., robotics), human cognitive modeling (e.g., decision-making and belief revision), etc., among which, of course, the theory of computation and computer science in general. In this, computational logic may be summarily concretized in the equation *Algorithm = Logic + Control*.

As a subfield of logic, computational logic has been motivated by the desire to carry out faster and more accurately (the "control" factor in the equation above) one of the essential tasks of logic: proofs. Proofs can by and large be divided into *direct* and *indirect*, with the division consisting roughly in showing that a formula is a theorem or its negation is inconsistent, respectively. The former apply a finite sequence of rules of inference until the formula / the conclusion (of an argument) to be proven is reached, and the latter work by *reductio ad absurdum*, i.e. if a contradiction can be derived from the negation of a formula / a conclusion, then the formula / the argument is proven to be valid. We refer to the indirect proof methods as *refutation* proofs.

Having the adequateness of classical logic in mind, an interesting aspect of the above division is that the direct methods are applied when testing for validity, while refutation is mostly applied when testing for satisfiability. The advantage of the latter is implicit in Proposition 3.4.33.1 below, making of refutation a *de-facto* decision procedure: If the negation of a formula (a conclusion) is shown to be unsatisfiable, then we have proven that the formula (the argument) is valid. To be sure, we can prove validity directly, but the methods for validity testing are more limited in both application and ease. For instance, truth tables are a decision procedure for classical propositional logic, but their applicability is limited to a small number of sub-formulae; moreover, they are simply inefficient for classical first-order logic. Add to this the fact that the other validity testing methods can be cumbersome, as they essentially lack the character of algorithmic procedures. Indeed, they usually require the invocation of assumptions or hypotheses, there being no precise rules for this invocation; one can invoke a giraffe, if one wants to, though this is unlikely to help one in testing for validity.

As far as computerization is concerned, this impacts on both complexity and computability: Truth tables, in particular, can be too complex to give an output in useful time, and the non-algorithmic character of the other validity testing methods creates difficulties for even small-sized arguments or theories. On the contrary, refutation methods have shown to be easily amenable to computerization, with the additional advantages of easy-to-read proofs and basically limitless applications (though, of course, there may be space and time issues). For these reasons, by *computational logic* with respect to classical proofs we mean the *automation* of the *resolution* and *analytic tableaux calculi.* These calculi we elaborate on in Chapter 8.

But the applicability of computational logic goes well beyond automated theorem proving; one can actually program using logical languages. *Logic programming* is an important programming paradigm based on classical first-order predicate logic. The essentially deductive nature of logic programming lies on the following principle: A goal (a fact) belongs to a program (a theory, or a database) if and only if it is a logical consequence thereof. In this case, the query whether the fact belongs to the program is answered with "true"; otherwise, the answer is: "false." In terms of formal languages, logic programming has the advantage of being Turing-complete: If a sentence is valid, then it is provable. Although Turing-completeness does not entail first-order decidability, as is well known, this gives to logic programming a "reliability" that other programming paradigms can only aspire to. This explains why this programming paradigm finds important applications in computational contexts in which safety and security are paramount concerns. On the other hand, logic programming is subject to the computational limitations and restrictions "intrinsic" to logical languages, well concretized in the satisfiability problem(s), and as such the programmer is confronted with obstacles that are unsolvable from the viewpoint of classical computing. Nevertheless, just as in the case of automated deduction, logic programming is object of research into refinements or like improvements, and it is only to be expected that the "control" factor in the equation *Algorithm = Logic + Control* might undergo future improvements. We explore logic programming in Chapter 9, both from a general theoretical viewpoint, using pure Prolog, and from the more practical viewpoint, with real Prolog and one of its "dialects," to wit, Datalog.

0.4. Classical deductive computing and its assumptions

We speak of *classical computing* whenever, given a computational problem, there is an exclusive mapping from (sets of) logical formulae to the two truth values of classical logic, to wit, truth and falsity. By "exclusive mapping," it is meant that a logical output (e.g., a formula, possibly empty, or a clause, also possibly empty) must be either true or false, and can never be both true and false. In a less intuitive vocabulary but equivalently, we say that a logical output must be mapped to either 0 or 1, and can never be simultaneously mapped to both 0 and 1. These are the well-known classical principles of bivalence and non-contradiction, respectively. The range of these functions is in fact irrelevant, as long as it has two and only two distinct elements: {false,true}, {0, 1}, {Yes,No} can all be treated as equivalent sets. Such parsimony hides a surprisingly complex notion of logical consequence (see Augusto, 2020b), and in fact, as will be elaborated on in Chapter 2 below, it supports results of computational complexity and decidability that are difficult to digest even as we write today. All thanks to the Turing machine.

Indeed, one cannot overstate the importance of the Turing machine for the field of computation in general and for classical computing in particular. In Turing (1937), the British scientist A. Turing not only defined the meaning of computing in the new light that would motivate the design of the digital computer, but also firstly stated the now well-established fact that there are limits for computers, digital or others, and these are theoretically established precisely by means of the Turing machine. Firstly, the Turing machine established the classical definition of *algorithm* and associated this to the fundamental notion in computability theory of *decision procedure* in the following way: There is a decision procedure for a computational problem–i.e. a computational problem has a "Yes/No" answer–if and only if there is an algorithm for it. In fact, these two notions coincide in the Turing machine.

Secondly, by *limits* it is meant that there are time and space issues to be considered in the carrying out of algorithms. Indeed, one might well have a clearly defined decision procedure but no solution in view. In computing jargon, a problem may be *decidable* (vs. *undecidable*), but be in practice hard to solve, or essentially *unsolvable* (vs. *solvable*). For instance, suppose you need to produce a timetable for a teaching organization in which no two different classes can take place in the same room at the same time; given a relatively small input of a thousand such classes, it would take a computer centuries to produce a timetable satisfying this scheduling problem. More relevant for our interests, such an apparently easy problem as that of deciding whether a set of formulae has some interpretation in which all the formulae are true–the *satisfiability problem*, or *SAT*–can in practice be essentially unsolvable for more than a few formulae. These problems have to do with *complexity*, and they are distinguished roughly between *easy* and *hard* problems.

Going back to the decidability issue, it is now also a well-established fact, also thanks to the Turing machine, that classical first-order logic is undecidable. This classical computing result was established by means of the so-called halting problem, i.e. we can establish an equivalence between this problem and the undecidability of first-order logic. In a similar way, the expressiveness of first-order logic was put into relation with the results on computational complexity via the Turing machine.

According to the view concretized in the Turing machine, computation is carried out in discrete steps in which states (configurations) are clearly distinguished as before, current, and after states in the computing system and/or process. This also entails that the state-space is predetermined, even if the Turing machine is a non-deterministic one; in other words, a Turing machine is a closed system. Moreover, a Turing computing system boasts infinite resources, both in memory and output (the tapes of the Turing machine are infinite, or can be designed to be so), and it consumes nothing in terms of power, as it is a mathematical object rather than a machine proper. Actually, more than a mathematical object, the Turing machine is a logical construct, as it can be conceived as computing the so-called characteristic function that maps logical formulae to the two truth values of classical logic. Finally, this view of computing professes that in essence there is only one algorithm that fits all computational problems, as any algorithm that can be carried out by a particular Turing machine can be carried out equally by the *universal* Turing machine. In sum, we have the view dictating that computation is a discrete, state predetermined, zero-power consumption, and resource-infinite affair. We call this view the *Turing paradigm of classical computing*.

These original assumptions of the Turing paradigm motivated further five paradigms of classical computing, all the six paradigms contributing to what we can now call the *assumptions of classical computing* (Stepney et al., 2005). Some of these further assumptions are, according to the indicated paradigms: Program execution is sequential and the program is static (the *von Neumann paradigm*); a program is a mathematical function that in practical terms is seen as a black box with a single well-defined output channel (the *output paradigm*); a program is a time- and space-bounded function–i.e. it runs in a computer that is largely a closed system with respect to the exterior world and that can be turned on and switched off at will–mapping initial input to final output and running with no randomness (the *algorithmic paradigm*); the specification, formulated in provable/disprovable terms, is the rigid cornerstone of all development and implementation (the *refinement paradigm*); computation is something carried out in an artifact (the hardware) that does not change during the process (the *"computer as artifact" paradigm*).

All these, and other, assumptions of classical computing can be seen in a critical light, and we shall indeed do so in the second volume of this book, but in the present volume we take them as positive, or just neutral, features of computation. Beyond logic-based (non-)classical computing other, non-discrete mathematical objects and structures are often required, and models of computation other than the Turing machine exist (e.g., artificial neural networks, cellular automata)–topics that we shall have the opportunity to touch in the second volume, but not exactly elaborate on, as they are not–at least at first sight–logic-related. In any case, this circumscription of classical computing justifies our choice of mathematical contents in Chapters 1 and 2 of the present volume.

In the first paragraph of this Section, we stated the fundamental criterion of classical computing as computing whose output can be mapped to the truth-value set containing only two elements denotative of truth and falsity. But this holds also for the input. Indeed, we consider here as classical the kind of computing that is *truth-preserving*. For instance, if all the sentences in a knowledge base are mapped to truth, then only sentences that preserve this truth–i.e. are true, too, in the knowledge base–can be considered as logical conclusions thereof. For this reason, logic programming (see Chapter 9) is seen here as a part of classical computing, whereas production systems are not, in spite of the fact that they were conceived in a computational context and are an important component of artificial intelligence based on a construct somehow related to the logical operator known as material conditional.[8] In effect, in a production-system rule of the form "If X_1 and X_2... and X_n, do Y", there is no semantics, bivalent or other, involved, as the "do Y" part is just a recommendation or a command for the system, often just a stimulus-response description. Differently, in a logic-program rule of the form "If X_1 and X_2... and X_n, then Y", the conditional *if ... then* is truth-preserving in the sense that Y holds, or is true, if and only if all the X_i are true (see Section 6.1). We call this truth-preservation in computing *deductive computation*, and though we do not often restate this fact henceforth, it should be kept always in (the back of) your mind when reading this text.[9]

---

[8]See Newell (1973, 1990); the first reference is mainly of historical interest, the second is a rather comprehensive elaboration from the viewpoint of computational cognitive modeling.

[9]Do not, however, equate deductive computation and classical computation, as truth-preserving computing may be carried out in non-classical computation.

Part II.

Mathematical Foundations

1. Mathematical and computational notions

As said in the Introduction, computational logic is applied mathematical logic, and evidently there is no mathematical logic without mathematics. Additionally, the classical theory of computation is also essentially a mathematical subject, even if it can be seen from a narrower perspective as a subject constituted by computational notions; these are, in effect, typically defined in a mathematical language and the associated results are proven with mathematical proof techniques. This said, we do not expect the reader to be literate in mathematical matters, and we assure the more reticent reader that the mathematics required for (this level of) computing with classical logic is, to put it kindly, not difficult. More specifically, an adequate grasping of the contents in this book requires basic notions in both *discrete mathematics* and *abstract mathematics*, the latter being important for mathematical proofs, which are ubiquitous in computational logic.

The present Chapter on mathematical notions for computational logic is not intended to substitute textbooks on set theory and discrete mathematics, being provided for the sake of notational and definitional uniformity, as well as self-containment. In particular, the examples and exercises in this Chapter are provided with the contents of this volume in mind; for instance, we focus on computability and countability with respect to functions and sets. For the same reason, our treatment of graphs and trees is very elementary, as in this volume we shall be using these essentially for models of computation, as well as for derivation, parse, and proof trees.

The reader requiring further, perhaps more elementary, examples and exercises should consult adequate textbooks (e.g., Gallier, 2011; Makinson, 2008). For mathematical proofs, we refer the reader who wishes to go beyond the contents of Section 1.3 to Bloch (2011).

1.1. Basic notions

1.1.1. Sets, relations, functions, and operations

The fundamental notion of mathematics is that of a set.

Definition. 1.1.1. A *set* is a collection of objects from a *universe* or *domain of discourse* **U**; these objects are its *members* or *elements*.[1]

1. The elements of a set can be specified in two ways, to wit, *intensionally* (e.g., $A = \{x \mid x$ is an even positive integer, $x \leq 10\}$), or *extensionally* (e.g., $A = \{2, 4, 6, 8, 10\}$). The order of the elements is irrelevant (e.g., $\{2, 4, 6, 8, 10\} = \{10, 4, 2, 8, 6\}$) and repeated elements are not considered (e.g., $\{a, c, b, a, c, a\} = \{a, c, b\}$).

2. If x is an element of a set A, we write $x \in A$; otherwise, we write $x \notin A$. The *complement of* A, denoted by $\overline{A}$, is the set $\{x \mid x \notin A\}$.

3. A set with a single element is a *singleton*. The *empty* set, i.e. the set with no elements, is denoted by $\emptyset$.

4. The *cardinality* of a set A, denoted by $|A|$, is the number of elements of A. A is finite if $|A| = n$, $n \geq 0$ is finite; otherwise, A is infinite. If A is infinite, we specify it extensionally by means of ellipses (e.g., $\{1, 2, 3, ...\}$ or $\{..., -2, -1, 0, 1, 2, ...\}$).

Example 1.1.1. The following infinite sets play an important role in discrete mathematics:

$$\mathbb{N} = \{0, 1, 2, 3, ...\} \text{ is the set of natural numbers}$$

$$\mathbb{Z} = \{..., -2, -1, 0, 1, 2, ...\} \text{ is the set of integers}$$

$$\mathbb{Z}^+ = \{1, 2, 3, ...\} \text{ is the set of positive integers}$$

$$\mathbb{Q} = \left\{ \frac{p}{q} \mid p, q \in \mathbb{Z}, q \neq 0 \right\} \text{ is the set of rational numbers}$$

Definition. 1.1.2. Let $\subseteq$ denote a *containment relation*.

---

[1] Sometimes it is useful to distinguish the universe, denoted by **U**, and (its) domains, denoted by $\mathscr{D}$. For instance, for **U** = *Numbers*, we may specify $\mathscr{D}_1$ = *Integers*, $\mathscr{D}_2$ = *Even numbers*, etc.

1. A set B is said to be a *subset* of a set A if $B \subseteq A$ (read "the set B is contained or is equal to the set A").

2. In turn, a set A is called a *superset* of B if $A \supseteq B$ (read "the set A contains or is equal to the set B").

3. If we have $B \subsetneq A$ ("B is contained in, but is not equal to, A") and $A \supsetneq B$ ("A contains, but is not equal to, B"), then we speak of *strict containment*. B is called a *proper subset* of A and A is called a *proper superset* of B if $B \subsetneq A$ or $A \supsetneq B$, respectively.[2]

4. We have the equality $A = B$ if and only if (abbr.: iff) $A \subseteq B$ and $B \subseteq A$.

5. Given a set C, the *power set* of C, denoted by 2^C, is the set of all the subsets of C. Let $C = \{x, y, z\}$; then

$$2^C = \{\emptyset, \{x\}, \{y\}, \{z\}, \{x, y\}, \{x, z\}, \{y, z\}, \{x, y, z\}\}.$$

The power set 2^C of a set C always contains $\emptyset$ and always contains C itself.

Example 1.1.2. We have the following containment relation for the sets of numbers:

$$\mathbb{Q} \supseteq \mathbb{Z} \supseteq \mathbb{N}$$

Let us define $\mathbb{N} = \{1, 2, 3, ...\}$.[3] We have $\mathbb{N} \subseteq \mathbb{Z}^+$ and $\mathbb{Z}^+ \subseteq \mathbb{N}$; hence, $\mathbb{N} = \mathbb{Z}^+$.

More strictly, we have the relation $\mathbb{Q} \supset \mathbb{Z} \supset \mathbb{N}$ for the sets of numbers. This strict containment relation helps us to define the set $\mathbb{R}$ of the real numbers in the following way: A real number r is rational if $r \in \mathbb{Q}$; otherwise, r is *irrational*. In effect, we have the strict containment relation:

$$\mathbb{R} \supset \mathbb{Q} \supset \mathbb{Z} \supset \mathbb{N}$$

$\sqrt{2} = 1.41421356...$ and $\pi = 3.14159265....$ are examples of irrational numbers.[4]

---

[2] Note that $B \subsetneq A$ is not necessarily equivalent to $B \subset A$, unless it is so stated. In effect, $B \subset A$ usually denotes that B is an arbitrary subset of A.

[3] If we define $\mathbb{N} = \{1, 2, 3, ...\}$, here denoted by $\mathbb{N}^+$ for disambiguation, then the inclusion of 0 in the natural numbers is denoted by $\mathbb{N} \cup \{0\}$.

[4] The set $\mathbb{R}$ plays a very limited role in discrete mathematics; hence, we discuss it

Definition. 1.1.3. Let A be a set. For $n \geq 1$, R is a *n-ary relation* over A if $R \subseteq A^n$, i.e. R is a subset of $\underbrace{A \times ... \times A}_{n}$. Thus, a *unary* relation R on A is a relation $R \subseteq A$, and a *binary* relation R on A is a relation $R \subseteq (A \times A)$. When $n = 0$, we may speak of a *degenerate* relation. Let A and B be sets. A relation R from A to B is a binary relation from A to B, i.e. $R \subseteq (A \times B)$.

Definition. 1.1.4. A *function* (or *mapping*) f defined on two sets A (the *domain* of f) and B (the *range* of f), denoted by $f : A \longrightarrow B$, is a subset $C \subseteq (A \times B)$ such that for each $x \in A$ there is one and only one *pair* in C of the form (x, y), $y \in B$, such that $f(x) = y$.

1. A function $f : A \longrightarrow B$ is said to be *defined at* $x \in A$, A is the *domain of definition* of f, if the value $f(x) = y \in B$ is calculable. Otherwise, f is said to be *undefined at* x.

2. Let $f : A \longrightarrow B$ be a function, $D \subseteq A$ and $E \subseteq B$. The *image of D under f* is the subset $f(D) = \{f(x) \,|\, x \in D\}$. The *inverse image of E under f* is the subset $f^{-1}(E) = \{x | f(x) \in E\}$.

3. For $n \geq 1$, we say that f is an *n-ary function* over a set A if $f : A^n \longrightarrow A$, i.e. if $f(x_1, ..., x_n) \in A$ whenever $x_1, ..., x_n \in A$. If $n = 0$, we may speak of a *degenerate* function. A 0-ary function over A denotes a fixed element of A, i.e. a *constant*.

4. We say that a function $f : A \longrightarrow B$ is *one-to-one*, or *injective*, when for all $x, y \in A$ if $x \neq y$, then $f(x) \neq f(y)$. An *injection* is also called a *monomorphism* (cf. Def. 1.2.3.4).

5. A function $f : A \longrightarrow B$ is said to be *onto*, or *surjective*, when for every $y \in B$ there is a $x \in A$ such that $f(x) = y$. A *surjection* is also called an *epimorphism* (cf. Def. 1.2.3.4).

6. A function is said to be *bijective*, or a *bijection*, if it is both one-to-one and onto. A bijection is also called a *one-to-one correspondence*, or an *isomorphism* (cf. Def. 1.2.3.4).

   a) The bijective function $\iota_A : A \longrightarrow A$ such that $\iota_A(x) = x$ for all $x \in A$ is called the *identity function*.

---

only briefly. The set $\mathbb{C}$ of the complex numbers contains all the other sets of numbers, but they play no role in the contents of this book, reason why we omit them above.

7. Given two sets A and B, a function f that assigns to each element $x \in A'$, $A' \subset A$, exactly one element $y \in B$ is called a *partial function on A* (and is denoted by φ). If f applies to the whole of A, then f is a *total function*. A (total) function f is said to be *computable* if there is an algorithm that, given input x, produces the value $f(x)$ in finitely many steps. A partial function φ is said to be *partially computable* if there is an algorithm that, on input x, produces the value $\varphi(x)$ in finitely many steps but produces no output if φ is not defined at x (e.g., the algorithm may not terminate).

8. Let $f : A \longrightarrow B$ and $g : B \longrightarrow C$. Then, the *composition of f and g* is the function $g \circ f : A \longrightarrow C$ defined as $(g \circ f)(x) = g(f(x))$ for all $x \in A$.

9. A function $f : A \longrightarrow A$ such that $f \circ f = f$ is said to be *idempotent*.

10. We say that a function $f : A \longrightarrow A$ is *iterated* or *iterative* if we have $f^n : A \longrightarrow A$ such that the f^n are defined recursively by $f^0(x) = \iota_A(x)$ and $f^n(x) = f \circ f^{n-1}(x)$ for some $x \in A$.

The finite or infinite cardinality of a set plays a central role in many results in the classical theory of computation. We accordingly expand on Definition 1.1.1.4.

Definition. 1.1.5. A set is *finite* if it is either empty or else it can be put in a one-to-one correspondence with the set $\{1, 2, ..., n\} \subseteq \mathbb{Z}^+$ (or the set $\{0, 1, 2, ..., n-1\} \subseteq \mathbb{N}$) such that its elements constitute the *sequence* $a_1, a_2, ..., a_n$ (or $a_0, a_1, a_2, ..., a_{n-1}$, respectively) with n *terms*. A set is *infinite* if it is not finite. A set is *denumerable*, or *countably infinite*, if its elements can be listed in a sequence indexed as $a_0, a_1, a_2, ...$ (or as $a_1, a_2, ...$), i.e. if it can be put in a one-to-one correspondence with the sets $\mathbb{N}$ or $\mathbb{Z}^+$, respectively.[5] We say that a set is *countable* if it is either finite or denumerable; otherwise, it is *uncountable*.

The following two theorems are central for the sets of numbers. The proof of Theorem 1.1.2 is by contradiction, a proof technique discussed below in Section 1.3.2.

Theorem 1.1.1. *The sets $\mathbb{N}$, $\mathbb{Z}$, and $\mathbb{Q}$ are countable.*

---

[5]In other words, a set is denumerable if it has the same cardinality as $|\mathbb{N}| = \aleph_0$ (or $|\mathbb{Z}^+| = \aleph_0$). The symbol $\aleph_0$ is read "aleph null."

Proof: Left as an exercise. (Hint for $\mathbb{Q}$: First list the elements p/q with $p + q = 2$, followed by those with $p + q = 3$, and so on, i.e. $1/1, 1/2, 2/1, 3/1, 2/2...$ Make each different p and q determine a table with entries a_{ij} where $i = p$ and $j = q$.)

Theorem 1.1.2. *The set $\mathbb{R}$ of the real numbers is uncountable.*

Proof: The proof is by contradiction and is known as *Cantor diagonalization argument*. Suppose that the set $\mathbb{R}$ is countable. Then, because any subset of a countable set is also countable (cf. Exercise 1.1.1.5), the subset of all real numbers in the interval $[0, 1]$ is also countable, and by Definition 1.1.5 we can list the numbers in this interval in some order, say, $r_1, r_2, ...$ Let d denote decimal representation and let d_{ij} represent the j-th number in the decimal representation of the i-th r, with $j \in \{0, 1, 2, ..., 9\}$. For instance, if $r_6 = 0.573421...$, then $d_{61} = 5$, $d_{62} = 7$, $d_{65} = 2$, etc. Then we have the list

$$r_1 = 0.d_{11}d_{12}d_{13}d_{14}d_{15}...$$

$$r_2 = 0.d_{21}d_{22}d_{23}d_{24}d_{25}...$$

$$r_3 = 0.d_{31}d_{32}d_{33}d_{34}d_{35}...$$

$$\vdots$$

Create now a real number $r = 0.d_1d_2d_3d_4d_5...$ according to the following rule for the decimal digits (choose any two different digits for α and β):

$$d_i = \begin{cases} \alpha & \text{if } d_{ii} \neq \alpha \\ \beta & \text{otherwise} \end{cases}$$

For instance, given the list

$$r_1 = 0.572819...$$

$$r_2 = 0.051689...$$

$$r_3 = 0.182543...$$

$$r_4 = 0.235164...$$

$$r_5 = 0.134638...$$

$$\vdots$$

and choosing $\alpha = 2$, $\beta = 3$, we have the number $r = 0.22322...$ But r is not in the list, as for each i it differs from the decimal expansion of r_i in the i-th place to the right of the decimal point. This shows that the assumption that every number in the interval $[0, 1]$ can be listed is false. Moreover, we have it that any set with an uncountable subset is itself uncountable. Therefore, we conclude that $\mathbb{R}$ is uncountable.[6] **QED**

Definition. 1.1.6. Let A be a set. A *n-ary operation* O on A, for $n \geq 1$, is a function f that assigns an element of A to every n-tuple of elements of A, i.e. $f : A^n \longrightarrow A$. When $n = 0$, we may speak of a *degenerate* operation. Let A and B be two sets; the following are the fundamental operations on the sets A and B:

1. The *intersection* of A and B is the set $A \cap B = \{x | x \in A \text{ and } x \in B\}$. Now let $\{A_i | i \in I\}$ be the family of sets A_i indexed by the set I; then the intersection of the sets A_i, written

$$\bigcap_{i \in I} A_i = \{x | \forall i \in I, x \in A_i\}$$

   is the set of objects that are members of every set A_i. A and B are *disjoint* sets if $A \cap B = \emptyset$ and a collection of sets $\{A_i | i \in I\}$ is disjoint if $\bigcap_{i \in I} A_i = \emptyset$.

2. The *union* of A and B is the set $A \cup B = \{x | x \in A \text{ or } x \in B\}$. Now let $\{A_i | i \in I\}$ be the family of sets A_i indexed by the set I; then the union of the sets A_i, written

$$\bigcup_{i \in I} A_i = \{x | \exists i \in I, x \in A_i\}$$

   is the set of objects that are members of at least one set A_i in the family.

3. We call the set $\{(a, b) | a \in A \text{ and } b \in B\}$, where (a, b) is an *ordered pair* or *2-tuple*, the *Cartesian product* of A and B. We denote this

---

[6]We actually have $|\mathbb{R}| = 2^{|\mathbb{N}|} = 2^{\aleph_0} = \mathfrak{c}$, and $\mathfrak{c} > \aleph_0$, given that–this is a theorem stated by Cantor–the cardinality of a set is always less than the cardinality of its power set. Moreover, $\mathfrak{c} = \aleph_1$, and $\aleph_0 < \aleph_1$, there being no cardinal number X between $\aleph_0$ and $\aleph_1$. This celebrated hypothesis by Cantor is known as *the continuum hypothesis*.

set operation by $A \times B$. The Cartesian product of $A_1, A_2, ..., A_n$ is the set

$$A_1 \times A_2 \times ... \times A_n = \prod_{i=1}^{n} A_i = \{(a_1, a_2, ..., a_n) \,|\, \forall i \,(a_i \in A_i)\}$$

of n-tuples whose i-th coordinate is from A_i.

4. The *set difference* is the set

$$A - B = A \backslash B = A \cap \overline{B} = \{x \,|\, x \in A \text{ and } x \notin B\}.$$

$A - B$ and $A \backslash B$ are alternative ways to write the same set operation.

Definition. 1.1.7. Let there be given a set A. If for every $a \in A$ it is the case that $a \in A_i$ for some $A_i \subseteq A$, $A_i \neq \emptyset$, then the collection $\mathscr{P}_A = \{A_1, ..., A_n\}$ such that

$$(i) \qquad \bigcup_{i=1}^{n} A_i = A$$

and

$$(ii) \qquad A_i \cap A_j = \emptyset$$

for $i \neq j$, is called a *partition* of A and each A_i is called a *part* (or *block* or *cell*) of $\mathscr{P}_A$.

Example 1.1.3. Let there be given the set $\mathbb{N}$.

- Then, for $n \in \mathbb{N}$, the sets $A_1 = \{n \,|\, n^2 < 10\} = \{0, 1, 2, 3\}$ and $A_2 = \{n \,|\, n^2 > 10\} = \{4, 5, 6, ...\}$ constitute a partition of $\mathbb{N}$, as $A_1, A_2 \subset \mathbb{N}$ or, equivalently $\mathbb{N} \supset A_1$ and $\mathbb{N} \supset A_2$ and, in fact,

$$A_1 \cup A_2 = \{0, 1, 2, 3, 4, 5, ...\} = \mathbb{N}$$

and moreover

$$A_1 \cap A_2 = \emptyset.$$

Conditions 1.1.7.i and 1.1.7.ii are satisfied and we write

$$\mathscr{P}_{\mathbb{N}} = \{A_1, A_2\}.$$

- On the contrary, the sets $B_1 = \{n \,|\, n < 6\}$ and $B_2 = \{n \,|\, n > 6\}$ do not constitute a partition of $\mathbb{N}$, as $6 \notin B_1 \cup B_2$ and thus $B_1 \cup B_2 \neq \mathbb{N}$, and condition 1.1.7.i is not satisfied.

- Let now $C_1 = \{n|n \leq 6\}$ and $C_2 = \{n|n \geq 6\}$; then, condition 1.1.7.ii is violated, and the sets C_1, C_2 do not constitute a partition of $\mathbb{N}$, either.

Exercises

Exercise 1.1.1.1. Show that, given two sets A and B, a partial function from A to B can be seen as a function $\dot{\varphi} : A \longrightarrow (B \cup \{u\})$, where $u \notin B$,

$$\dot{\varphi}(a) = \begin{cases} \varphi(a) & \text{if } a \in A' \\ u & \text{if } \varphi \text{ is undefined at } a \end{cases}$$

A' is the domain of definition of φ.

Exercise 1.1.1.2. Show that the following functions are computable:

1. $f(x) = x - 1$, for $x > 0$ and $f(0) = 0$.

2. $f(x, y) = y$.

Exercise 1.1.1.3. Show that $\emptyset$ is computable.

Exercise 1.1.1.4. Prove that for every set S, (i) $\emptyset \subseteq S$, and (ii) $S \subseteq S$.

Exercise 1.1.1.5. Prove that every infinite set has a countably infinite subset, and every subset of a countable set is countable.

Exercise 1.1.1.6. Show that the set $\mathbb{N} \times \mathbb{N}$ is countable.

Exercise 1.1.1.7. Show that if S_i is a countable set for every $i \in \mathbb{N}$, then

$$S = \bigcup_{i=0}^{\infty} S_i$$

where ∞ denotes an infinite sequence, is also countable.

Exercise 1.1.1.8. Determine whether, and show why, the following sets are countable or uncountable:

1. $\{A|A \subset \mathbb{N}, A \text{ has three elements}\}$ (consider all such sets A)

2. $\{B|B \subset \mathbb{N}, B \text{ is finite}\}$ (consider all such sets B)

3. $\{C_i|C_i \subset \mathbb{N}, i = 1, ..., n, C_i \cap C_j = \emptyset \text{ for all } i \neq j, \bigcup_{i=1}^n C_i = \mathbb{N}\}$

  4. $\{f|f : \mathbb{N} \longrightarrow \{0,1\}\}$ (consider all such functions f)

  5. $\{g|g : \mathbb{N} \longrightarrow \mathbb{N}\}$ (consider all such functions g)

  6. $\{h|h : \mathbb{N} \longrightarrow \mathbb{N}, h \text{ is non-decreasing}\}$ (consider all such functions h)

Exercise 1.1.1.9. Prove that if S is a countably infinite set, then 2^S is uncountable. Apply the diagonalization method.

Exercise 1.1.1.10. Research into *recursive functions*. After giving a precise definition of a recursive function, elaborate on the following notions and state their relation to *recursion*:

  1. Ackermann function.

  2. Fibonacci sequence.

  3. Towers of Hanoi.

  4. Wrapper function.

  5. Tail-recursive function.

  6. μ-recursive function.

  7. Primitive recursive function.

Exercise 1.1.1.11. Is an iterative function the same as a recursive function? Elaborate on this topic and illustrate your elaboration with relevant examples.

Exercise 1.1.1.12. Prove Theorem 1.1.1.

1.1.2. Binary relations and ordered sets

Given a set A, it is often important to order its elements in a specific way and to specify relations among them. Binary relations are central in classical computing.

Definition. 1.1.8. A binary relation R over a set A is a relation $R \subseteq A \times A$. We say that a binary relation R is

  1. *reflexive* if for every $x \in A$ we have it that xRx.

2. *irreflexive* if for every $x \in A$ it is not the case that xRx.

3. *symmetric* if for every $x, y \in A$ we have it that if xRy, then yRx.

4. *anti-symmetric* if for every $x, y \in A$ we have it that if xRy and yRx, then $x = y$.

5. *transitive* if for every $x, y, z \in A$ we have it that if xRy and yRz, then xRz.

6. an *equivalence relation* if it is reflexive, symmetric, and transitive.

Example 1.1.4. Let the set $\mathbb{Z}$ be given.

- For $x \in \mathbb{Z}$, the relation $=$ is reflexive, as we have $x = x$, while the relation $<$ is irreflexive, because $x \not< x$.

- For $x, y \in \mathbb{Z}$, the relation "is immediately next to" is symmetric, as for any two integers x and y we have the pairs (x, y) and (y, x) (e.g., $(1, 2)$ and $(2, 1)$), but $x \neq y$.

- Let $m \neq 0, n \in \mathbb{Z}$; we say that m divides n and n is divisible by m, denoted by $m|n$, if there is $p \in \mathbb{Z}$ such that $n = mp$, or equivalently, if $\frac{n}{m} \in \mathbb{Z}$. For $m, n \in \mathbb{Z}^+$, the relation $|$ is antisymmetric, as whenever we have $m|n$ and $n|m$, then $m = n$.

- Let us consider $m \in \mathbb{Z}$, $m > 1$. We say that x is congruent to y modulo m, and we write $x \approx y \, (\bmod \, m)$, if $x - y$ is divisible by m. The relation $R = x \approx y \, (\bmod \, m)$ is an equivalence relation on $\mathbb{Z}$. In effect, for every $x \in \mathbb{Z}$ we have $x \approx x \, (\bmod \, m)$, as $x - x$ is divisible by m, and thus R is reflexive. R is also symmetric, as for any $x, y \in \mathbb{Z}$, if $x - y$ is divisible by m, then $-(x - y)$ is also divisible by m. Finally, R is transitive, because if for $x, y, z \in \mathbb{Z}$ we have it that $x - y$ and $y - z$ are both divisible by m, then $x - z = (x - y) + (y - z)$ is also divisible my m.

Definition. 1.1.9. A binary relation $\leq$ on a set A that is reflexive and transitive is called a *preorder*, and $\mathcal{R} = (A, \leq)$ is correspondingly called a preorder.

- $\mathcal{R}$ is a *partial order*, or a *poset*, if $\leq$ is reflexive, anti-symmetric, and transitive.

- If $\leq$ is irreflexive and transitive, then $\mathcal{R}$ is a *strict partial order* (often denoted by $<$ or $\leq$). Let $\mathcal{R}$ be a strict partial order; then for $x, y, z \in A$ we say that y *covers* x in A if $x < y$ and for $x \leq z \leq y$ it is the case that $z = x$ or $z = y$.

- A poset $\mathcal{R}$ is *totally ordered* if every $x, y \in A$ are comparable, i.e. we have $x \leq y$ or $y \leq x$. We then speak of a *total* or *linear order*.

A convenient way to represent graphically a poset is a *Hasse diagram* (also: *poset diagram*). A Hasse diagram is a graph, namely a simple graph, a structure which will be defined bellow (Section 1.2.3); however, its highly intuitive character sanctions its use in examples and exercises after a brief informal definition. Essentially, the elements of a poset are depicted in a Hasse diagram both vertically and horizontally in such a way that if $a \leq b$ then a is located below b, and if $a \not\leq b$ then both a and b may be placed at the same level; additionally, neither reflexivity nor transitivity are explicitly represented, i.e. there are neither loops (edges connecting a vertex a to itself), nor edges of the form $a\_\_c$ whenever $a\_\_b$ and $b\_\_c$ are in the diagram, respectively.

Example 1.1.5. Let the set A be given whose members and partial order are those of the Hasse diagram of Figure 1.1.1. Clearly, the pairs $(\square, \spadesuit)$ and $(\triangle, @)$ are incomparable. Comparable are, for instance, $\square < \blacklozenge$ and $\triangle > \spadesuit$.

Definition. 1.1.10. Given a poset $\mathcal{R} = (A, \leq)$ and a subset $B \subseteq A$, we have the following order relations for $x, y \in A$ and $z \in B$:

1. x is a *minimal element* of A, written $min(A) = x$, if there is no $y \in A$ such that $y \leq x$.

2. x is a *maximal element* of A, written $max(A) = x$, if there is no $y \in A$ such that $x \leq y$.

3. x is the *least element* of A if for every $y \in A$ we have $x \leq y$.

4. x is the *greatest element* of A if for every $y \in A$ we have $y \leq x$.

5. x is a *lower bound* of B if for every $z \in B$ we have $x \leq z$.

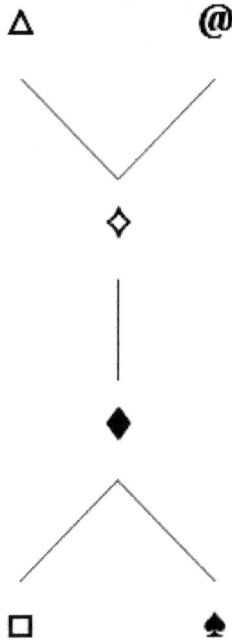

Figure 1.1.1.: A partially ordered set.

6. x is an *upper bound* of B if for every $z \in B$ we have $z \leq x$.

7. x is a *greatest lower bound (glb)*, or *infimum*, of B, written $glb\,(B) = x$ or $inf\,(B) = x$, if x is the lower bound of B and $x' \leq x$ for any other lower bound x' of B.

8. x is a *least upper bound (lub)*, or *supremum*, of B, written $lub\,(B) = x$ or $sup\,(B) = x$, if x is the upper bound of B and $x \leq x'$ for any other upper bound x' of B.

Example 1.1.6. The Hasse diagram of Figure 1.1.2 shows a poset $\mathcal{R} = (A, \leq)$ where $A = \{a, b, ..., m\}$.

- We have $min\,(A) = \{a, b, c\}$ and $max\,(A) = \{l, m\}$. This shows that $|min\,(A)| \geq 1$ and $|max\,(A)| \geq 1$ for some set $A \neq \emptyset$; on the contrary, the least and the greatest elements of a set A, if they exist, are unique. In the case at hand, there is neither a least nor a greatest element of A.

- With respect to the subset $B \subset A$, $B = \{d, k, f\}$, we have the following order relations: There is neither a least element nor a lower bound, but there are two minimal elements, d and f; k is the maximal element, the greatest element, and the lub of B; there is no glb, i.e. $glb\,(B) = \emptyset$; the upper bound of B is the subset $\{k, l, m\}$.

- Consider now the subset $D \subset A$, $D = \{a, b, c\}$. We have $max\,(D) = min\,(D) = D$; there is neither a least nor a greatest element; the upper bound of D is the subset $\{k, l, m\}$ but there is no lower bound; $lub\,(D) = \emptyset$ and $glb\,(D) = \{k\}$.

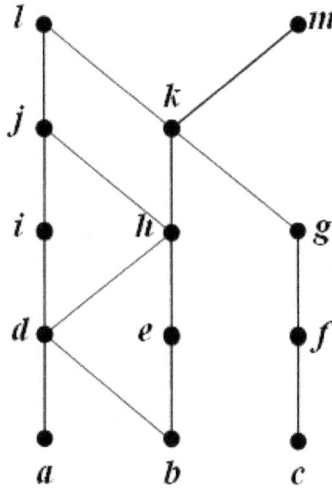

Figure 1.1.2.: Hasse diagram of a poset.

Definition. 1.1.11. Let R be a binary relation on a set A. The *transitive closure* of R is the connectivity relation R^* associated with R such that $R^* = \bigcup_{i=1}^{\infty} R^i$ of all the positive powers of R. In other words, for a relation R, the transitive closure of R is the smallest transitive relation containing R.

Definition. 1.1.12. Let $\mathcal{R} = (A, \leq)$ be a poset.

1. The *interval* between $x, y \in A$ is defined as:

$$[x, y] = \{z \in A | x \leq z \leq y\}$$

$$(x, y) = \{z \in A | x < z < y\}$$
$$[x, y) = \{z \in A | x \leq z < y\}$$
$$(x, y] = \{z \in A | x < z \leq y\}$$

2. An *ideal*, or *downset*, in $\mathcal{R}$ is a subposet S such that if $x \in S$ and $y < x$, then $y \in S$. The *ideal generated by an element* x in a poset $\mathcal{R}$ is the downset

$$\downarrow \{x\} = \{y \in \mathcal{R} | y \leq x\}.$$

The *ideal generated by a subset* A in $\mathcal{R}$ is the downset

$$\mathcal{D}_A = \bigcup_{x \in A} \downarrow \{x\}.$$

3. A *filter*, or *upset*, in $\mathcal{R}$ is a subposet S such that if $x \in S$ and $y > x$, then $y \in S$. The *filter generated by an element* x in a poset $\mathcal{R}$ is the upset

$$\uparrow \{x\} = \{y \in \mathcal{R} | y \geq x\}.$$

The *filter generated by a subset* A in $\mathcal{R}$ is the upset

$$\mathcal{U}_A = \bigcap_{x \in A} \uparrow \{x\}.$$

Exercises

Exercise 1.1.2.1. Consider the set $A = \{a, b, c\}$ and the following five relations on A:

$R_1 = \{(a, a), (a, b), (a, c), (c, c)\}$
$R_2 = \{(a, a), (a, b), (b, a), (b, b), (c, c)\}$
$R_3 = \{(a, a), (a, b), (b, b), (b, c)\}$
$R_4 = A \times A$
$R_5 = \emptyset$

1. Determine which of these relations are reflexive, irreflexive, symmetric, anti-symmetric, transitive.

2. Is any of these an equivalence relation?

Exercise 1.1.2.2. Consider the following relations on $\mathbb{Z}$:

$$R_1 = \{(a, b) \,|\, a < b\}$$
$$R_2 = \{(a, b) \,|\, a \geq b\}$$
$$R_3 = \{(a, b) \,|\, a = b\}$$
$$R_4 = \{(a, b) \,|\, a + b > 5\}$$
$$R_5 = \{(a, b) \,|\, a - b \leq 0\}$$

Which of these relations contain the following pairs?

1. $(3, 4)$

2. $(1, 1)$

3. $(-1, 1)$

4. $(2, 1)$

5. $(-2, -1)$

6. $(3, 3)$

Exercise 1.1.2.3. Consider the poset in Figure 1.1.2. Find all the order relations listed in Definition 1.1.10 for the following subsets:

1. $\{b, h, f\}$.

2. $\{l, m\}$.

3. $\{d\}$.

Exercise 1.1.2.4. Consider the poset with the Hasse diagram of Figure 1.1.3. Find, if they exist,

1. the upper bounds and the lower bounds of $\{\alpha, \beta, \gamma\}$.

2. the upper bounds and the lower bounds of $\{\iota, \theta\}$.

3. the upper bounds and the lower bounds of $\{\alpha, \beta, \gamma, \varsigma\}$.

4. $glb\,(\{\beta, \delta, \eta\})$ and $lub\,(\{\beta, \delta, \eta\})$.

Exercise 1.1.2.5. Let S be the set of the 26 letters of the Latin alphabet and let R be a relation on S such that aRb iff $|a| = |b|$, for $|x|$ the length of string x. Is R an equivalence relation?

Exercise 1.1.2.6. Let $\mathscr{C}$ be any class of sets. Show that the relation $\subseteq$ is a partial order on $\mathscr{C}$.

Exercise 1.1.2.7. Let the set of all people be given and let R be defined as

$$R = \{(a, b) \,|\, a \text{ has met } b\}.$$

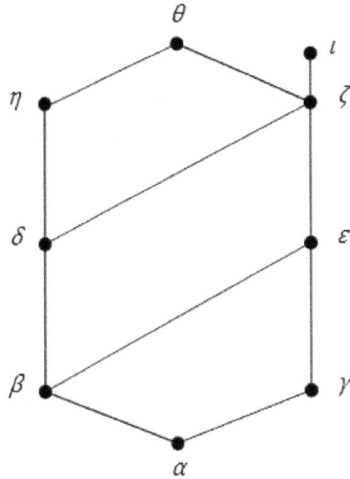

Figure 1.1.3.: Hasse diagram of a poset.

1. Determine R^n for $n > 1$.

2. Determine R^*.

Exercise 1.1.2.8. Let the set of all European countries be given and let R be defined as

$$R = \{(a, b) \,|\, a \text{ and } b \text{ share a border}\}.$$

1. Determine R^n for $n > 0$.

2. Determine R^*.

1.2. Discrete structures

1.2.1. Algebraic structures

Algebraic structures, and in particular Boolean algebras, are fundamental structures for an adequate understanding of both classical computation and computational logic. We give here the basic notions of these structures.

1.2.1.1. Basic algebraic notions

Definition. 1.2.1. An *algebraic structure* (or *system*) is a triple $3 = (A, O, R)$ consisting of a non-empty set A, a set O of operations on A, and a set R of relations on A. If $O = \emptyset$, 3 is called a *model*, or *relational structure*. If $R = \emptyset$, 3 is called an *algebra*.

Example 1.2.1. For $\mathbb{Z}$ the set of integers, the structure $(\mathbb{Z}, +)$ is an algebra, the structure $(\mathbb{Z}, \leq)$ is a model.

Definition. 1.2.2. An *(abstract) algebra* is a pair

$$\mathfrak{A} = (U, O) = (U, \{o_i | i \in I\})$$

where U is a non-empty set, called the *universe* or *carrier* of $\mathfrak{A}$, and for each $i \in I$, o_i is a *basic operation* on U. If U is a singleton, then $\mathfrak{A}$ is said to be *degenerate*. $\mathfrak{A}$ is a *finite* algebra if I, the *index set* of $\mathfrak{A}$, is finite; otherwise, it is *infinite*. The *type* of $\mathfrak{A}$ is a function $t : I \longrightarrow \mathbb{N}$ where $t(i)$ equals the arity of the operation o_i.

Definition. 1.2.3. Let $\mathfrak{A} = (U, \{o_i | i \in I\})$ be an algebra.

1. An algebra $\mathfrak{B} = (Z, \{o_i | i \in I\})$ is a *subalgebra* of $\mathfrak{A}$ iff $Z \subseteq U$ and, for $i \in I$, Z is closed with respect to each o_i.

2. An algebra $\mathfrak{C} = (U, \{o_i | i \in K\})$ is a *reduct* of $\mathfrak{A}$ if $K \subsetneq I$; if, on the contrary, $K \supsetneq I$, then we say that $\mathfrak{C}$ is an *expansion* of $\mathfrak{A}$.

3. An algebra $\mathfrak{D} = (U, \{p_k | k \in K\})$ is *similar* to the algebra $\mathfrak{A}$ iff $I = K$ and $o_i^m = p_i^m$ for each $i \in I$ and m the arity of o_i, p_i. In other words, two algebras are similar iff they have the same type.

4. A *homomorphism* of $\mathfrak{A}$ into a similar algebra $\mathfrak{D} = (Y, \{p_i | i \in I\})$, denoted by $Hom(\mathfrak{A}, \mathfrak{D})$, is a mapping

$$h : U \longrightarrow Y$$

such that for each $i \in I$ and for arbitrary $u_1, ..., u_{m(i)} \in U$ we have

$$h\left(o_i\left(u_1, ..., u_{m(i)}\right)\right) = p_i\left(h\left(u_1\right), ..., h\left(u_{m(i)}\right)\right).$$

A one-to-one homomorphism is a *monomorphism* or *embedding* of $\mathfrak{A}$ into $\mathfrak{D}$. A homomorphism of $\mathfrak{A}$ into itself is an *endomorphism*.

If $h : \mathfrak{A} \longrightarrow \mathfrak{D}$ is a homomorphism such that $h(U) = Y$, then h is an *epimorphism*. An epimorphism that is also a monomorphism is an *isomorphism* of $\mathfrak{A}$ and $\mathfrak{D}$.

5. If $Z \subseteq U$ is non-empty, then there exists the least subalgebra $\mathfrak{E} = (X, \{o_i | i \in I\})$ of $\mathfrak{A}$ such that $Z \subseteq X$, i.e. the intersection of all subalgebras of $\mathfrak{A}$ with universes that contain Z.

6. $Z \subseteq U$, $Z \neq \emptyset$, is a set of *generators* of $\mathfrak{A}$ if the least subalgebra of $\mathfrak{A}$ containing Z in its universe is $\mathfrak{A}$ itself.

7. $\mathfrak{A}$ is an *absolutely free* algebra if there is a set $Z \subseteq U$ of generators of $\mathfrak{A}$ such that every mapping $s : Z \longrightarrow Y$ of a similar algebra $\mathfrak{D} = (Y, \{p_k | k \in K\})$ can be extended to a homomorphism from $\mathfrak{A}$ into $\mathfrak{D}$. Then, the elements of Z are said to be *free generators* of $\mathfrak{A}$ and $\mathfrak{A}$ is said to be *freely generated* by Z.

8. Let $\mathscr{C}$ be a class of similar algebras. The algebra $\mathfrak{A} \in \mathscr{C}$ is said to be a *$\mathscr{C}$-free algebra* iff it has a set of generators G such that any mapping $f : G \longrightarrow Z$, for Z the universe of an arbitrary algebra $\mathfrak{B} \in \mathscr{C}$, can be extended to a homomorphism $h : \mathfrak{A} \longrightarrow \mathfrak{B}$. If $\mathfrak{A}$ is $\mathscr{C}$-free, then G is said to be a set of *$\mathscr{C}$-free generators* for the algebra $\mathfrak{A}$.

Definition. 1.2.4. *Semigroups and monoids* – An algebraic structure $\mathfrak{X} = (X, \star)$ such that X is a non-empty set and $\star$ is an associative binary operation is a semigroup. If X is a poset, then $\mathfrak{X}$ is called a partially-ordered semigroup. A (partially-ordered) semigroup with the distinguished element 1 or 0 (denoted by ι), $\mathfrak{X} = (X, \star, \iota)$, is a monoid.

Example 1.2.2. The algebraic structure $\mathfrak{A} = (B, -, +, \cdot, 0, 1)$ is an algebra of type $(1, 2, 2, 0, 0)$, because it has one binary operation (denoted by 1), two binary operations (denoted by 2), and two 0-ary operations (denoted by 0) on the elements of B. If these operations are seen as the additive inverse, addition and multiplication, and the additive and multiplicative identities, respectively, then $\mathfrak{A}$ is called a *ring*. Any algebra of type $(1, 2, 2, 0, 0)$ is said to be similar to a ring. A *group* is an algebra $\mathfrak{E} = (B, -, \star, 0)$ of type $(1, 2, 0)$, where $\star$ denotes some binary operation. If we specify a group as $\mathfrak{E} = (\mathbb{Z}, -, +, 0)$, i.e. as the group of integers under negation and addition, and do so for a monoid as $\mathfrak{X} = (\mathbb{Z}, +, 0)$, the monoid of integers under addition, then $\mathfrak{X}$ is a reduct of $\mathfrak{E}$ and this is an expansion of $\mathfrak{X}$. Note that a monoid is an algebra of type $(2, 0)$, but the monoid $\mathfrak{X} = (\mathbb{N}, +, 0)$ is *not* a reduct of a group. Thus, attention is required in defining both the carriers and the operations of algebras.

1.2.1.2. Boolean algebra

The many relations between Boolean algebra and classical deductive computing with classical logic require a good grasping of the following contents.

Definition. 1.2.5. An algebra $\mathfrak{B} = (B,',\wedge,\vee,0,1)$ of type $(1,2,2,0,0)$ is a *Boolean algebra* if the following conditions hold for the operations of *complementation* $(')$, *meet* $(\wedge)$, and *join* $(\vee)$, and for the distinguished elements 0 and 1:

1. If $|B| = 0$, i.e. $\mathfrak{B} = (\emptyset,',\wedge,\vee,0,1)$, then $',\wedge,\vee$ are constant and $0 = 1$. $\mathfrak{B}$ is called a degenerate Boolean algebra.

2. If $|B| = 2$, then we have the two-element Boolean algebra $\mathbf{2} = (\{0,1\},',\wedge,\vee,0,1)$ with elements 0 and 1, $0 \leq 1$, such that we have the following equalities:[7]

   a) $0' = 1$ and $1' = 0$

   b)
$\wedge$	1	0
1	1	0
0	0	0

$\vee$	1	0
1	1	1
0	1	0

   $(1 \wedge 1 = 1, 1 \wedge 0 = 0,$ etc.$)$

3. Let $x,y \in B$. The operations $\wedge$ and $\vee$ are inter-definable by means of $'$ as
$$x \wedge y = (x' \vee y')'$$
$$x \vee y = (x' \wedge y')'$$

4. For elements $x,y \in B$ we have the following equalities:
$$x \wedge y = min(x,y)$$
$$x \vee y = max(x,y)$$
$$x' = 1 - x$$

---

[7] $\mathbf{2}$ abbreviates $\mathfrak{B}_2$.

Definition. 1.2.6. Given two Boolean algebras $\mathfrak{A}$ and $\mathfrak{B}$, a *Boolean homomorphism* is a function $h : \mathfrak{A} \longrightarrow \mathfrak{B}$ such that for all $x, y \in \mathfrak{A}$ we have

$$h\left(x \wedge y\right) = h\left(x\right) \wedge h\left(y\right)$$

$$h\left(x \vee y\right) = h\left(x\right) \vee h\left(y\right)$$

and

$$h\left(x'\right) = h\left(x\right)'.$$

Although in this definition $\mathfrak{A}$ and $\mathfrak{B}$ have exactly the same operations, this need not be so. For instance, we could have the set of operations $O_{\mathfrak{B}} = \{-, \cap, \cup\}$ such that

$$h\left(x \wedge y\right) = h\left(x\right) \cap h\left(y\right)$$

etc., and we still have a Boolean homomorphism.

Definition. 1.2.7. Let h be a Boolean homomorphism. Then, the *kernel* of h is the set

$$ker\left(h\right) = h^{-1}\left[\{0\}\right]$$

and the *hull* of h is the set

$$hull\left(h\right) = h^{-1}\left[\{1\}\right].$$

Proposition. 1.2.8. *Let* $\mathfrak{B} = (B, ', \wedge, \vee, 0, 1)$ *be a non-degenerate Boolean algebra. Then, for* $x_1, ..., x_n \in B$ *we have the equalities*

$$1 = x_1 \vee ... \vee x_n = \bigvee_{i=1}^{n} x_i$$

$$0 = x_1 \wedge ... \wedge x_n = \bigwedge_{i=1}^{n} x_i$$

Proof: Left as an exercise.

When in the context of Boolean logic, it is important to consider sets of formulas.

Definition. 1.2.9. Let $\mathfrak{L} = (F, ', \wedge, \vee, 0, 1)$ be an algebra of formulas, in which F is a set of formulas. A variable $x_i \in F_0 \subseteq F$ in a finite sequence $x_1, ..., x_n$ is said to be a *Boolean variable* if there is a *Boolean function of degree n*

$$f : \{0, 1\}^n \longrightarrow \{0, 1\}$$

such that we have $f(x_i) = 0$ or $f(x_i) = 1$, spoken of as *truth-value assignments*. Additionally, we have:

$$[f(x_i) = \neg x_i] = 1 - f(x_i)$$

$$\left[f(x_1, ..., x_n) = \bigwedge_{i=1}^{n} x_i\right] = min\,(f(x_1), ..., f(x_n))$$

$$\left[f(x_1, ..., x_n) = \bigvee_{i=1}^{n} x_i\right] = max\,(f(x_1), ..., f(x_n))$$

We often abbreviate and write simply $x_i = 1$ or $x_i = 0$, as well as:

$$\neg x = 1 - x$$

$$\bigwedge_{i=1}^{n} x_i = min\,(x_1, ..., x_n)$$

$$\bigvee_{i=1}^{n} x_i = max\,(x_1, ..., x_n)$$

Definition. 1.2.10. A *Boolean expression* ϕ in the individual variables $x_1, ..., x_n$, written $\phi(x_1, ..., x_n)$, is an expression defined inductively in the following way:

1. the constants $0, 1$, and all variables x_i are Boolean expressions in $x_1, ..., x_n$;

2. if ϕ and ψ are Boolean expressions in the variables $x_1, ..., x_n$, then so are ϕ', $\phi \wedge \psi$, and $\phi \vee \psi$.

The constants 0 and 1 are the logical constants known as *falsum* (false) and *verum* (truth), respectively, in Boolean logic. Definition 1.2.10.2 "translates" directly to the *well-formed formulas* of Boolean logic $\neg \phi$, $\phi \wedge \psi$, and $\phi \vee \psi$.[8]

Example 1.2.3. The following are examples of Boolean expressions:

- $\phi(p) = \neg p$

---

[8]For convenience, we shall write indifferently $\neg\phi$ or ϕ' in the framework of Boolean logic.

- $\omega\,(p,q,r) = (p \vee q) \wedge (\neg p \vee r)$

- $\chi\,(x,w,y,z) = (x \wedge \neg w) \vee \neg y \vee z$

- $\psi\,(x_1, x_2, x_3) = \neg x_1 \wedge (\neg x_2 \vee x_1) \wedge (\neg x_2 \vee \neg x_3)$

Definition. 1.2.11. An expression of the form

$$\phi\,(p_1, ..., p_n)$$

where the p_i are propositional Boolean variables, is called a *propositional Boolean formula*.

Definition. 1.2.12. An expression of the form

$$\blacklozenge_1 x_1 ... \blacklozenge_n x_n\,(\phi\,(x_1, ..., x_n))$$

where the x_i are individual Boolean variables and $\blacklozenge_i \in \{\forall, \exists\}$, is called a *quantified Boolean formula* (QBF).

Both expressions, propositional and quantified, constitute the basic elements of *Boolean logic*. This will be frequently put into relation with classical logic in Parts III-V of this book.

Exercises

Exercise 1.2.1.1. Show that, given an arbitrary set B, the class of all subsets of B is a Boolean algebra $\mathfrak{B} = (B,', \wedge, \vee, 0, 1)$.

Exercise 1.2.1.2. Show that in a Boolean algebra $\mathfrak{B} = (B,', \wedge, \vee, 0, 1)$, if $|\mathfrak{B}| = \emptyset$, then the operations $', \wedge, \vee$ are constant and $0 = 1$.

Exercise 1.2.1.3. Let $\mathfrak{B} = (B,', \wedge, \vee, 0, 1)$ be a Boolean algebra. Show that if $B = \{x\}$, then

$$\bigwedge \{x\} = \bigvee \{x\} = x.$$

Exercise 1.2.1.4. Show that a Boolean algebra obeys the *duality principle*: *Every Boolean algebra has an isomorphic dual by interchanging* 0 *for* 1 *and* $\wedge$ *for* $\vee$.

Exercise 1.2.1.5. Prove that a mapping between Boolean algebras that preserves join and complement is a Boolean homomorphism.

Exercise 1.2.1.6. Show that a homomorphism $h : \mathfrak{A} \longrightarrow \mathfrak{A}'$ is one-to-one iff either the kernel of h is trivial or the hull of h is.

Exercise 1.2.1.7. A *Boolean ring* is a structure $\mathfrak{A} = (B, -, +, \cdot, 0, 1)$ with identity element 1 and in which every element $x \in B = \mathbb{R}$ is idempotent, i.e. $x^2 = x$. Show that $\mathfrak{A}$ is a Boolean algebra $\mathfrak{B} = (B,', \wedge, \vee, 0, 1)$. (Note that there are two directions to this proof.)

Exercise 1.2.1.8. A *field* (or *algebra*) *of sets with unit* U is a structure $\mathfrak{F} = (\Psi, -, \cap, \cup, U, \emptyset)$ of type $(1, 2, 2, 0, 0)$ where $\Psi \supseteq 2^U$ and Ψ is closed under the operations $-, \cap, \cup$. Show that $\mathfrak{F}$ is isomorphic to a non-degenerate Boolean algebra $\mathfrak{B} = (B,', \wedge, \vee, 0, 1)$.

Exercise 1.2.1.9. For arbitrary elements x and y, the following are the *De Morgan's laws* :

$$(x \wedge y)' = x' \vee y'; (x \vee y)' = x' \wedge y'$$

Show that these laws hold in a Boolean algebra $\mathfrak{B} = (B,', \wedge, \vee, 0, 1)$.

Exercise 1.2.1.10. Show that every finite Boolean algebra has 2^n elements.

Exercise 1.2.1.11. Show that every Boolean function can be expressed as a Boolean formula.

Exercise 1.2.1.12. Show that for all Boolean functions f and g the following statements are equivalent:

1. $f \leq g$

2. $f \vee g = g$

3. $\bar{f} \vee g = 1$

4. $f \wedge g = f$

5. $f \wedge \bar{g} = 0$

Exercise 1.2.1.13. Prove Proposition 1.2.8.

1.2.2. Lattices

Lattices are important discrete structures for both classical logic and classical computing with it. Although they can be discussed as algebraic structures, a separate discussion is pedagogically useful.

Definition. 1.2.13. A poset $\mathcal{R} = (A, \leq)$ in which every two elements $x, y \in A$ have a $lub\,(x, y) := x \vee y = y$, called the *join* of x and y, or a $glb\,(x, y) := x \wedge y = x$, called the *meet* of x and y, is called a *semi-lattice*. If $\mathcal{R}$ has only a meet (a join), then it is a *lower* (respectively, *upper*) *semi-lattice* or a *meet semi-lattice* (a *join semi-lattice*, respectively). A poset $\mathcal{R}$ that is both an upper and a lower semi-lattice is a *lattice*; in turn, every lattice is a poset $\mathcal{R}$. A lattice $\mathcal{L}' = ((A', \leq), \vee, \wedge)$ is a *sublattice* of the lattice $\mathcal{L} = (\mathcal{R}, \vee, \wedge)$ if $A' \subseteq A$ and $\mathcal{L}'$ has the same operations as $\mathcal{L}$. A lattice is *complete* if every subset of A has both a join (a supremum, or $\bigvee A$) and a meet (an infimum, or $\bigwedge A$).

Theorem 1.2.1. *Any non-empty finite lattice* $\mathcal{L} = (\mathcal{R}, \vee, \wedge)$ *is complete.*

Proof: Left as an exercise.

Theorem 1.2.2. *The lattices of all transitive relations and of all equivalence relations on a set are complete lattices.*

Proof: Left as an exercise.

Proposition. 1.2.14. *Let for any* $x, y \in A$, $lub\,(x, y) := x \cup y$ *and* $glb\,(x, y) := x \cap y$. *Let further the pair* $\mathcal{S} = (2^A, \subseteq)$ *be given. Then the structure* $(\mathcal{S}, \cup, \cap)$ *is a complete (distributive) lattice.*

Proof: Left as an exercise.

Proposition. 1.2.15. *The lattice* $\mathcal{L} = (\mathcal{S}, \cup, \cap)$ *satisfies the following properties for every* $x, y, z \in A$:

1. *Commutativity:* $x \cup y = y \cup x$, $x \cap y = y \cap x$.

2. *Associativity:* $(x \cup y) \cup z = x \cup (y \cup z)$, $(x \cap y) \cap z = x \cap (y \cap z)$.

3. *Idempotency:* $x \cup x = x$, $x \cap x = x$.

4. *Absorption:* $x \cup (x \cap y) = x$, $x \cap (x \cup y) = x$.

Proof: Left as an exercise.

Example 1.2.4. Let $A = \{a, b, c\}$. Then,

$$2^A = \{\emptyset, \{a\}, \{b\}, \{c\}, \{a, b\}, \{a, c\}, \{b, c\}, \{a, b, c\}\}.$$

We show the join and meet tables of $(\mathcal{S}, \cup, \cap)$ (Fig.s 1.2.1-2). Figure 1.2.3 shows the Hasse diagram of the same structure. Obviously, $\mathcal{L} = (\mathcal{S}, \cup, \cap)$ is a complete lattice with $\bigwedge A = \emptyset$ and $\bigvee A = \{a, b, c\}$.

Definition. 1.2.16. A lattice $\mathcal{L} = (\mathcal{R}, \vee, \wedge)$ is *distributive* iff the following are true for all $x, y, z \in \mathcal{R}$:

1. $x \wedge (y \vee z) = (x \wedge y) \vee (x \wedge z)$

2. $x \vee (y \wedge z) = (x \vee y) \wedge (x \vee z)$

Proposition. 1.2.17. *If a lattice is non-distributive, it must contain a sub-lattice isomorphic to one of the lattices in Figure 1.2.4.*

Proof: Left as an exercise.

Definition. 1.2.18. A lower bound in a lattice $\mathcal{L}$ is an element $0 \in \mathcal{R}$ such that $0 \wedge x = 0$ (equivalently, $0 \leq x$) for all $x \in \mathcal{R}$. An upper bound in $\mathcal{L}$ is an element $1 \in \mathcal{R}$ such that $1 \vee x = 1$ (equivalently, $x \leq 1$) for all $x \in \mathcal{R}$. A lattice $\mathcal{L}$ is *bounded* if it contains a lower bound and an upper bound. An element x in a bounded lattice $\mathcal{L}$ such that $0 < x$ and there is no $y \in \mathcal{L}$ such that $0 < y < x$ is called an *atom* of $\mathcal{L}$.

Definition. 1.2.19. A lattice $\mathcal{L}$ is *complemented* if

1. it is bounded;

2. for each $x \in \mathcal{R}$ there is a $y \in \mathcal{R}$, called the complement of x, such that $x \vee y = 1$ and $x \wedge y = 0$.

Definition. 1.2.20. A complemented distributive lattice is called a *Boolean lattice*.

Definition. 1.2.21. Let the structure $\mathcal{B} = (A, ', \wedge, \vee, 0, 1)$ be a Boolean lattice. Then, the following properties are satisfied for all $x, y, z \in A$ and for the distinguished elements 0 and 1:

$\cup$	$\emptyset$	$\{a\}$	$\{b\}$	$\{c\}$	$\{a,b\}$	$\{a,c\}$	$\{b,c\}$	$\{a,b,c\}$
$\emptyset$	$\emptyset$	$\{a\}$	$\{b\}$	$\{c\}$	$\{a,b\}$	$\{a,c\}$	$\{b,c\}$	$\{a,b,c\}$
$\{a\}$	$\{a\}$	$\{a\}$	$\{a,b\}$	$\{a,c\}$	$\{a,b\}$	$\{a,c\}$	$\{a,b,c\}$	$\{a,b,c\}$
$\{b\}$	$\{b\}$	$\{a,b\}$	$\{b\}$	$\{b,c\}$	$\{a,b\}$	$\{a,b,c\}$	$\{b,c\}$	$\{a,b,c\}$
$\{c\}$	$\{c\}$	$\{a,c\}$	$\{b,c\}$	$\{c\}$	$\{a,b,c\}$	$\{a,c\}$	$\{b,c\}$	$\{a,b,c\}$
$\{a,b\}$	$\{a,b\}$	$\{a,b\}$	$\{a,b\}$	$\{a,b,c\}$	$\{a,b\}$	$\{a,b,c\}$	$\{a,b,c\}$	$\{a,b,c\}$
$\{a,c\}$	$\{a,c\}$	$\{a,c\}$	$\{a,b,c\}$	$\{a,c\}$	$\{a,b,c\}$	$\{a,c\}$	$\{a,b,c\}$	$\{a,b,c\}$
$\{b,c\}$	$\{b,c\}$	$\{a,b,c\}$	$\{b,c\}$	$\{b,c\}$	$\{a,b,c\}$	$\{a,b,c\}$	$\{b,c\}$	$\{a,b,c\}$
$\{a,b,c\}$	$\{a,b,c\}$	$\{a,b,c\}$	$\{a,b,c\}$	$\{a,b,c\}$	$\{a,b,c\}$	$\{a,b,c\}$	$\{a,b,c\}$	$\{a,b,c\}$

Figure 1.2.1.: Join table of 2^A.

∩	∅	{a}	{b}	{c}	{a,b}	{a,c}	{b,c}	{a,b,c}
∅	∅	∅	∅	∅	∅	∅	∅	∅
{a}	∅	{a}	∅	∅	{a}	{a}	∅	{a}
{b}	∅	∅	{b}	∅	{b}	∅	{b}	{b}
{c}	∅	∅	∅	{c}	∅	{c}	{c}	{c}
{a,b}	∅	{a}	{b}	∅	{a,b}	{a}	{b}	{a,b}
{a,c}	∅	{a}	∅	{c}	{a}	{a,c}	{c}	{a,c}
{b,c}	∅	∅	{b}	{c}	{b}	{c}	{b,c}	{b,c}
{a,b,c}	∅	{a}	{b}	{c}	{a,b}	{a,c}	{b,c}	{a,b,c}

Figure 1.2.2.: Meet table of 2^A.

a.	$x \wedge y = y \wedge x$	*Commutativity of $\wedge$*
	$x \vee y = y \vee x$	*Commutativity of $\vee$*
b.	$x \wedge (y \wedge z) = (x \wedge y) \wedge z$	*Associativity of $\wedge$*
	$x \vee (y \vee z) = (x \vee y) \vee z$	*Associativity of $\vee$*
c.	$x \wedge (y \vee z) = (x \wedge y) \vee (x \wedge z)$	*Distributivity of $\wedge$*
	$x \vee (y \wedge z) = (x \vee y) \wedge (x \vee z)$	*Distributivity of $\vee$*
d.	$0' = 1$	*Complementation*
	$1' = 0$	*Complementation*
e.	$(x')' = x$	*Complementation*
f.	$x \wedge x' = 0$	*Complementation*
	$x \vee x' = 1$	*Complementation*
g.	$x \wedge 1 = x$	*Identity element of $\wedge$*
	$x \vee 0 = x$	*Identity element of $\vee$*

1. From the properties of distributivity, complementation, and identity it follows that in a Boolean lattice we have:

$$x = x \wedge y \text{ iff } x \vee y = y$$

2. Then, the relation $x \leq y$ is a partial order with least element 0 and greatest element 1, and with respect to $\leq$ we have for the operations of *meet* ($\wedge$) and *join* ($\vee$), respectively:

$$x \wedge y = inf(x, y)$$

$$x \vee y = sup(x, y)$$

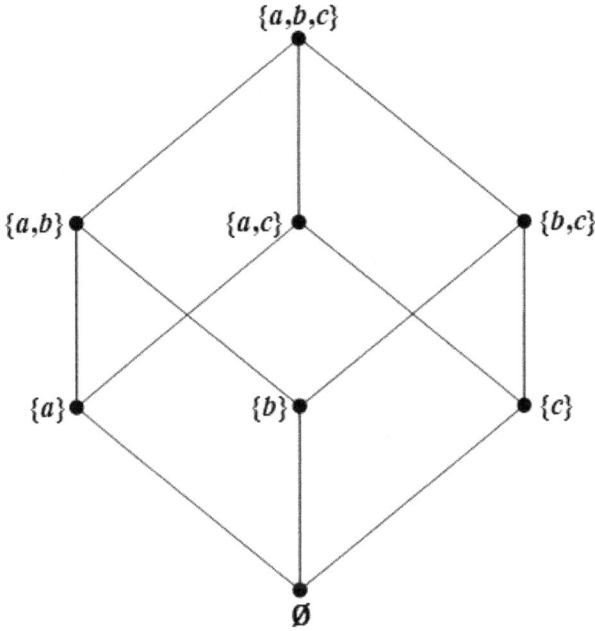

Figure 1.2.3.: The lattice $(\mathcal{S}, \cup, \cap)$.

Proposition. 1.2.22. *The power set of any non-empty set forms a Boolean algebra with the two operations meet and join defined as intersection ($\cap$) and union ($\cup$), respectively.*

Proof: Left as an exercise.

Proposition. 1.2.23. *Every finite lattice is bounded. Given the poset $\mathcal{R} = (\{x_1, ..., x_n\}, \leq)$, then we have the following equalities:*

$$1 = x_1 \vee ... \vee x_n = \bigvee_{i=1}^{n} x_i$$

$$0 = x_1 \wedge ... \wedge x_n = \bigwedge_{i=1}^{n} x_i$$

Proof: Left as an exercise.

Definition. 1.2.24. Let $\mathcal{L}$ be a lattice and A, B sets of elements of $\mathcal{L}$.

45

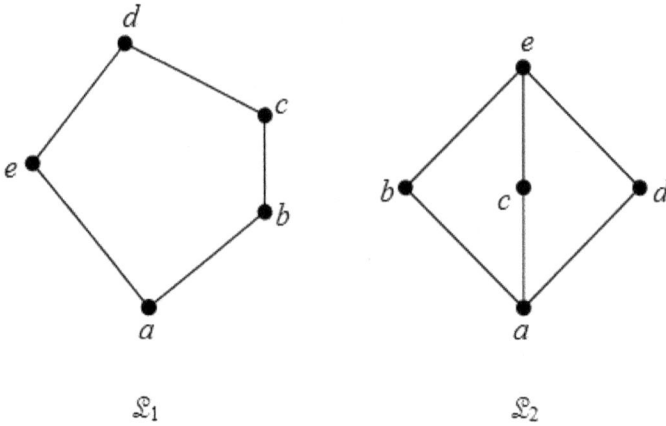

Figure 1.2.4.: The non-distributive lattices $\mathcal{L}_1$ and $\mathcal{L}_2$.

1. A is called an *ideal* of $\mathcal{L}$ if:

- $x, y \in A$ implies that $x \cup y \in A$, and
- $x \in \mathcal{L}$, $y \in A$ and $x \leq y$ imply that $x \in A$.

a) A is said to be a *proper ideal* of $\mathcal{L}$ if $A \neq \mathcal{L}$. A proper ideal A is *maximal* if for any ideal A

$$X \subseteq A \subseteq \mathcal{L} \quad \Rightarrow \quad X = A \text{ or } A = \mathcal{L}.$$

2. B is called a *filter* of $\mathcal{L}$ if:

- $x, y \in B$ implies that $x \cap y \in B$, and
- $x \in \mathcal{L}$, $y \in B$ and $x \geq y$ imply that $x \in B$.

a) B is said to be a *proper filter* of $\mathcal{L}$ if $B \neq \mathcal{L}$. A proper filter B is *maximal* (or an *ultrafilter*) if for any filter B

$$B \subseteq X \subseteq \mathcal{L} \quad \Rightarrow \quad B = X \text{ or } X = \mathcal{L}.$$

Definition. 1.2.25. Let $\mathcal{L}$ be a lattice.

1. For each $x \in \mathcal{L}$, $\downarrow \{x\}$ (cf. Def. 1.1.12.2) is an ideal of $\mathcal{L}$. This particular ideal is called the *principal ideal generated by x*.

2. For each $x \in \mathcal{L}$, $\uparrow \{x\}$ (cf. Def. 1.1.12.3) is a filter of $\mathcal{L}$. This particular filter is called the *principal filter generated by x*.

Proposition. 1.2.26. *In a finite lattice, every filter and every ideal is principal.*

Proof: Left as an exercise.

Proposition. 1.2.27. *Let $\mathcal{L}$ be a lattice. Then, the following holds for all $x, y \in \mathcal{L}$:*

$$[x, y] = (\uparrow \{x\}) \cap (\downarrow \{y\})$$

Proof: Left as an exercise.

Definition. 1.2.28. A subset X of a lattice $\mathcal{L}$ is said to have the *finite-meet property* if whenever $x_1, ..., x_n \in X$, then $\bigwedge_{i=1}^{n} x_i \neq 0$.

Exercises

Exercise 1.2.2.1. Prove by means of the properties in Definition 1.2.21 that the relation $\leq$ is a partial order.

Exercise 1.2.2.2. Show that under the operations of join and meet every Boolean algebra is a distributive lattice.

Exercise 1.2.2.3. Prove that, in every lattice, the validity of the property of the distributivity of $\wedge$ ($\vee$) implies the validity of the property of the distributivity of $\vee$ ($\wedge$, respectively).

Exercise 1.2.2.4. Prove that in a distributive lattice with 0 and 1, if there is a complement $'$, then it is unique.

Exercise 1.2.2.5. Show that the following property, known as *absorption law*, holds in a Boolean algebra:

$$x \wedge (x \vee y) = x \vee (x \wedge y) = x.$$

Exercise 1.2.2.6. Show that the kernel (hull) of any Boolean homomorphism is a Boolean ideal (filter, respectively).

Exercise 1.2.2.7. Prove that B is an ultrafilter iff B is the hull of a homomorphism $h : B \longrightarrow 2$.

Exercise 1.2.2.8. Consider the posets of Figures 1.1.1-3.

1. Are these posets lattices? Account for your answer.

2. What lattices can you extract from their subsets? List them all.

Exercise 1.2.2.9. Prove the theorems and propositions in this Section.

Exercise 1.2.2.10. Research into the *lattice model of information flow* and give its main features.

1.2.3. Graphs and trees

Definition. 1.2.29. *Graphs* – Let $V = \{v_0, v_1, ..., v_n\}$ be a set of vertices and $E = \{e_0, e_1, ..., e_n\}$, $e_i = \{v_j, v_k\}$, be a set of edges. A *graph* $\mathfrak{G}$ is a pair (V, E) such that all the endpoints v_j, v_k of E (i.e., nodes) are contained in V.

1. In a *directed graph* (or *digraph*) $\vec{\mathfrak{G}} = \left(V, E, \vec{f}\right)$, V is as for $\mathfrak{G}$, E is a set of arcs, i.e. directed edges ("arrows"), and $\vec{f}$ is an incidence function assigning to each member of E a tail and a head, indicating the origin and the end of the edges, respectively.

2. A *multi-edge* is a set of ≥ 2 edges with the same endpoints.

3. A *loop* is an edge or arc joining a vertex to itself.

4. A *simple* graph (digraph) is a graph $\mathfrak{G}$ with no loops or multi-edges (a digraph $\vec{\mathfrak{G}}$ with no self-loops and no pair of arcs with the same head and tail, respectively).

5. The edge $e_i = \{v_j, v_k\}$ *connects* the vertices v_j, v_k.

6. A *path* ε in a graph $\mathfrak{G}$ is an alternating sequence of vertices and edges of the form $v_0, e_1, v_1, e_2, ..., e_{n-1}, v_{n-1}, e_n, v_n$ in which each e_i contains both v_{i-1} and v_i.

7. A graph $\mathfrak{G}$ is *connected* if each pair of vertices is joined by a path. The number of edges is the *length* of the path.

8. If $v_0 = v_n$ in a path, then the path is said to be *closed*.

9. A closed path is *directed* if $v_0, v_1, ..., v_n$ are pairwise distinct vertices of a digraph $\vec{\mathfrak{G}}$.

10. A *cycle* is a closed path of length ≥ 3.

11. A cycle is *directed* if it is a closed directed path.

Example 1.2.5. Figure 1.2.5 shows a simple graph. Figure 2.2.3 shows a labeled digraph, which is a state diagram of a finite automaton. Every automaton, including Turing machines, can be modeled by means of a labeled digraph (see Section 2.2).

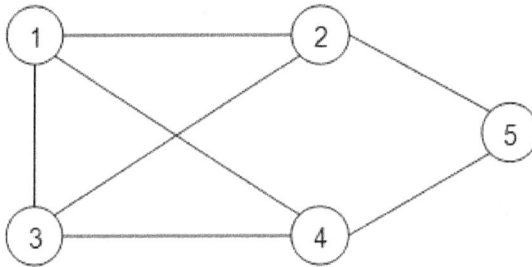

Figure 1.2.5.: A simple graph.

Definition. 1.2.30. *Trees* – A *tree* is a connected graph containing no cycles.

1. A *subtree* is a subgraph of a graph that is also a tree.

2. A tree is said to be *finite* if it has a finite number of vertices (*nodes*) and edges (*branches*). Otherwise, it is said to be *infinite*.

3. A tree is *labeled* if it has labels attached to its vertices.

4. A tree is *rooted* if it has a least node s_0.

5. In a rooted tree, the *ancestor* (the *descendant*) of a vertex v is either v itself or any vertex that is the *predecessor* (the *successor*, respectively) of v on a path from the root. A *proper ancestor* (*proper descendant*) of a vertex v is any ancestor (descendant, respectively) except v itself.

6. In a rooted tree, the *child* of a vertex v is a vertex such that v is its immediate ancestor. A vertex that is the immediate predecessor of v on the unique path from the root to v is called the *parent* of vertex v.

7. In a rooted tree, a *leaf* is a vertex that has no children. A vertex that is not a leaf is an *interior* vertex.

8. A tree is *ordered* if it is rooted and the children of each internal vertex are linearly ordered.

9. The *depth* of a vertex v in a rooted tree is the number of edges in the unique path from the root to v.

10. A tree is said to be *binary* if it is an ordered rooted tree and each of its vertices has at most two children, a right and a left child.

11. A binary tree is said to be *complete* if every parent thereof has two children and all leaves are at the same depth.

Example 1.2.6. Figures 7.3.2-3, as well as Figure 8.1.1, are examples of labeled binary trees. The analytic tableaux calculus is based on labeled binary trees (see, e.g., Fig. 8.2.2): A proof in this calculus is a complete (closed) labeled binary tree. Figure 2.1.1 shows a string derivation tree, i.e. an ordered labeled tree. A parse tree (Fig. 2.1.3) is also an ordered labeled tree.

Exercises

Exercise 1.2.3.1. Given the graph $\mathfrak{G} = (V, E)$, a $n \times n$ *adjacency matrix* $A_{\mathfrak{G}} = (a_{ij})$ is a matrix in which each a_{ij} entry satisfies

$$a_{ij} = \begin{cases} 1 & \text{if } (v_i, v_j) \in E \\ 0 & \text{otherwise} \end{cases}.$$

Draw the graphs with the following adjacency matrices:

1.

$$A_{\mathfrak{G}_1} = \begin{pmatrix} 0 & 0 & 1 & 1 \\ 0 & 0 & 1 & 0 \\ 1 & 1 & 0 & 1 \\ 1 & 1 & 1 & 0 \end{pmatrix}$$

2.

$$A_{\mathfrak{G}_2} = \begin{pmatrix} 0 & 1 & 1 \\ 1 & 0 & 0 \\ 1 & 0 & 0 \end{pmatrix}$$

3.

$$A_{\mathfrak{G}_3} = \begin{pmatrix} 1 & 1 & 0 & 1 \\ 0 & 0 & 0 & 1 \\ 1 & 1 & 0 & 1 \\ 1 & 0 & 0 & 1 \end{pmatrix}$$

Exercise 1.2.3.2. Describe formally the indicated graphs. (Check the Table of Figures for the page numbers.)

1. Figure 2.2.3.

2. Figure 2.2.5.2.

3. Figure 2.2.10.

4. Figure 2.2.13.

Exercise 1.2.3.3. Represent by means of labeled rooted trees the following statements:

1. $\left[\left(a + \frac{b}{2}\right) \cdot c\right] - d$

2. $(\neg p \vee q) \rightarrow [\neg (r \wedge (s \leftrightarrow \neg q))]$

3. $b_i [3 * [\S a]]$

4. $(6x - 5) + (2y/3)(2x)^2$

5. $\sqrt{\frac{a - 6b^3}{c}}$

Exercise 1.2.3.4. Let there be given a graph $\mathfrak{G} = (V, E)$ with $|V_{\mathfrak{G}}| = n > 1$. Show that the following statements are equivalent:

1. $\mathfrak{G}$ is a tree.

2. $\mathfrak{G}$ has no cycles and $|E_{\mathfrak{G}}| = n - 1$.

3. $\mathfrak{G}$ is connected and $|E_{\mathfrak{G}}| = n - 1$.

Exercise 1.2.3.5. Show that a finite tree with n vertices must have $n - 1$ edges.

Exercise 1.2.3.6. Describe formally the indicated trees. (Check the Table of Figures for the page numbers.)

1. Figure 2.1.5.

2. Figure 5.3.2.

3. Figure 7.3.2.

4. Figure 8.1.1.

5. Figure 8.2.2.

1.3. Mathematical proof

As a mathematical subject, the theory of classical computation requires mastery of the basic proof techniques, which are *mathematical induction* and *proof by contradiction* (often also spoken of as *reductio ad absurdum*). In computational logic, these two techniques have a flavor of their own, but it is useful to know how to use them in the context of discrete mathematics. We duly provide the essential notions of these techniques; a more comprehensive discussion of these and other proof techniques can be found in Bloch (2011).

Of course, mathematical logic and logic in general are all about proving theorems, and computational logic is all about doing so by computational means, but in fact mathematical induction is applied at the metalanguage level rather than at the object-language level–contrarily to proofs by contradiction, which also feature at the latter level. In effect, though mathematical induction is an inference rule characteristic of deductive reasoning, logical proofs proper do not apply it (see Chapters 5, 8 and 9), but in order to be assured that the expressions of an object-language are well-formed, and that the relation of logical consequence between such expressions holds, we need to apply it at the metalanguage level, i.e. the level at which the objects of logic and their relations are analyzed from the viewpoint of their mathematical properties. As for proof by contradiction, the reader will have the opportunity of applying this technique frequently in Parts IV and V. For further aspects of logical proofs, see Augusto (2019).[9]

---

[9]Sometimes, constructive proofs are also called for, it being understood by such proofs that actually an object must be constructed. We use this proof technique sparingly, and it should be obvious when employed. We give some notes on further methods of mathematical proof in the appropriate places.

1.3.1. Mathematical induction

In a proof by *mathematical induction* (MI), we require two main steps: the *basis (step)* and the *induction step*. Say that we have a string of n symbols; the idea is that if we can prove that this string is legal for $n = 1$ (or some other $n \in \mathbb{N}^+$;[10] the basis), then we can prove (the induction step) that a string formed with arbitrarily many n symbols is equally legal by proving that if a string formed with m symbols is legal–the *induction hypothesis*–then a string with $m + 1$ symbols is also legal. MI proper is restricted to the set of natural numbers, but we can generalize it to any well-founded mathematical structure (e.g., lattices, trees), in which case we speak of *structural induction.*[11] It is this latter generalization that is of interest to mathematical logic and computer science, but it helps to understand thoroughly MI proper.

Example 1.3.1. The following is an often used example to introduce MI. We wish to prove that the following statement $P(n)$ holds for all positive natural numbers n.[12]

$$(P(n)) \qquad \sum_{i=1}^{n} i = \frac{n(n+1)}{2}.$$

Induction basis: We show that $P(1)$ holds. In effect, we have

$$1 = \frac{1(1+1)}{2}.$$

Induction step: We now show that if $P(m)$ holds, then $P(m+1)$ holds, too. We begin by assuming that $P(m)$ holds for arbitrary $m \in \mathbb{N}$ and then show that $P(m+1)$ also holds:

$$(P(m+1)) \qquad \sum_{i=1}^{m+1} i = \sum_{i=1}^{m} i + (m+1) =$$

$$= \frac{m(m+1)}{2} + (m+1) = \frac{m^2 + 3m + 2}{2} = \frac{(m+1)(m+2)}{2} =$$

$$= \frac{(m+1)((m+1)+1)}{2}$$

---

[10]More rarely, $n = 0$ for the basis. Even more rarely, $n \in \mathbb{Z}$.

[11]Throughout this text, when we write "by induction" with respect to proofs we often mean "by structural induction." When we write "define inductively"–which we often do below–, we could equally well–or (more) correctly, according to some views–write "define recursively."

[12]$\sum_{i=1}^{n} i$ is an abbreviation for the *summation* $1 + 2 + ... + n$, $1 \leq i \leq n$.

and what was to be proven has been proven (abbreviated **QED**, for the Latin expression *"quod erat demonstrandum"*).

We formalize the above:

Definition. 1.3.1. *Principle of mathematical induction:* Let P be a proposition defined on the set $\mathbb{N}^+$. Then, for each $n \in \mathbb{N}^+$, $P(n)$ either holds or does not hold. Suppose that $P(1)$ holds and $P(m+1)$ holds whenever $P(m)$ holds. Then, $P(n)$ holds for arbitrary $n \in \mathbb{N}^+$.

Below, we often abbreviate the proofs by structural induction, namely for logical formulae. A non-abbreviated proof by structural induction for logical formulae is given in Example 1.3.2, where property P is well-formedness (compare with Definition 3.1.3, for instance).

Example 1.3.2. Let P be a certain property such that:
(i) *Induction basis:* The symbols $\top$ and $\bot$, and every propositional atom satisfy P.
(ii) *Induction steps:*

- ($\neg$) Suppose ϕ satisfies P (*induction hypothesis*) and let $\neg$ be a unary operator. Then $\neg\phi$ satisfies P.

- ($\heartsuit$) Suppose ϕ and ψ satisfy P (*induction hypothesis*) and let $\heartsuit$ be a binary operator. Then $\phi\heartsuit\psi$ satisfies P.

- ($\blacklozenge$) Suppose ϕ satisfies P (*induction hypothesis*); let $\blacklozenge$ be a quantifier and x be a variable symbol. Then $\blacklozenge x(\phi)$ satisfies P.

The principle of induction for logical formulae states that if all the conditions above are satisfied, then every first-order logical formula satisfies property P.

1.3.2. Proof by contradiction

Proof by contradiction is a method of mathematical proof theoretically based on the following properties of Boolean expressions (cf. Section 1.2.1): For any (propositional) variable p, it holds that

$$\text{(LEM)} \qquad p \vee \neg p = 1$$

and

$$\text{(LC)} \qquad p \wedge \neg p = 0.$$

In logical jargon, we say that for every proposition p, either p is true or its negation (i.e. $\neg p$) is, a principle known as *law of excluded middle* (LEM), and p cannot be both true and false, known as *law of contradiction* (LC).

These two laws become the basis of a proof by contradiction in the following way: Of a given proposition p, we may assume that it is either true or false. Let us assume that p is false, so that $\neg p$ is true. If by the end of our reasoning we have reached a proposition of the form $q \wedge \neg q$, for some proposition q, then we have clearly–though indirectly[13]–reached a contradiction, and the only way out of it is to retract the truth of $\neg p$ and accept the truth of p instead.[14]

We formalize this:

Definition. 1.3.2. A proof by contradiction has the following structure:

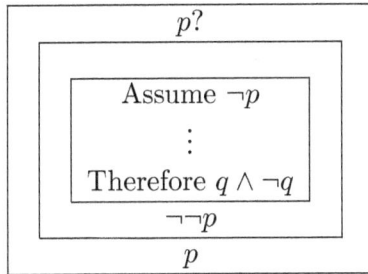

$$
\boxed{
\begin{array}{c}
p? \\[4pt]
\boxed{
\begin{array}{c}
\boxed{
\begin{array}{c}
\text{Assume } \neg p \\
\vdots \\
\text{Therefore } q \wedge \neg q
\end{array}
} \\
\neg\neg p
\end{array}
} \\
p
\end{array}
}
$$

Example 1.3.3. This is a classic proof by contradiction of the irrationality of the number $\sqrt{2}$. Let there be given the proposition

The number $\sqrt{2}$ is irrational.

We assume, for the sake of contradiction, that this proposition is false, i.e. we assume that $\sqrt{2}$ is rational. If $\sqrt{2}$ is rational, then there are

---

[13]Reason why proofs by contradiction are a kind of *indirect proof*.

[14]Although proof by contradiction is often seen as equivalent to proof by *reductio ad absurdum*, it actually is a particular kind of the general form of proofs that involve an absurd, or impossible, conclusion. In fact, *reductio ad absurdum* is more properly used for argumentation, any form thereof resulting (supposedly) in absurdity qualifying as such. For example, the reasoning

If that's so, then pigs can fly.

fits into the broadest sense of *reductio ad absurdum*. This notwithstanding, it is also very frequent in mathematical reasoning to refer to a proof by contradiction as a proof by *reductio (ad absurdum)*. See Augusto (2019b) for a discussion of this topic.

1. Mathematical and computational notions

integers a and b, $b \neq 0$, for which

$$(2.1) \qquad \sqrt{2} = \frac{a}{b}.$$

We additionally assume that $\frac{a}{b}$ is reduced to its simplest terms, which entails that not both of a and b can be even. So, one or both of a and b must be odd. From 2.1, it follows that

$$(2.2) \qquad 2 = \left(\frac{a}{b}\right)^2 = \frac{a^2}{b^2}$$

or

$$a^2 = 2b^2$$

and the square of a is an even number. Hence, a itself is an even number. Let us now replace $a = 2k$ in 2.2; we have

$$2 = \frac{(2k)^2}{b^2} = \frac{4k^2}{b^2}$$

and hence

$$2b^2 = 4k^2$$

or equivalently

$$b^2 = 2k^2$$

and we have it that b is itself even. So we have it that both of a and b are even, but we know that not both of a and b can be even, so that we have reached a contradiction. We reject the statement that $\sqrt{2}$ is a rational number, as being false, and accept the irrationality of $\sqrt{2}$. **QED**

Exercises

Exercise 1.3.1. Prove the following statements by applying MI:

1. $\sum_{i=0}^{n} 2^i = 2^{n+1} - 1$ (for all $n \in \mathbb{N}$).

2. $\sum_{i=1}^{n} i = n^2$ (for all $n \in \mathbb{N}^+$, n is odd).

3. $n < 2^n$ (for all $n \in \mathbb{N}$).

4. $n^3 - n$ is divisible by 3 (for $n \in \mathbb{N}^+$).

5. If $|S| = n$ for a set S and finite n, then S has 2^n subsets.

6. For $n \geq 2$ and $A_i \in U$, U is some universal set, $\overline{\bigcap_{i=1}^{n} A_i} = \bigcup_{i=1}^{n} \overline{A_i}$.

Exercise 1.3.2. Prove by MI that every formula ϕ has the same number of open and closed parentheses.

Exercise 1.3.3. Comment further on the ubiquity of MI in mathematical logic and computer science. (Hints: Parameters; infinity.)

Exercise 1.3.4. Recall from Section 1.1 the set of the reals $\mathbb{R}$. Why is it the case that in MI proofs $n \notin \mathbb{R}$?

Exercise 1.3.5. Prove the following statements by applying the method of proof by contradiction:

1. If $3n + 2$ is odd, then n is odd.

2. If A and B are sets, then $A \cap (B - A) = \emptyset$.

3. At least 4 of 22 chosen days fall on the same day of the week.

4. The sum of an irrational number and a rational number is an irrational number.

1.4. Algorithms and programs

Finally, no textbook on computational logic would be complete without an introduction to algorithms and programs, mathematical constructs that will be required throughout this text. We shall not focus on programming in general or programming languages other than those in the logic-programming paradigm, and we refer the reader to Sebesta (2012) and Scott (2009) for these foci. We refer the reader interested in a comprehensive general study of algorithms to Cormen et al. (2009).

We begin by giving the essential features of an algorithm and elaborate then briefly on additional aspects.

Definition. 1.4.1. Given a specific input I, an *algorithm* Ψ is a finite sequence of precise instructions for a procedure Ψ leading to a solution as the output O. Formally, we have the triple $\Psi = (I, \Psi, O)$.

A procedure Ψ is said to be an *effective procedure* if, besides being an unambiguous and well-ordered set of instructions, it is carried out in a discrete, deterministic way. In general terms, the solution is a "Yes/No" answer if the effective procedure terminates. If the effective procedure (eventually) ends in a "Yes" answer, then it, together with the input and the output, is said to be an algorithm to solve a particular (class of)

problem. More often than not, the "Yes" answer is actually some specific
output, in particular when Ψ is a function. In any case, an algorithm is
generally constituted by an *input*, an *output*, and the *procedure* between
both. This is so even when the algorithm appears not to be so formally
describable: For instance, a burnt light-bulb might be seen as the input
to an algorithm to change a defective light-bulb, the working light-bulb
as the output, and actions such as using a ladder to reach the defective
light-bulb, unscrewing it, checking whether it is indeed defective (e.g.,
burnt), etc., are the *steps* constituting the procedure intermeadiating
between the input and the output.

   We shall not be concerned with such algorithms as for changing a
burnt light-bulb. Specifically, the algorithms we shall be interested in are
called *computational algorithms*, as they arise in the context of classical
computation. In other words, they fall in the algorithmic tradition that
started with D. Hilbert's *Entscheidungsproblem* and A. Turing's famous
negative answer to this problem via the Turing machine. In this narrow
context, a computation is formally denoted by

$$q_0, I \vdash_M^* O, q_f$$

where I and O denote the input and the output, respectively, q_0 is the
initial state of computing machine M, q_f is its final state, $\vdash$ denotes a
single step, and $\vdash_M^*$ denotes that M requires zero or more ($*$) individual
steps to perform the computation, or carry out procedure Ψ, so that in
fact we have

$$q_0, \underbrace{I \vdash_M^* O}_{\Psi}, q_f$$

$$\underbrace{}_{\Psi}$$

for an algorithm Ψ_M for machine M.[15] When this computation is carried
out in a classical computing device or medium (see Introduction) in such
a way that the intermediate procedure does not change (say, it does not
evolve), we speak of a *classical algorithm*.

   In order for a computing machine like a digital computer to carry out
an algorithm it requires a program, which, in turn, requires a programming language:

Definition. 1.4.2. Let L denote a precisely-defined language, i.e. a
formal language. Given an algorithm $\Psi = (I, \Psi, O)$ such that $I, O \in L$
and $\Psi \subseteq L$, then L is a *programming language* and $\Psi_L = (I_L, \Psi_L, O_L) =$

---

[15]This formal representation allows for the carrying-out of algorithms to be precisely
   measured in terms of complexity, i.e. we can measure the temporal and/or spatial
   costs of an algorithmic procedure Ψ, a topic we discuss in Section 2.3 below.

$\Pi_{M,L}$ is called a *(computer) program (written for M in L)* if there is a computing machine M that can *implement* Ψ_L so that, given input I_L, it outputs O_L by carrying out Ψ_L.

This definition entails that Ψ_L is a sequence of *instructions* rigorously formulated in a machine-readable language L, called *code*. In effect, we now have

$$\underbrace{\underbrace{q_0, I_L \vdash_M^* O_L, q_f}_{\Psi_L}}_{\Pi_{M,L}}$$

where it is evident that states q_0, q_f of machine M are considered part of the program. In effect, a computer carries out a specific computation, or algorithm, by following the instructions that amount to the correct sequence of configurations it has to realize as a–physical or abstract–machine. This said, programmers do not in fact need to write the code for the machine, as this is an exceedingly difficult task that is better accomplished by a compiler, a piece of software that translates the high-level languages used by programmers, such as Java, C, Pascal, Prolog, etc., into the low-level, or machine, languages. For this reason, henceforth we concentrate on high-level languages, in which we include *pseudo-code*, a "programming" language that is very similar to English and whose use aims at allowing humans to read the algorithm more or less easily.[16]

Example 1.4.1. The *Euclidean algorithm* is one of the best–and oldest–algorithms known for finding the greatest common divisor of two integers a and b, denoted by $\gcd(a,b)$ and defined as

$$\gcd(a,b) = \begin{cases} a & \text{if } b = 0 \\ \gcd(b, a \bmod b) & \text{if } b > 0 \end{cases}.$$

The operation of finding $\gcd(a,b)$ is computable by means of an algorithm, because the greatest common divisor of two integers–not both

---

[16] As stated in the Preface, for the sake of generality we shall not stick to any programming language, but we shall also not stick to a particular pseudo-code. The aim of the algorithms provided in this book is the carrying-out "by hand" of specific computations and our belief is that this requires a mathematical–in any case notational–understanding of the input, the output, and the intermediate procedure; for this reason, our algorithms are frequently just mathematical descriptions of these elements. This said, we use expressions such as **if ... then ... else, while ... do**, etc., which are ubiquitous in computing.

zero–exists, as the set of their common divisors is finite and non-empty. Although there are other methods to compute gcd (a, b) (e.g., prime factorization), the Euclidean algorithm is a very efficient procedure. For pedagogical reasons, we restrict our example to non-negative integers. The input to the algorithm is constituted by the two integers a and b, and the output is gcd $(a, b) = d$, d is the greatest integer such that $d|a$ and $d|b$ (see Example 1.1.4). The simplest algorithm for this computation, which is recursion-based, is as follows:

Input: integers a, b such that $a > 0$ and $b \geq 0$
Output: gcd (a, b)

if $b = 0$
    return a
else return gcd $(b, a \bmod b) =$ gcd (b, r)

This is actually a program, as the algorithm is formulated in pseudo-code. In other words, the computing "machine" is here a human. However, it often pays off for a programmer to write a program first in pseudo-code, and then "translate" this into real code. Note in this example how the procedure in pseudo-code corresponds to the mathematical definition (given above) of the function for the greatest common divisor. It is evident that in case $b \neq 0$ one needs to compute gcd $(b, a \bmod b)$, which is only possible if one knows how to carry out a modulo operation. In particular, one needs to know the theoretical basis for this requirement, which is given by the result (a lemma) "if $a = bq + r$, a, b, q, r are all integers, then gcd $(a, b) =$ gcd (b, r)." For instance, we apply the above definition to find gcd $(723, 320)$:

$$
\begin{aligned}
723 &= (320 \times 2) + 83 \\
320 &= (83 \times 3) + 71 \\
83 &= (71 \times 1) + 12 \\
71 &= (12 \times 5) + 11 \\
12 &= (11 \times 1) + 1 \\
11 &= (1 \times 11) + 0
\end{aligned}
$$

We obtain the result gcd $(723, 320) = 1 =$ gcd $(1, 0)$. Because more often than not $b \neq 0$, then a more explicit algorithm is required.

Example 1.4.2. The following is a more explicit version of the Euclidean algorithm:

Input: integers a, b such that $a > 0$ and $b \geq 0$
Output: $\gcd(a, b)$

make
    $x = a$
    $y = b$
while $y \neq 0$
    $r = x \bmod y$
    $x = y$
    $y = r$
    return x

From the short discussion above, it should be clear that the following properties are required for any classical algorithm:

- *Finiteness*: No matter how large the number of steps between the input and the output might be, the set of steps constituting an algorithmic procedure is finite.

- *Generality*: The procedure should be applicable to any given values of a specific problem, not just to particular input values thereof.

- *Effectiveness*: Each step should be completable in finite time. Otherwise, the algorithm is inefficient.

While the two first properties are easily attainable when designing an algorithm, the last one depends on many factors. In particular, an algorithm Ψ_A can be classified as *more* or *less efficient* than another algorithm Ψ_B for the same input-output depending on the physical resources, namely time or space, it requires to carry out the intermediate procedure Ψ. Hence, the classification by their complexity is a fundamental one for the design and analysis of classical algorithms, and we discuss this topic in Section 2.3 below. For other types of algorithm classification (method, task, etc.), we provide some research guidelines in the Exercises.

Exercises

Exercise 1.4.1. Bézout's theorem states that if a and b are positive integers, then there exist integers s and t such that $\gcd(a, b) = sa + tb$. From this theorem, which defines $\gcd(a, b)$ as a linear combination in which s and t are the (Bézout) coefficients of a and b, respectively, an algorithm to compute gcd (a, b) can be designed. Design it using a pseudo-code similar to that in Examples 1.4.1-2.

Exercise 1.4.2. Consider this algorithm to compute gcd (a, b):

Input: integers a, b such that $a > 0$ and $b \geq 0$
Output: gcd (a, b)

if $b = 0$
    return $(a, 1, 0)$
else $(d', x', y' = \gcd(b, a \bmod b))$
    $(d, x, y) = (d', y'x' - \lfloor a/b \rfloor y')$
    return (d, x, y)

1. Find out the mathematical result(s) it was designed upon.

2. Compare it with the Euclidean algorithm in Example 1.4.2: Which do you think is the most efficient? Account for your answer.

Exercise 1.4.3. Research into the following types of algorithms as according to a very broad classification:

1. Divide-and-conquer algorithms.

2. Greedy algorithms.

3. Dynamic programming.

4. Linear programming.

Exercise 1.4.4. A more fine-grained classification of algorithms is given below. For each type, (a) give a definition, (b) give the most important instances, and (c) indicate relevant applications in classical computing:

1. Sort algorithms.

2. Search algorithms.

3. String matching and parsing algorithms.

4. Hashing.

Part III.

Classical Computing

2. Fundamentals of classical computing

In the large field of computing, we may today speak of a great divide between *classical* and *non-classical computing*. This divide, in turn, corresponds roughly to a distinction between *hard* and *soft computing*. Examples of the latter are artificial neural networks, Bayesian networks, and evolutionary algorithms. The former is exceptionally physically realized in the so-called *digital computer*, typically abbreviated as simply *computer*. In this book, and unless otherwise stated, a *computer* is any machine, physical or abstract, that carries out classical, or hard, computing; equally, we shall often abbreviate *classical computing* simply as *computing*.

Computers essentially *make logical decisions* and *calculate functions*, and they do so by manipulating strings of symbols. In terms of some formal language, this entails *recognizing* its strings and *accepting* them, often in order to *generate* a string as output. Thus, generation, recognition, and acceptance of strings of a formal language are at the root of the various *models of classical computation* (henceforth just *models of computation*). Of these, we are particularly interested in *Turing machines*, given the fundamental results concerning complexity and computability that were obtained thanks to them.

But the Turing machine is only one–though the most powerful–of a kind of mathematical objects associated to models of computation known as *automata*, and, in face of the Chomsky hierarchy and its importance for programming languages, we need to know about less powerful automata. Accordingly, we provide in this Chapter an introductory, yet extensive, treatment of the fundamentals of classical computing, to wit, *formal languages*, their associated *grammars*, and their associated *models of computation*. Because computation is time- and space-bounded models thereof need to address the fundamental topics of *computability* (what can and cannot be computed in principle) and *complexity theory* (what can and cannot be computed in practice).

We give here only the essential aspects of these topics highly relevant for anyone working in the field of computation, including computational logic. A book-length further elaboration on this Chapter is Augusto

(2021). In this, we provide discussions on further topics such as, for example, grammar cleaning, normal form transformations, minimization algorithms, as well as a comprehensive discussion on finite transducers. For "classics" on these topics from a more comprehensive perspective, the reader is referred to Davis & Weyuker (1983) and Hopcroft, Motwani, & Ullman (2013), the first of which contains extensive material on classical logic from a computational perspective.

2.1. Formal languages and grammars

Classical computation is performed over a language that is unequivocally specified.[1] Such a language is called a *formal language*, and this unequivocal specification is provided by a *formal grammar*. A formal language consists of strings of symbols of a specified alphabet that are well formed according to specified formation rules.

Definition. 2.1.1. Let Σ be a non-empty set of symbols. Then, a *string*, or *word*, w on the set Σ is a finite sequence of symbols over Σ. A set Σ over which strings or words are built is called an *alphabet* and the members of Σ are its *letters*.[2]

Example 2.1.1. Let $\Sigma = \{a, b, c\}$. Then, $u = abba = ab^2a$ and $v = cbabbaaac = cbab^2a^3c$, where a^n is an abbreviation for $\overbrace{a...a}^{n}$, are strings on Σ. The sequence $affe$ is *not* a string on Σ, because $f, e \notin \Sigma$.

Definition. 2.1.2. Let w be a string on an alphabet Σ.

1. The *length* of w, denoted by $|w|$, is the number of letters in w. If $|w| = 0$, then we speak of the *empty string*, and we denote it by ϵ.

2. The *frequency* of a letter a in a word w, denoted by $|w|_a$, is the number of occurrences of a in w.

3. A *substring* w' of w is a sequence of consecutive letters occurring in w. A substring that starts (ends) at the leftmost (rightmost, respectively) letter is a *prefix* (*suffix*, respectively).

---

[1] Although this does not guarantee that it is unambiguous, as it will be seen.
[2] Actually, one can distinguish *word* from *string* by defining the former as a finite *or countably infinite* sequence of symbols. Unless otherwise stated, we shall not distinguish between both; moreover, henceforth we shall speak of string.

Example 2.1.2. Let v be the string in Example 2.1.1. Then, $|v| = 9$, $|v|_a = 4$, $|v|_b = 3$, and $|v|_c = 2$. We have $|v|_a + |v|_b + |v|_c = |v|$. Two substrings of v are, for instance, $v' = cbab$ and $v'' = ba^3c$, a prefix and a suffix of v, respectively. The string $cbbb$ is *not* a substring of v.

The following are the basic operations on strings:

Definition. 2.1.3. Let $u = a_1...a_n$ and $v = b_1...b_m$ be two strings on the alphabet $\Sigma = \{a, b\}$.

1. The *concatenation* of u and v, written uv, is the string constituted by the string u followed by the string v, i.e. $a_1...a_nb_1...b_m$. We have $|uv| = |u| + |v|$.

2. Given a string u, we define the *i-th power* of u as follows: $u^0 = \epsilon$, $u^1 = u$ and, for $i > 1$, $u^i = u^{i-1}u$. The length of u^i is $i|u|$.

3. A string w is the *reverse* or *mirror image* of u, denoted by u^R, if $w = a_n...a_1$. A string that is identical to its reverse is a *palindrome*.

4. Let now $v = b_1...b_n$. The *shuffle* of u and v, denoted by $u \,\square\, v$, is the string $a_1b_1...a_nb_n$.

Example 2.1.3. Let Σ be the 26-letter Latin alphabet adopted for English. Then, *bookkeeper* is the concatenation of the strings *book* and *keeper*, and *says* is the concatenation of the strings *say* and *s*. $says^0 = \epsilon$, $says^1 = says$, $says^2 = sayssays$, etc. The proper nouns *Eve*, *Otto*, and *Hannah* are palindromes. The shuffle of *goat* and *milk* is *gmoialtk*.

Proposition. 2.1.4. *The empty string ϵ is an identity element for concatenation, as for some string* u *we have* $\epsilon u = u\epsilon = u$.

Concatenation is generally not a commutative operation when $|\Sigma| > 1$, though it is associative:

1. unless v and u are identical strings we have $uv \neq vu$;

2. given strings u, v, and w, we have $(uv)\,w = u\,(vw)$.

Definition. 2.1.5. A *formal language* is a pair $L = (\Sigma, G)$ where Σ is a (possibly infinite) set of symbols called the *letters* (of Σ),[3] and G is a finite set of *rules* for making *well-formed*, or *legal*, strings of L. Σ is called the *alphabet* of L and G is called the *grammar* (or *syntax*) of L.

---

[3]In practice, alphabets are typically finite sets.

Example 2.1.4. Clearly, a natural language like English is a formal language. The English alphabet (in this formal sense) is constituted by lower and upper case letters of the Latin alphabet, the digits 0,1,...,9, the blank symbol, and various punctuation symbols (e.g., ; and !). The grammar of English is the set of (many!) rules to make well-formed strings. Prolog (see Chapter 9) is a formal language with far fewer rules than English, but much of what can be expressed in English can be so in Prolog as well.

Thus, both natural languages and computer languages are formal languages. Definition 2.1.5 can be further specified as follows:

Definition. 2.1.6. A *(formal) language L over an alphabet* Σ is a collection of strings w on Σ. Let now Σ^* denote the set of all strings (including the empty string) on Σ. Then, a (formal) language L over Σ just is a subset of Σ^* (and any subset of Σ^*, including $\emptyset$, is a language).[4] More formally:

$$L = \{w \in \Sigma^* | w \text{ has property } P\}$$

Example 2.1.5. Let $\Sigma = \{a, b\}$. The following are languages over Σ:

1. $L_1 = \{w \in \Sigma^* | w = a^m b^m, m > 0\} = \{ab, a^2b^2, a^3b^3, ...\}$

2. $L_2 = \{w \in \Sigma^* | w = ab^n, n \geq 0\} = \{a, ab, ab^2, ...\}$

3. $L_3 = \{w \in \Sigma^* | w = w^R\}$, i.e. w is a palindrome

4. $L_4 = \{w \in \Sigma^* | w \text{ has an even number of } a\} = \{aa, aab, bbbaaaa, ...\}$

5. $L_5 = \{w \in \Sigma^* | w \text{ ends with } bb\} = \{bb, abb, babaabb, ...\}$

It should be obvious that if Σ contains at least one letter, then Σ^* is infinite, and because formal languages L over Σ just are subsets of Σ^*, they are in infinite number, too. However, a language L over an alphabet Σ can be finite or infinite.

Example 2.1.6. Let $\Sigma = \{a, b, c, ..., y, z\}$, $|\Sigma| = 26$. Then, the following are languages over Σ, L_{1-3} are infinite languages, and L_{4-7} are finite languages:

1. $L_1 = \Sigma^*$

2. $L_2 = \{w \in \Sigma^* | w \text{ consists solely of vowels}\}$

---

[4]We henceforth often omit the adjective "formal" and speak only of languages.

3. $L_3 = \{w \in \Sigma^* | w = w^R\}$

4. $L_4 = \{w \in \Sigma^* | w$ is a single-letter string$\}$

5. $L_5 = \{my, your, his, her, its, our, their\}$

6. $L_6 = \{w \in \Sigma^* | w$ occurs in Shakespeare's plays$\}$

7. $L_7 = \emptyset$

Given Proposition 2.1.4 above, we have the following mathematical characterization of a language L:

Definition. 2.1.7. Let Σ be an alphabet. Then, Σ^* is a *free monoid* with domain the set of all strings over Σ and with string concatenation as the semigroup operation $\star$. The set $\Sigma^* - \{\epsilon\}$ is the *ϵ-free semigroup* Σ^+ on the alphabet Σ with string concatenation as the semigroup operation $\star$.

This allows us to reformulate Definition 2.1.6 in a more mathematical framework:

Definition. 2.1.8. A *(formal) language* on an alphabet Σ is a subset L of the free monoid Σ^*. An *ϵ-free language* on Σ is a subset of the ϵ-free semigroup Σ^+.

We focus now on operations on languages:

Definition. 2.1.9. Let $L_{(i)}$, $i = 1, 2, ...$, be languages over an alphabet Σ.

1. The *reverse* of a language L, denoted by L^R, is the set
$$L^R = \{w \in \Sigma^* | w^R \in L\}.$$

2. The *concatenation* $L_1 L_2$ of L_1 and L_2 is the set
$$L_1 L_2 = \{w \in \Sigma^* | w = uv, u \in L_1, v \in L_2\}.$$

3. The *shuffle* $L_1 \sqcup L_2$ of L_1 and L_2 is the set
$$L_1 \sqcup L_2 = \{w \in \Sigma^* | w = u \sqcup v, u \in L_1, v \in L_2\}.$$

4. The *i-th power* of the language L is defined as follows: $L^0 = \{\epsilon\}$ for $i = 0$, $L^1 = L$ for $i = 1$, $L^{i+1} = L^i L$ for $i > 1$.

5. The *Kleene closure* (or *Kleene star*) of a language L, denoted by L^*, is the (infinite) union of all its powers, i.e. $\bigcup_{i \geq 0} L^i$. The *positive closure* of L, denoted by L^+, is the same as L^* without the empty string, i.e. $L^+ = \bigcup_{i \geq 1} L^i$.

6. The *union* of two languages L_1 and L_2 is the set operation $L_1 \cup L_2$.

7. The *intersection* of two languages L_1 and L_2 is the set operation $L_1 \cap L_2$.

8. The *complement* of a language L over an alphabet Σ is the set

$$\overline{L} = \{w \in \Sigma^* | w \notin L\}.$$

Example 2.1.7. Let $L_1 = \{yaw, czt\}$ and L_2 be the set of the English possessive adjectives, i.e. $L_2 = \{my, your, his, her, its, our, their\}$. We have the following languages:

1. $L_2^R = \{ym, ruoy, sih, reh, sti, ruo, rieht\}$.

2. $L_1 L_2 = \{yawmy, yawyour, yawhis, ..., cztour, czttheir\}$.

3. $L_1^0 = \{\epsilon\}$, $L_1^1 = \{yaw, czt\}$, $L_1^2 = \{yawyaw, yawczt, cztczt, cztyaw\}$,
   ...

4. $L_1 \cup L_2 = \{yaw, czt, my, your, his, her, its, our, their\}$.

5. $L_1 \cap L_2 = \emptyset$.

Definition. 2.1.10. Let Σ be an alphabet and let $\mathscr{L}$ be the set of all the finite subsets of Σ^*. Then $\mathscr{L}$ is a *language class*. Let $\Diamond$ be some operation on languages. A language class $\mathscr{L}$ is said to be *closed under* $\Diamond$ iff, whenever $L_1, L_2 \in \mathscr{L}$, then $\Diamond L_{i=1,2} \in \mathscr{L}$ and $(L_1 \Diamond L_2) \in \mathscr{L}$.

Recall the general definition of a formal language (Def. 2.1.5) as the pair (Σ, G), where G is a finite set of rules (a grammar) for forming well-formed strings over the alphabet Σ. Because a formal language is generated by a *formal grammar* (sometimes also called a *phrase-structure grammar*) we can identify the former with the latter–and, as we shall see later, we can identify these with (a mathematical abstraction of) a *machine*.

Definition. 2.1.11. A *formal grammar* is a 4-tuple $G = (V, T, S, P)$ where $V \neq \emptyset$ is a finite set of *variables* (or *non-terminal symbols*), $T \neq \emptyset$ is a finite set of *terminal symbols*, $V \cap T = \emptyset$, $S \in V$ is the *start variable*, and P is a finite set of *grammar*, or *production*, *rules* (also: *productions*).[5]

1. A production (rule) is a pair of strings (α, β), written $\alpha \to \beta$, where α must contain at least one non-terminal symbol.[6] In the production (rule) $\alpha \to \beta$, α is called the *antecedent*, and β the *consequent*.

2. Let $\alpha, \beta \in [(V \cup T)^* = (V \cup T \cup \{\epsilon\})]$ be strings. We say that β is *directly derivable* from α (or α *yields* β) with respect to G, and write $\alpha \underset{G}{\Longrightarrow} \beta$, if there is a production $\gamma \to \delta$ and strings $\zeta_1, \zeta_2 \in (V \cup T)^*$ such that $\alpha = \zeta_1 \gamma \zeta_2$ and $\beta = \zeta_1 \delta \zeta_2$.

3. A *derivation* of the string β–called the *yield*–from the string α is a sequence of n direct derivations or steps

$$\alpha \underset{G}{\Longrightarrow} \gamma_1, \gamma_1 \underset{G}{\Longrightarrow} \gamma_2, ..., \gamma_n \underset{G}{\Longrightarrow} \beta = \alpha \underset{G}{\overset{n}{\Longrightarrow}} \beta$$

or $\alpha \underset{G}{\overset{*}{\Longrightarrow}} \beta$, for *zero or more* steps. $\underset{G}{\overset{n}{\Longrightarrow}}$, or $\underset{G}{\overset{*}{\Longrightarrow}}$, is the *reflexive and transitive closure* of $\underset{G}{\Longrightarrow}$, as we have (i) $\alpha \underset{G}{\overset{*}{\Longrightarrow}} \alpha$, and (ii) if $\alpha \underset{G}{\overset{*}{\Longrightarrow}} \beta$ and $\beta \underset{G}{\overset{*}{\Longrightarrow}} \gamma$, then we have $\alpha \underset{G}{\overset{*}{\Longrightarrow}} \gamma$.

   a) A derivation $\alpha \Longrightarrow \beta$ is called a *leftmost derivation*, denoted by $\alpha \Longrightarrow_l \beta$, if at each step a production is applied to the leftmost variable.

   b) A derivation $\alpha \Longrightarrow \beta$ is called a *rightmost derivation*, denoted by $\alpha \Longrightarrow_r \beta$, if at each step a production is applied to the rightmost variable.

4. A string $\alpha \in (V \cup T)^*$ is called a *sentential form* if $S \underset{G}{\overset{*}{\Longrightarrow}} \alpha$. If $S \underset{G}{\overset{*}{\Longrightarrow}}_l \alpha \left(S \underset{G}{\overset{*}{\Longrightarrow}}_r \alpha \right)$, then α is called a *left-sentential form* (a *right-sentential form*, respectively).

---

[5] We often abbreviate "(non-)terminal symbols" as "(non-)terminals."

[6] Recall that a string can be composed of a single letter, or no letter at all (the empty string). The use of Greek lowercase letters emphasizes the fact that a formal grammar is in many ways similar to an axiom system (see below).

5. The *language $L \subseteq T^*$ generated by the grammar G*, denoted by $L(G)$, is defined as:[7]

$$L(G) = \left\{ w \in T^* | S \underset{G}{\overset{*}{\Longrightarrow}} w \right\}$$

   a) The *leftmost language generated by the grammar G*, denoted by $L_l(G)$, is the language of strings of terminals with leftmost derivations from the start variable S.

   b) The *rightmost language generated by the grammar G*, denoted by $L_r(G)$, is the language of strings of terminals with rightmost derivations from the start variable S.

6. The grammars $G_1 = (V_1, T_1, S_1, P_1)$ and $G_2 = (V_2, T_2, S_2, P_2)$ are *equivalent* iff $L(G_1) = L(G_2)$.

Unless otherwise stated, we shall use the following conventions:

- $A, B, C, ...$ denote variables;

- $a, b, c, ...$ denote terminals;

- $\alpha, \beta, \gamma, \delta, \zeta$ denote strings of variables and/or terminals;

- $u, v, ..., z$ denote strings of terminals;

- X denotes a symbol that may be either a variable or a terminal.

It should be noted that productions are, in practical terms, rules for *rewriting*, or, to be more precise, *replacement*. This will become clear in the many examples below. See Example 2.1.11 for an example of the derivation of a string from a given production set.

From the grammars to be discussed below, the regular grammars and the context-free grammars are often required to be *clean*, i.e. they cannot have productions that are ill-defined or non-generating (productions of the kind $A \overset{*}{\Longrightarrow} \emptyset$), and there cannot be inaccessible productions, or productions with a variable in the antecedent that cannot be reached from S. Grammar cleaning is an algorithmic procedure (see Algorithm 2.1).

Although detecting unreachable and ill-defined symbols is crucial to the process of cleaning a grammar, in fact we also want to avoid circularity and superfluous derivations, so that productions of the kind $A \rightarrow B$,

---

[7]This explains the meaning of "non-terminal" symbols: they do not occur in the generated language.

called *unitary* (or *unit productions*) and of which the *recursive* production $A \to A$ is a special case, should also be eliminated in the process. Evidently, productions of the kind $A \to \epsilon$ should also be eliminated. Thus, before even starting the process of cleaning a grammar, these two kinds of productions–known as *renaming* (or *copying*) and *empty productions* (or *ϵ-productions*), respectively–must be removed while leaving the grammar unchanged. To do this, it suffices to (i) add to the production set P the productions in which A has been rewritten as ϵ or as the consequent(s) of B, and (ii) remove the respective empty and unit productions. See Exercises 2.1.6-7.

Exercises

Exercise 2.1.1. Show that for any strings u, v over some alphabet Σ:

1. $|uv| = |u| + |v|$.

2. $|uv| = |vu|$.

Exercise 2.1.2. Let $\Sigma = \{a, b, c\}$ be an alphabet and let $u = abbcbb$, $v = baaacb$, and $w = ccc$ be strings over Σ. Find:

1. all substrings of v.

2. uv.

3. wu^2.

4. ϵu.

5. $v \epsilon w$.

6. $v \square u$.

7. v^3.

8. $|v^3|$.

9. $|v|_a$, $|v|_b$, $|v|_c$.

10. $|vw|_a$, $|vw|_b$, $|wv|_c$.

Exercise 2.1.3. Let $\Sigma = \{0, 1\}$ be an alphabet. Give examples of strings accepted by the following languages over Σ:

Algorithm 2.1 Grammar cleaning

Input: A grammar $G = (V, T, S, P)$ possibly with inaccessible and ill- or non-defined symbols
Output: A clean grammar $G' = (V', T, S, P')$

STEP 1

1. Compute the set $WDF \subseteq V$ defined as

$$WDF = \{A| (A \to x) \in P, x \in T^*\}.$$

2. Compute the set WDF' defined as

$$WDF' = WDF \cup \{B| (B \to C_1 C_2 ... C_n) \in P\}$$

where every $C_i \in WDF$.

3. Compute
$$V - WDF' = \overline{WDF}.$$

These symbols are ill or non-defined and must be eliminated from the grammar along with the productions in which they occur on the left side.

STEP 2

1. Compute the set of accessible symbols $ACCS \subseteq V$ defined as

$$ACCS = \{S, A| (S \to \alpha A \beta) \in P\}.$$

2. Compute the set $ACCS'$ defined as

$$ACCS' = ACCS \cup \{B| (A \to \alpha B \beta) \in P, A \in ACCS\}.$$

3. Compute
$$V - ACCS' = \overline{ACCS}.$$

These are inaccessible symbols and must be eliminated from the grammar along with the productions in which they occur on the left side.

TERMINATE

---

1. $L = \{0^2 1^l 0^2 | l > 0\}$.

2. $L = \{(01)^n \,|n > 0\}$.

3. $L = \{1^l 01^m 01^n | l > 0, m, n \geq 0\}$.

Exercise 2.1.4. Let $L_1 = \{pit, tip, pip\}$ and $L_2 = \{la, di, da\}$ be two languages over the alphabet $\Sigma = \{a, b, c, ..., y, z\}$. Find:

1. L_1^0.

2. L_2^1.

3. L_1^2.

4. L_1^R.

5. $L_1^1 L_2^R$

6. $L_2 L_1$.

7. $L_1 \,\square\, L_2$.

8. $L_1 \cup L_2$.

Exercise 2.1.5. Find the language generated by the grammar $G = (V, T, S, P)$

1. with $V = \{S, A\}$, $T = \{a, b, c\}$, and the productions

$$P = \left\{ \begin{array}{c} S \to aSb \\ Aab \to c \\ aS \to Aa \end{array} \right\}.$$

2. with $V = \{S, A, B\}$, $T = \{0, 1\}$, and the productions

$$P = \left\{ \begin{array}{c} S \to 0B \\ A \to 0B \\ B \to 1A \\ B \to 1 \end{array} \right\}.$$

3. with $V = \{S, A, B\}$, $T = \{a, b\}$, and the productions

$$P = \left\{ \begin{array}{c} S \to aA \\ A \to aAB \\ A \to a \\ B \to b \end{array} \right\}.$$

4. with $V = \{S, A, B\}$, $T = \{0, 1\}$, and the productions

$$P = \left\{ \begin{array}{c} S \rightarrow 0AB \\ AB \rightarrow 0 \\ A \rightarrow 1 \\ B \rightarrow AB \end{array} \right\}.$$

Exercise 2.1.6. Remove the empty production from the following production set while leaving the respective grammar unchanged:

$$P = \left\{ \begin{array}{c} S \rightarrow aAB \mid Ab \\ A \rightarrow cBd \mid \epsilon \\ B \rightarrow cBA \mid a \end{array} \right\}$$

Exercise 2.1.7. Remove the unit productions from the following production set while leaving the respective grammar unchanged:

$$P = \left\{ \begin{array}{c} S \rightarrow aAB \mid A \mid C \\ A \rightarrow cBd \mid Bc \mid b \\ B \rightarrow cBA \mid a \\ C \rightarrow cAd \mid c \end{array} \right\}$$

Exercise 2.1.8. Clean the given grammars:

1. $G = (\{S, A, B, C, D, E\}, \{a, b, c, d\}, S, P)$ with

$$P = \left\{ \begin{array}{c} S \rightarrow S \mid aA \mid Ba \mid aaa \\ A \rightarrow aa \\ B \rightarrow aa \\ C \rightarrow b \mid DE \\ D \rightarrow EC \mid c \\ E \rightarrow CD \mid d \end{array} \right\}.$$

2. $G = (\{S, A, B, C\}, \{a, b\}, S, P)$ with

$$P = \left\{ \begin{array}{c} S \rightarrow A \mid AB \mid aSB \mid C \\ A \rightarrow aAB \\ B \rightarrow \epsilon \mid Bb \end{array} \right\}.$$

3. $G = (\{S, A, B, C, D, E, F\}, \{0, 1\}, S, P)$ with

$$P = \left\{ \begin{array}{c} S \to 0A1 \mid 1DA \mid AE1 \\ A \to 00A \mid C01A \mid DC10 \\ B \to 1AB \mid 00D \\ C \to A0 \mid \epsilon \\ D \to 1010 \mid AD \\ E \to E11CB \\ F \to 11B \mid \epsilon \mid 00A \end{array} \right\}.$$

2.1.1. Regular languages

It will be useful, in the discussion of regular languages, to put the definition above of formal grammars (Def. 2.1.11) in the back of our minds. We shall bring it back to our frontal cortex at the end of this Section.

Definition. 2.1.12. Let Σ be an alphabet. The set of *regular expressions* over Σ is defined as follows:

1. $\emptyset$ and ϵ are regular expressions;

2. Each letter $a \in \Sigma$ is a regular expression;

3. If r is a regular expression, then so is (r^*);

4. If r_1 and r_2 are regular expressions, then so are $(r_1 + r_2)$ and $(r_1 \cdot r_2)$;

5. Nothing else is a regular expression over Σ.

Example 2.1.8. Let $\Sigma = \{a, b, c, ..., y, z\}$ be an alphabet. The following are regular expressions over Σ:[8]

1. $\emptyset$

2. ϵ

3. f

4. $((c \cdot u) \cdot p) = (c \cdot u \cdot p)$

5. $(((y \cdot a) \cdot (w^*)) \cdot n) = (y \cdot a \cdot w^* \cdot n)$

---

[8]Note the omission of superfluous parentheses on the right side of the equalities.

6. $(p + (f + (e + (r + d)))) = (p + f + e + r + d)$

7. $(a + (b \cdot (b^*))) = (a + bb^*)$

Definition. 2.1.13. The language $L(r)$ over Σ defined by a regular expression r over Σ is as follows:

1. $L(\emptyset) = \emptyset$ and $L(\epsilon) = \{\epsilon\}$;

2. $L(a) = \{a\}$, where $a \in \Sigma$;

3. $L(r^*) = (L(r))^*$, the Kleene closure of $L(r)$;

4. $L(r_1 + r_2) = L(r_1) \cup L(r_2)$;

5. $L(r_1 \cdot r_2) = L(r_1) L(r_2)$.

Definition. 2.1.14. Let L be a language over an alphabet Σ. Then, L is called a *regular language* over Σ if there is a regular expression r over Σ such that $L = L(r)$.

Example 2.1.9. The following are the regular languages corresponding to the regular expressions of Example 2.1.8:

1. $L(\emptyset) = \emptyset$

2. $L(\epsilon) = \{\epsilon\}$

3. $L(f) = \{f\}$

4. $L(c \cdot u \cdot p) = \{cup\}$

5. $L(y \cdot a \cdot w^* \cdot n) = \{yaw^*n\} = \{yan, yawn, yawwn, ...\}$

6. $L(p + f + e + r + d) = \{p, f, e, r, d\}$

7. $L(a + bb^*) = \{a, bb^*\} = \{a, b^+\} = \{a, b, bb, bbb, ...\}$

Proposition. 2.1.15. *Every finite language is regular.*

Proof: Left as an exercise.

But is an arbitrary infinite language regular? This is a relevant question, as it turns out that most interesting and useful languages are infinite. The following theorem, known as *pumping lemma for regular languages*, provides the means for a proof by contradiction: If a language L does not satisfy the specified conditions, then it is not a regular language. These conditions are necessary, but not sufficient: a language may satisfy the conditions and be non-regular. The typical method for proving the regularity of a language L is by constructing a regular expression based on this lemma. Another method is by constructing a finite-state automaton; in fact, it is by means of such an automaton that we prove this lemma (see next Section).

Theorem 2.1.1. *Pumping lemma for regular languages – Let L be a regular language. Then, there is a constant m such that if z is any word in L and $|z| \geq m$ (the "pumping length"), we can write $z = uvw$ in such a way that*

1. $|uv| \leq m$;

2. $|v| \geq 1$;

3. *for all $i = 0, 1, 2, ...$, $z_i = uv^i w \in L$.*

Proof: See next Section. See also Example 2.1.10.

Informally, the pumping lemma states that if a regular language L contains a (sufficiently) long string z, then it contains an infinite set of strings of the form $uv^i w$. v is the substring that can be "pumped," i.e. repeated at will or removed with z remaining a string in L.

Example 2.1.10. Consider language L_1 of Example 2.1.5. This language is not regular. We can easily prove this by contradiction. (i) Let the string given be $z = a^m b^m$. Using condition 1 of the pumping lemma, v must be a and $u = \epsilon$. If $v = a$, then by "pumping" a we end up with more instances of a than instances of b, so that $m \neq m$, a contradiction. (ii) Let now $z = a^k b^k$, $k \leq 1 < m$. We can have $v = a$ or $v = b$, but in any case we end up with more instances of one letter so that $k \neq k$, a contradiction. Let now $|a^k b^k| \leq m$ so that we can have $v = ab$; then z_i will have *as* following *bs*, i.e. $z_i = \underbrace{abab...ab}_{i \text{ times}}$. In either case, $z_i \notin L_1$, which is a contradiction. Thus L_1 is not a regular language. See below (Theorem 2.2.5) for a proof using a finite automaton.

2. *Fundamentals of classical computing*

Recall now Definition 2.1.13. The following proposition follows from points 3-5 of this Definition:

Proposition. 2.1.16. *Regular languages are closed under Kleene closure, union, and concatenation.*

Proof: Left as an exercise.

Additionally, regular languages are closed under all other operations in Definition 2.1.9, as well as under further operations (e.g., homomorphisms).

Recall the definition of a formal grammar (Def. 2.1.11). A regular language is generated by a specific formal grammar known as *type-3* grammar.

Definition. 2.1.17. A grammar $G = (V, T, S, P)$ is *regular* (or of *type 3*) if in each rule $r \in P$, $r = \alpha \to \beta$, it is the case that $\alpha \in V$ and $\beta \in (V \cup T \cup \{\epsilon\})$. A language $L \subseteq T^*$ is a *regular language* if there is a regular grammar G such that $L = L(G)$.

Although the treatment above of regular languages essentially neglected this definition, we bring it back to our frontal cortex now, because the study of context-free languages (next Section), as well as of the remaining languages of the Chomsky hierarchy (see below), does not dispense with the notion of a formal grammar.

Exercises

Exercise 2.1.1.1. Let $\Sigma = \{a, b\}$. Describe the language $L(r)$ with:

1. $r = ab^*a^*$

2. $r = aa + ab + bb$

3. $r = a^* + b^*$

4. $r = abb^*a$

5. $r = a^*ba^*ba^*$

6. $r = (aa + bb + (ab + ba)(aa + bb)^*(ab + ba))^*$

7. $r = b^*a(a + ab^*a)^*$

8. $r = (b + ab^*a)^*ab^*$

9. $r = (a + bbb)^*$

Exercise 2.1.1.2. Prove that, given an alphabet $\Sigma = \{a, b\}$:

1. $\{\epsilon\}$ is a regular language.

2. $(aa + aaa)^* = \epsilon + aaa^*$.

3. $(a + b)^* ab (a + b)^* + b^* a^* = (a + b)^*$.

Exercise 2.1.1.3. Show that the following languages on the alphabet $\Sigma = \{a, b\}$ are not regular:

1. $L = \{a^n b^{2n} | n > 0\}$

2. $L = \{a^n b^m | 0 < n \leq m\}$

3. $L = \left\{a^{n^2} | n \geq 0\right\}$

Exercise 2.1.1.4. A regular grammar can be *right-* or *left-linear* if it has rules of the form $\alpha \rightarrow \beta A$ or $\alpha \rightarrow A\beta$, respectively, for $A \in V$. A language is regular iff it has both a right- and a left-linear grammar that generate it. The language $L = \{a (ba)^*\}$ is regular. Determine the productions of each of the right- and left-linear grammars generating L.

Exercise 2.1.1.5. Prove (Complete the proof of) the theorems and propositions in this Section that were left as an exercise (with a sketchy proof, respectively).

2.1.2. Context-free languages

Regular languages are too simple and limited for, say, a high-level programming language. In order to be able to generate complex languages of this kind we need more complex grammars. A context-free grammar (CFG) is just such a grammar.

Definition. 2.1.18. A grammar $G = (V, T, S, P)$ is *context-free* (or of *type 2*) if each production $r \in P$ has the form $\alpha \rightarrow \beta$, where $\alpha \in V$, $|\alpha| = 1$, and $\beta \in (V \cup T)^*$. A language $L \subseteq T^*$ is a *context-free language* (CFL) if there is a CFG G such that $L = L(G)$.

As a matter of fact, a CFG G may be greatly simplified, having only two types of productions.

Definition. 2.1.19. A CFG G in which every production is of the two forms $A \to BC$ and $A \to a$, where $A, B, C \in V$ and $a \in T$, is said to be in *Chomsky normal form*.

Theorem 2.1.2. *For every CFG G there is a CFG G' in Chomsky normal form such that $L(G') = L(G) - \{\epsilon\}$.*

Proof: The reader may try it as an exercise, in which case the first step is to prove that if one eliminates from a CFG G all ϵ-productions and all productions of the form $A \to B$ (called *unit productions*), the resulting grammar G' is a CFG. Next, show that one can convert the remaining rules into Chomsky normal form. (Hint: Algorithm 2.2.)

Another important normal form for CFGs is the *Greibach normal form*, conversion into which requires that a CFG G be already in Chomsky normal form. See Algorithm 2.3.

Before we begin elaborating on the CFGs and the corresponding CFLs, we should contrast these with the context-sensitive grammars (CSGs).

Definition. 2.1.20. A grammar $G = (V, T, S, P)$ is *context-sensitive* (or of *type 1*) if each production $r \in P$ has the form $\gamma A \delta \to \gamma X \delta$, where $A \in V$, $\gamma, \delta \in (V \cup T)^*$, and $X \in (V \cup T)^+$ (i.e. $X \neq \epsilon$). Put differently, every production in a CSG is of a form $\alpha \to \beta$ such that $|\beta| \geq |\alpha|$, i.e. no production is length-decreasing.[9] A language $L \subseteq T^*$ is a *context-sensitive language* (CSL) if there is a CSG G such that $L = L(G)$.

The informal explanation for this label is that the production $\gamma A \delta \to \gamma X \delta$ allows the substitution of A by X only in the *context* "$\gamma \cdots \delta$".

We retake the CSLs in Section 2.1.4. We now move on with our elaboration on the CFLs.

Example 2.1.11. We begin with a very simple example of a CFG. Consider the language $L = \{a^m b^m | m > 0\}$. This is language L_1 of Example 2.1.5 that was proven to be non-regular in Example 2.1.10. L is a context-free language. In effect, consider the grammar $G = (V, T, S, P)$ with $V = \{S\}$, $T = \{a, b\}$, and $P = \{r_1, r_2\}$, where $r_1 = S \to aSb$ and $r_2 = S \to ab$. By applying $m - 1$ times production r_1 and then r_2 we obtain the derivation

---

[9]The production $S \to \epsilon$ is allowed only in case S does not occur in the consequent of any of the remaining productions.

Algorithm 2.2 Chomsky-normal-form transformation

Input: A clean CFG $G = (V, T, S, P)$
Output: A CFG $G' = (V', T, S, P')$ in Chomsky normal form (ChNF)

INITIATE with P

STEP I: Given a rule $r_i \in P$ of the form $\alpha \to \beta_1 \mid \beta_2 \mid ... \mid \beta_n$:

1. If r_i is a production with body $|\beta_j| = 1, \beta \in T$, then **output** $P_{I.1} = P$ and go to step 2.

2. If r_i is a production with body $|\beta_j| > 1$:

    a) for $|\beta_j| = 2$, $\beta \in V$, **output** $P_{I.2.a} = P$.

    b) for each symbol a in β_j such that $a \in T$, create a new symbol $U_a \in V$. Add a rule $U_a \to a$ to G and replace a in the bodies β_j of rules r_i by U_a. **Output** $P_{I.2.b}$.

    c) for $|\beta_j| \geq k, k \geq 3, \beta \in V$, such that β_j has the form $A \to B_1 B_2 ... B_k$, introduce $k - 2$ new variables $W_1, W_2, ..., W_{k-2}$ and replace rule r_i by $k - 1$ rules r' such that

$$A \to B_1 W_1 \qquad\qquad\qquad\qquad\qquad\qquad \left(r_1' \right)$$
$$W_1 \ \to B_2 W_2 \qquad\qquad\qquad\qquad\qquad \left(r_2' \right)$$
$$\vdots \qquad\qquad\qquad\qquad\qquad\qquad\qquad\qquad \vdots$$
$$W_{k-3} \ \to B_{k-2} W_{k-2} \qquad\qquad \left(r_{k-2}' \right)$$
$$W_{k-2} \ \to B_{k-1} B_k \qquad\qquad\quad \left(r_{k-1}' \right)$$

and **output** $P_{I.2.c}$.

STEP II: Introduce a new start symbol S_0 and add the rule $S_0 \to S$ to $P_{I.2.c}$. **Output** P'.

TERMINATE

Algorithm 2.3 Greibach-normal-form transformation

Input: A CFG $G = (V, T, S, P = P_0)$ in ChNF
Output: A CFG $G' = (V', T' = T, A_1, P' = P_4)$ in Greibach normal form (GNF)

START with P_0

1. Relabel all the k variables of P_0 as $A_1, A_2, ..., A_k$ making $S = A_1$ and ordering the rules such that $r_1 = A_1 \to \beta, ..., r_k = A_k \to \beta$. **Output** P_1.

2. Verify that for every rule $r_i \in P_1$ of the form $A_i \to A_j \alpha$, $|\alpha| = n - 1$ if $n > 0$, it is the case that $i < j$. If not, substitute A_j in the body of r_i by the body of A_j. Repeat until $j \geq i$ or $A_j \to \sigma \in T$. **Output** P_2.

3. If for any $r \in P_2$ we have a rule of the form

$$A_i \to A_i \alpha_1 \,|\, A_i \alpha_2 \,|\, ... \,|\, A_i \alpha_k \,|\, \beta_1 \,|\, \beta_2 \,|\, ... \,|\, \beta_l$$

for $|\alpha| = n - 1$ if $n > 0$, and $\beta = \sigma_1 \sigma_2 ... \sigma_n \in (T \cup V)^*$ and $\sigma_1 \neq A_i$, then replace r by

a) $Z \to \alpha_j \,|\, \alpha_j Z$ for $1 \leq j \leq k$, Z is a new variable, and
b) $A_i \to \beta_j \,|\, \beta_j Z$ for $1 \leq j \leq l$.
c) **Output** P_3.

4. If for any $r_i \in P_3$ we have a rule of the form $A_i \to \alpha$, for $A_i, \alpha \in V$, replace the initial variables of productions in the body of r_i until r_i is of the form $A_i \to a_1 \alpha \,|\, a_2 \alpha \,|\, ... \,|\, a_m \alpha$, where $a_j \in T$. Proceed in the order $i = n, n - 1, n - 2, ..., 1$ for the A_i. **Output** P_4.

TERMINATE

$$\underbrace{S \Longrightarrow aSb \Longrightarrow a^2Sb^2 \Longrightarrow ... \Longrightarrow a^{m-1}Sb^{m-1}}_{(m-1)r_1} \underbrace{\Longrightarrow a^m b^m}_{r_2}.$$

Note that r_2 is a final production, i.e. once it is applied, no S remains in the resulting string, and so this resulting string is terminal.

Theorem 2.1.3. *Let L_1 and L_2 be CFLs. Then, $L_1 \cup L_2$, $L_1 L_2$, and $(L_1)^*$ are also CFLs.*

Proof: We prove that $L_1 \cup L_2$ is also a CFL and leave the other two remaining cases as an exercise. Suppose that the CFGs $G_1 = (V_1, T, S_1, P_1)$ and $G_2 = (V_2, T, S_2, P_2)$ generate the CFLs L_1 and L_2, respectively. Assume further $V_1 \cap V_2 = \emptyset$. Design a new CFG $G = (V, T, S, P)$ with $V = V_1 \cup V_2 \cup \{S\}$ and $P = P_1 \cup P_2 \cup \{S \to S_1, S \to S_2\}$. We can derive a string $x \in L(G)$ by starting with either $S \to S_1$ or $S \to S_2$ and continuing the derivation in either G_1 or G_2, so that $L(G) = L(G_1) \cup L(G_2)$. In effect, suppose that $S \overset{*}{\underset{G}{\Longrightarrow}} x$. Then, the first step in the derivation must be either $S \Longrightarrow S_1$ or $S \Longrightarrow S_2$. Because $V_1 \cap V_2 = \emptyset$, in the first case we remain in G_1 and $x \in L_1$; similarly for $x \in L_2$. Thus $L_1 \cup L_2 = L(G)$. **QED**

Theorem 2.1.3 has more than its intrinsic mathematical interest; it comes in handy in practical terms. In effect, to construct a CFG may be a non-trivial task. Because a CFL can be a union of simpler CFLs, one can break a CFG into simpler parts and then construct CFGs for these parts. Similarly, practical applications can be found for concatenation and iteration.

The CFG of Example 2.1.11 is a rather simple example in the sense that it is easy to determine the language generated by looking at the productions. This is, however, not the norm. One way to determine more complex languages is by resorting to derivation, or parse, trees.

Definition. 2.1.21. Let $G = (V, T, S, P)$ be a CFG. We say that an ordered labeled tree is a *derivation*, or *parse*, *tree* for a string $w \in G$, denoted by $\mathcal{T}_w$ (abbreviating $\mathcal{T}_{G_w}$), iff:

1. The root of $\mathcal{T}_w$ is S;

2. Each and every interior vertex of $\mathcal{T}_w$ is labeled with an element from V;

3. Each and every leaf of $\mathcal{T}_w$ is labeled with an element from $T \cup \{\epsilon\}$;

4. If a vertex has a label $A \in V$ and its children are labeled from left to right with $X_1, X_2, ..., X_n$, then P must contain a production of the form $A \to X_1 X_2 ... X_n$;

5. A leaf labeled with ϵ has no siblings.

Definition. 2.1.22. A derivation tree $\mathcal{T}_w$ is called a *partial derivation tree*, denoted by $\mathcal{T}_w^i$, $i = 0, 1, ..., n$, if $S = \mathcal{T}_w^0$ is the root and every leaf is labeled with an element from $V \cup T \cup \{\epsilon\}$.

This should be distinguished from a *subtree*, in which the root can be any particular vertex of the derivation tree (cf. Def. 1.2.27.1). We denote a subtree of a string w by $\mathcal{T}_w'$.

Example 2.1.12. Let the following be the productions of a CFG $G = (V, T, S, P)$ with $V = \{S, A, B\}$ and $T = \{a, b, c\}$:

$$P = \left\{ \begin{array}{l} S \to aAB \\ A \to Bba \\ B \to bB \\ B \to c \end{array} \right\}$$

Note that for the antecedent B there are two *alternative* consequents. Given this, we can write a single production rule where "|" denotes "or":

$$B \to bB \mid c$$

The productions of this grammar generate the string $w = acbabc$ as a result of the derivation (we omit the subscript G in $\Longrightarrow$)

$$S \Longrightarrow aAB \Longrightarrow a\,(Bba)\,B \Longrightarrow acbaB \Longrightarrow acba\,(bB) \Longrightarrow acbabc$$

Figure 2.1.1 shows the different partial derivation trees and the final derivation tree. In this, the string is read in the leaves from left to right. Other strings of this grammar are any of $ab^n cbab^m c$ for $n, m \geq 0$, so we have the CFL $L = \{ab^n cbab^m c \mid n, m \geq 0\}$.

Theorem 2.1.4. Let $G = (V, T, S, P)$ be a CFG. Then, $S \overset{*}{\underset{G}{\Longrightarrow}} \alpha$ iff there is a derivation tree in G with yield α.

Proof: Left as an exercise. (Hint: Prove that for any $A \in V$, $A \overset{*}{\underset{G}{\Longrightarrow}} \alpha$ iff there is a subtree with vertex A with α as the yield. Then make $A = S$.)

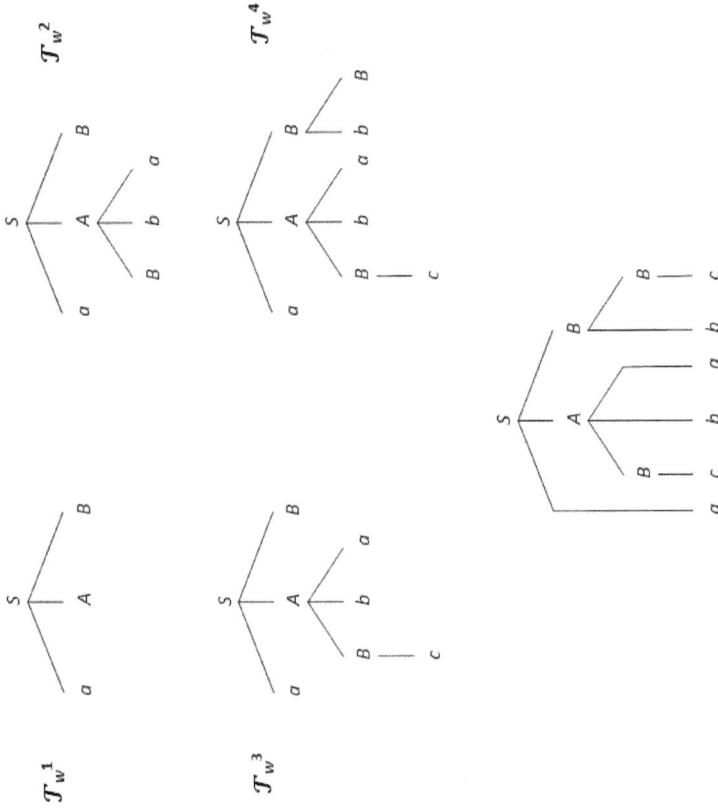

Figure 2.1.1.: Derivation tree of the string $w = acbabc \in L(G)$ with the corresponding partial derivation trees.

However, given a CFG G, there may be more than one derivation (tree) for a single string, and they may be unequally appropriate. The following example illustrates this problem.

Example 2.1.13. Let $G = (\{S\}, \{a, +, *, (,)\}, S, P)$ be a CFG with the following productions in P:

$$S \to a \mid S + S \mid S * S \mid (S)$$

Figure 2.1.2 shows that the string $a + a * a$ has two leftmost derivations.

The problem exhibited in Example 2.1.13 is called *ambiguity*.

Definition. 2.1.23. A CFG G is said to be *ambiguous* if there is a string $w \in L(G)$ with two or more different leftmost derivations in G. Otherwise, G is *unambiguous*.

This is more rigorously captured in the following theorem:

Theorem 2.1.5. *Let G be a CFG and $w \in L(G)$. Then, the following statements are equivalent:*

1. *w has more than one derivation tree.*

2. *w has more than one leftmost derivation.*

3. *w has more than one rightmost derivation.*

Proof: Left as an exercise. (Hint: given two derivation trees $\mathcal{T}_1$ (abbreviating $\mathcal{T}_{w1}$) and $\mathcal{T}_2$, specify the first step where the two corresponding derivations differ such that, given an occurrence of a variable A, there will be two nodes with different children such that $\mathcal{T}_1 \neq \mathcal{T}_2$.)

The ambiguity exhibited in Example 2.1.13 resides in the following possible *interpretations*: Derivation tree 1 interprets this string as a sum, whereas derivation tree 2 interprets the string as a product. Suppose $a = -2$; then we have $a + (a * a) = 2$ and $(a + a) * a = 8$.

This ambiguity is as highly undesirable in programming as it is in arithmetic, as programs are required to have a unique interpretation. This problem can be solved by agreeing, whenever a step has more than one variable occurrence, to use the leftmost one when *parsing* the string. In effect, a string is ambiguous if it has more than one parse, or derivation, tree. And parsing a string–i.e. building its parse tree–actually superimposes a structure on it. In fact, parsing is the only efficient way to recognize a string of a CFL.

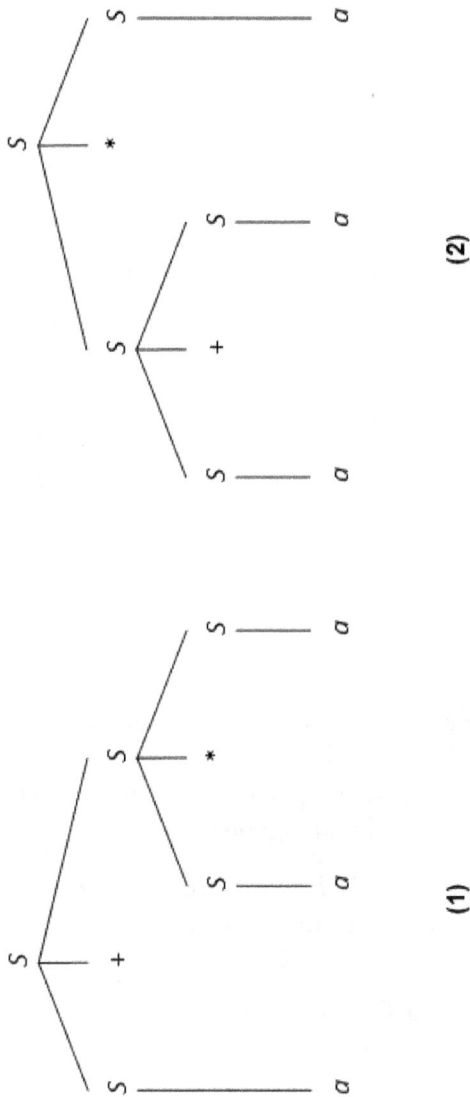

Figure 2.1.2.: Two leftmost derivations of the string $a + a * a$.

Example 2.1.14. Let us retake the CFG of Example 2.1.13. Note that above we did not apply the production rule $S \rightarrow (S)$. By doing it now, and by agreeing to use the leftmost variable, we verify that there is only one leftmost derivation corresponding to the string $a + (a * a)$. Figure 2.1.3 shows the derivation, or parse tree, of this string.

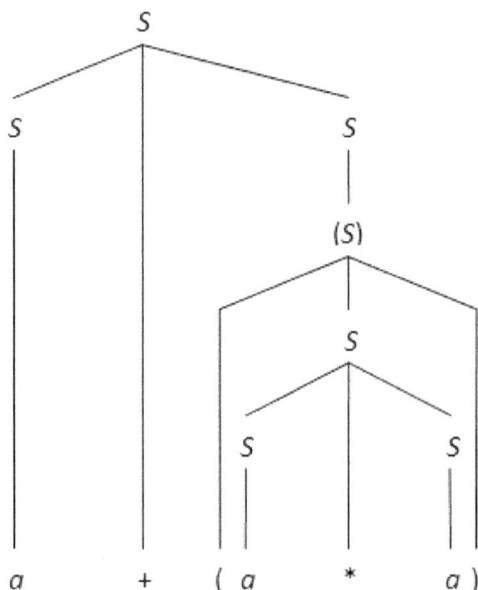

Figure 2.1.3.: Parse tree of an unambiguously derived string.

However, this convention does not always work, because there are languages that are inherently ambiguous.

Definition. 2.1.24. A CFL L is said to be *inherently ambiguous* if every CFG that generates it is ambiguous.

Example 2.1.15. The CFL $L = \{a^i b^j c^k | i = j \text{ or } j = k\}$ is inherently ambiguous. We leave the proof as an exercise.

We often need to find out whether a language is, or is not, a CFL. Just as in the case of regular languages, there is a *pumping lemma for CFLs* that allows us to detect non-CFLs. Informally, this lemma states that in a given CFL there are always two short substrings close together that can be "pumped" at will, i.e. can both be repeated the same number (zero or more) of times.

Theorem 2.1.6. *Pumping lemma for CFLs – Let L by any CFL. Then there is an integer n depending only on L such that, if $z \in L$ and $|z| \geq n$, then we can write $z = uvwxy$ satisfying the following three conditions:*

1. $|vx| \geq 1$;

2. $|vwx| \leq n$;

3. $uv^i wx^i y \in L$ for all $i \geq 0$.

Before we prove this theorem, we need to state and prove an additional lemma:

Lemma 2.1.7. *Let $S \overset{*}{\underset{G}{\Rightarrow}} z$, where G is in Chomsky normal form. Suppose that $\mathcal{T}_z$ is a derivation tree for z in G and that no path in $\mathcal{T}_z$ contains more than k nodes. Then $|z| \leq 2^{k-2}$.*

Proof: First, suppose that $z = a$. Then, $\mathcal{T}_z$ has only two nodes, labeled S and a respectively, and a single edge (see Fig. 2.1.4.1), i.e. $|z| = 2^{2-2} = 1$. Clearly, $|z| \leq 2^{k-2}$. Otherwise, because G is in Chomsky normal form, the root of $\mathcal{T}_z$ must have exactly two immediate successors, α and β, labeled by variables (e.g., A and B, respectively, for a production $S \rightarrow AB$). We now have two subtrees $\mathcal{T}_z' = \mathcal{T}_\alpha$ and $\mathcal{T}_z'' = \mathcal{T}_\beta$ with roots A and B, respectively (see Fig. 2.1.4.2). In each of $\mathcal{T}_\alpha, \mathcal{T}_\beta$ the longest path must contain $\leq k - 1$ nodes. By induction, we may assume that each of $\mathcal{T}_\alpha, \mathcal{T}_\beta$ has $\leq 2^{k-3}$ leaves, so that we have

$$|z| \leq 2^{k-3} + 2^{k-3} = 2^{k-2}.$$

QED

We can now easily prove Theorem 2.1.6.

Proof of Theorem 2.1.6: (Sketch) Clearly, given a grammar G in Chomsky normal form, if we start with a derivation

$$(D) \qquad S \overset{*}{\underset{G}{\Rightarrow}} uAy \overset{*}{\underset{G}{\Rightarrow}} uvAxy \overset{*}{\underset{G}{\Rightarrow}} uvwxy = z$$

we can end up with a yield $uv^i wx^i y$ by the sequence

$$(D') \quad S \overset{*}{\underset{G}{\Rightarrow}} uAy \overset{*}{\underset{G}{\Rightarrow}} uvAxy \overset{*}{\underset{G}{\Rightarrow}} uv^2 Ax^2 y \overset{*}{\underset{G}{\Rightarrow}} \dots$$

$$\dots \overset{*}{\underset{G}{\Rightarrow}} uv^i Ax^i y \overset{*}{\underset{G}{\Rightarrow}} uv^i wx^i y.$$

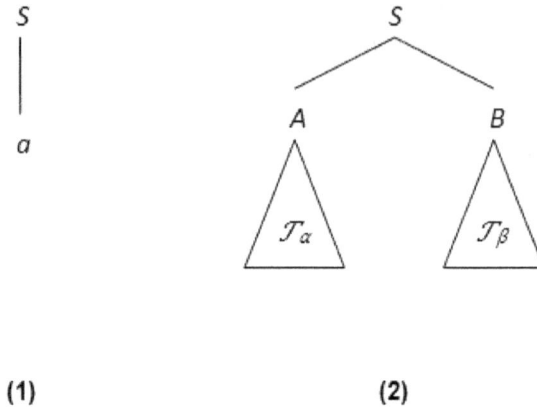

Figure 2.1.4.: Parse trees for productions (1) $S \to a$ and (2) $S \to AB$.

That is, a sufficiently long derivation in G must contain a recursive (or self-embedded) variable A such that w can be derived from any of the occurrences of A and thus any of the strings uwy, $uvwxy$, uv^2wx^2y, ..., can be derived from S. Recall that this grammar G must generate a language $L(G) - \{\epsilon\}$ in Chomsky normal form. This means that every derivation step (i) either introduces a terminal by the application of a production of the form $A \to a$, (ii) or augments the length of the sentential form by means of a production $A \to BC$ (cf. Def. 2.1.19).

Because this grammar is in Chomsky normal form, any derivation tree must be a binary tree of the basic form shown in Figure 2.1.5, and because z is a long string, the parse tree for this string must have a long path. As a matter of fact, so long that there must be some variable that occurs more than once on this path–and near the bottom of the path, for that matter. In terms of Lemma 2.1.7, we let $n = 2^k$ for k the number of variables in G. If $z \in L(G)$ and $|z| \geq n$, then any parse tree $\mathcal{T}_z$ must have a path with length $\geq k + 1$, since $|z| > 2^{k-1}$. But this path has $\geq k + 2$ vertices, all labeled with variables with the exception of the last one. Hence, by Lemma 2.1.7 we have

$$|vwx| = |v \, [\![\mathcal{T}_\beta]\!] \, x| = |[\![\mathcal{T}_\alpha]\!]| \leq 2^n$$

where $[\![\mathcal{T}_\gamma]\!]$ denotes the string consisting of the labels of the leaves of the tree $\mathcal{T}_\gamma$. **QED**

Example 2.1.16. The language $L = \{a^n b^n c^n | n > 0\}$ is not a CFL. In effect, suppose $L = L(G)$ is a CFL with G a grammar in Chomsky

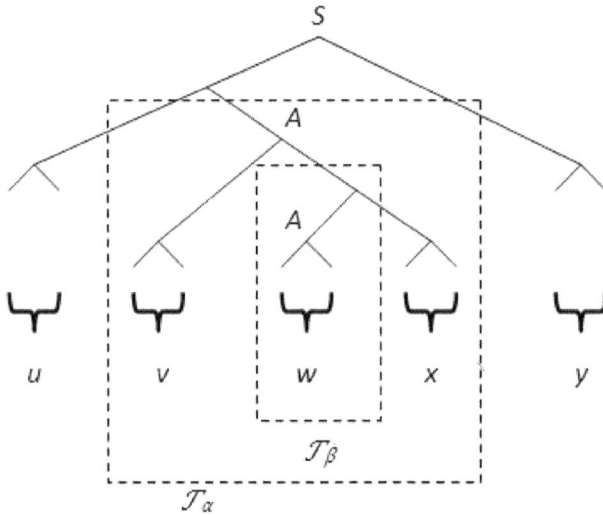

Figure 2.1.5.: Parse tree for $z = uv^i wx^i y$.

normal form with n variables. Now choose k so large that $|a^k b^k c^k| > 2^n$, i.e. $k > 2^n/3$. We have $a^k b^k c^k = uvwxy$. Set $z_i = uv^i wx^i y$, so that we have $z_i \in L$ for $i = 0, 1, 2, \dots$ For $i = 2$, we thus have $z_2 = uv^2 wx^2 y$. But v and x must each contain only one of a, b, or c, so that one of these three letters must be left out. For $i = 3, 4, 5, \dots$, z_i contains more and more copies of v and x; moreover, $vx \neq \epsilon$. Hence, it is not possible for z_i to have the same number of occurrences of a, b, and c. Thus, $z_i \notin L$, and we reached a contradiction.

However, some non-CFLs evade detection by means of the pumping lemma for CFLs. For instance, $L = \{a^i b^j c^k d^l | i = 0 \text{ or } j = k = l\}$ is not a CFL, but this lemma fails to determine this. This is so because this pumping lemma is not strong enough, namely in the sense that it does not tell us much about the location of the substrings v and x, i.e. the substrings that are pumped. The following lemma, known as *Ogden's lemma*, allows the designation of some positions of z as "distinguished," so that some of the distinguished positions are guaranteed to appear in the pumped substrings. As a matter of fact, the pumping lemma for CFLs just is the special case of Ogden's lemma in which all the positions in z are distinguished.

Theorem 2.1.8. *Ogden's lemma – Let L be a CFL. Then there is an integer n such that, if $z \in L$ and $|z| \geq n$, and we mark n or more positions of z as "distinguished," then we can write $z = uvwxy$ such that*

1. *the string vx has at least one distinguished position;*

2. *the string vwx has at most n distinguished positions;*

3. $uv^iwx^iy \in L$ for all $i \geq 0$.

Proof: Left as an exercise. (Hints: Let us call vertices of a derivation tree both of whose children have distinguished descendants *branch points*. Construct a path ε in the tree beginning at the root only with branch points. It should follow that there are at least $k+1$ branch points in ε. Apply the pigeonhole principle.)

Example 2.1.17. Ogden's lemma allows us to determine that the language $L = \{a^i b^j c^k | i \neq j, j \neq k, i \neq k\}$ is not a CFL. We leave the explanation as an exercise.

Further properties of CFLs are as follows. We leave the proofs of the following propositions as exercises.

Proposition. 2.1.25. *The class of CFLs is closed under substitution and homomorphism. This class is not closed under intersection, complementation, and difference.*

Proposition. 2.1.26. *Let L_1 and L_2 be a CFL and a regular language, respectively. Then, $L_1 \cap L_2$ and $L_1 - L_2$ are CFLs.*

Many questions are relevant with respect to CFLs. However, not all relevant questions have an answer; those that do not fail to have an algorithm. In effect, for arbitrary CFGs G_1 and G_2, the questions whether (i) $L(G_1) = L(G_2)$, (ii) $L(G_1) \cap L(G_2) = \emptyset$, (iii) $L(G_1) \subseteq L(G_2)$, and (iv) $L(G_i) = \Sigma^*$ are undecidable.

We present now theorems for the decidable questions (proofs are left as exercises).

Theorem 2.1.9. *Given a CFG G, there are algorithms to determine whether (i) $L(G) = \emptyset$, (ii) $L(G)$ is finite, and (iii) $L(G)$ is infinite.*

Theorem 2.1.10. *Let G be a CFG and w be a string. Then there is an algorithm to determine whether $w \in L(G)$.*

Exercises

Exercise 2.1.2.1. Determine the CFLs given the indicated productions:

1. $P = \{S \rightarrow SaS \mid b \mid \epsilon\}$.

2. $P = \{S \rightarrow aS \mid bS \mid \epsilon\}$.

3. $P = \left\{ \begin{array}{l} S \rightarrow aA \mid bA \\ A \rightarrow aS \mid bS \end{array} \right\}$.

4. $P = \{S \rightarrow aS \mid aSbS \mid \epsilon\}$.

5. $P = \{S \rightarrow ab \mid bS \mid a \mid b\}$.

Exercise 2.1.2.2. Obtain the derivation trees of the following strings given the indicated productions of CFGs:

1. $w = abaabaa$ given the productions $S \rightarrow aAS \mid a$ and $A \rightarrow bS$.

2. $w = a^n$ given the production $S \rightarrow SS \mid a$.

3. $w = a^n b^m, n, m \geq 1$, given the productions $S \rightarrow aS \mid aA$ and $A \rightarrow bA \mid b$.

Exercise 2.1.2.3. Let $G = (V, T, S, P)$ with $V = \{A, B\}$ and $T = \{a, b\}$ be a CFG. The productions of G are as follows:

$$P = \left\{ \begin{array}{l} S \rightarrow aB \mid bA \\ A \rightarrow aS \mid bAA \mid a \\ B \rightarrow bS \mid aBB \mid b \end{array} \right\}$$

Obtain the string $aaabbabbba$ by (i) a leftmost derivation and (ii) a rightmost derivation.

Exercise 2.1.2.4. Show that the CFGs G with the following productions are ambiguous:

1. $P = \left\{ \begin{array}{l} S \rightarrow a \mid aAb \mid abSb \\ A \rightarrow aAAb \mid bS \end{array} \right\}$.

2. $P = \{S \rightarrow SS \mid a \mid b\}$.

Exercise 2.1.2.5. Explain why the following languages are inherently ambiguous:

1. $L = \{a^i b^j c^k | i = j \text{ or } j = k\}$ (Example 2.1.15)

2. $L = \{a^n b^m c^m d^n | n, m > 0\} \cup \{a^n b^n c^m d^m | n, m > 0\}$

Exercise 2.1.2.6. Determine whether the following languages are CFLs:

1. $L = \left\{ a^{n^2} | n \geq 1 \right\}$

2. $L = \{w \in \{a, b, c\}^* \, \| w|_a = |w|_b = |w|_c\}$

Exercise 2.1.2.7. Show why the pumping lemma for CFLs fails in the case of the language $L = \{a^i b^j c^k d^l | i = 0 \text{ or } j = k = l\}$.

Exercise 2.1.2.8. Show why Ogden's lemma allows us to determine that the language $L = \{a^i b^j c^k | i \neq j, j \neq k, i \neq k\}$ is not a CFL.

Exercise 2.1.2.9. Transform the grammars of Exercise 2.1.8.1-3 into:

1. Chomsky normal form.

2. Greibach normal form.

Exercise 2.1.2.10. Prove (Complete the proof of) the theorems and propositions above that were left as an exercise (with a sketchy proof, respectively).

2.1.3. Recursively enumerable languages

The languages that we approach next are fundamental for computing theory in more than one way. In this Section, we discuss briefly the basic features of these languages as far as formal grammars are concerned, and we elaborate further on them below, namely in Section 2.3.

Definition. 2.1.27. A grammar $G = (V, T, S, P)$ is an *unrestricted grammar* (UG; also: *type-0 grammar*) if it has productions of the form $\alpha \to \beta$ for strings α, β such that $\alpha, \beta \in (V \cup T)^*$, with $|V(\alpha)| \geq 1$. A language $L \subseteq T^*$ is a *recursively enumerable language* (REL) iff there is a UG G such that $L = L(G)$.

Contrarily to CFGs, UGs allow for productions in which a variable may *depend on the context*. For instance, the production $Bc \rightarrow bc$ allows the substitution of the variable B by the terminal b only if B is immediately followed by c on the antecedent. In particular, given a UG G, the assumption $S \overset{*}{\underset{G}{\Rightarrow}} vAx \overset{*}{\underset{G}{\Rightarrow}} z$ no longer implies that $z = vwx$ for some string w. Nevertheless, just as in the case of regular languages and CFLs, a UG production is terminal when its right side has no variables.

Example 2.1.18. Recall from Example 2.1.16 that the language $L = \{a^n b^n c^n | n > 0\}$ is not a CFL. But this language is a REL. We show how a UG can generate it. We begin by creating a variable L denoting "left end of a string." We now require productions that (i) allow us to obtain strings of the form $L(ABC)^n$, and (ii) allow for an alphabetical (re)arrangement of the variables A, B, and C. These productions are as follows:

$$(i) \quad S \rightarrow SABC \mid LABC$$

$$(ii) \quad BA \rightarrow AB; CB \rightarrow BC; CA \rightarrow AC$$

Now, of course, we need productions that (iii) allow us to replace variables that are already in alphabetic order by terminals:

$$(iii) \quad LA \rightarrow a; aA \rightarrow aa; aB \rightarrow ab; bB \rightarrow bb; bC \rightarrow bc; cC \rightarrow cc$$

Recall from the discussion of CFLs that a few questions with respect to these languages are undecidable. In fact, the problem of undecidability is even more acute for the RELs: the questions of equivalence, inclusion, membership, finiteness, and emptiness are all *undecidable*. (See Section 2.3 for a discussion on decidability issues.) Furthermore, the class of RELs is not closed under complementation. Nevertheless, we have the following result:

Theorem 2.1.11. *Let L_1 and L_2 be RELs. Then, $L_1 \cup L_2$ and $L_1 \cap L_2$ are also RELs.*

Proof: The proof requires the notion of a Turing machine, so it is left as an exercise in Section 2.2.3.

Exercises

Exercise 2.1.3.1. Despite being called unrestricted, UGs do not allow any production of the form $\epsilon \rightarrow \alpha$, for $\alpha \in (V \cup T)^*$. Explain why.

Exercise 2.1.3.2. Show that for each UG G there is a UG G_1 all of whose productions are either of the form

$$\alpha \to \beta$$

with $\alpha, \beta \in (V_1 \cup T_1)^+$ and $|\alpha| \leq |\beta|$, or of the form

$$A \to \epsilon$$

for $A \in V_1$.

Exercise 2.1.3.3. Find a UG generating the following RELs:

1. $L = \left\{ a^{2^n} \mid n \in \mathbb{N} \right\}$

2. $L = \{ xxx \mid x \in \{a, b\}^* \}$

3. $L = \{ a^n b^n a^n b^n \mid n \geq 0 \}$

Exercise 2.1.3.4. Find the REL generated by the following productions:

$$S \to ABCS \mid ABC$$

$$AB \to BA; BC \to CB; CA \to AC; AC \to CA; BA \to AB; CB \to BC$$

$$A \to a; B \to b; C \to c$$

Exercise 2.1.3.5. Find the REL generated by the following productions:

$$S \to S_1 B; S_1 \to a S_1 b; a S_1 b \to aa; bB \to bbbB; B \to \epsilon$$

Exercise 2.1.3.6. Show that the language $L = \{ a^n b^n c^n \mid n > 0 \}$ is a CSL. (Hint: Alter the consequent $LABC$ in production (i); cf. Example 2.1.18.)

2.1.4. The Chomsky hierarchy (I)

Recall the definition of a formal grammar (Def. 2.1.11). Note that the definition of a UG (Def. 2.1.27) is the closest one to this general definition of a formal grammar. Actually, they are one and the same definition. In effect, this type-0 grammar provides the "framework" in which the other studied grammars must fit. N. Chomsky, the designer of this 0-3 typology, established the following strict-containment hierarchy known as the *Chomsky hierarchy* (Chomsky, 1956; 1959), in which we abbreviate "regular language" as "RGL" and indicate between brackets the type of grammar that generates each class of language:

$$\mathscr{RGL}\ (3) \subset \mathscr{CFL}\ (2) \subset \mathscr{CSL}\ (1) \subset \mathscr{REL}\ (0)$$

The following theorem states this in a different formulation.

Theorem 2.1.12. *There are regular languages that are not CFLs, and there are CFLs that are not CSLs, and CSLs that are not RELs.*

Proof: This is a strict containment relation, which means that there is at least one language generated by a grammar of type j that is not generated by a grammar of type i, where $j < i$, i covers j, and $i, j = 0, 1, 2, 3$. In effect, the language $L = \{a^n b^n | n > 0\}$ is not regular (cf. Example 2.1.10), but it is a CFL (cf. Example 2.1.11). The language $L = \{a^n b^n c^n | n > 0\}$ is not a CFL (cf. Example 2.1.16), but it is a CSL (cf. Exercise 2.1.3.6). Because RELs are generated by UGs, there are CSLs that are not RELs, as the former are generated by grammars with restrictions, namely by CSGs. **QED**

We now analyze this hierarchy from the viewpoint of its implications. Firstly, this is obviously a hierarchy from the most restricted to the less restricted–or most general–grammars, and it is thus a hierarchy with practical implications: If, given a language L, there is an *algorithm* that decides that L is a regular language, then we need search no more, because the class of these languages is included in all other classes of formal languages. In other words, the more restricted a grammar is, the simpler a program can be written to decide on the type of language. We summarize the main features of this hierarchy in Table 2.1.1 and provide a very simple algorithm for deciding on the language class (Algorithm 2.1). (We leave the algorithm for deciding on the type of grammar as an exercise.)

The importance of this hierarchy will become clearer in the following Section, as it is the case that each class of languages is associated to a mathematical object–an *abstract machine*–that is in fact a model of computation implementing a specific decision program. This has, again, practical importance: the more restricted a language is, the simpler it is to devise a decision program for it. Anticipating on the material to follow, programming a Turing machine, the abstract machine associated to RELs, can be a Herculean task for even very elementary functions (see Example 2.2.16).

But the importance of this hierarchy lies also precisely in what is left out from it. That is to say that there are (infinitely many) languages

---
Algorithm 2.4 Language class by grammar type

Input: A grammar $G = (V, T, S, P)$
Output: Language class $\mathscr{C}$ such that $G \in \mathscr{C}$

```
{
if G is regular then
   return "regular"
} else {
  if G is context-free then
     return "context-free"
  } else {
    if G is context-sensitive then
       return "context-sensitive"
    } else {
      if G is unrestricted then
          return "recursively enumerable"
}
```

---

that are not RELs. This and further consequences of this hierarchy from the computational perspective are discussed below in Section 2.3.

Exercises

Exercise 2.1.4.1. Algorithm 2.4 is too simple, lacking the details that allow the identification of the different grammar types. Complete it with these details. (Hint: Table 2.1.1.)

Exercise 2.1.4.2. Design an algorithm as complete as possible for, given a language L, determining the type of grammar G that generates $L(G)$ according to the Chomsky hierarchy.

2.2. Models of computation

With a formal language at hand, we often need to know what kind of a language it is–or is not. The most efficient way to solve this problem is by describing the language by means of a mathematical abstraction

Type	Name of G	$\alpha \to \beta$
0	Unrestricted (UG)	$\alpha \in (V \cup T)^+, \lvert V(\alpha) \rvert \geq 1$ $\beta \in (V \cup T)^*$
1	Context-sensitive (CSG)	$\alpha, \beta \in (V \cup T)^+, \lvert \beta \rvert \geq \lvert \alpha \rvert$ $\alpha = \gamma A \delta$ $\beta = \gamma X \delta$ $\gamma, \delta \in (V \cup T)^*$ $X \in (V \cup T)^+$
2	Context-free (CFG)	$\alpha \in V, \lvert \alpha \rvert = 1$ $\beta \in (V \cup T)^*$
3	Regular	$\alpha \in V, \lvert \alpha \rvert = 1$ $\beta = Av$ or vA $A \in V^*, v \in T^*$

Table 2.1.1.: The Chomsky hierarchy.

of a machine. This machine, in turn, has associated with it a computer model. The simplest of such machines are merely recognizers or acceptors in the sense that, given an input string w, their computation is limited to a "Yes/No" answer. More complex models can actually produce some output, and the most complex machines–the Turing machines–can both decide languages and compute functions. Memory is associated with this growing complexity in the following way: From basically no memory for the simplest language recognizers, memory increases until it reaches virtually infinity in the Turing machine.

2.2.1. Finite-state machines

Finite-state machines are the simplest machines associated with language recognition/acceptance and they recognize/accept the simplest (non-trivial) formal languages, i.e. the regular languages. A finite-state machine is either a finite-state recognizer (or acceptor) or a finite-state transducer. We begin with the former, which are typically called finite automata, and we then elaborate briefly on the latter. A more comprehensive discussion of all these machines can be found in Augusto (2021).

Definition. 2.2.1. A *finite-state recognizer* or *acceptor*, or *finite automaton* (abbr.: FA), is a 5-tuple

$$M = (Q, \Sigma, q_0, A, \delta)$$

where $Q = \{q_0, ..., q_n\}$ is a finite set of *states*, Σ is a finite *input alphabet*, $q_0 \in Q$ is the *initial state*, $A \subseteq Q$ is the set of *accepting states*, and $\delta : (Q \times \Sigma) \longrightarrow Q$ is the *transition function*.

The *computer model* for a FA (Fig. 2.2.1) consists of a logic box programmed by the transition function δ: whenever a read-only head examines a letter $\sigma \in \Sigma$ on a cell of the input tape in state $q \in Q$, the computer switches into state $\delta(q, \sigma)$ and moves on to read the letter on the next cell (so this is single-direction tape motion).

Definition. 2.2.2. Let $M = (Q, \Sigma, q_0, A, \delta)$ be a FA. The transition function for a string of input symbols, denoted by $\hat{\delta}$, is called the *extended transition function*. Clearly,

$$\hat{\delta} : (Q \times \Sigma^*) \longrightarrow Q$$

which is defined as,

1. for every $q \in Q$ and ϵ,
$$\hat{\delta}(q, \epsilon) = q$$

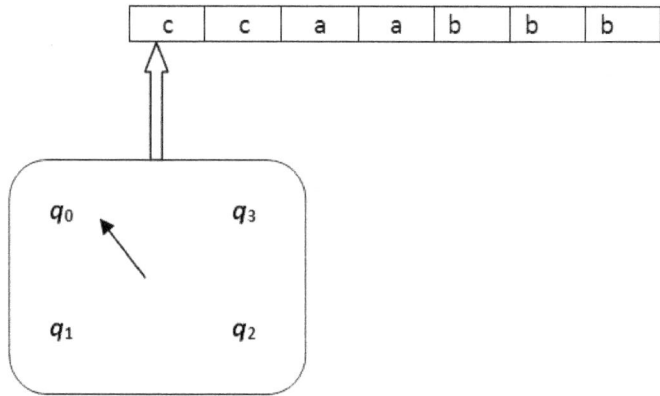

Figure 2.2.1.: Computer model for a FA.

2. for every $q \in Q$, every $y \in \Sigma^*$, and every $\sigma \in \Sigma$,

$$\hat{\delta}\left(q, y\sigma\right) = \delta\left(\hat{\delta}\left(q, y\right), \sigma\right).$$

In point 2 of the Definition above, our interest falls obviously on $x = y\sigma$ for some $x \in \Sigma^*$, as we want to evaluate $\hat{\delta}\left(q, x\right)$. We can do this, because we know what state the FA is in after having started in q and having finished reading the letters in y; then, we apply the last transition, corresponding to σ. Similarly, we can evaluate the concatenation of two strings $x, y \in \Sigma^*$ by the equality

$$\hat{\delta}\left(q, xy\right) = \hat{\delta}\left(\hat{\delta}\left(q, x\right), y\right).$$

Definition. 2.2.3. We say that a FA $M = (Q, \Sigma, q_0, A, \delta)$ *accepts* a string $x \in \Sigma^*$ if $\hat{\delta}\left(q, x\right) \in A$; otherwise, M is said to *reject* x.

Definition. 2.2.4. A *configuration* for a FA M is a pair $C = (q, w)$ such that $q \in Q$ and $w \in \Sigma^*$ denoting that M is at state q with the read-only head positioned at a letter of the string w.[10]

1. A *computation* for M is a sequence $C_0, C_1, ..., C_n$ of configurations, $C_0 = (q_0, w_0)$ and $C_n = (q_n, w_n)$, such that $C_{i-1} \vdash_M C_i$, read "configuration C_{i-1} yields configuration C_i in one step," for $i = 1, ..., n$.

---

[10]Some texts use the expression "instantaneous description" instead of "configuration."

2. A configuration (q, w) *yields* a configuration (q', w'), denoted by $(q, w) \vdash_M^* (q', w')$, where $\vdash_M^*$ denotes the reflexive and transitive closure of the step relation $\vdash_M$, if there is a computation of (q', w') from (q, w) in zero or more steps.

3. A string $w \in \Sigma^*$ is *accepted* by a FA M if, starting at state q_0 with the first letter, its transition sequence terminates in a state $q_f \in A$ such that
$$(q_0, w) \vdash_M^* (q_f, \epsilon).$$
Otherwise, w is *rejected*.

4. A string $w \in \Sigma^*$ is *recognized* by a FA M if the final state $q_f \in Q$ is some accepting state $q_i \in A$, $0 \le i \le n$.

Another way to define a FA $M = (Q, \Sigma, q_0, A, \delta)$ is by means of its state diagram $D = \vec{\mathfrak{D}}(M)$.

Definition. 2.2.5. The *state diagram* $D = \vec{\mathfrak{D}}(M)$ of a FA M is a labeled digraph with the following characteristics:

1. The vertices of $\vec{\mathfrak{D}}(M)$ are the states $q_i \in Q$, $i = 0, 1, ..., n$.

2. For each state $q \in Q$ and each letter $\sigma \in \Sigma$ there is an arc from vertex q to vertex $\delta(q, \sigma)$ labeled with the letter σ. If $\delta(q_i, \sigma) = q_j$, then if M is in state q_i and the current input letter is σ, M will move to state q_j, $i \gtreqless j$.

3. The starting state q_0 is designated by an incoming arc and an accepting state is indicated by two concentric circles.

Example 2.2.1. Figure 2.2.2 shows the FAs accepting the regular expressions of Examples 2.1.8-9. Note that the first FA accepts the language $L(M) = \{\epsilon\}$; in fact, a FA whose initial state coincides with an accepting state (i.e. $q_0 \in A$) accepts the empty string. Differently, a FA accepting the language $L(M) = \emptyset$ has solely the initial state q_0 (and perhaps a final state q_f, but there is no arc connecting both states).

Example 2.2.2. Let $M = (\{q_0, q_1, q_2\}, \{0, 1\}, q_0, \{q_0, q_1\}, \delta)$ be a FA with the transition function as specified in the following transitions:

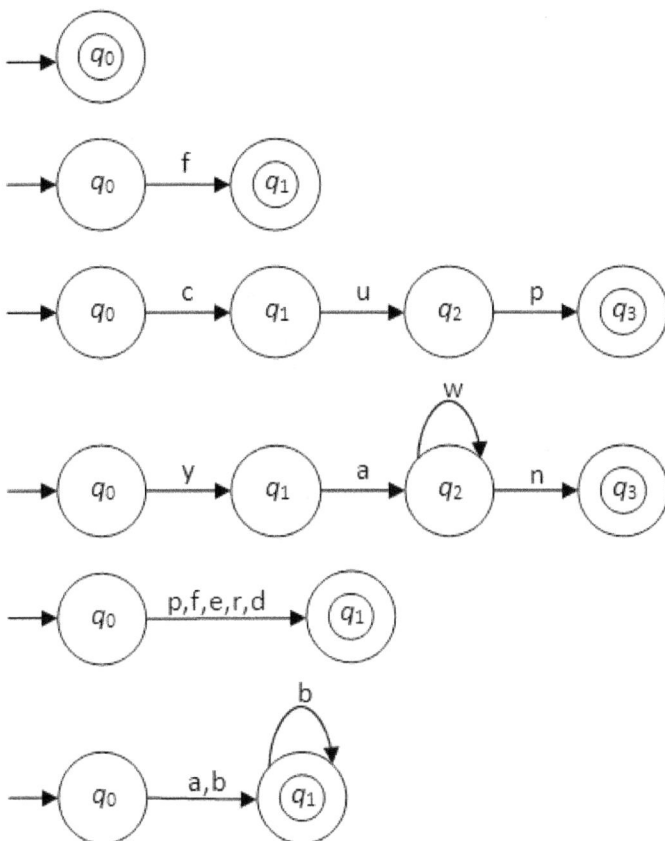

Figure 2.2.2.: State diagrams of FAs.

initial q	input σ	$\delta(q,\sigma)$
q_0	0	q_0
q_0	1	q_1
q_1	0	q_0
q_1	1	q_2
q_2	0	q_2
q_2	1	q_2

This gives the following *transition table*:

	$\delta(q,0)$	$\delta(q,1)$
q_0	q_0	q_1
q_1	q_0	q_2
q_2	q_2	q_2

This FA accepts all the strings that can be obtained from the regular expression $(0+10)^*(1+\epsilon)$, i.e. the language $L = \{0, 10\}^* \{\epsilon, 1\}$. The state diagram of this FA is shown in Figure 2.2.3. Note that if there are two successive 1, then the FA enters state q_2, which it cannot leave; thus q_2 is a non-accepting state–known as a *trapping state*, because the single arc leaving it is a loop.

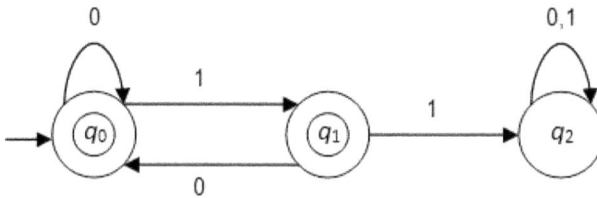

Figure 2.2.3.: A FA with two accepting states and one rejecting state.

Example 2.2.2 shows exactly what it is that FAs *recognizing a language* $L(M)$ do: they distinguish those strings w belonging to $L(M)$ from those that do not by accepting the former and rejecting all other strings. In other words, they provide a "Yes/No" (accept/reject) output. If an FA accepts no strings whatsoever, it still recognizes one language, to wit, the language $L(\emptyset) = \emptyset$.

As a matter of fact, FAs must be distinguished according to whether they are deterministic or non-deterministic. All the examples above are of the latter. We give the definition of a non-deterministic FA and the deterministic version is defined by contrast with this. Whenever necessary, henceforth we specify that a FA is deterministic by the abbreviation DFA.

Definition. 2.2.6. A *non-deterministic finite automaton* (NDFA) is a 5-tuple

$$M = (Q, \Sigma, q_0, A, \delta)$$

where Q, Σ, q_0, A are just as in a DFA and δ is the transition function $(Q \times \Sigma) \longrightarrow 2^Q$, or $(Q \times (\Sigma \cup \{\epsilon\})) \longrightarrow 2^Q$ if it includes ϵ-*transitions*, i.e. transitions $\delta(q, \epsilon) = q'$ (see Fig. 2.2.4).[11]

The computer model for a NDFA is like that of a DFA, but whenever it reads a string w on the input file while on state q, it switches into *any* state of $\hat{\delta}(q, w)$ and moves on to read the next letter. The main features of a NDFA is that it allows zero, one, or more transitions from a state on a single input letter, as well as a transition on no input at all. These account for the non-determinism of these automata.

Definition. 2.2.7. Given a NDFA, a state q is said to be in the ϵ-*closure* of $P \subseteq Q$, denoted by $\epsilon(P)$, if it is an element of P or it can be reached from an element thereof by means of applying $n \geq 1$ ϵ-transitions. The extended transition function $\hat{\delta} : (Q \times \Sigma^*) \longrightarrow 2^Q$ is defined in the following way:

1. For every $q \in Q$:

$$\hat{\delta}(q, \epsilon) = \epsilon(\{q\})$$

2. For every $q \in Q$, every $y \in \Sigma^*$, and every $\sigma \in \Sigma$:

$$\hat{\delta}(q, y\sigma) = \epsilon \left(\bigcup \left\{ \delta(p, \sigma) \mid p \in \hat{\delta}(q, y) \right\} \right)$$

Example 2.2.3. Figure 2.2.4 shows a NDFA that accepts the language $L = \{001\}^* \{0, 010\}^*$ by allowing an ϵ-transition.

Definition. 2.2.8. A *finite automaton* M is either a DFA or a NDFA.

---

[11] Alternatively, one can, instead of δ, consider a *transition relation* Δ, i.e. a finite subset of $Q \times \Sigma^* \times 2^Q$.

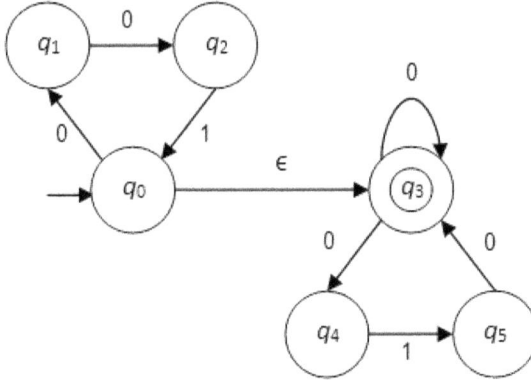

Figure 2.2.4.: A NDFA accepting the language $L = \{001\}^* \{0, 010\}^*$.

Definition. 2.2.9. The *language accepted* by a FA $M = (Q, \Sigma, q_0, A, \delta)$, denoted by $L(M)$, is the set of all strings accepted by M. Formally,

$$L(M) = \{w \in \Sigma^* | w \text{ is accepted by } M\}$$

or equivalently, appealing to the extended transition function,

$$L(M) = \left\{w \in \Sigma^* | \hat{\delta}(q_0, w) \in A\right\}.$$

Definition. 2.2.10. The FAs M_1 and M_2 are said to be *equivalent* if $L(M_1) = L(M_2)$.

Theorem 2.2.1. *If the language L is accepted by a NDFA with ϵ-transitions M, then it is accepted by a NDFA without ϵ-transitions M'.*

Proof: (Sketch) Let $M = (Q, \Sigma, q_0, A, \delta)$ and $M' = (Q, \Sigma, q_0, A', \delta')$ where for every $q \in Q$ and every $\sigma \in \Sigma$ we have $\delta'(q, \epsilon) = \emptyset$ and $\delta'(q, \sigma) = \hat{\delta}(q, \sigma)$. Let now a string $x \in \Sigma^*$ be accepted by M if $\hat{\delta}(q_0, x) \cap A \neq \emptyset$. Furthermore, define:

$$A' = \begin{cases} A \cup \{q_0\} & \text{if } \epsilon \in L \\ A & \text{otherwise} \end{cases}$$

This definition, according to which $A \subseteq A'$, assures us that M' accepts ϵ in case $\epsilon \in L$ but $q_0 \notin A$, and the definition above of $\hat{\delta}$ (cf. Def. 2.2.7) means that, for every $q \in Q$ and every $x \in \Sigma^*$, $\hat{\delta}(q, x)$ is the set of

108

states reachable by M by using the letters $\sigma \in x$ *and* ϵ-transitions. In turn, our definition above of δ' is intended to assure us that $\hat{\delta}'(q, x) = \hat{\delta}(q, x)$(why?). So, first of all we have to prove by induction on x that this indeed holds for every q and every x for which $|x| \geq 1$. Begin with the obvious step for $x = a \in \Sigma$. Next, let σ be an arbitrary element of Σ and let $\hat{\delta}'(q, y) = \hat{\delta}(q, y)$ for every q and some y for which $|y| \geq 1$. From

$$\hat{\delta}'(q, y\sigma) = \bigcup \left\{ \delta'(p, \sigma) \,|p \in \hat{\delta}'(q, y) \right\}$$

obtain (how?)

$$\hat{\delta}'(q, y\sigma) = \bigcup \left\{ \hat{\delta}(p, \sigma) \,|p \in \hat{\delta}(q, y) \right\}$$

which is a special case of the formula

$$\hat{\delta}(q, yz) = \bigcup \left\{ \hat{\delta}(p, z) \,|p \in \hat{\delta}(q, y) \right\}.$$

Now all that is left to do is to verify that $L = L(M) = L(M')$, which is done by checking for ϵ and $|x| \geq 1$ with respect to $L(M)$ and $L(M')$. (Hint: Check when $A = A'$ and $A \subseteq A'$.) **QED**

Example 2.2.4. Another way to prove Theorem 2.2.1 would be by applying the notion of ϵ-closure. We show this directly by means of an example. First of all, it is convenient to make, for $P \subseteq Q$,

$$\epsilon(P) = \bigcup_{q \in P} \epsilon(q).$$

Define the extended transition function as in Definition 2.2.7 and extend both δ and $\hat{\delta}$ to sets of states R as, for $w \in \Sigma^*$ and $\sigma \in \Sigma$,

$$\delta(R, \sigma) = \bigcup_{q \in R} \delta(q, \sigma)$$

$$\hat{\delta}(R, w) = \bigcup_{q \in R} \hat{\delta}(q, w)$$

Let $M = (\{q_0, q_1, q_2\}, \{a, b, c\}, q_0, \{q_2\}, \delta)$ be a NDFA with the following transition table for $\delta(q, \sigma)$ and $\hat{\delta}(q, \sigma)$:

q	$\delta(q,a)$	$\delta(q,b)$	$\delta(q,c)$	$\delta(q,\epsilon)$	$\hat{\delta}(q,a)$	$\hat{\delta}(q,b)$	$\hat{\delta}(q,c)$
q_0	$\{q_0\}$	$\emptyset$	$\emptyset$	$\{q_1\}$	$\{q_0,q_1,q_2\}$	$\{q_1,q_2\}$	$\{q_2\}$
q_1	$\emptyset$	$\{q_1\}$	$\emptyset$	$\{q_2\}$	$\emptyset$	$\{q_1,q_2\}$	$\{q_2\}$
q_2	$\emptyset$	$\emptyset$	$\{q_2\}$	$\emptyset$	$\emptyset$	$\emptyset$	$\{q_2\}$

We show how we obtained $\hat{\delta}(q,\sigma)$. First, we computed

$$\hat{\delta}(q_0,\epsilon) = \epsilon(\{q_0\}) = \{q_0,q_1,q_2\}$$

With this result, we next computed

$$
\begin{aligned}
\hat{\delta}(q_0,a) &= \epsilon\left(\delta\left(\hat{\delta}(q_0,\epsilon)\right),a\right) \\
&= \epsilon(\delta(\{q_0,q_1,q_2\},a)) \\
&= \epsilon(\delta(q_0,a)\cup\delta(q_1,a)\cup\delta(q_2,a)) \\
&= \epsilon(\{q_0\}\cup\emptyset\cup\emptyset) \\
&= \epsilon(\{q_0\}) = \{q_0,q_1,q_2\}
\end{aligned}
$$

and

$$
\begin{aligned}
\hat{\delta}(q_0,b) &= \epsilon\left(\delta\left(\hat{\delta}(q_0,\epsilon)\right),b\right) \\
&= \epsilon(\delta(\{q_0,q_1,q_2\},b)) \\
&= \epsilon(\delta(q_0,b)\cup\delta(q_1,b)\cup\delta(q_2,b)) \\
&= \epsilon(\emptyset\cup\{q_1\}\cup\emptyset) \\
&= \epsilon(\{q_1\}) = \{q_1,q_2\}
\end{aligned}
$$

and so on for all the remaining states and letters. Note that $\hat{\delta}(q,\sigma)$ is not necessarily equal to $\delta(q,\sigma)$, because the latter includes only the states reachable from q by arcs labeled σ, whereas the former includes all states reachable from q by arcs labeled σ (*including arcs labeled* ϵ). From the complete transition tables above it is an easy matter to build (left as an exercise) the state diagrams of these associated NDFA with and without ϵ-transitions.

Recall what was said above (note to Def. 2.2.6) about the main features of a NDFA. A DFA is a special case of a NDFA in which for each state there is a single transition on each letter. But at the practical level they are very different: Whereas we can transform the transition table or the state diagram of a DFA into a computer program, we cannot do this (as readily) for a NDFA. Thus, we need to be able to construct a DFA out of a NDFA. The following theorem provides the mathematical foundation for this *simulation*.

Theorem 2.2.2. *Any language that can be accepted by a NDFA M can be accepted by a DFA M'. Thus, for every NDFA M we can construct an equivalent DFA M'.*

Proof: (Sketch) A DFA $M' = \left(Q', \Sigma, q_0', A', \delta'\right)$ simulates a NDFA $M = (Q, \Sigma, q_0, A, \delta)$ by means of a correspondence between the states of the DFA and *sets of states* of the NDFA, i.e. by making $Q' = 2^Q$. More specifically, for every $q \in Q'$ and $\sigma \in \Sigma$, setting

$$\delta'(q, \sigma) = \bigcup \{\delta(p, \sigma) \,|\, p \in q\}$$

gives us

$$\delta'([q_1, q_2, ..., q_i], \sigma) = [p_1, p_2, ..., p_j]$$

where $[q_1, q_2, ..., q_i]$ is a single state, from

$$\delta(R, \sigma) = \delta(\{q_1, q_2, ..., q_i\}, \sigma) = \{p_1, p_2, ..., p_j\}$$

where R is a set of states. Set now

$$q_0' = [q_0]$$

and for every $x \in \Sigma^*$

$$\hat{\delta}'\left(q_0', x\right) = \hat{\delta}(q_0, x)$$

so that we have

$$\hat{\delta}'\left(q_0', x\right) = [q_1, q_2, ..., q_i]$$

iff

$$\hat{\delta}(q_0, x) = \{q_1, q_2, ..., q_i\}.$$

Make the basis $|x| = 0$, and the proof follows by induction on the length m of the inputs. The proof is complete when we show that a string x is accepted by M' exactly if $\hat{\delta}'\left(q_0', x\right) \in A'$. **QED**

Example 2.2.5. The proof of this theorem provides the algorithmic means to convert a NDFA into a DFA known as *subset construction* (cf. Algorithm 2.5).[12] Let $M = (|Q| = 5, \{a, b\}, q_0, \{q_2, q_4\}, \delta)$ be a NDFA (cf. Fig. 2.2.5.1) whose transition function δ is given by the following table:

---

[12]This method can also be applied to convert a NDFA with, into one without, ϵ-transitions.

2. *Fundamentals of classical computing*

Algorithm 2.5 Subset construction

Input: NDFA $M = (Q, \Sigma, q_0, A, \delta)$

Output: DFA $M' = \left(Q', \Sigma, q_0', A', \delta' \right)$ such that $L(M) = L(M')$

---

1. Set $Q' = 2^Q$

2. For each $\sigma \in \Sigma$ and each subset $R \in 2^Q$ obtain $\delta'([R], \sigma) = \bigcup \{ \delta(p, \sigma) \, | p \in R \}$ such that $\delta' \left(\underbrace{[q_1, q_2, ..., q_i]}_{1 \text{ state}}, \sigma \right) = [p_1, p_2, ..., p_j]$

3. Set $q_0' = [q_0]$

4. Make $A' = \{ S \in Q' | [S] \supseteq q_i \in A \}$

	$\delta(q, a)$	$\delta(q, b)$
$\rightarrow q_0$	$\{q_0, q_1\}$	$\{q_0, q_3\}$
q_1	$\{q_2\}$	$\emptyset$
q_2	$\emptyset$	$\emptyset$
q_3	$\emptyset$	$\{q_4\}$
q_4	$\emptyset$	$\emptyset$

We obtain the FA $M' = \left(Q', \{a, b\}, q_0', A', \delta' \right)$ in which $Q' = 2^Q = \{[\emptyset], [q_1], ..., [q_0, q_1, ..., q_4]\}$, $q_0' = [q_0]$, $A' = \{[q_0, q_1, q_2], [q_0, q_3, q_4]\}$, and $\delta'([q], \sigma)$ is obtained as, for instance,

$$
\begin{aligned}
\delta'([q_0, q_1], a) &= \delta(q_0, a) \cup \delta(q_1, a) \\
&= \{q_0, q_1\} \cup \{q_2\} \\
&= \{q_0, q_1, q_2\} \\
&= [q_0, q_1, q_2]
\end{aligned}
$$

The states that interest us are those which can be reached from the initial state, with those that do not satisfy this requirement being omitted. We obtain this result by computing the transition table for δ for all the subsets of Q:

	$\delta\left(q,a\right)$	$\delta\left(q,b\right)$
$\rightarrow \{q_0\}$	$\{q_0,q_1\}$	$\{q_0,q_3\}$
$\{q_1\}$	$\{q_2\}$	$\emptyset$
$\vdots$	$\vdots$	$\vdots$
$\{q_0,q_1\}$	$\{q_0,q_1,q_2\}$	$\{q_0,q_3\}$
$\vdots$	$\vdots$	$\vdots$
$\emptyset$	$\emptyset$	$\emptyset$

The table for δ' is as follows:

	$\delta'\left([q],a\right)$	$\delta'\left([q],b\right)$
$\rightarrow [q_0]$	$[q_0,q_1]$	$[q_0,q_3]$
$[q_0,q_1]$	$[q_0,q_1,q_2]$	$[q_0,q_3]$
$[q_0,q_3]$	$[q_0,q_1]$	$[q_0,q_3,q_4]$
$[q_0,q_1,q_2]$	$[q_0,q_1,q_2]$	$[q_0,q_3]$
$[q_0,q_3,q_4]$	$[q_0,q_1]$	$[q_0,q_3,q_4]$

Figure 2.2.5.2 shows the state diagram of M'. For graphical convenience, we abbreviate q_i as i. Verify that $L\left(M\right) = L\left(M'\right)$, both automata accepting strings ending in a^2 or b^2. Note that, as expectable, there are more transitions in M' than in M.

Proposition. 2.2.11. *The class of languages $\mathscr{L}$ accepted by FAs is closed under Kleene closure, union, concatenation, complementation, and intersection.*

Proof: The proof is left as an exercise. (Hint: See the next theorems.)

Theorem 2.2.3. *Let $M_1 = (Q_1, \Sigma, q_1, A_1, \delta_1)$, $M_2 = (Q_2, \Sigma, q_2, A_2, \delta_2)$ be FAs accepting the languages L_1 and L_2, respectively. Let further $M = (Q, \Sigma, q_0, A, \delta)$ be a FA where $Q = Q_1 \times Q_2$, $q_0 = (q_1, q_2)$, $A = A_1 \times A_2$, and where the transition function is defined, for every p, q, σ such that $q \in Q_1$, $p \in Q_2$, and $\sigma \in \Sigma$, by*

$$\delta\left((q,p),\sigma\right) = \left(\delta_1\left(q,\sigma\right), \delta_2\left(p,\sigma\right)\right).$$

Then, (i) if $A = \{(q,p) \mid q \in A_1 \text{ or } p \in A_2\}$, M accepts the language $L_1 \cup L_2$; (ii) if $A = \{(q,p) \mid q \in A_1 \text{ and } p \in A_2\}$, M accepts the language $L_1 \cap L_2$; and (iii) if $A = \{q,p \mid q \in A_1 \text{ and } p \notin A_2\}$, M accepts the language $L_1 - L_2$.

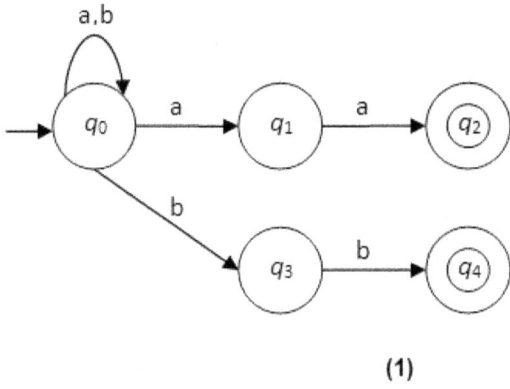

(1)

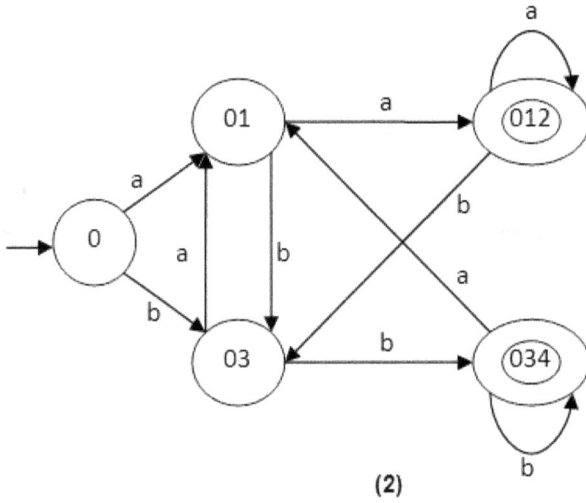

(2)

Figure 2.2.5.: Equivalent NDFA (1) and DFA (2).

Proof: Left as an exercise. (Hint: Consider the formula

$$\hat{\delta}(q_0, x) = \left(\hat{\delta}_1(q_1, x), \hat{\delta}_2(q_2, x)\right)$$

and apply structural induction on x.)

Theorem 2.2.4. *(Kleene's theorem; Kleene, 1956) A language L over an alphabet Σ is regular iff $L = L(M)$, i.e. L is the language accepted by some FA M.*

Proof: (Sketch) Recall the definitions of regular expression and regular language (Def.s 2.1.12 and 2.1.13, respectively). The proof of this theorem is by induction on $r = (), \epsilon, a \in \Sigma$, which constitutes the basis for the induction (i.e. there are no operators involved). For the induction, i.e., for $L(M_1) \cup L(M_2)$, $L(M_1)L(M_2)$, and $L(M_1)^*$, let $M_1 = (Q_1, \Sigma, q_1, \{f_1\}, \delta_1)$ and $M_2 = (Q_2, \Sigma, q_2, \{f_2\}, \delta_2)$ where f_i abbreviates q_{f_i}, for q_f denoting final state. Let q_0 be a new initial state and f_0 a new final state. In the first case, we have to construct a NDFA

$$M = (Q_1 \cup Q_2 \cup \{q_0, f_0\}, \Sigma, q_0, \{f_0\}, \delta)$$

where $Q_1 \cap Q_2 = \emptyset$. In the second case, we have to construct a NDFA

$$M = (Q_1 \cup Q_2, \Sigma, q_1, \{f_2\}, \delta).$$

For the third case, we have to construct a NDFA

$$M = (Q_1 \cup \{q_0, f_0\}, \Sigma, q_0, \{f_0\}, \delta).$$

The correctness of the proof falls on the right definitions of the several $\delta(\cdot, \cdot)$ involved; Figure 2.2.6 provides clues thereto. Then apply Theorems 2.2.1-2. **QED**

As language recognizers, FAs are useful to prove non-regular languages, as well as for distinguishing finite from infinite languages. Recall the pumping lemma (Theorem 2.1.1); the proof of this lemma is constructed precisely by appealing to FAs.[13] In effect, we can complete Theorem 2.1.1 in the following way:

Theorem 2.2.5. *Pumping lemma – Let $L = L(M)$, where M is a FA with m states. Let $z \in L$, $|z| \leq m$. Then we can write $z = uvw$ such that z satisfies the three conditions of Theorem 2.1.1.*

---

[13]Yet another way to prove that a language L is not regular is by means of the *Myhill-Nerode theorem*. We skip this theorem, as it is not essential and its proof is rather complex. See, e.g., Davis & Weyuker (1983).

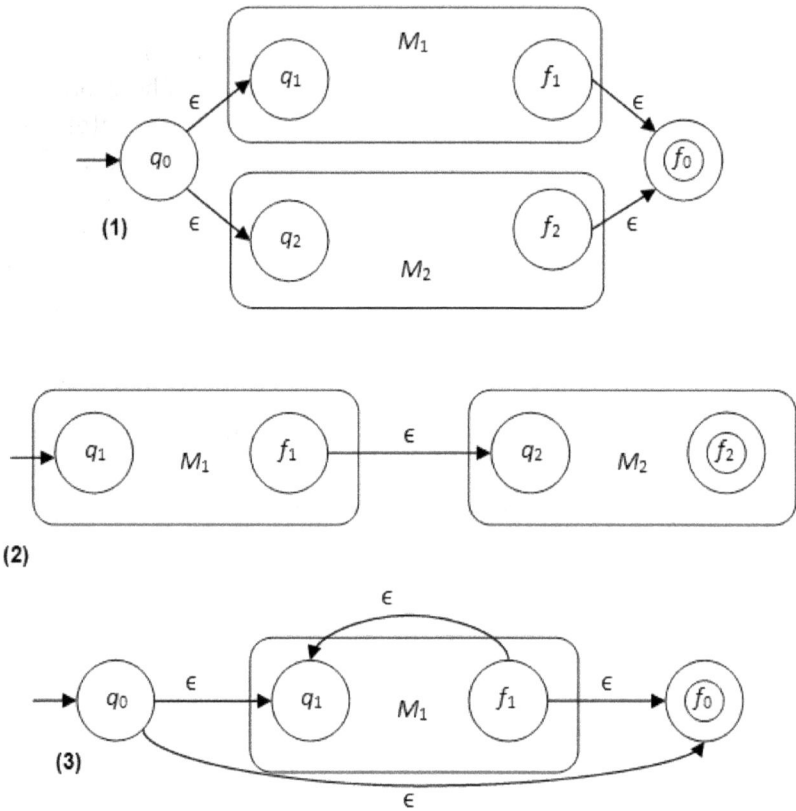

Figure 2.2.6.: Schematic diagrams for FAs accepting (1) $L_1 \cup L_2$, (2) $L_1 L_2$, and (3) $(L_1)^*$.

Proof: If a language L is regular, then there is a FA M with a definite number of states that accepts L. Let us suppose that M has m states. Consider a string with m or more letters $a_1 a_2 ... a_n$, $n \geq m$ and let $\delta(q_0, a_1 a_2 ... a_i) = q_i$ for $i = 1, 2, ..., n$. This includes the initial state q_0, but since there are only m states, it is not possible for the $m + 1$ states $q_0, q_1, ..., q_m$ to be distinct. By appealing to the pigeonhole principle,[14] we conclude that M must be in at least one state more than once, i.e. there must be two integers j and k, $0 \leq j < k \leq m$ such that $q_j = q_k$. In other words, there must be a loop. We have $|a_{j+1}...a_k| \geq 1$, because $j < k$, and because $k \leq m$ we also have $|a_{j+1}...a_k| \leq m$. If $a_1 a_2 ... a_n$ is in $L(M)$, then $a_1 a_2 ... a_j a_{k+1} a_{k+2} ... a_n$ is also in $L(M)$, as there is a path from q_0 to q_n that goes directly through q_j without doing the loop $a_{j+1}...a_k$. But we could do this loop as many times as we want to, so that $a_1 a_2 ... a_j (a_{j+1}...a_k)^i a_{k+1} a_{k+2} ... a_n$ is also in $L(M)$ for $i = 0, 1, ...$ Simplifying by the equalities $u = a_1 a_2 ... a_j$, $v = a_{j+1}...a_k$, and $w = a_{k+1} a_{k+2} ... a_n$, and by applying the extended transition function, we have

$$\hat{\delta}(q_0, u) = q_j$$

$$\hat{\delta}(q_j, v) = q_j$$

$$\hat{\delta}(q_j, w) = q_n \in A$$

so that we have in fact

$$\hat{\delta}(q_0, uvw) = \hat{\delta}(q_0, uv^i w) = q_n \in A$$

and $z = uv^i w \in L(M)$ for $i \geq 0$. Figure 2.2.7 shows the FA M that accepts the regular expression $z = uv^{(i)}w$. **QED**

The acceptance or rejection of a language L by a FA is a *decision problem* (cf. Section 2.3). Two further decision problems that can be answered by a FA are whether L is non-empty and whether L is infinite. The following two theorems (the proofs are left as exercises; hint: the pumping lemma features in the proofs) actually provide algorithms–albeit inefficient ones–to answer these questions.

Theorem 2.2.6. *Let M be a FA with m states. If $L(M) \neq \emptyset$, there is a string $z \in L(M)$ such that $|z| < m$.*

Theorem 2.2.7. *Let M be a FA with m states. Then $L(M)$ is infinite iff $L(M)$ contains a string z such that $m \leq |z| < 2m$.*

---

[14]The *pigeonhole principle* states that if $(n+1)$ objects are distributed among n sets, then at least one of the sets must contain at least two objects.

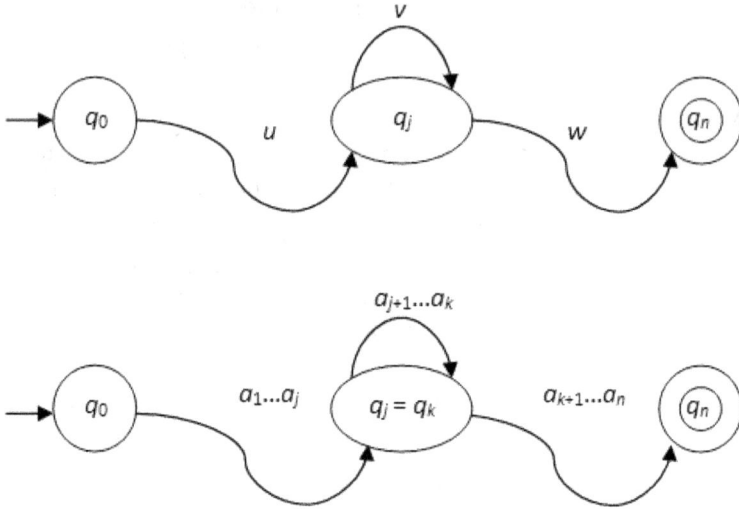

Figure 2.2.7.: A FA M for the pumping lemma.

It is rather obvious that a machine needs to produce some sort of output, in order to be useful for computation in practical terms. A finite transducer is such a machine.

Definition. 2.2.12. A *(deterministic) finite transducer* is a 6-tuple

$$M = (Q, \Sigma_I, \Sigma_O, q_0, \delta, \tau)$$

where Q, q_0 are as for a FA, Σ_I is the *input alphabet*, Σ_O is the *output alphabet*, $\delta : (Q \times \Sigma_I) \longrightarrow Q$ is the transition function, and $\tau : (Q \times \Sigma_I) \longrightarrow \Sigma_O^*$ is the *output function*. Specifications for the output function define two kinds of finite transducer:

1. A finite transducer whose output function produces a single letter for each input letter, i.e. $\tau : (Q \times \Sigma_I) \longrightarrow \Sigma_O$, is called a *Mealy machine*.

2. A finite transducer whose output function depends only on the current state, i.e. $\tau : Q \longrightarrow \Sigma_O$, is called a *Moore machine*.

Example 2.2.6. The following transition and output table is for a Moore machine with alphabet $\Sigma_I = \Sigma_O = \{0, 1\}$ (the arrow indicates the initial state). Figure 2.2.8.1 shows this Moore machine.

	$\delta(q,0)$	$\delta(q,1)$	$\tau(q,0) = \tau(q,1)$
$\rightarrow q_0$	q_1	q_2	0
q_1	q_1	q_3	1
q_2	q_0	q_2	0
q_3	q_2	q_3	1

Figure 2.2.8.2 shows a Mealy machine with alphabet $\Sigma_I = \Sigma_O = \{0,1\}$ and with the following transition and output table:

	$\delta(q,0)$	$\delta(q,1)$	$\tau(q,0)$	$\tau(q,1)$
$\rightarrow q_0$	q_1	q_2	1	0
q_1	q_1	q_3	1	1
q_2	q_0	q_2	0	0
q_3	q_2	q_3	0	1

This Mealy machine is equivalent to the Moore machine in Figure 2.2.8.1. It should be easy to see that the conversion from a Moore to an equivalent Mealy machine is straightforward.[15] In this example, the input alphabet and the output alphabet are the same, but this need not be so. Note the distinct ways to indicate the output in the two machines: in a Mealy machine, the output is indicated on the edges in the form *input/output*, whereas in a Moore machine it is indicated as *state/output* on the vertices of the diagram.

Every Moore machine with output alphabet $\Sigma_O = \{0,1\}$ can be considered as a FA by taking outputs "1" as accepting states. A FA, in turn, can be seen as a Moore machine by considering that it enters an accepting state when it prints "1", and enters some other state when printing "0." Thus, a transition and output table for a Moore machine can be turned into a transition table for a FA by proceeding in the following way: in the final column of the former, replace 1 by, say, $*$ and 0 by a blank space. See Example 2.2.7.

Example 2.2.7. The table below is the transition table for a FA equivalent to the Moore machine in Example 2.2.6.

	$\delta(q,0)$	$\delta(q,1)$	
$\rightarrow q_0$	q_1	q_2	
q_1	q_1	q_3	$*$
q_2	q_0	q_2	
q_3	q_2	q_3	$*$

---

[15]But the reverse conversion is not so straightforward. See Augusto (2021) for the conversion algorithm.

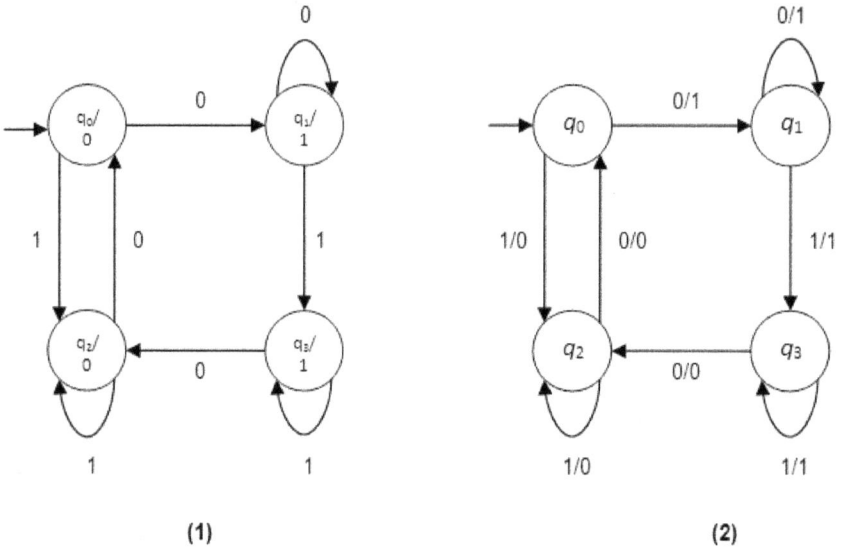

Figure 2.2.8.: Moore (1) and Mealy (2) machines.

Definition. 2.2.13. A *finite-state machine* is a FA or a finite transducer.

Exercises

Exercise 2.2.1.1. Let $\Sigma = \{0, 1\}$. Construct a FA M accepting precisely those strings over Σ / the language

1. with an even number of 0.

2. that end in a single 1.

3. that end in an odd number $n > 1$ of 1.

4. with the prefix 10.

5. $L = \{0, 1\}^* \{000, 111\} \{0, 1\}^*$.

6. $L = \{0, 1\}^* \{00\}$.

7. $L = \{010\}^* \{11, 0110\}^*$.

8. $L = \{00, 001\}^* \{1\}$.

9. $L = \{1, 01\}^* \{0\}$.

10. $L = \{010, 101\} \{101, 010\}^*$.

Exercise 2.2.1.2. Describe a FA recognizing all (partial) anagrams of the string *amber*.

Exercise 2.2.1.3. Describe a FA with the Latin-based alphabet for input alphabet Σ that recognizes strings with the 5 vowels occurring in the usual order, i.e. *aeiou* (e.g., the English word *facetious*). Describe its behavior when given the string *milk* as input.

Exercise 2.2.1.4. Construct a NDFA M accepting the language $L = \{0, 11\}^* \{10^*, \epsilon\}$ that is actually a combination of two NDFAs M_1 and M_2.

Exercise 2.2.1.5. Convert the NDFA with ϵ-transitions of Example 2.2.3 into a NDFA without ϵ-transitions.

Exercise 2.2.1.6. Construct the NDFA with ϵ-transitions of Example 2.2.4. Construct the equivalent NDFA without ϵ-transitions.

Exercise 2.2.1.7. The NDFA M' without ϵ-transitions equivalent to one with such transitions M can have (many) more states than this. Construct the NDFA M' without ϵ-transitions equivalent to the NDFA $M = (\{q_0, q_1, q_2, q_3, q_4\}, \{0, 1\}, q_0, \{q_0\}, \delta)$ with the following transition table:

q	$\delta(q, 0)$	$\delta(q, 1)$	$\delta(q, \epsilon)$	$\hat{\delta}(q, 0)$	$\hat{\delta}(q, 1)$
q_0	$\emptyset$	$\emptyset$	$\{q_1\}$	$\{q_1, q_2\}$	$\emptyset$
q_1	$\{q_1, q_2\}$	$\emptyset$	$\emptyset$	$\{q_1, q_2\}$	$\emptyset$
q_2	$\emptyset$	$\{q_3\}$	$\emptyset$	$\emptyset$	$\{q_0, q_1, q_3\}$
q_3	$\emptyset$	$\{q_4\}$	$\{q_0\}$	$\{q_1, q_2\}$	$\{q_4\}$
q_4	$\{q_3\}$	$\emptyset$	$\emptyset$	$\{q_0, q_1, q_3\}$	$\emptyset$

Exercise 2.2.1.8. Formulate the two algorithms based on Theorems 2.2.6 and 2.2.7 in some pseudo-code.

Exercise 2.2.1.9. Build the state diagram of the finite transducer with the following transition and output table given $\Sigma_I = \Sigma_O = \{0, 1\}$:

	$\delta(q, 0)$	$\delta(q, 1)$	$\tau(q, 0)$	$\tau(q, 1)$
$\rightarrow q_0$	q_1	q_0	1	0
q_1	q_3	q_0	1	1
q_2	q_1	q_2	0	1
q_3	q_2	q_1	0	0

Exercise 2.2.1.10. Build the state diagram of the FA in Example 2.2.7.

Exercise 2.2.1.11. Show that the following languages are not regular:

1. $L = \{a^n b^m \in \Sigma^* | n \neq m\}$

2. $L = \{ww^R | w \in \Sigma^*\}$

3. $L = \{w \in \Sigma^* | |w|_a < |w|_b\}$

Exercise 2.2.1.12. Give informal accounts of the machines accepting the languages in Exercise 2.2.1.11.

Exercise 2.2.1.13. As seen above, a finite machine does not accept the language $L_1 = \{a^m b^m \in \Sigma^* | m \geq 0\}$. However, it does accept the language $L_2 = \{a^m b^n \in \Sigma^* | m, n \geq 1\}$. Explain this contrast by appealing to the computation abilities of finite machines.

Exercise 2.2.1.14. Prove (Complete the proof of) the theorems and propositions in this Section that were left as an exercise (with a sketchy proof, respectively).

Exercise 2.2.1.15. Give informal accounts of the behavior of the FAs in the proofs of Exercise 2.2.1.14.

2.2.2. Pushdown automata

Pushdown automata (PDAs) are associated to CFLs in the same way that finite-state machines are associated to regular languages: a language $L(G)$ can be generated by a CFG G iff it can be accepted by a PDA M such that $L(G) = L(M)$. But unlike finite-state machines, non-determinism cannot always be eliminated in PDAs; if non-determinism can be eliminated in a PDA, then we have a *parser* for a certain CFL. Parsers thus accept only a subset of the CFLs, precisely that subset in which the syntax of most programming languages is included.

Recall the definition of a formal grammar (Def. 2.1.11). A PDA is essentially a (possibly non-deterministic) finite-state machine with an auxiliary stack that provides an unlimited amount of memory for P. The operation of a stack is basically on a last-in-first-out rule: adding a symbol to the memory is interpreted as *pushing* it onto a stack so that it becomes the top symbol; this explains the coinage of these machines.

As said, the memory may be unlimited, but the PDA has access only to the symbol at the top of the stack, i.e. the most recently added symbol. Let X be the symbol currently on top of the stack; then, $\alpha = YX$, where α is a string, denotes the fact that a new symbol Y is added to the stack and is now the top symbol. But the PDA can simply leave the stack unchanged, denoted by $\alpha = X$, or *pop X off* the stack (i.e. delete X), an action denoted by $\alpha = \epsilon$. Visualizing a tray dispenser at a cafeteria may help gaining an intuitive idea of a PDA; the formal description of a computer model of a PDA (see below) lends rigor to this idea.

Definition. 2.2.14. A pushdown automaton is a 7-tuple

$$M = (Q, \Sigma, \Gamma, q_0, A, Z_0, \delta)$$

where Q is a finite set of states, $q_0 \in Q$ is the initial state, Σ is a finite set of symbols called the *input alphabet*, Γ is a finite set of symbols called the *stack alphabet*, A is the set of accepting states, $Z_0 \in \Gamma$ is the *start stack symbol*, and $\delta : (Q \times (\Sigma \cup \{\epsilon\}) \times \Gamma) \longrightarrow (Q \times \Gamma^*)$ is the transition function.[16]

The computer model for a PDA (Fig. 2.2.9) consists of a logic box programmed by the transition function δ and equipped with a read-only head and a stack. The read-only head reads an input tape moving in a single direction. When this head reads a substring u while in state q with substring α at the top of the stack, the machine selects a corresponding entry from δ and performs the indicated transition.

Definition. 2.2.15. A *configuration* C for a PDA M is a triple $C = (q, u, \alpha)$ where $q \in Q$, $u \in \Sigma^*$, and $\alpha \in \Gamma^*$.

1. We say that the configuration (q, u, α) yields the configuration (p, v, β) in one *move*, and write $(q, u, \alpha) \vdash_M (p, v, \beta)$, if there is a transition $((q, u, \alpha), (p, \beta))$. More exactly, we have $(q, u, X\gamma) \vdash_M (p, v, \xi\gamma)$, for $u = \lambda v$, $\lambda \in \Sigma \cup \{\epsilon\}$,[17] and where if $\alpha = X\gamma$ for some $X \in \Gamma$ and some $\gamma \in \Gamma^*$, then $\beta = \xi\gamma$ for some string ξ for which $(p, \xi) \in \delta(q, \lambda, X)$.

2. A *computation* for a PDA M is a finite sequence of configurations $C_0, C_1, ..., C_n$ such that $C_{i-1} \vdash_M C_i$, $1 \leq i \leq n$.

---

[16] Alternatively, one can consider a transition relation $\Delta = Q \times (\Sigma \cup \{\epsilon\}) \times \Gamma \times Q \times \Gamma^*$.
[17] Clearly, $u = v$ if $\lambda = \epsilon$. Alternatively, we can write $(q, uv, X\gamma) \vdash_M (p, v, \xi\gamma)$.

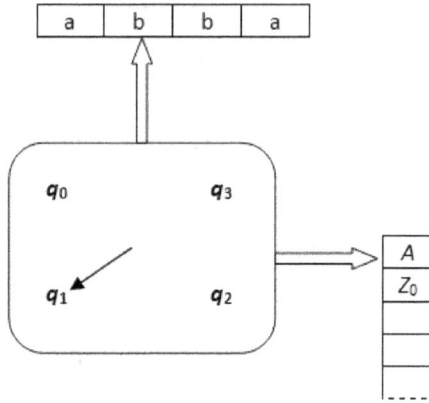

Figure 2.2.9.: Computer model for a PDA.

3. A configuration C yields a configuration C' if there is a computation of C' from C in n moves, denoted by $C \vdash_M^n C'$, or in zero or more moves, denoted by $C \vdash_M^* C'$.

This can be restated in terms of the transition function:

Definition. 2.2.16. Given a PDA M, we write

$$\delta(q, a, Z) = \{(p_1, \alpha_1), (p_2, \alpha_2), ..., (p_n, \alpha_n)\}$$

where $q, p_{1 \leq i \leq n} \in Q$, $a \in \Sigma$, $Z \in \Gamma$, and $\alpha_{1 \leq i \leq n} \in \Gamma^*$, to indicate that in state q, reading input symbol a, and with Z as the top stack symbol, M can enter state p_i, replace symbol Z by string α_i and advance the input head one symbol.[18] We write

$$\delta(q, \epsilon, Z) = \{(p_1, \alpha_1), (p_2, \alpha_2), ..., (p_n, \alpha_n)\}$$

to indicate that M at state q and with Z as the top symbol on the stack can enter any state p_i and replace Z by α_i independently of the input symbol being read. Note that in this case the head does not advance.

By convention, the leftmost symbol of α_i is placed highest on the stack, and the rightmost symbol of α_i is placed lowest on the stack.

Definition. 2.2.17. Let there be given a PDA M. An *accepting configuration* of M is any configuration in which the state $p \in A$, and we

---

[18]It is not possible to choose state p_i and string α_j such that $i \neq j$ in a single move.

write[19]

$$(q_0, uv, Z_0) \vdash_M^* (p, v, \alpha).$$

Example 2.2.8. Recall the CFL $L = \{a^m b^m | m \geq 0\}$ of Example 2.1.11. The PDA for L is $M = (Q, \Sigma, \Gamma, q_0, A, Z_0, \delta)$ where $Q = \{q_0, q_1, q_2, q_3\}$, $A = \{q_0, q_3\}$, $\Sigma = \{a, b\}$ and the transitions are as follows:

State $q \in Q$	Input $(a, b \in \Sigma) \cup \{\epsilon\}$	Stack symbol $A, Z_0 \in \Gamma$	Move(s)
q_0	a	Z_0	(q_1, AZ_0)
q_1	a	A	(q_1, AA)
q_1	b	A	(q_2, ϵ)
q_2	b	A	(q_2, ϵ)
q_2	ϵ	Z_0	(q_3, Z_0)

The sequence of configurations that result in the acceptation of the string *aabb* is

$$(q_0, aabb, Z_0) \vdash (q_1, abb, AZ_0) \vdash (q_1, bb, AAZ_0) \vdash (q_2, b, AZ_0)$$

$$\vdash (q_2, \epsilon, Z_0) \vdash (q_3, \epsilon, Z_0)$$

This is translated in terms of δ as

$$\begin{aligned}
\delta(q_0, a, Z_0) &= \{(q_1, AZ_0)\} \\
\delta(q_1, a, A) &= \{(q_1, AA)\} \\
\delta(q_1, b, A) &= \{(q_2, \epsilon)\} \\
\delta(q_2, b, A) &= \{(q_2, \epsilon)\} \\
\delta(q_2, \epsilon, Z_0) &= \{(q_3, Z_0)\}
\end{aligned}$$

Definition. 2.2.18. The state diagram $D = \vec{\mathfrak{D}}(M)$ of a PDA M is a labeled digraph as for a FA (cf. Def. 2.2.5, items 1 and 3) such that for each transition $\delta(q, u, X) = (p, \alpha)$ there is an arc from vertex q to vertex p labeled $u; X/\alpha$ (also: $u, X/\alpha$), interpreted as "read u; replace X by α."

Example 2.2.9. We can actually simplify the PDA in Example 2.2.8 by making the set $Q = \{q_0, q_1, q_2\}$ and $A = \{q_2\}$. Figure 2.2.10 shows this PDA accepting (by final state) the language $L(M) = \{a^m b^m | m \geq 0\}$.

---

[19]Note that only u has been accepted, not the input string uv, and the stack is not empty. Thus, this is not a terminal accepting configuration. We shall shortly specify terminal accepting configurations.

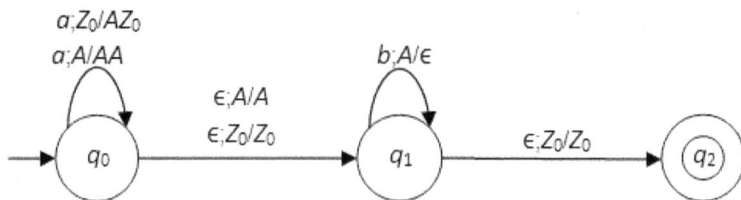

Figure 2.2.10.: A PDA M accepting the language $L(M) = \{a^m b^m | m \geq 0\}$.

We give a precise definition of string and language acceptance by a PDA:

Definition. 2.2.19. A string w is *accepted* by a PDA M if, starting in state q_0 at the first symbol, the transition sequence of M terminates in two possible ways:

1. Termination in an accepting state $q_f \in A$. We denote this by

$$(q_0, w, Z_0) \vdash^*_M (q_f, \epsilon, \alpha)$$

where $\alpha \in \Gamma^*$. Accordingly, we have the *language accepted by final state*, denoted by $L(M)$, defined as the language

$$L(M) = \{w \in \Sigma^* | (q_0, w, Z_0) \vdash^*_M (q_f, \epsilon, \alpha)\}.$$

2. Termination in an empty stack after M has read the last symbol. We denote this by

$$(q_0, w, Z_0) \vdash^*_M (q, \epsilon, \epsilon)$$

where $q \in Q$. Accordingly, we have the *language accepted by empty* (or *null*) *stack*, denoted by $N(M)$, defined as the language

$$N(M) = \{w \in \Sigma^* | (q_0, w, Z_0) \vdash^*_M (q, \epsilon, \epsilon)\}.$$

The first mode of acceptance uses the internal memory (state), whereas the second uses the external memory, i.e. the stack. Unless it is necessary to specify that an accepted language is of the $L(M)$ or $N(M)$ kind, we shall designate a language accepted by a PDA M indifferently by $L(M)$. The following theorem supports this simplification:

126

Theorem 2.2.8. *For each PDA M, one can construct a PDA M' such that $L(M) = N(M')$, and vice-versa $N(M) = L(M')$.*

Proof: (Sketch) We have to show that (I) given a PDA M that accepts a language L by final state, we can construct a PDA M' that simulates M by accepting L by empty stack, and (II) given a PDA M that accepts L by empty stack, we can construct a PDA M' that simulates M by accepting L by final state.

(I) The idea is to let M' have the option to erase its stack whenever M enters a final state. For this end, we provide M' with a state q_e. Moreover, we provide M' with a bottom-of-stack marker X_0 intended to bar M' from accidentally accepting should M empty its stack without entering a final state. Let $M = (Q, \Sigma, \Gamma, q_0, Z_0, A, \delta)$. Now, let

$$M' = \left(Q \cup \left\{q_0', q_e\right\}, \Sigma, \Gamma \cup \{X_0\}, q_0', X_0, \emptyset, \delta'\right)$$

where δ' is defined as follows: (1) $\delta'\left(q_0', \epsilon, X_0\right) = \{(q_0, Z_0 X_0)\}$; (2) for all $q \in Q$, all $a \in \Sigma \cup \{\epsilon\}$, and all $Z \in \Gamma$, $\delta'(q, a, Z)$ includes the elements of $\delta(q, a, Z)$; (3) for all $q \in A$ and $Z \in \Gamma \cup \{X_0\}$, $\delta'(q, \epsilon, Z)$ contains (q_e, ϵ); (4) for all $Z \in \Gamma \cup \{X_0\}$, $\delta'(q_e, \epsilon, Z)$ contains (q_e, ϵ). Figure 2.2.11.1 shows a schematic state diagram of this simulation.

(II) The idea is to make M' detect when M empties its stack, entering a final state when and only when this occurs, so that no additional symbol is "consumed." M' detects that M has emptied its stack when a stack symbol X_0 appears. Let $M = (Q, \Sigma, \Gamma, q_0, Z_0, \emptyset, \delta)$. Now, we let

$$M' = \left(Q \cup \left\{q_0', q_f\right\}, \Sigma, \Gamma \cup \{X_0\}, q_0', X_0, \{q_f\}, \delta'\right)$$

where δ' is defined as follows: (1) $\delta'\left(q_0', \epsilon, X_0\right) = \{(q_0, Z_0 X_0)\}$; (2) for all $q \in Q$, all $a \in \Sigma \cup \{\epsilon\}$, and all $Z \in \Gamma$, we have $\delta'(q, a, Z) = \delta(q, a, Z)$; (3) for all $q \in Q$, $\delta'(q, \epsilon, X_0)$ contains (q_f, ϵ). Figure 2.2.11.2 shows a schematic state diagram of this simulation.

Now, let I.(1)–(4) and II.(1)–(3) be actually rules and let x be in $L(M)$ or $N(M)$, accordingly. For case I, if $x \in L(M)$, then for some $q \in A$ we have $(q_0, x, Z_0) \vdash_M^* (q, \epsilon, \alpha)$. Now we consider M' with input x: By rule I.(1),

$$\left(q_0', x, X_0\right) \vdash_{M'}^* (q_0, x, Z_0 X_0).$$

Do the same (i.e. give the configurations and computations) for rules I.(2)–(4). Then work the other way round. Do the same for case II.

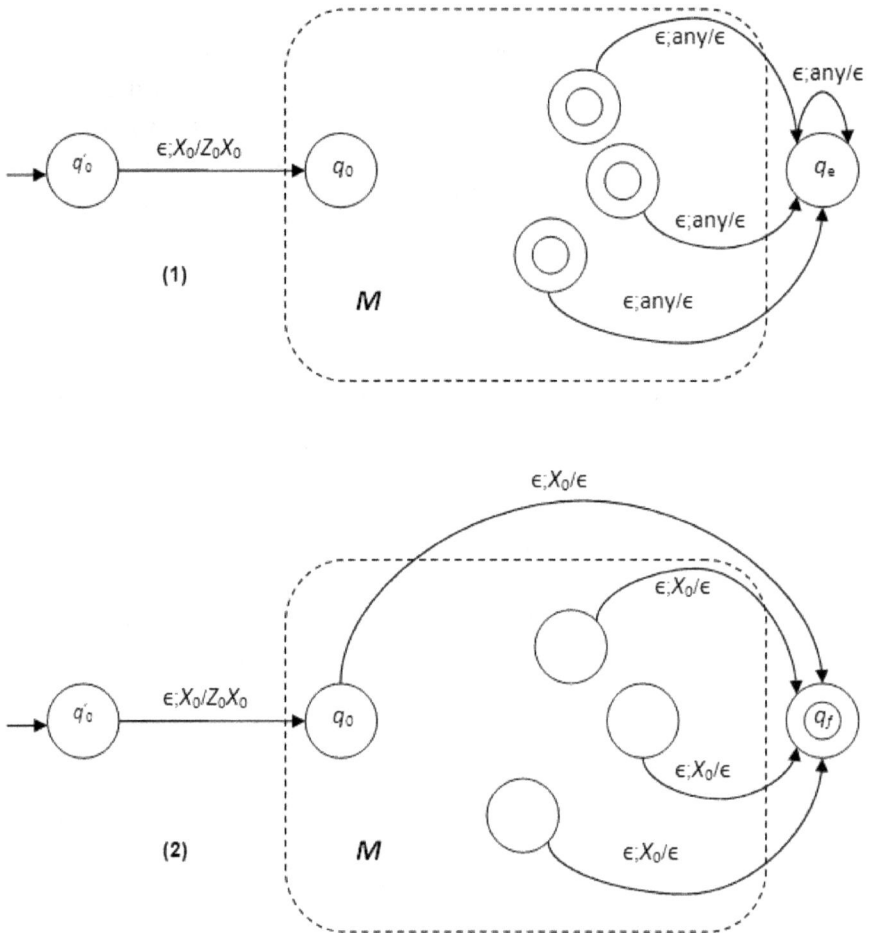

Figure 2.2.11.: Proving the equivalence of $L\left(M\right) = N\left(M\right)$.

QED

We began this Section by stating that there is an equivalence between CFLs and PDAs. The following two theorems state this in a formal way. The proofs of both theorems are highly convoluted, and we leave them as exercises, providing only the essential idea behind them. The examples that follow the sketches provide further clues for the complete proofs.

Theorem 2.2.9. *If L is a CFL, then there is some PDA M such that $L = L(M)$.*

Proof: (Sketch) If L is a CFL, then there is a CFG $G = (V, T, S, P)$ that generates it. Let $w \in L(G)$. Then, there is a derivation $S \overset{*}{\underset{G}{\Rightarrow}} w$ where S is the start symbol and each step of this derivation yields an intermediate string of variables and terminals. In particular, there is a leftmost derivation $S \overset{*}{\underset{G}{\Rightarrow}}_l w$ and $L = L(M)$. A PDA M given input w starts by writing S on its stack and then goes through a sequence of intermediate strings, non-deterministically selecting at each step a production for a particular variable and making the corresponding substitutions. M eventually arrives at a string w' with only terminals; if $w' = w$, then M accepts $L(G)$. M needs the feature of being able to store the intermediate strings; the stack requires an adaptation for this, as an intermediate string can have both variables and terminals, but M needs to detect a variable to make the corresponding substitution. The solution is to keep only a part of the intermediate string on the stack, namely the part starting with the leftmost variable symbol; terminal symbols anteceding this variable symbol are matched with symbols in the input string directly. In this way, the symbol at the top of the stack is always associated to a variable in an intermediate string, and the derivation is a leftmost derivation. **QED**

Example 2.2.10. In designing a PDA M that accepts a CFL L, it helps if the grammar that generates L is in Greibach normal form, i.e. a CFG in which all the productions have the form $A \rightarrow aX$ where $a \in T$ and $X \in V$ is a variable string (see Algorithm 2.3). Let $G = (\{S, A, B, C\}, \{a, b, c\}, S, P)$ be a CFG in Greibach normal form with the following productions in P:

$$P = \begin{cases} S \rightarrow aA \\ A \rightarrow aABC \mid bB \mid a \\ B \rightarrow b \\ C \rightarrow c \end{cases}$$

2. Fundamentals of classical computing

The PDA $M = (\{q_0, q_1, q_2\}, \Sigma, \Gamma, q_0, Z_0, \{q_2\}, \delta)$, where $\Sigma = T$ and $\Gamma = V \cup Z_0$, accepts L if its transitions are as follows:

$$
\begin{aligned}
\delta(q_0, \epsilon, Z_0) &= \{(q_1, SZ_0)\} \\
\delta(q_1, a, S) &= \{(q_1, A)\} \\
\delta(q_1, a, A) &= \{(q_1, ABC), (q_1, \epsilon)\} \\
\delta(q_1, b, A) &= \{(q_1, B)\} \\
\delta(q_1, b, B) &= \{(q_1, \epsilon)\} \\
\delta(q_1, c, C) &= \{(q_1, \epsilon)\} \\
\delta(q_1, \epsilon, Z_0) &= \{(q_2, Z_0)\}
\end{aligned}
$$

Given the input $aaabc$, we have the leftmost derivation

$$S \Longrightarrow aA \Longrightarrow_l aaABC \Longrightarrow_l aaaBC \Longrightarrow_l aaabC \Longrightarrow_l aaabc.$$

Theorem 2.2.10. *If, given a PDA M and a language L, $L = N(M)$, then L is a CFL.*

Proof: (Sketch) We now need a CFG G that simulates the moves of M. In particular, we need a leftmost derivation in G of some string x to be a simulation of M when given input x. Because this is a leftmost derivation, the processed prefix is the terminal part of x, to the right of which lies the variable part. Thus, we require that the symbols in the stack of M be matched to the unprocessed suffixes of x; when the stack is empty, x is a string of terminals. To this end, we need special variables. Let $M = (Q, \Sigma, \Gamma, q_0, Z_0, \emptyset, \delta)$ be a PDA. We need a grammar $G = (V, T, S, P)$ where $T = \Sigma \cup \{\epsilon\}$ and V is a set of objects of the form $[qAp]$ for $A \in \Gamma$ and $q, p \in Q$. The productions in P are of the form

$$S \rightarrow [q_0 Z_0 q]$$

for every $q \in Q$, and

$$[qAq_{m+1}] \rightarrow a\,[q_1 B_1 q_2]\,[q_2 B_2 q_3] \dots [q_m B_m q_{m+1}]$$

for every $q, q_1, q_2, \dots q_{m+1} \in Q$, every $a \in \Sigma \cup \{\epsilon\}$, and $A, B_1, B_2, \dots B_m \in \Gamma$, such that $\delta(q, a, A) \supseteq (q_1, B_1 B_2 \dots B_m)$. If $m = 0$, then we have $[qAq_1] \rightarrow a$. The idea is to have

$$[qAp] \underset{G}{\overset{*}{\Longrightarrow}}_l x \text{ iff } (q, x, A) \vdash_M^* (p, \epsilon, \epsilon).$$

That is, we want $[qAp]$ to derive x in a leftmost derivation iff x causes M to erase an A from its stack by a sequence of zero or more moves that starts in state q and ends in state p; at each move, the stack content is increased or decreased by a single symbol, until it is wholly empty at state p.

($\Rightarrow$) First we want to prove that if $(q, x, A) \vdash^i_M (p, \epsilon, \epsilon)$, then $[qAp] \underset{G}{\overset{*}{\Rightarrow}}_l x$. The proof is by induction on i. If $i = 1$, then $\delta(q, x, A) \supseteq (p, \epsilon)$, and thus $([qAp] \rightarrow x) \in G$. For $i > 1$, let $x = ay$, so that we have the computation

$$(q, ay, A) \vdash (q_1, y, B_1 B_2 ... B_n) \vdash^{i-1}_M (p, \epsilon, \epsilon).$$

Let $y = y_1 y_2 ... y_n$, where each y_j causes B_j to be popped off the stack, so that we have (how?)

$$[qAp] \underset{G}{\overset{*}{\Rightarrow}}_l ay_1 y_2 ... y_n = x.$$

($\Leftarrow$) We now want to prove that if $[qAp] \underset{G}{\overset{i}{\Rightarrow}}_l x$, then $(q, x, A) \vdash^*_M (p, \epsilon, \epsilon)$. The proof is again by induction on i. (We leave the details as an exercise. Hint: write $x = ax_1 x_2 ... x_n$ so that the first move of M is $(q, x, A) \vdash_M (q_1, x_1 x_2 ... x_n, B_1 B_2 ... B_n)$.)

Finally, if we let $A = Z_0$ and $q = q_0$, then we obviously have

$$[q_0 Z_0 p] \underset{G}{\overset{*}{\Rightarrow}}_l x \text{ iff } (q_0, x, Z_0) \vdash^*_M (p, \epsilon, \epsilon)$$

and thus

$$S \underset{G}{\overset{*}{\Rightarrow}}_l x \text{ iff } (q_0, x, Z_0) \vdash^*_M (p, \epsilon, \epsilon).$$

Hence, $x \in L(G)$ iff $x \in N(M)$. **QED**

Example 2.2.11. Let δ as given below be the transition function for the PDA $M = (\{q_0, q_1\}, \{a, b\}, \{A, Z_0\}, q_0, Z_0, \emptyset, \delta)$:

$$
\begin{aligned}
\delta(q_0, a, Z_0) &= \{(q_0, AZ_0)\} \\
\delta(q_0, a, A) &= \{(q_0, AA)\} \\
\delta(q_0, b, A) &= \{(q_1, \epsilon)\} \\
\delta(q_1, b, A) &= \{(q_1, \epsilon)\} \\
\delta(q_1, \epsilon, A) &= \{(q_1, \epsilon)\} \\
\delta(q_1, \epsilon, Z_0) &= \{(q_1, \epsilon)\}
\end{aligned}
$$

The construction of the CFG G that generates $N(M)$ follows the following steps: Begin by setting the set V with the objects of the form

$[q_i A q_j]$, $i, j = 0, 1$. Next, determine the productions with S as an antecedent. Then, work out the productions for $[q_0 Z_0, q_i]$ (e.g., $[q_0 Z_0 q_0] \to a [q_0 A q_0] [q_0 Z_0 q_0]$). Do the same for the variables required by each transition. Determine then the variables that are superfluous, i.e. variables for which there are no productions or from which no terminal string can be derived. Delete these variables. *Et voilà!*

As said above, the CFLs accepted by a deterministic PDA constitute an important subset. We now focus on this topic.

Definition. 2.2.20. An automaton $M = (Q, \Sigma, \Gamma, q_0, Z_0, A, \delta)$ is said to be a *deterministic PDA* (DPDA) if the following conditions are satisfied:

1. For every $q \in Q$, every $a \in \Sigma \cup \{\epsilon\}$, and every $X \in \Gamma$, $|\delta(q, a, X)| \leq 1$.

2. For every $q \in Q$, every $a \in \Sigma$, and every $X \in \Gamma$, if $|\delta(q, \epsilon, X)| \neq \emptyset$, then $|\delta(q, a, X)| = \emptyset$.

A language L is called a *deterministic CFL* (DCFL) if there is a DPDA accepting it.

Condition 2 above hinders there being a choice on the same input, whereas condition 1 hinders there being a choice between a ϵ-transition and a transition using the next input. Because there are now no choices, the transitions are written simply

$$\delta(q, a, X) = (p, \alpha).$$

The following proposition specifies a *normal form* for the DPDAs: the only stack operations of a DPDA are erasing the top symbol or pushing a symbol. The proofs for conditions 1 and 2 are left as exercises.

Proposition. 2.2.21. *Every DCFL L is a $L(M)$ for a DPDA M such that if $\delta(q, a, X) = (p, \alpha)$,*

1. *then $|\alpha| \leq 2$, and*

2. *then $\alpha = \epsilon$, $\alpha = X$, or $\alpha = YX$.*

Proof: (Sketch) With respect to condition 1, we want M to push only one symbol per move. So, if $\delta\left(q, a, X\right) = (r, \alpha)$ and $\alpha > 2$, let $\alpha = Y_1 Y_2 ... Y_n$, $n \geq 3$, and create states $p_1, p_2, ..., p_{n-2}$. These are non-accepting states. Define then $\delta\left(q, a, X\right) = (p_1, Y_{n-1} Y_n)$. What else do we need so that, in state q, on input a, X is at the top of the stack, this DPDA still replaces X with α and then enters state r, but now does so in $n-1$ moves? As for condition 2, let M' be a DPDA satisfying condition 1. We need some DPDA M to simulate M'. Define M and δ so that it can be shown by induction on the number of moves that

$$\left(q_0', w, X_0\right) \vdash_{M'}^* (q, \epsilon, X_1 X_2 ... X_n)$$

iff

$$\left(\left[q_0' X_0\right], w, Z_0\right) \vdash_M^* ([q X_1], \epsilon, X_2 X_3 ... X_n Z_0)$$

and $L\left(M\right) = L\left(M'\right)$. **QED**

Further results are as follows:

Proposition. 2.2.22. *The class of DCFLs is closed under (i) complementation and (ii) intersection with $\mathscr{RGL}$.*

Proof: (Sketch) (i) We need a DPDA that reads the entire input string and whose accepting states can only be reading states; then, we swap acceptance and non-acceptance with respect to these states. (ii) (Hint: You will need a DPDA M_1 and a deterministic FA M_2.) **QED**

Proposition. 2.2.23. *The class of DCFLs is not closed under (i) union, (ii) concatenation, or (iii) Kleene closure.*

Proof: (Sketch) Let $L_1 = \left\{a^i b^i c^j | i, j \geq 0\right\}$ and $L_2 = \left\{a^i b^j c^j | i, j \geq 0\right\}$. For (i), $L_1 \cup L_2$ is equivalent to $L_3 = \left\{a^i b^j c^k | i \neq j, j \neq k\right\}$, but L_3 is not a DCFL (why?). (ii) Show why $L_4 = 0 L_1 \cup 0 L_2$ is not a DCFL. (iii) $L_5 = \{0\} \cup (0 L_1 \cup L_2)$ is a DCFL, but $(L_5)^*$ is not (why?). **QED**

The simplest way to define a DCFL is to say that it is generated by a CFG with restrictions. In particular, the restricted CFGs known as LR(k)–abbreviating "left-to-right scan of the input producing a rightmost derivation and using k symbols of look-ahead on the input"–grammars generate *exactly* the DCFLs. These grammars are particularly important for compilation. We elaborate on LR(k) grammars for $k = 0, 1$ in Augusto (2021). The reader wishing more depth of treatment of these grammars–and of parsing in general–can consult, for instance, Sippu & Soisalon-Soininen (1990) and Grune & Jacobs (2010).

Exercises

Exercise 2.2.2.1. Design a PDA M that accepts the language

1. $N(M) = \{a^n b^m a^n | m, n \geq 1\}$.

2. $L(M) = \{a^n b^m a^n | m, n \geq 1\}$.

3. $N(M) = \{a^m b^m c^n | m, n \geq 1\}$.

4. $N(M) = \{0^n 1^{2n} | n \geq 1\}$.

5. $L(M) = \{0^m 1^n | m > n\}$.

Exercise 2.2.2.2. Let the PDA $M = (Q, \Sigma, \Gamma, q_0, Z_0, A, \delta)$ with $|Q| = 3$, $\Sigma = \{a, b\}$, $\Gamma = \{A, B, Z_0\}$, and $A = \{q_2\}$ be given. The following are the transitions by M accepting the language L:

$$
\begin{aligned}
\delta(q_0, a, Z_0) &= \{(q_1, A), (q_2, \epsilon)\} \\
\delta(q_1, b, A) &= \{(q_1, B)\} \\
\delta(q_1, b, B) &= \{(q_1, B)\} \\
\delta(q_1, a, B) &= \{(q_2, \epsilon)\}
\end{aligned}
$$

1. Determine L.

2. Draw the state diagram of M.

Exercise 2.2.2.3. A PDA accepting $L = \{u \in \Sigma^* | u = ww^R\}$ must guess correctly where the middle of the string u is. Given the PDA $M = (Q, \Sigma, \Gamma, q_0, Z_0, A, \delta)$ with $|Q| = 3$, $\Sigma = \{a, b\}$, $\Gamma = \{X, Y, Z_0\}$ and $A = \{q_2\}$, give M's computation for input string $abba$.

Exercise 2.2.2.4. In Example 2.2.10, account for the first transition function, i.e. $\delta(q_0, \epsilon, Z_0) = \{(q_1, SZ_0)\}$.

Exercise 2.2.2.5. Construct a PDA M that corresponds to the CFGs with the following production sets:

1.

$$
P = \left\{ \begin{array}{c} S \to aABB \,|\, aAA \\ A \to aBB \,|\, a \\ B \to bBB \,|\, A \end{array} \right\}
$$

2.

$$P = \{S \to aSbb \mid aab\}$$

3.

$$P = \left\{ \begin{array}{l} S \to aAA \mid A \\ A \to aS \mid bS \mid a \end{array} \right\}$$

4.

$$P = \left\{ \begin{array}{c} S \to b \mid d \mid aSB \mid bA \\ A \to bA \mid b \\ B \to c \end{array} \right\}$$

Exercise 2.2.2.6. Solve the exercise proposed in Example 2.2.11.

Exercise 2.2.2.7. Is the PDA of Example 2.2.8 deterministic?

Exercise 2.2.2.8. Research further into LR(k) grammars and research also into LL(k) grammars. For each of these grammars:

1. Define it precisely.

2. Find its main applications.

Exercise 2.2.2.9. Prove (Complete the proof of) the theorems, propositions, etc. in this Section that were left as an exercise (with a sketchy proof, respectively).

2.2.3. Turing machines

The Turing machine, so coined after its conceiver A. Turing (1937), provides the most encompassing and most powerful model of computation; so much so that it actually is at the root of the modern digital computer. In effect, this particular automaton is equivalent, in definitional terms, to the fundamental notion of a computable function (see next Section). This, in turn, is connected to the RELs in the sense that a Turing machine can enumerate the elements of such a language, which action equates with accepting it. Thus, in the same way that finite machines are associated with regular languages and PDAs are so with CFLs, Turing machines are associated with RELs.

Definition. 2.2.24. A *Turing machine* (TM) is a 7-tuple

$$M_T = (Q, \Gamma, \#, \Sigma, q_0, A, \delta)$$

where Q is a finite set of states, Γ is the *tape alphabet*, $\# \in \Gamma$ is the *blank symbol*, $\Sigma \subseteq (\Gamma - \#)$ is the *input alphabet*, $A \subseteq Q$ is the set of *accepting states*, and $\delta : (Q \times \Gamma) \longrightarrow (Q \times \Gamma \times \{L, R\})$, where L, R denote direction (left and right, respectively), is the *transition function.*

There are (many) variations to this definition. For instance, we may have $\# \in \Sigma$. In particular, we can consider two *halting states*, to wit, q_a and q_r for acceptation and rejection, respectively, so that the transition function is $\delta : (Q \times \Gamma) \longrightarrow ((Q \cup \{q_a, q_r\}) \times \Gamma \times \{L, R\})$. Also alternatively, the set A may be empty, acceptation being the case when the machine processes a string completely, or computes a function, and no more subsequent moves are specified. Another variation is the addition of a stop or no-action symbol, S or N, respectively, to $\{L, R\}$. However, these variations do not change the computing power of a Turing machine.

As said above, a Turing machine is a surprisingly parsimonious machine if we take into consideration its computational power. Its computer model (Fig. 2.2.12) reflects this parsimony and the examples and exercises below illustrate its computational power.

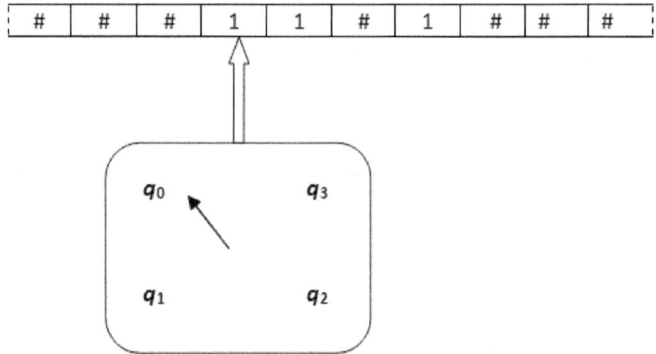

Figure 2.2.12.: Computer model for a Turing machine.

The *computer model* for a Turing machine $M_T = (Q, \Gamma, \#, \Sigma, q_0, A, \delta)$ consists of a logic box programmed by δ equipped with a read-and-write head. This head reads an input string starting on the left of an input tape that is infinite to the right.[20] When the machine reads a symbol a while in state q, it switches into state p, replaces symbol $a \in \Gamma$ by symbol $b \in \Gamma$, and moves one tape cell in direction $D = \{L, R\}$, so that

---

[20]The input tape can be infinite also to the left, but this does not change the computing power of the Turing machine.

we have

$$\delta\left(q,a\right)=\left(p,b,D\right).$$

Definition. 2.2.25. A *configuration* for a Turing machine M_T is a pair $C=(q,u\underline{a}v)$ indicating that M_T is in state q with the present tape string uav and reading the symbol a.

1. The M_T configuration $(q,u\underline{a}v)$ *yields* the configuration $(p,x\underline{b}y)$ *in one step* (or *move*), denoted by

$$\underbrace{(q,u\underline{a}v)}_{C_i}\vdash_T\underbrace{(p,x\underline{b}y)}_{C_{i+1}}$$

   if and only if the transition $\delta\left(q,a\right)$ changes configuration $(q,u\underline{a}v)$ to configuration $(p,x\underline{b}y)$.

2. A *starting configuration* C_0 for M_T is a configuration of the form $(q_0,\epsilon\underline{a}v)$, or $(q_0,\#\underline{a}v)$, indicating that a is the symbol in the left-most cell of the input tape or string, respectively.

3. An *accepting configuration* C_n for M_T has the form $(q,u\underline{a}v)$ indicating that uav is the output, the read-and-write head is positioned at a, and $q\in A$ or $q=q_a$. If $q=q_r$ in C_n, then $(q,u\underline{a}v)$ is a *rejecting configuration*.

4. A *hanging configuration* C_n for M_T is a configuration of the form $(q,\epsilon\underline{a}v)$ such that the transition function $\delta\left(q,a\right)$ instructs the machine to move left (i.e. off the tape), or any configuration from which the machine is instructed to make an impossible move.

5. We write $C_i\vdash_T^n C_j$ or $C_i\vdash_T^* C_j$, $i,j=0,1,...,n$, to denote that there is a sequence of $n\geq 1$ or $n\geq 0$ moves, respectively, taken by M_T to go from configuration C_i to configuration C_j.

6. A *computation* for a Turing machine M_T on input $w\in\Sigma^*$ is any of four possibilities:

   a) A finite sequence $C_0,C_1,...,C_n$ of configurations (where possibly $n=0$) such that

   $$\underbrace{q_0,w}_{C_0}\vdash_T^*\underbrace{q_i,z}_{C_n}$$

   where $q_i\in A$ or $q_i=q_a$ and $z\in\Gamma^*$, and C_n is an accepting configuration;

b) A finite sequence $C_0, C_1, ..., C_n$ of configurations (where possibly $n = 0$) such that

$$\underbrace{q_0, w}_{C_0} \vdash_T^* \underbrace{q_i, z}_{C_n}$$

where $q_i \notin A$ or $q_i = q_r$ and $z \in \Gamma^*$, and C_n is a rejecting configuration;

c) A finite sequence $C_0, C_1, ..., C_n$ of configurations (where possibly $n = 0$) such that

$$\underbrace{q_0, w}_{C_0} \vdash_T^* \underbrace{q_i, z}_{C_n}$$

where $q_i \notin A$ and $z \in \Gamma^*$, and C_n is a hanging configuration;

d) An infinite sequence $C_0, C_1, ...$ of configurations such that

$$\underbrace{q_0, w}_{C_0} \vdash_T^* \infty.$$

7. An infinitely repeated computation $C_i \vdash_T^n C_i$, $n \geq 2$, is an *infinite loop*. In practical terms, we have $C_0 \vdash_T^* \infty$.[21]

8. A string $w \in \Sigma^*$ is accepted by the Turing machine M_T if M_T *halts* on input w and accepts w, i.e. if we have

$$q_0, w \vdash_T^* q_i, z$$

for some $q_i \in A$ (or $q_i = q_a$) and $z \in \Gamma^*$. Otherwise, if M_T *halts* but does not accept w (for instance, we have $q_0, w \vdash_T^* q_r, z$), then w is rejected by M_T.[22] A language $L \subseteq \Sigma^*$ is accepted by a Turing machine M_T if

$$L = L(M_T) = \{w \in \Sigma^* | M_T \text{ accepts } w\}.$$

9. A language L is *Turing-machine acceptable* if there is a Turing machine M_T that accepts it.

---

[21] This explains why an infinite sequence (Def. 4.3.3.6.d) is more often than not called an infinite loop, when in fact no loop proper is the case.

[22] It is important to notice here that a TM can *halt* in either an accepting or a rejecting state. Contrast this with the cases when a TM does *not* halt (cf. Defs. 2.2.25.6.d and 2.2.25.7).

In order to define a transition table for a Turing machine the following definition is required:

Definition. 2.2.26. An $m \times n$ *Turing machine* M_T is a Turing machine with m states and n symbols.

Definition. 2.2.27. A *transition table* for an $m \times n$ Turing machine M_T is a table with m rows and n columns such that the entry in row q and column a is $\delta(q, a)$.

Recall that Turing machines also compute functions, besides recognizing formal languages.[23]

Definition. 2.2.28. In effect, we say that a Turing machine M_T *defines* a function $f(x) = y$ for strings $x, y \in \Sigma^*$ and some $q_i \in A$ if

$$q_0 x \vdash_T^* q_i y.$$

We begin by showing how Turing machines compute functions, because their behavior appears more intuitive here than when recognizing languages. However, programming a Turing machine for any but the most basic functions is typically a Herculean task.[24] We thus stick for the time being to very basic functions from integers to integers.

Example 2.2.12. The following is the transition table for a Turing machine $M_T = (\{q_0, q_1, q_2, q_3\}, \{1, \#\}, \#, \{1\} \cup \{\#\}, q_0, \{q_3\}, \delta)$ adding two positive integers (the integers are represented in unary notation $1^n = \underbrace{111...1}_{n}$; e.g. $1^3 = 111 = 3$):

	1	#
q_0	$q_0 1R$	$q_1 1R$
q_1	$q_1 1R$	$q_2 \# L$
q_2	$q_3 \# R$	–
q_3	–	–

On input $11\#111\#\#...$, M_T behaves as follows (we omit the subscript T in $\vdash$ and the commas between states and strings):[25]

$$q_0 \underline{1}1\#111\# \vdash q_0 1\underline{1}\#111\# \vdash q_0 11\underline{\#}111\# \vdash q_1 111\underline{1}11\# \vdash q_1 11111\underline{1}\#$$

---

[23] As we shall see, these are actually one and the same function for Turing machines.
[24] Contrast the following example with Example 2.2.16 below.
[25] In this input string, we have $\Sigma = \{1\} \cup \{\#\}$ because the symbol $\#$ between 1s is part of the string, whereas the symbol $\#$ after the final 1 simply indicates the first empty cell after the input string.

$$\vdash q_1 111111\underline{1}\# \vdash q_1 111111\underline{\#} \vdash q_2 111111\underline{1}\# \vdash q_3 11111\#\underline{\#}$$

$$\approx q_0 11\#111\# \vdash^8 q_3 11111\#$$

The behavior of the machine is as follows: Given the input $11\#111\#\#...$, which corresponds to the two integers 2 and 3, in order to add both strings M_T has firstly, after reading the two initial 1s, to delete the blank space between them, replace it with 1, and then keep moving right until it finds the last symbol 1 and the first empty cell after it; M_T moves one cell to the left, deletes this 1, moves to the right and halts in an accepting state. The output string is 11111 followed by infinitely many blank symbols. Figure 2.2.13 shows the state diagram of this M_T. More formally, we say that M_T defines the function $f(m, n) = m + n$, $m, n \in \mathbb{Z}^+$, for an arbitrary input string $1^m \# 1^n$ if

$$q_0, 1^m \# 1^n \vdash_T^{(m+n)+3} q_3, 1^m 1^n.$$

Finally, because there are no transitions from state q_3 we can simply consider it as a halting state for acceptance q_a; this allows us to simplify the transition table as follows:

	1	#
q_0	$q_0 1R$	$q_1 1R$
q_1	$q_1 1R$	$q_2 \# L$
q_2	$q_a \# R$	$-$

We can then write the computation

$$q_0, 1^m \# 1^n \vdash_T^{(m+n)+3} q_a, 1^m 1^n$$

where termination in an accepting state is made explicit.

Just like finite machines and PDAs, Turing machines can be defined by means of state diagrams.

Definition. 2.2.29. Given a Turing machine M_T, its state diagram is a labeled digraph $D = \vec{\mathfrak{D}}(M_T)$ as for a FA such that for each transition $\delta(q, a) = (p, b, D)$ there is an arc from vertex q to vertex p labeled with "a/bD," read "if M_T reads symbol a, then replace a by b and move in the direction D."

Clearly, if we have a label of the form "a/aD," then we simply leave a untouched and mind only the direction instruction.

Example 2.2.13. Figure 2.2.13 shows the state diagram of the Turing machine in Example 2.2.12.

1/1R   1/1R

q_0 #/1R q_1 #/#L q_2 1/#R q_3

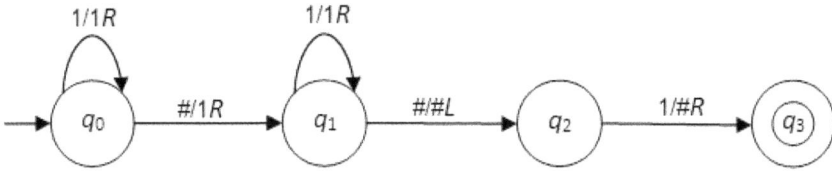

Figure 2.2.13.: A Turing machine that computes the function $f(m, n) = m + n$ for $m, n \in \mathbb{Z}^+$.

Example 2.2.14. We show now a Turing machine accepting a language. We consider the language $L = \{0^n 1^n | n > 0\}$. A Turing machine M_T accepting this language has the following sets: $Q = \{q_0, q_1, q_2, q_3, q_4\}$, $\Sigma = \{0, 1\}$, $\Gamma = \{0, 1, X, Y, \#\}$, and $A = \{q_4\}$. The behavior of the machine is as follows: Given the input $0^n 1^n \#\#...$, M_T replaces the leftmost 0 by X, moving then right to the leftmost 1 and replacing it by Y; next, M_T moves left to the rightmost X, moves one cell to the leftmost 0, and repeats the cycle. If after changing a 1 to a Y the machine finds no more 0, it checks for any 0 left on the tape; if there is none, it accepts the input string. We show the transition table for M_T:

	0	1	X	Y	#
q_0	$q_1 X R$	$-$	$-$	$q_3 Y R$	$-$
q_1	$q_1 0 R$	$q_2 Y L$	$-$	$q_1 Y R$	$-$
q_2	$q_2 0 L$	$-$	$q_0 X R$	$q_2 Y L$	$-$
q_3	$-$	$-$	$-$	$q_3 Y R$	$q_4 \# R$
q_4	$-$	$-$	$-$	$-$	$-$

The computation carried out by M_T given an input string 0011 is as follows (we omit the subscript T in $\vdash$ and the commas between states and strings):

$$q_0 \underline{0} 011 \vdash q_1 X \underline{0} 11 \vdash q_1 X 0 \underline{1} 1 \vdash q_2 X \underline{0} Y 1 \vdash q_2 \underline{X} 0 Y 1 \vdash q_0 X \underline{0} Y 1 \vdash q_1 X X \underline{Y} 1$$

$$\vdash q_2 X X \underline{Y} Y \vdash q_2 X \underline{X} Y Y \vdash q_0 X X \underline{Y} Y \vdash q_3 X X Y \underline{Y} \vdash q_4 X X Y Y \underline{\#}$$

Theorem 2.2.11. *For every UG G, there is a Turing machine M_T such that $L(M_T) = L(G)$.*

Proof: Left as an exercise. (Hint: describe a (non-deterministic)[26] Turing machine M_T that accepts $L(G)$.)

---

[26]See Def. 2.2.30 below.

There are many variations to the computer model of the Turing machine above. In effect, there are Turing machines with $n > 1$ tapes, with a two-way infinite tape, with $k > 1$ heads, with a read-only input tape, etc. For our purposes, the distinction between a *deterministic* and a *non-deterministic* Turing machine is the most relevant.

Definition. 2.2.30. A Turing machine M_T is said to be *non-deterministic* if, given a state q and a symbol a being read, the machine has a finite number of choices for the next move. Otherwise, i.e. if there is exactly one move for any pair (q, a), M_T is said to be *deterministic*.

Theorem 2.2.12. *If a language L is accepted by a non-deterministic Turing machine M_{T1}, then it is accepted by some deterministic Turing machine M_{T2}.*

Proof: Left as an exercise. (Hints: The transition function for M_{T1} must be $\delta : (Q \times \Gamma) \to 2^{(Q \times \Gamma \times \{L,R\})}$. M_{T2} requires a means to keep all of M_{T1}'s configurations on its tape, in order to simulate M_{T1}. Language acceptance must be addressed.)

As a matter of fact, the variations above are to some extent superfluous, because there is a Turing machine capable of simulating any other Turing machine. In order to introduce this machine we require the notion of an encoding; this notion will prove relevant also in Sections below.

Definition. 2.2.31. Let a Turing machine M_T be given; then, there is an *encoding function* $e : M_T \longrightarrow \{0, 1\}^*$ that is a unique description of M_T.[27] In detail, given a Turing machine M_T, a generic move $\delta(q_i, \sigma_j) = (q_k, \sigma_l, D_m)$ is *encoded* by the binary string

$$\xi = 0^i 10^j 10^k 10^l 10^m$$

and a *binary code* for M_T is

$$111 \ \text{code}_1 \ 11 \ \text{code}_2 \ 11...11 \ \text{code}_n \ 111 = \langle M_T \rangle$$

where every code_i has the form of ξ.

As said, this encoding is unique, i.e. every string ξ is interpreted as the code for at most one Turing machine, and a Turing machine has many codes.

---

[27] Turing (1937) referred to this encoding that, for practical ends, just is the program of a Turing machine, as the *Standard Description*. Of course, a Turing machine just is its program, or Standard Description.

Definition. 2.2.32. A *universal Turing machine* is a machine M_{TU} such that, given an arbitrary Turing machine M_T and $z \in \Gamma^* \subset M_T$, and an encoding function e, upon receiving an input string of the form $\xi^* = e\,(M_T)\,e\,(z) = \langle M_T, z \rangle$

1. M_{TU} accepts ξ^* iff M_T accepts z.[28]

2. M_{TU} produces output $\langle y \rangle$ if M_T accepts z and produces output y.

Example 2.2.15. Let $q_{i,k}, i, k \geq 0$, be coded as $\langle q_0 \rangle = 0$, $\langle q_1 \rangle = 00$, etc. Let the following be the binary coding for the tape symbols: $\langle 1 \rangle = 00$ and $\langle \# \rangle = 0$. Finally, encode the directions $\langle L \rangle = 0$ and $\langle R \rangle = 00$. Then, Figure 2.2.14 shows the encoding of the Turing machine $M_T = (\{q_0, q_1, q_2, q_3\}, \{1, \#\}, \#, \{1\}, q_0, \{q_3\}, \delta)$ of Example 2.2.12, as well as the encoding of this Turing machine with the input string $z = 11\#111$ with $\langle 1 \rangle = 1$ and $\langle \# \rangle = 0$, so that $\langle z \rangle = 110111$.

We now present the fundamental theorem associating UGs with RELs. The inverse of this theorem is far more difficult to prove, and we leave it as an optional exercise.

Theorem 2.2.13. *Given a UG G, if $L = L\,(G)$, then L is a REL.*

Proof: Design a non-deterministic Turing machine M_T with two tapes, one (tape 1) for the input string w and the other (tape 2) for the sentential form $\alpha \in G$. Let $|\alpha| = n$. M_T initializes α to S and repeatedly (i) selects a position i in α, $1 \leq i \leq n$, (ii) selects a production $\beta \rightarrow \gamma$ of G, (iii) replaces β by γ if β begins in position i, and (iv) compares this new sentential form with the input on tape 1: M_T accepts w if the two tapes match; otherwise, M_T returns to (i). Thus, only, and all, the sentential forms of G appear on tape 2. Hence, $L\,(G) = L\,(M_T) = L$, L is a REL by Definition 2.1.27. **QED**

We conclude this discussion of Turing machines with an illustration of the statement above that it can be quite challenging to program a Turing machine for even very elementary functions.

Example 2.2.16. Figure 2.2.15 shows the state diagram of a Turing machine that computes the function $f\,(n, m) = 2n + 3m$ for $n, m \in \mathbb{Z}^+$. Figure 2.2.16 shows the program in a notation that can be implemented

---

[28]Obviously, if $\Gamma \subseteq \{0, 1\}^*$, then for a string $z \in \Gamma^*$ we have $e\,(z) = \langle z \rangle = z$.

$$\langle M_T \rangle = 1110100101001001101010010010010010011001001001100100100110010100100100110101001000101011000100100010010100111$$

$$\langle M_T, z \rangle = 1110100101001001101010010010010010011001001001100100100110010100100100110101001000101011000100100010010100111110111$$

Figure 2.2.14.: The encodings $\langle M_T \rangle$ and $\langle M_T, z \rangle$.

in the Turing machine simulator to be found at http://ironphoenix.org/tril/tm/. In this specific simulator:

- we use a tape alphabet $\{0,1\}$ with the additional symbol X, i.e. $\Gamma = \{0,1\} \cup X$;

- on the tape file only 1 or 0 can be found initially;

- if there is no 1 or X on a cell, then there is a 0;

- state q_0 in the state diagram (Fig. 2.2.15) corresponds to state 1 in the program, state q_1 to state 2, etc.;

- right is denoted by $>$ and left by $<$;

- the initial string on the tape is $1^n\_1^m$, where $\_$ denotes a blank cell.

The program can be easily adapted to any other simulator.

Now that we have a good grasp of the Turing machines (for RELs), we can approach a particular type of Turing machine that is associated to the CSLs (cf. Def. 2.1.20).

2.2.33. A *linear-bounded automaton* (LBA) is a non-determi-nistic Turing machine

$$M_{[\![T]\!]} = (Q, \Gamma, \#, \Sigma, q_0, A, \delta)$$

with two additional tape symbols $[\![,]\!] \in \Gamma$, called *left* and *right endmark-ers*, respectively.

Definition. 2.2.34. The *initial configuration* of a LBA $M_{[\![T]\!]}$ with input w is $q_0[\![w]\!] = q_0 [\![x_1 x_2 ... x_n]\!]$. A string $w \in [(\Sigma - \{[\![,]\!]\})^* = \Sigma^\bullet]$ is accepted by a LBA $M_{[\![T]\!]}$ if it halts on input w, i.e. if we have the computation

$$q_0[\![w]\!] \vdash^*_{[\![T]\!]} q_i z$$

for some $q_i \in A$ and $z \in \Gamma^*, |z| \geq |w|$. A language $L \subseteq \Sigma^\bullet$ is accepted by a Turing machine $M_{[\![T]\!]}$ if

$$L = L\left(M_{[\![T]\!]}\right) = \left\{w \in \Sigma^\bullet | M_{[\![T]\!]} \text{ accepts } w\right\}.$$

The additional tape symbols $[\![,]\!] \in \Gamma$ around w cannot be printed over and the read-and-write head cannot move to the right of $]\!]$ nor to the left of $[\![$ during the computation. That is, no production in a CSG can be length-decreasing. Another way to put the above is to say that

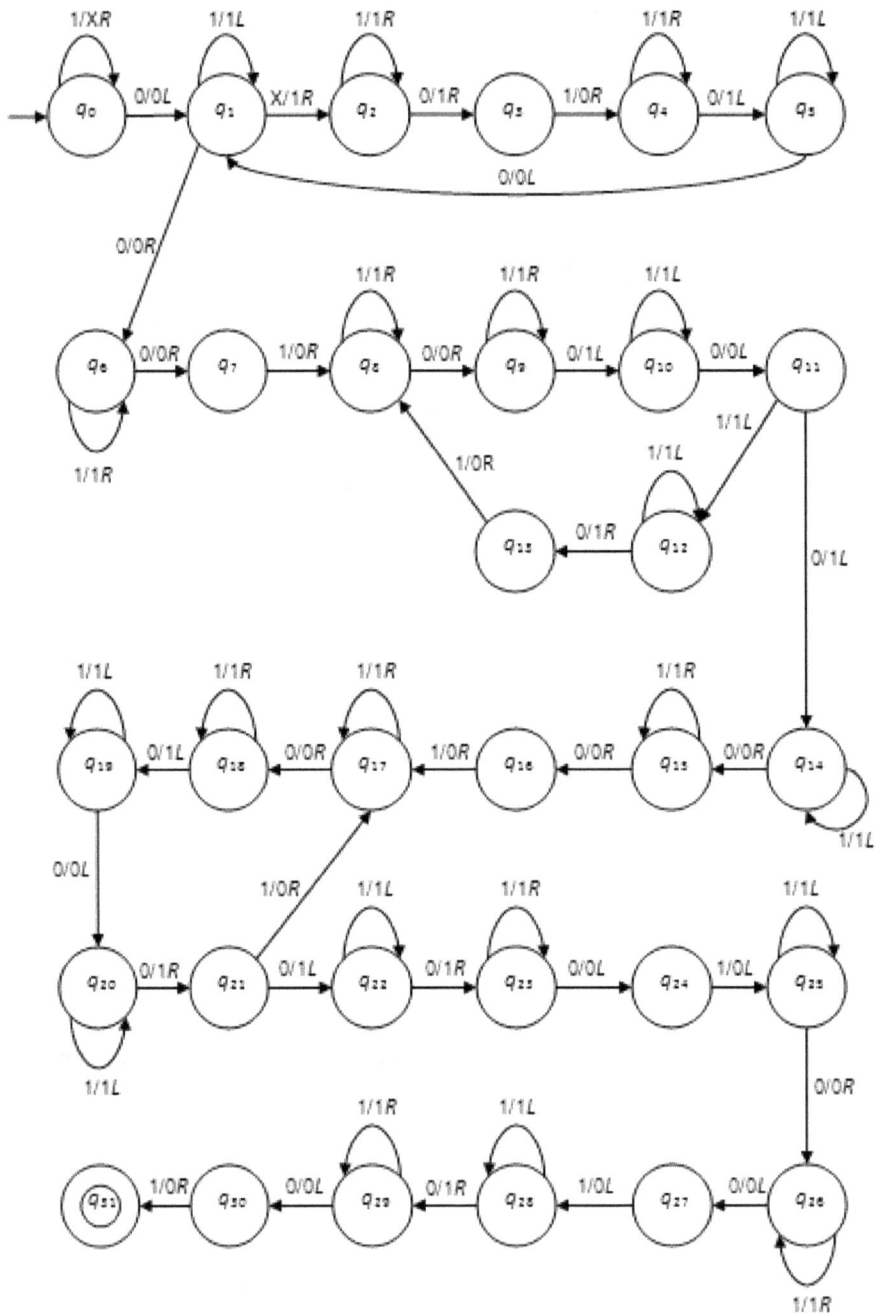

Figure 2.2.15.: A Turing machine that computes the function $f(n, m) = 2n + 3m$ for $n, m \in \mathbb{Z}^+$.

1,1 1,x,>	11,1 11,1,<	21,_ 22,1,>
1,_ 2,_,<	11,_ 12,_,<	22,1 18,_,>
2,1 2,1,<	12,1 13,1,<	22,_ 23,1,<
2,x 3,1,>	12,_ 15,1,<	23,1 23,1,<
2,_ 7,_,>	13,1 13,1,<	23,_ 24,1,>
3,1 3,1,>	13,_ 14,1,>	24,1 24,1,>
3,_ 4,1,>	14,1 9,_,>	24,_ 25,_,<
4,1 5,_,>	15,1 15,1,<	25,1 26,_,<
5,1 5,1,>	15,_ 16,_,>	26,1 26,1,<
5,_ 6,1,<	16,1 16,1,>	26,_ 27,_,>
6,1 6,1,<	16,_ 17,_,>	27,1 27,1,>
6,_ 2,_,<	17,1 18,_,>	27,_ 28,_,<
7,1 7,1,>	18,1 18,1,>	28,1 29,_,<
7,_ 8,_,>	18,_ 19,_,>	29,1 29,1,<
8,1 9,_,>	19,1 19,1,>	29,_ 30,1,>
9,1 9,1,>	19,_ 20,1,<	30,1 30,1,>
9,_ 10,_,>	20,1 20,1,<	30,_ 31,_,<
10,1 10,1,>	20,_ 21,_,<	31,1 32,_,>
10,_ 11,1,<	21,1 21,1,<	

Figure 2.2.16.: Program for a Turing machine computing the function $f(n, m) = 2n + 3m$ for $n, m \in \mathbb{Z}^+$.

a LBA is a non-deterministic Turing machine with a finite tape whose finiteness is determined by $cn + 2$, where c is a specific constant for each individual LBA, n is the length in tape cells of the input string w, and 2 corresponds to the cells for the left and right endmarkers, one cell for each. This finiteness with respect to the tape places LBAs on the side of finite machines and PDAs, rather than on the side of the Turing machines, which have an infinite tape.

Theorem 2.2.14. *If $L \subseteq \Sigma^\bullet$ is a CSL, then there is a LBA that accepts L.*

Proof: We conceive a LBA M (abbreviating $M_{[\![T]\!]}$) with a second track on its input tape. On the input string $[\![w]\!]$ on its tape, M writes the symbol S below the leftmost symbol of w on a second track. If $w = \epsilon$, M rejects w. Otherwise, M proceeds by guessing a production and a position in the sentential form on the second track. Applying the production, M moves the part of the sentential form to the right whenever the sentential form increases in length, but stops in a rejecting state if the new sentential form is longer than w, as given the non-decreasing feature of CSG productions there can be no derivation $S \overset{*}{\Longrightarrow} \alpha \overset{*}{\Longrightarrow} w$ in which the sentential form α is longer than w. M rejects thus empty strings and productions in which the antecedent is longer than the consequent. In other words, M accepts only strings generated by a CSG and all strings generated by a CSG. **QED**

The use of two tape tracks allows for M to be operational with n cells instead of $2n$. In fact, by using k tracks a LBA M can simulate a computation of a Turing machine requiring kn cells, where kn is a linear bound on the input length. This explains the label "linear-bounded" for these automata.

Exercises

Exercise 2.2.3.1. Given the transition tables and the tape strings, determine the computing behavior of each Turing machine and, if any, its output. Draw their state diagrams.

1. Tape string: 11; transition table:

	1	#
q_0	$q_0 1R$	$q_1 \# L$
q_1	$q_2 \# R$	$q_1 1R$
q_2	$q_2 \# R$	$q_a 1R$

2. Tape string: 111; transition table:

	1	#
q_0	$q_1\#R$	$q_5\#L$
q_1	$q_1 1R$	$q_2\#R$
q_2	$q_2 1R$	$q_3 1L$
q_3	$q_3 1L$	$q_4\#L$
q_4	$q_4 1L$	$q_0 1R$
q_5	$q_5 1L$	$q_a\#R$

3. Tape string: 0011; transition table:

	0	1	#
q_0	$q_1 1R$	$q_1 0R$	$-$
q_1	$q_1 1L$	$q_1 0R$	$q_a\#L$

4. Tape string: 0000; transition table:

	0	a	b	#
q_0	$q_1 aR$	$-$	$q_a bR$	$-$
q_1	$q_1 0R$	$-$	$q_1 bR$	$q_2 bL$
q_2	$q_2 0L$	$q_0 aR$	$q_2 bL$	$-$

Exercise 2.2.3.2. Construct the state diagram of the Turing machine accepting the language L of Example 2.2.14.

Exercise 2.2.3.3. Design Turing machines accepting the following languages:

1. $L = \{a^n b^n c^n | n \geq 1\}$

2. $L = \{a^m b^n | m < n\}$

Exercise 2.2.3.4. Design a Turing machine that computes the function

$$f(x,y) = \begin{cases} x-y & \text{if } x > y \\ 0 & \text{otherwise} \end{cases}.$$

Hint: You will need a combination of Turing machines (see Fig. 2.2.17).

Exercise 2.2.3.5. Determine the function computed by the Turing machine in Figure 2.2.18. (Hint: Input ...$\#\#11\#\#$...)

Exercise 2.2.3.6. Describe succinctly the behavior of the Turing machine in Example 2.2.16.

Exercise 2.2.3.7. Describe the behavior of a Turing machine accepting the following languages:

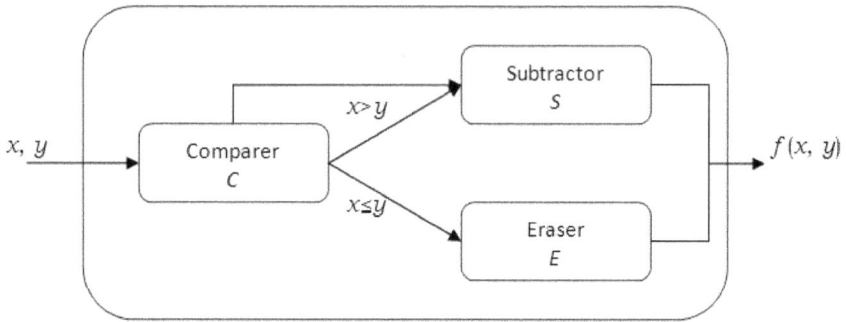

Figure 2.2.17.: A combination of Turing machines.

1. $L = \{w\#w | w \in \{0,1\}^*\}$

2. $L = \{0^{2^n} | n \geq 0\}$

3. $L = \{\#x_1\#x_2\#...\#x_n | x_i \in \{0,1\}^*, x_i \neq x_j \text{ for every } i \neq j\}$

Exercise 2.2.3.8. Identify the languages over $\Sigma = \{a,b,c\}$ accepted by the Turing machines M_1 (Fig. 2.2.19) and M_2 (Fig. 2.2.20) by testing them with the given inputs. Trace their moves for both acceptance and rejection.

1. *aaaabbcccc*

2. *aabbbcccccc*

3. *aabbcccc*

4. *aaabbbccc*

Exercise 2.2.3.9. Prove Theorems 2.2.11 and 2.2.12.

2.2.4. The Chomsky hierarchy (II)

As seen above, the computation of both finite machines and PDAs has to do with accepting (or rejecting) and generating a language, whereas the computation of a Turing machine is more powerful: it computes functions and it can decide membership in recursive sets. By "is more powerful," it is intended to express the fact that all Turing machines are FAs and PDAs, though the reverse does not hold. In effect, Turing

150

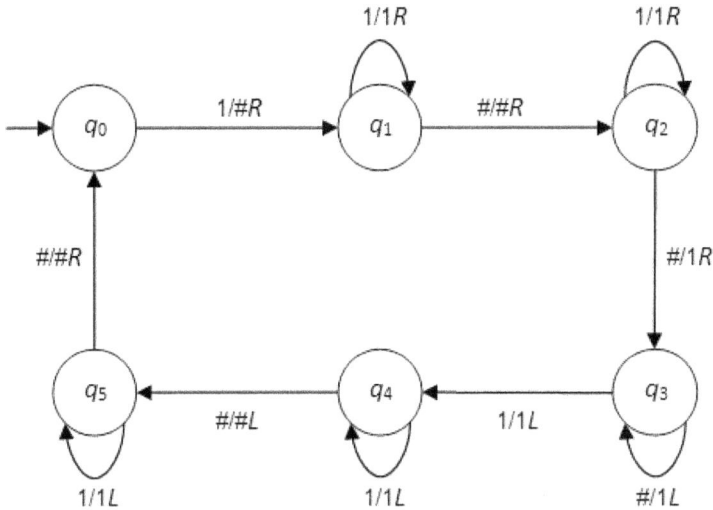

Figure 2.2.18.: A Turing machine.

machines have an enhanced output capability with respect to these two other kinds of automata. But they are also more powerful than the LBAs, as these accept only a subset of the RELs, i.e. the CSLs.

As a matter of fact, the CSLs are important in terms of the Chomsky hierarchy, because this class is strictly or properly contained in the class of the *recursive languages*. These, in turn, distinguish themselves from the RELs according to the following important definition (recall from Def. 2.2.25 that a Turing machine M_T with input alphabet Σ *accepts* a language $L \subseteq \Sigma^*$ if $L = L(M_T)$, and $L(M_T)$ is a REL):

Definition. 2.2.35. A Turing machine M_T *decides* a language L if M_T computes the characteristic function $\chi_L : \Sigma^* \longrightarrow \{0,1\}$ defined as follows for some string x:

$$\chi_L(x) = \begin{cases} 1 & \text{if } x \in L \\ 0 & \text{otherwise} \end{cases}.$$

L is a *recursive language* if there is a Turing machine M_T that decides L.

Theorem 2.2.15. *Every recursive language is recursively enumerable.*

Proof: Given a string $x \in \Sigma^*$ and a Turing machine M_T that decides $L \subseteq \Sigma^*$, on input x M_T halts and produces an output 1 ($\chi_L(x) = 1$) or 0 ($\chi_L(x) = 0$); if the output is 1, accept, otherwise reject. **QED**

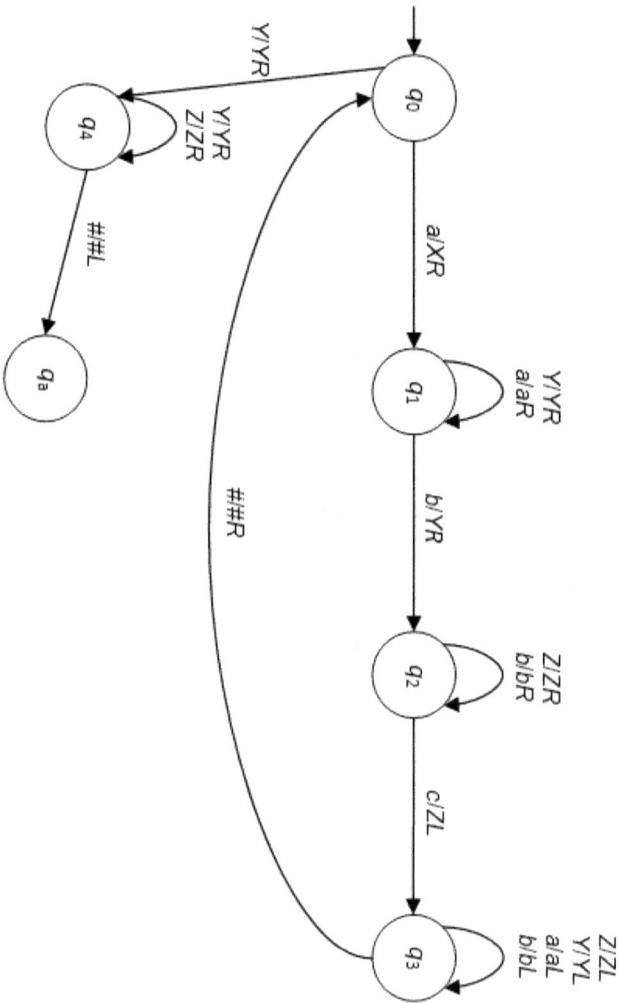

Figure 2.2.19.: Turing machine M_1 accepting a language over $\Sigma = \{a, b, c\}$.

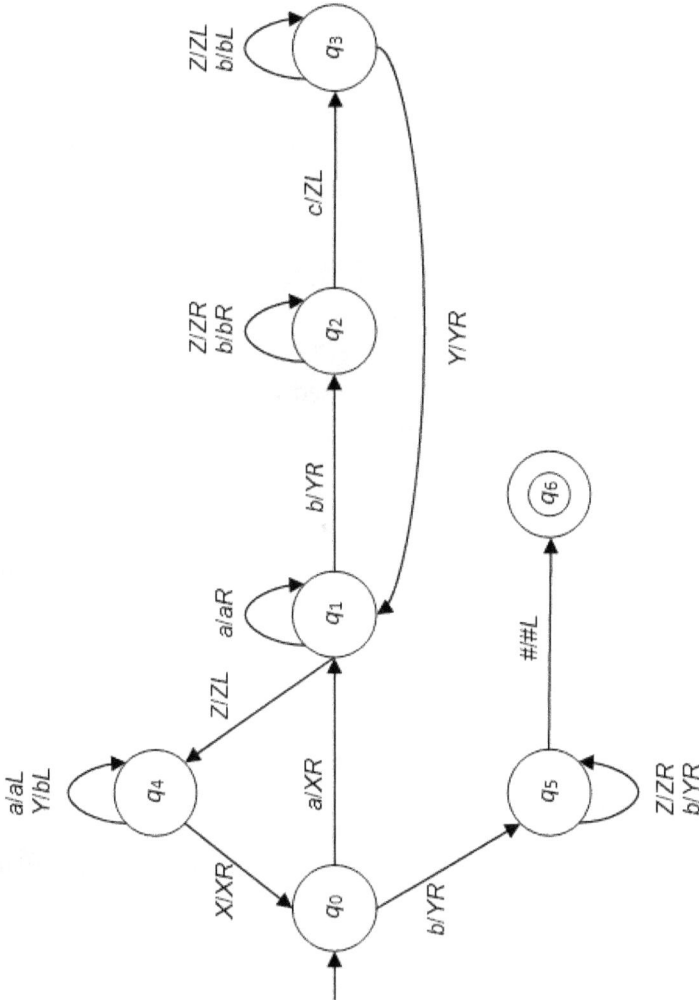

Figure 2.2.20.: Turing machine M_2 accepting a language over $\Sigma = \{a, b, c\}$.

Theorem 2.2.16. *If $L \subseteq \Sigma^*$ is accepted by a total Turing machine M_T, i.e. a Turing machine that halts on every input string x, then L is a recursive language.*

Proof: M_T halts on every input string x. Then, the characteristic function $\chi_L(x) : \Sigma^* \longrightarrow \{0,1\}$ constitutes an algorithm for deciding L in the following way: if M_T accepts x, return 1, otherwise, i.e. if M_T rejects x, then return 0. **QED**

   In the next Section, we shall see more clearly how deciding a language is different from accepting a language. In the meantime, in Table 2.2.1 we show the extended[29] Chomsky hierarchy with both the formal languages and the machines associated to them. Just to complete the hierarchy first given in Section 2.1.4, we have

$$\mathcal{RGL}\ (3) \subset \mathcal{CFL}\ (2) \subset \mathcal{CSL}\ (1) \subset \mathcal{RL} \subset \mathcal{REL}\ (0)$$

where we abbreviate "recursive language" as RL. Note in Table 2.2.1 that there is no specific grammar associated to the recursive languages; in effect, no such grammar is known.

Exercises

   Exercise 2.2.4.1. Show that if A and B are recursive sets, then $A \cup B$, $A \cap B$, and $\overline{A}$ are also recursive. (Hint: Begin with $\chi_{\overline{A}}(x) = 1 - \chi_A(x)$, x is a string.)

   Exercise 2.2.4.2. Show that CSLs are closed under union, intersection, concatenation, and positive closure.

   Exercise 2.2.4.3. A PDA with two stacks is as powerful as a Turing machine. Explain why.

   Exercise 2.2.4.4. A LBA is not as powerful as an unrestricted Turing machine. Explain why, expanding on the reason given above.

   Exercise 2.2.4.5. Prove the following statement: Let $M_1, M_2, \ldots$ be an enumeration of some set of Turing machines halting on all inputs. Then, there is some recursive language that is not $L(M_i)$ for any i (M_i abbreviates M_{T_i}).

---

[29] Because Chomsky did not include the recursive languages in his hierarchy. Note that currently the Chomsky hierarchy is actually larger.

Grammar	Language	Computer Model
0 (UG)	Recursively enumerable (REL)	Turing machines (TMs)
	Recursive	Total TMs (Deciders)
1 (CSG)	Context-sensitive (CSL)	Linear-bounded automata (LBA)
2 (CFG)	Context-free (CFL)	Pushdown automata (PDA)
3 (Regular)	Regular	Finite machines

Table 2.2.1.: The extended Chomsky hierarchy: Languages and associated computer models.

Exercise 2.2.4.6. Prove that there is a recursive language that is not a CSL. (Hint: Use the statement in Exercise 2.2.4.5.)

2.3. Computability and complexity

2.3.1. The decision problem and Turing-decidability

It should be obvious by now that formal languages are at the very core of the theory of computation. In effect, whatever can be computed, can so only with and over a formal language. But this is still too general: in fact, whatever can be computed, can so only with and over *a language that is decidable*. In other words, we are speaking here of a language for which there is an algorithm for deciding whether an arbitrary string x belongs to the language. Thus, the question whether such an algorithm exists is paramount. This is known as the *decision problem*.

Definition. 2.3.1. Given an arbitrary string x and a language L, it is asked whether $x \in L$. The *decision problem* consists in the following: is there an algorithm to answer the question whether $x \in L$?

The decision problem is formulated in terms of set membership, and deciding with respect to a subset of a language whether or not a given string x belongs to it is in no way different from deciding with respect to a subset of the natural numbers whether a given number x is an element thereof.

Definition. 2.3.2. A set $S \subseteq \mathbb{N}$ is said to be *recursive, computable*, or *decidable* if there is an algorithm Ψ that terminates *after a finite amount of time* and correctly decides with respect to a given number x whether $x \in S$. More formally, $S \subseteq \mathbb{N}$ is a recursive (or computable or decidable) set if there is a total computable function $f : \mathbb{N} \longrightarrow \{0, 1\}$ such that for a given number x we have $f(x) = 1$ if $x \in S$ and $f(x) = 0$ if $x \notin S$.

Hence, for a set to be classified as computable, it must be the case that algorithm Ψ *also* outputs the answer "No" whenever $x \notin S$. In the specific case of a formal language, *decidability* is formulated in the following way:

Definition. 2.3.3. A formal language L is said to be *(Turing-)decidable* if there is an algorithm guaranteed to determine whether an arbitrary string x belongs to L.

As seen above, this algorithm is the computation of the characteristic function $\chi_L(x) : \Sigma^* \longrightarrow \{0,1\}$ by a total Turing machine. This explains why saying *decidable* just is the same as saying *Turing-decidable* with respect to a recursive language. If ambiguity may arise, then the term "decidable" should be used for languages that are decidable by other kinds of Turing machine (e.g. a non-deterministic Turing machine; abbr.: NDTM). In any case, this shows the fundamental importance of the Turing machines for the whole of the field of computation, well expressed in the following statement known as *Church-Turing Thesis*:

2.3.4. *A function that is effectively calculable is a function that can be computed by a Turing machine.*

This statement, which essentially defines a computable function in terms of the Turing machine,[30] is not a theorem, as it cannot be proved, thus remaining a thesis. This said, so far no one came up with an algorithmic procedure that cannot be implemented by a Turing machine. Whether the Turing machine is an *efficient* model of computation, that is altogether another issue, as it can be extremely difficult to program a Turing machine for even very elementary functions, as Example 2.2.16 shows.

This thesis is named after A. Church, too, because it was shown by Turing (in Turing, 1937) that a Turing machine is equivalent to Church's lambda calculus in defining an effectively calculable function, i.e. Turing machines and the lambda calculus are equivalent models of computation. But the intuitive power of the Turing machine makes it an "easier" model than the other equivalent models of computation.[31] Indeed, to say of a language that it is recursive or computable is the same as saying that its elements can be *enumerated*–or, more informally, *listed*. And the easiest way to define this is by using a multi-tape Turing machine.[32]

Definition. 2.3.5. Let M_T be a k-tape Turing machine, $k \geq 1$, and let $L \subseteq \Sigma^*$ be a formal language. We say that M_T *enumerates* L if its operation satisfies the following conditions with respect to tape 1:

---

[30]In effect, from the Church-Turing thesis as formulated above, we can rigorously define a computable function as a function $f : \Sigma^* \longrightarrow \Sigma^*$ such that on every input w a Turing machine halts with only $f(w)$ on its tape. Compare this with Def. 1.1.4.7.

[31]See also Church (1936a, b). Besides the already mentioned lambda calculus, there are (many) more equivalent models of computation: e.g., register machines, Post machines, and μ-recursive functions. Incidentally, and as a justification for our focusing exclusively on Turing machines, K. Gödel considered that the Turing machine was sufficient to define an effectively calculable function (cf. Gödel, 1964).

[32]Recall, however, that variations such as multiple tapes do not change the computing power of the Turing machine.

2. Fundamentals of classical computing

1. the tape head never moves to the left, and no non-blank symbol printed on it is subsequently erased or printed over, i.e. tape 1 is a one-way write-only tape;

2. for every $x \in L$, at some point the contents of this tape will be

$$x_1 \# x_2 \# ... \# x_n \# x \#$$

for some $n \geq 0$, $x_1, x_2, ..., x_n \in L$, $x_1, x_2, ..., x_n, x$ are all distinct, and there is nothing printed after $\#$ if L is finite.

Let us now say that a set of strings is listed in *canonical order* if shorter strings precede longer strings and strings of the same length are alphabetically ordered. Then, we have the following theorem with respect to the distinction between RELs and recursive languages:

Theorem 2.3.1. *For every language $L \subseteq \Sigma^*$, L is recursively enumerable iff there is a Turing machine M_T that enumerates L, and L is recursive iff there is a Turing machine M_T that enumerates the elements of L in canonical order.*

Proof: Left as an exercise.

Exercises

Exercise 2.3.1.1. Show that the set Σ^* of all strings over Σ is countable.

Exercise 2.3.1.2. Show that the following languages are decidable:

1. $ACPT_{FA} = \{\langle M, w \rangle \,|\, M$ is a FA that accepts string $w\}$

2. $ACPT_{LBA} = \{\langle M, w \rangle \,|\, M$ is a LBA that accepts string $w\}$

3. $REG_{r,w} = \{(r, w) \,|\, r$ is a regular expression that generates string $w\}$

4. $REG_{CFG} = \{(G, w) \,|\, G$ is a CFG that generates string $w\}$

5. $EMPTY_{FA} = \{\langle M \rangle \,|\, M$ is a FA and $L(M) = \emptyset\}$

Exercise 2.3.1.3. Prove Theorem 2.3.1.

Exercise 2.3.1.4. Give at least three more formulations of the Church-Turing thesis, i.e. statements equivalent to §2.3.4.

2.3.2. Undecidable problems and Turing-reducibility

The importance of the recursive languages for computing is that they
mark the endpoint of the *decidable languages* and, as will be seen, of the
problems solvable by a computer. In practical terms as far as decidability is concerned, we have it that there are easily formulated problems
that actually have no algorithmic solution. For instance, although programs and program specifications are precisely defined mathematical
objects, the general problem of program verification cannot be solved
by a computer.

This means that the fundamental question whether $x \in L$ has no
conclusive answer for the languages beyond the recursive ones in the
extended Chomsky hierarchy, which means that for the RELs the best
scenario one may get in the case that $x \notin L$ is that a Turing machine
will not answer that $x \in L$, though it is not certain whether or not it
will halt. In effect, there may be input strings $x \notin L$ that may cause a
Turing machine to loop forever. This means that the best we can get
for RELs is *semi-decidability*.

Definition. 2.3.6. A set $S \subseteq \mathbb{N}$ is said to be *recursively enumerable*
or *semi-decidable* if there is an algorithm that correctly decides when a
given number $x \in S$, but the algorithm may give no answer for $x \notin S$.
Formally formulated, there is a partial computable function φ such that

$$\varphi(x) = \begin{cases} 1 & \text{if } x \in S \\ \text{undefined, or does not halt} & \text{if } x \notin S \end{cases}.$$

Recall from above the definition of a computable set (cf. Def. 2.3.2).

Example 2.3.1. Consider the set

$$S = \{p | p \text{ is a polynomial over } x \text{ with an integral root}\}.$$

We ask whether S is decidable. Given a polynomial p over x such
as $2x^3 + 6x^2 - x + 5$, we ask if $p \in S$. In other words, we want to
know if $x = n, n \in \mathbb{Z}$. There is indeed an algorithm to test for this
decision problem: evaluate p for the values $0, 1, -1, 2, -2, 3, ...$ one at
a time successively; if for some $n \in \mathbb{Z}$ it is the case that $p = 0$, then
$\chi_S(p) = 1$. Nevertheless, we have no guarantee that this algorithm will
ever terminate: it will only terminate if indeed $p \in S$; otherwise, it is
obvious that it will run forever, because $\mathbb{Z}$ is an infinite set. Thus, this
set is in fact not computable, though it is recursively enumerable.[33]

---

[33] S is in fact called a *Diophantine set*. Put more rigorously, it can be shown that

Example 2.3.1 shows in a clear way that the property of being recursively enumerable may not be of much help in practical terms. The importance of these results cannot be overstated: they mark the limits of what computing devices can do, namely with respect to the decision problem (see above). These limits are notably illustrated by the following examples.

Example 2.3.2. *The Acceptance Problem (ACPT)* – Let the following set be given (M abbreviates M_T and TM does so for Turing machine):

$$ACPT_{TM} = \{\langle M, w\rangle \,|\, M \text{ is a TM and } M \text{ accepts } w\}.$$

Although it is recursively enumerable, this set is undecidable. In effect, given input w, M can halt in either an accepting or a rejecting state, but M may loop on w, and M has no way to determine that it is looping (i.e. not halting) on a given input.

Example 2.3.3. *The Halting Problem (HALT)* – The set

$$HALT_{TM} = \{\langle M, w\rangle \,|\, M \text{ is a TM and } M \text{ halts on } w\}$$

is undecidable. That is, the problem whether, given an arbitrary program for a Turing machine and an arbitrary string, the program halts has no algorithmic solution. This was actually one of the first decision problems to be proven undecidable, and it was so famously in Turing (1937) by means of the Turing machine. We give the sketch of a proof and leave the details as an exercise. Let $\langle M\rangle$ be the binary encoding of a TM $M_T = (Q, \Gamma, \#, \Sigma, q_0, A, \delta)$ and $w \in \{0,1\}^+$ be some string. Assume now that there is a Turing machine H (abbreviating H_T) that solves HALT. H has two final states, q_Y and q_N for "Yes" and "No" answers, respectively, such that on input $(\langle M\rangle, w)$ machine H ends in state q_Y if M halts on w, and H ends in state q_N if M does not halt on w. We construct a Turing machine H' from H such that H' halts if M ends in state q_N and H' does not halt (i.e. loops forever) if M ends in state q_Y. We then construct another Turing machine H'' from H' that makes a copy of $\langle M\rangle$ and behaves like H'. We have it that H'' with input $\langle M\rangle$

---

there is a polynomial $p(n, x_1, ..., x_n)$ with integer coefficients such that the set of values of $n \in \mathbb{Z}$ for which the equation $p(n, x_1, ..., x_n) = 0$ has solution is not computable. The direct consequence of this result is that Hilbert's Tenth Problem is unsolvable. See Matiyasevich (1993) for a comprehensive elaboration on this result in the framework of Hilbert's Tenth Problem and Diophantine equations; for a discussion more immediately in the context of computability theory (typically with the restriction $n \in \mathbb{N}$), see Leary & Christiansen (2015).

runs H' with input $(\langle M \rangle, \langle M \rangle)$, and we reach a contradiction when we run H'' on input $\langle H'' \rangle$: If H'' halts, then it loops for ever; if it does not halt, then it halts.[34]

Note that in Examples 2.3.2-3 both sets $ACPT_{TM}$ and $HALT_{TM}$ are actually languages. This means that, in face of a problem requiring a "Yes/No" answer, we can formulate it in terms of whether a particular language is–or is not–recursive. Suppose now you suspect that problem B is hard to solve and you have an unsolvable problem A formulated as a language that cannot be decided; if you can reduce A to B, then you know that B is also undecidable. In turn, if you think that A is easy to solve and in fact you have a decidable problem B, then if you can reduce A to B you know that A is also decidable. This is known as *reducibility*.

Definition. 2.3.7. Given two languages A and B, we say that there is a *mapping reduction from A to B*, denoted $A \preceq B$, if there is a computable function $f : \Sigma^* \longrightarrow \Sigma^*$ such that, for every w, we have

$$(f_{red}) \qquad w \in A \quad \text{iff} \quad f(w) \in B.$$

The function f is the *reduction from A to B*.

One can actually speak of *Turing-reducibility*:[35] it is obvious that the reduction above actually reduces questions of membership (to a language) in A to membership in B, it being the case, for example, that there is a total Turing machine M that is a *decider* for B. Then, the proofs involve combining this decider with other Turing machines, so as to form "composite" Turing machines with one tape for each of the individual machines. The basic idea is to show that language A is recursive in language B. In other words, the algorithm for deciding membership in B can be converted into an algorithm for deciding membership in A.

---

[34]Equivalently, but perhaps more intuitively, we can think in terms of functions in the following way: Given a function $f(x)$, which we name as f_x, we define a function:

$$halt(x) = \begin{cases} 1 & \text{if } f_x \text{ halts on input } x \\ 0 & \text{otherwise} \end{cases}$$

We next define a new function:

$$g(x) = \begin{cases} \text{loop forever} & \text{if } halt(x) = 1 \\ 0 & \text{otherwise} \end{cases}$$

Derive now a contradiction from defining

$$g(n) = f_n(n)$$

for n some natural number.

[35]Strictly conceived, Turing-reducibility is a generalization of mapping reducibility.

Theorem 2.3.2. *Let A and B be two languages. If $A \preceq B$ and B is decidable, then A is decidable.*

Proof: (Sketch) We assume that M (abbreviating M_T) is a decider for B and f is the reduction from A to B. Then, we can define a Turing machine M' such that, given input w, M' computes $f(w)$ and runs M on input w and outputs whatever is the output of M. Hence, we have $\chi_A(x) = \chi_B(f(x))$, for some string $x \in A$. **QED**

By contraposition,[36] we have the following result:

Corollary 2.3.3. *If $A \preceq B$ and A is undecidable, then B is undecidable.*

Example 2.3.4. The *State-Entry Problem (STENTRY)*, defined as

$$STENTRY_{TM} = \{\langle M, w, q \rangle \,|M \text{ enters state } q \text{ on input } w\}$$

can be shown to be undecidable by reduction. We reduce HALT to STENTRY. Suppose we have an algorithm A that solves the latter. Then, there is a Turing machine M (abbreviating M_T) that enters state q on input w. By Theorem 2.3.2, there is also an algorithm to solve HALT. We modify M into a Turing machine M' that on input w halts in state q iff M does. On input w, M halts in state q only if some transition $\delta(q_i, \sigma)$ for some state q_i and some symbol σ is undefined. In order to obtain M' we change every undefined transition to $\delta(q_i, \sigma) = (q, \sigma, R)$, where q is the only final state. We run algorithm A on $\langle M', w, q \rangle$: if M' enters state q, then M halts on input w; otherwise, M does not halt. But we know that there is no algorithm for (M', w, q), i.e. HALT is undecidable. Therefore, because HALT$\preceq$STENTRY, by Corollary 2.3.3, STENTRY must also be undecidable.

As seen, there is indeed an algorithm for the recursive languages. However, there is no such algorithm for the whole class of the RELs. In the light of Theorems 2.2.15-16, the extended Chomsky hierarchy signifies that all languages up to and including the recursive languages are decidable, and that there are RELs that are not decidable, though they can be accepted by a Turing machine (otherwise, they would not be RELs). In effect, beyond the recursively enumerable sets, we have the *non-computable* or *undecidable* sets. In terms of the Chomsky hierarchy, we have the undecidable languages. And, impressively, the set of the languages that are not RELs is uncountable.

---

[36]Cf. ⊩ 7 in Proposition 4.3.1.

Theorem 2.3.4. *The set of languages over $\{0,1\}$ that are not RELs is uncountable.*

Proof: Left as an exercise. Let $\{0,1\}^* = S$. First, prove that $|S| = |\mathbb{N}|$. Next, prove that 2^S is uncountable (cf. Exercise 1.1.1.9).

On the other hand, we have the following theorem:

Theorem 2.3.5. *The set of Turing machines is countable.*

Proof: (Sketch) The set Σ^* of all strings over Σ is countable (cf. Exercise 2.3.1.1). Hence, we can list all the strings of Σ^*, for example, ordering them by length. For each of these strings there is an encoding $\langle M_T \rangle$ of a Turing machine, so that, if we omit all illegal encodings of Turing machines, we have a list of all Turing machines. **QED**

The result follows: There are more formal languages than Turing machines. This poses the question what the languages that are not RELs are like. Unfortunately, we have only an indirect glimpse into them.

Theorem 2.3.6. *There exists a REL L whose complement $\overline{L}$ is not recursively enumerable.*

Proof: We abbreviate M_{T_i} as M_i. By Theorem 2.3.5, given an input alphabet $\Sigma = \{a\}$, there are Turing machines $M_1, M_2, \dots$ such that there is some M_i associated to each REL $L(M_i)$ over Σ. Consider now a new language L such that, for each $i \geq 1$, with respect to the string a^i we have

$$a^i \in L \quad \text{iff} \quad a^i \in L(M_i).$$

Consider now the complement of L:

$$(1) \qquad \overline{L} = \left\{ a^i | a^i \notin L(M_i) \right\}.$$

Both L and $\overline{L}$ are well defined, but the latter is not recursively enumerable. This can be shown by contradiction. We assume that $\overline{L}$ is recursively enumerable. Then, there must be some M_j such that

$$(2) \qquad \overline{L} = L(M_j).$$

Let us consider now a string a^j. If $a^j \in \overline{L}$, then by (2) $a^j \in L(M_j)$ which entails by (1) that $a^j \notin \overline{L}$; if $a^j \in L$, then $a^j \notin \overline{L}$, which by (2) entails that $a^j \notin L(M_j)$, which by (1) entails that $a^j \in \overline{L}$. In either case, we reach a contradiction, so we conclude that $\overline{L}$ is not recursively enumerable. Proving that L is a REL is easy and we leave this part of the proof as an exercise. (Hint: Think universal!) **QED**

In the proof above, we remark that both L and $\overline{L}$ are well defined. However, the fact that $\overline{L}$ is not recursively enumerable shows us that there are well-defined languages that are not RELs. This, in turn, shows that there are well-defined languages for which no membership algorithm can be conceived, and thus no effective decision procedure can be obtained.

But, in fact, there are specific problems with respect to the recursive languages that are not computable in practice, either. This topic is approached in Section 2.3.4.

Exercises

Exercise 2.3.2.1. Prove the undecidability of ACPT by means of the diagonalization method.

Exercise 2.3.2.2. Give a full proof of the undecidability of HALT from the sketch in Example 2.3.3. (Hint: Use configurations.)

Exercise 2.3.2.3. Solve Exercise 1.1.1.1 again now by making use of Turing machines.

Exercise 2.3.2.4. Show that if HALT were decidable, then every REL would be recursive.

Exercise 2.3.2.5. Explain why HALT is theoretically decidable for linear-bounded automata or deterministic machines with a finite memory. Explain also why only theoretically so.

Exercise 2.3.2.6. Show that the complement of a recursive language is recursive. (Hint: See Fig. 2.3.1, where M abbreviates M_T.)

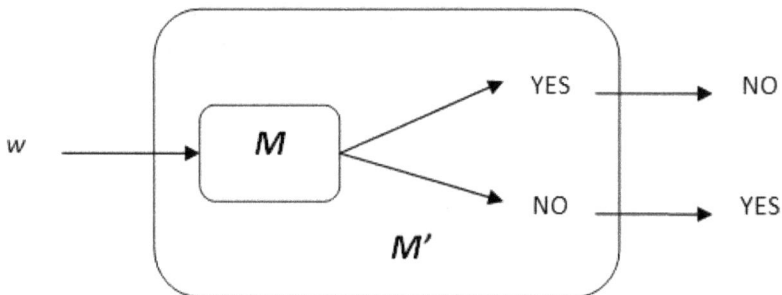

Figure 2.3.1.: A combination of Turing machines.

Exercise 2.3.2.7. Show by combining two Turing machines M_1 and M_2 into a Turing machine M that the class of recursive languages is closed under union and intersection.

Exercise 2.3.2.8. Show by combining two Turing machines M_1 and M_2 into a Turing machine M that the class of RELs is closed under union and intersection.

Exercise 2.3.2.9. Show by Turing-reducibility that the following languages are undecidable:

1. $REG_{TM} = \{\langle M \rangle \,|\, M$ is a TM and $L(M)$ is regular$\}$

2. $EMPTY_{TM} = \{\langle M \rangle \,|\, M$ is a TM and $L(M) = \emptyset\}$

3. $EMPTY_{LBA} = \{\langle M \rangle \,|\, M$ is a LBA and $L(M) = \emptyset\}$

4. $EQUAL_{TM} = \{\langle M_1, M_2 \rangle \,|\, M_1, M_2$ are TMs and $L(M_1) = L(M_2)\}$

Exercise 2.3.2.10. Research into the decision problem known as *Busy Beaver* and give its main aspects, i.e. describe the problem, formalize it, and provide the details of a proof of its undecidability.

Exercise 2.3.2.11. Research into *Rice's theorem* and give its main points.

Exercise 2.3.2.12. Research into *Post's Correspondence Problem* and give its main points.

Exercise 2.3.2.13. Research into *Hilbert's Tenth Problem* and elaborate on the proof of its undecidability found by Y. Matiyasevich.

Exercise 2.3.2.14. Search in the literature for other important computational problems that have been shown unsolvable by means of Turing-reducibility.

Exercise 2.3.2.15. Complete the proof of Theorem 2.3.6.

2.3.3. The Chomsky hierarchy (III)

We ended last Section with an indirect glimpse into the languages beyond the RELs, i.e. the complements of RELs. As seen, the RELs are merely *Turing-recognizable*. By this it is meant that there is a Turing machine M_T that recognizes a REL L, but M_T may enter an infinite loop for strings that do not belong to L. $ACPT_{TM}$ and $HALT_{TM}$ are notorious examples of Turing-recognizable languages, and the only glimpse that we get of their complements, to wit, $\overline{ACPT_{TM}}$ and $\overline{HALT_{TM}}$, respectively, is that they are not even Turing-recognizable. If we use the Turing-decidable languages (i.e. the recursive languages) as the border between decidability and Turing-recognizability, the extended Chomsky hierarchy can be visualized as in Figure 2.3.2. (Recall that the decidable languages include the regular languages, the CFLs, and the CSLs.)

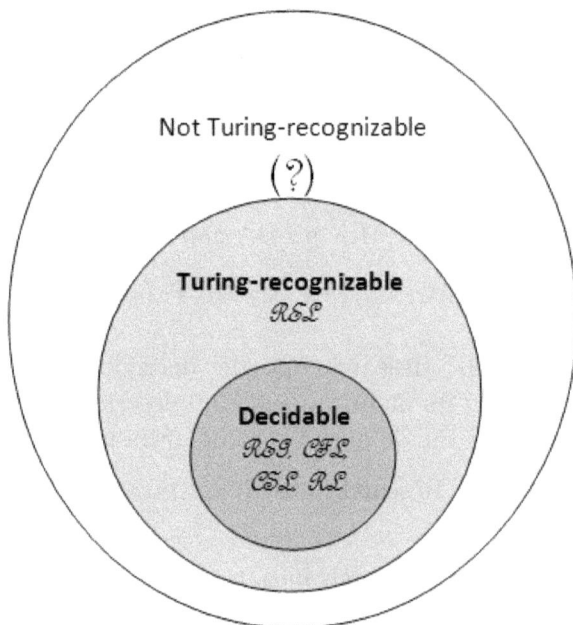

Figure 2.3.2.: The Chomsky hierarchy and beyond: Decidable, Turing-recognizable, and not-Turing-recognizable languages.

Knowledge of the material above will easily lead to the realization that just about all questions of interest with respect to RELs are undecidable. Moreover, as we go up in the Chomsky hierarchy, undecidability grows. Table 2.3.1 shows the decidability properties of the grammar classes in the Chomsky hierarchy.

Exercises

Exercise 2.3.3.1. With respect to Table 2.3.1, some of the proofs of (un)decidability have already been discussed/proposed as exercises above. Formalize the remaining decision problems on it and prove their (un)decidability status.

	$\mathcal{REG}$ Type 3	$\mathcal{CFL}$ Type 2	$\mathcal{CSL}$ Type 1	$\mathcal{REL}$ Type 0
Membership	Yes	Yes	Yes	No
Finiteness	Yes	Yes	No	No
Emptiness	Yes	Yes	No	No
Inclusion	Yes	No	No	No
Equivalence	Yes	No	No	No

Table 2.3.1.: Decidability ("Yes") and undecidability ("No") of some properties of interest for the Chomsky hierarchy.

2.3.4. Computational complexity

Strictly defined, complexity theory is concerned with classifying the degree of difficulty in testing for membership in the various formal languages. Less strictly defined, it classifies the amount of resources required to carry out a computational algorithm. This is a large field, so large indeed that a comprehensive listing of complexity classes can contain more than 400 entries. For our purposes, luckily, only a few such classes are required.

In the analysis of computational problems, there are resources other than time and space that can be taken into consideration, and models of computation other than the Turing machines. However, we shall concentrate on time and space resources, and our model of computation will be the Turing machine. We begin with a very general, abstract approach, and go then into more specific issues and classifications in the theory of computational complexity. This abstract approach is based on M. Blum's axioms for computational measures (Blum, 1967). Although they have lost some of their initial spark–due to the fact that many measures today relevant in computing theory do not satisfy them–, they will be fine for our purposes. In later Chapters, further specifications are made.

The following, rather informal, definition serves our purposes well:

Definition. 2.3.8. A *computational problem* is an infinite collection of *(problem) instances* with a *solution* for every instance.

The best way to tackle a computational problem is by reducing it. This may be a reduction to a *decision problem* in which input strings–typically, but not always, over the alphabet $\Sigma = \{0,1\}$–are problem instances.[37] Another possible reduction is to a *function problem*. Clearly, outputs in the latter reduction are more complex than the output "Yes/No" of decision problems, but function problems can be recast in terms of decision problems (and vice-versa). Other, less frequent, reductions are possible.

Example 2.3.5. Recall from Section 1.2.3 the definition of a graph and from Section 2.2.3 the notion of encoding. Given an undirected graph $\mathfrak{G}$, a Hamiltonian path in $\mathfrak{G}$ is a simple path that goes through each vertex exactly once. We want to test whether a graph $\mathfrak{G}$ contains a Hamiltonian path connecting two specified vertices, say, u and v, which constitutes a triple $(\mathfrak{G}, u, v)$. This is known as *The Hamiltonian Path Problem*, and it can be formulated in terms of a decision problem as follows with $\langle \cdot \rangle$ denoting an appropriate encoding:

$HAMPATH = \{\langle \mathfrak{G}, u, v \rangle \, | \, \mathfrak{G}$ has a Hamiltonian path connecting $u, v\}$.

We can encode $\mathfrak{G}$ over the alphabet $\Sigma = \{0,1\}$ by means of an adjacency matrix (cf. Exercise 1.2.3.1). Let $\mathfrak{G}$ be the graph of Figure 1.2.5. Then, we have the adjacency matrix

$$A_{\mathfrak{G}} = \begin{pmatrix} 0 & 1 & 1 & 1 & 0 \\ 1 & 0 & 1 & 0 & 1 \\ 1 & 1 & 0 & 1 & 0 \\ 1 & 0 & 1 & 0 & 1 \\ 0 & 1 & 0 & 1 & 0 \end{pmatrix}$$

which can be easily transformed into an array, which for our purposes constitutes a string w over the alphabet $\Sigma = \{0,1\}$ such that $w = \langle \mathfrak{G} \rangle$.

In practical terms, faced with a computational problem we want to know how much *effort* must be put into solving it, or what its *cost* in terms of resources is. In effect, once we determine that a problem is computable, this does not mean that we can actually compute it–efficiently, or at all. This is so because any computation takes up resources, such as (running) time and (utilized) space. Time and space might be infinite, but we are short-living and time and space have monetary costs for us; so, we require algorithms that both terminate in useful time and do not

---

[37]This *binary encoding* is especially relevant when we stipulate that the size n of the input is measured in bits, as it assures us that each character uses a constant number of bits and each integer $i > 0$ uses at most $c \log i$ bits for some constant $c > 0$. To avoid in any case is *unary encoding*, which uses the single symbol "1"; just think of representing any integer greater than, say, 20 in this system.

take up too much space. Thus, we need to restrict these resources appropriately. In order to do this we require a clear notion of computational measure; for time and space resources, the Blum axioms (Blum, 1967) provide a general approach.

Definition. 2.3.9. *The Blum axioms* – Let $M_0, M_1, ...$ be an effective enumeration of a (class) of machines. With each M_i there are associated two functions: (i) the partial recursive function $\varphi_i(n)$ for input n, and (ii) a partial recursive function $\Phi_i(n)$ such that $\Phi_i(n) = m$ is the *computational complexity of M_i on input n* iff the following axioms are satisfied:

- *Axiom 1:* $\varphi_i(n)$ converges iff $\Phi_i(n)$ converges.

- *Axiom 2:* The function

$$R(i, n, m) = \begin{cases} 1 & \text{if } \Phi_i(n) = m \\ 0 & \text{otherwise} \end{cases}$$

is (total) recursive.

Axiom 1 can be reformulated in terms of definition, i.e. $\Phi_i(n)$ is defined iff $\varphi_i(n)$ is defined (i.e. M_i halts on input n); reformulated in terms of Axiom 2, we have

$$\varphi_i(n) \text{ converges iff } \exists m\, [R(i, n, m) = 1].$$

Blum called the function $\Phi_i(n)$ the *step-counting function*, but it may be interpreted as counting either the number of steps or the amount of tape used by M_i given input n. So, the resources here envisaged are exclusively time and space computed as m steps or tape cells, respectively, in terms of complexity.

Blum (1967) does not specify the kind of (Turing) machine, so as to guarantee the generality of the axioms. We concentrate on two (rather general) types of Turing machine to specify complexity measures of time and space: deterministic and non-deterministic Turing machines, abbreviated as DTM and NDTM, respectively (see Section 2.2.3). We also focus on *worst-case analysis*, i.e. we are interested in longest running-time and largest storage-space requirements for inputs of length n.

Definition. 2.3.10. *Space complexity* – Let M be a Turing machine. If, for every input string of length n, M scans *at most* $S(n)$ storage tape cells, then M is said to be a *Turing machine of space complexity* $S(n)$ (or a $S(n)$ *space-bounded Turing machine*). Because every Turing machine uses at least one cell on any input, we may assume $S(n) \geq 1$ for all n. Moreover, we may assume that

$$S(n) = \max(1, \lceil S(n) \rceil)$$

where $\lceil S(n) \rceil$ denotes the ceiling for M given input n.[38] The language recognized by M is also said to be of space complexity $S(n)$.

Definition. 2.3.11. *Time complexity* – Let M be a Turing machine. If, for every input string of length n, M takes *at most* $T(n)$ steps (moves) before halting, then M is said to be a *Turing machine of time complexity* $T(n)$ (or a $T(n)$ *time-bounded Turing machine*). Because every Turing machine will require at least $n + 1$ steps to read the input and certify itself that the end of the string has been reached (by reading the first blank symbol on the right of the string), we may assume $T(n) = n + 1$ for all n. Moreover, we may assume that

$$T(n) = \max(n + 1, \lceil T(n) \rceil)$$

where $\lceil T(n) \rceil$ denotes the ceiling for M given input n. The language recognized by M is also said to be of time complexity $T(n)$.

Example 2.3.6. Consider the language

$$L = \{wcw^R | w \in \{0, 1\}^*\}$$

where c is some symbol not belonging to $\Sigma = \{0, 1\}$. A Turing machine M with two tapes makes at most $n + 1$ moves for an input of length n: firstly, M copies the half of the input that is to the left of c (i.e. w) to the second tape; when M finds c, M *simultaneously* moves the head to the right on the input tape (reading w^R) and moves the head to the left on the second tape, comparing the symbols on both tapes. We say that L is of time complexity $T(n) = n + 1$.

Definition. 2.3.12. Definitions 2.3.10-11 hold for DTMs.[39] We thus have the two *families* of space and time complexity classes DSPACE $(S(n))$ and DTIME $(T(n))$, respectively.

But these definitions also hold for NDTMs once we formulate some restrictions:

---

[38]The *ceiling* of a, denoted by $\lceil a \rceil$, is the smallest integer greater than or equal to a. E.g., $\lceil 3.6 \rceil = 4$; $\lceil 4 \rceil = 4$.

[39]With the proviso that if we use DTMs with any fixed $n \geq 1$ of tapes, we might not be able to consider all languages accepted in time $T(n)$.

Definition. 2.3.13. A NDTM is of space complexity $S(n)$ if it cannot scan more than $S(n)$ tape cells, regardless of how many choices it may have. We denote this family of space complexity by NSPACE $(S(n))$. A NDTM is of time complexity $T(n)$ if it cannot make more than $T(n)$ moves, regardless of how many choices it may have. We denote this family of time complexity by NTIME $(T(n))$.

Because it is more often than not the case that we are faced with complex expressions, especially so on large inputs, we actually use what is called *big-O notation*.

Definition. 2.3.14. Consider the functions $f, g : \mathbb{N} \longrightarrow \mathbb{R}^+$, where $\mathbb{R}^+$ is the set of the positive real numbers. We say that $f(n)$ *is of order* $g(n)$, and write $f(n) = \mathcal{O}(g(n))$, if there exist $c, n_0 \in \mathbb{Z}^+$ such that, for $n \geq n_0$, c is a constant,

$$f(n) \leq cg(n).$$

When $f(n) = \mathcal{O}(g(n))$, we say that $g(n)$ is an *(asymptotic) upper bound* for $f(n)$.

Clearly, $\mathcal{O}(g(n))$ is an approximation, or estimation, of $f(n)$. In practical terms, what we do is, given a polynomial $p(x)$ of degree n, to suppress the coefficient of the highest-order term in $p(x)$ and disregard all other terms to obtain $p(x) = \mathcal{O}(x^n)$.[40]

Example 2.3.7. Given the polynomial $3x^3 + 7x^2 + x + 36$, we have

$$3x^3 + 7x^2 + x + 36 = \mathcal{O}(x^3).$$

Example 2.3.8. Consider the language $L = \{0^m 1^m | m \geq 0\}$ and a single-tape Turing machine M (abbreviating M_T). We analyze the algorithm for deciding this language given the input string $|0^k 1^k| = n$.

1. M scans across the tape to verify that the input is of the form $0^* 1^*$. This takes n steps.[41] M then places the tape head again at the leftmost end of the tape. This, too, takes n steps, so that in total M takes $2n = \mathcal{O}(n)$ steps.

2. M repeatedly scans across the tape, crossing off a 0 and a 1 on each scan. Each scan uses $\mathcal{O}(n)$ steps. But M crosses off two symbols on each scan, so that actually there are at most $n/2$ scans. The total time for stage 2 is thus $\frac{n}{2}\mathcal{O}(n) = \mathcal{O}(n^2)$.

---

[40]More correctly, we have $p(x) \in \mathcal{O}(x^n)$.

[41]We simplify here, as M actually takes $n+1$ steps to reach the first blank symbol after the input string.

3. M scans the tape only once, in order to decide whether to accept or reject. This single scan takes at most $\mathcal{O}(n)$ steps.

Hence, the total time taken by M is

$$(\natural) \qquad \mathcal{O}(n) + \mathcal{O}(n^2) + \mathcal{O}(n) = \mathcal{O}(n^2).$$

In Example 2.3.8, we say that the Turing machine M has $\mathcal{O}(n^2)$-time complexity, or, more generally, M has polynomial-time complexity. We have the equality $\natural$, because we are concerned with upper bounds and $\mathcal{O}(n^2)$ *dominates* (over) $\mathcal{O}(n)$. In effect, for the standard functions of interest (see Table 2.3.2) we have, for $k, m \in \mathbb{N}$,

$$\mathcal{O}(\log n) \subset \mathcal{O}(n) \subset \mathcal{O}(n \log n) \subset \mathcal{O}\left(n^k\right) \subset \mathcal{O}(2^n)$$

and

$$\mathcal{O}\left(n^k\right) \subset \mathcal{O}\left(n^{k+1}\right) \subset \mathcal{O}\left(n^{k+2}\right) \subset \ldots \subset \mathcal{O}\left(n^{k+(m-1)}\right) \subset \mathcal{O}\left(n^{k+m}\right).$$

Thus, if given some function $f(n)$ we have $f = \mathcal{O}(n^2)$, then $f = \mathcal{O}(n^3)$, too, as $\mathcal{O}(n^3)$ is still an upper bound on f, but $f \neq \mathcal{O}(n)$.

Definition. 2.3.15. A Turing machine M has *polynomial-time complexity* if there exists a polynomial function $p(n) = \mathcal{O}(n^k)$, $k \in \mathbb{N}$, such that $T(x) \leq p(n)$ for $n = 0, 1, \ldots$. M has *polynomial-space complexity* if there is a polynomial function $p(n) = \mathcal{O}(n^k)$, $k \in \mathbb{N}$, such that $S(x) \leq p(n)$ for $n = 0, 1, \ldots$. M has *exponential-time complexity* if there is an exponential function $ex(n) = \mathcal{O}\left(2^{n^k}\right)$, $k \in \mathbb{N}$, such that $T(x) \leq ex(n)$ for $n = 0, 1, \ldots$. M has *exponential-space complexity* if there is an exponential function $ex(n) = \mathcal{O}\left(2^{n^k}\right)$, $k \in \mathbb{N}$, such that $S(x) \leq ex(n)$ for $n = 0, 1, \ldots$.

When we speak of computational complexity, we are actually referring to rates of growth of a function given an input n. To really appreciate the difference between a polynomial and an exponential function in terms of resources expenditure or requirements, check Table 2.3.2. Many interesting computational problems can be solved in polynomial time. Exponential-time algorithms are typically associated to *brute-force search*, i.e. an exhaustive search of the solution space, but this can often be replaced by an algorithm running in polynomial time. However, for many problems there has been found as yet no polynomial-time algorithm, whether because the problem has not been fully grasped or for

n	$\log n$	n	$n \log n$	n^2	n^3	2^n
5	3	5	15	25	125	32
10	4	10	40	100	10^3	10^3
100	7	100	700	10^4	10^6	10^{30}
1000	10	1000	10^4	10^6	10^9	10^{300}

Table 2.3.2.: Rates of growth of some standard functions.

other reasons. Therefore, familiarity with the classes of both polynomial-and exponential-time running algorithms is essential.

Definition. 2.3.16. We now define the complexity classes with respect to time and space complexity:[42]

1. The complexity class **P** contains all languages that are decidable by a DTM with polynomial-time complexity, i.e.

$$\mathbf{P} = \bigcup_{k \in \mathbb{N}} \mathrm{DTIME}\left(n^k\right).$$

2. The complexity class **PSPACE** contains all languages that are decidable by a DTM with polynomial-space complexity, i.e.

$$\mathbf{PSPACE} = \bigcup_{k \in \mathbb{N}} \mathrm{DSPACE}\left(n^k\right).$$

3. The complexity class **NP** contains all languages that are decidable by a NDTM with polynomial-time complexity, i.e.

$$\mathbf{NP} = \bigcup_{k \in \mathbb{N}} \mathrm{NTIME}\left(n^k\right).$$

4. The complexity class **NPSPACE** contains all languages that are decidable by a NDTM with polynomial-space complexity, i.e.

$$\mathbf{NPSPACE} = \bigcup_{k \in \mathbb{N}} \mathrm{NSPACE}\left(n^k\right).$$

---

[42] Although we denote classes elsewhere in this book by means of formal script font (e.g., $\mathscr{C}$, $\mathscr{L}$), it is usual to denote the complexity classes by bold fonts. We keep to this convention.

5. The complexity class **EXPTIME** contains all languages that are decidable by a DTM with exponential-time complexity, i.e.

$$\textbf{EXPTIME} = \bigcup_{k \in \mathbb{N}} \text{DTIME}\left(2^{n^k}\right).$$

6. The complexity class **EXPSPACE** contains all the languages that are decidable by a DTM with exponential-space complexity, i.e.

$$\textbf{EXPSPACE} = \bigcup_{k \in \mathbb{N}} \text{DSPACE}\left(2^{n^k}\right).$$

7. The complexity class **NEXPTIME** contains all the languages that are decidable by a NDTM with exponential-time complexity, i.e.

$$\textbf{NEXPTIME} = \bigcup_{k \in \mathbb{N}} \text{NTIME}\left(2^{n^k}\right).$$

8. The complexity class **NEXPSPACE** contains all the languages that are decidable by a NDTM with exponential-space complexity, i.e.

$$\textbf{NEXPSPACE} = \bigcup_{k \in \mathbb{N}} \text{NSPACE}\left(2^{n^k}\right).$$

Given this classification, we say that a computational problem is in a specific complexity class if, given an appropriate encoding of instances of the problem, the encoded language of Yes-instances–i.e., encoded instances for which an algorithm halts and outputs "Yes"–is a member of this class.

With respect to Definition 2.3.16, we have the following important remarks:

Remark. **2.3.17.** In fact,

$$\textbf{PSPACE} = \textbf{NPSPACE}$$

by *Savitch's theorem*, which states that for any function $f : \mathbb{N} \longrightarrow \mathbb{R}^+$ with $f(n) \geq n$,

$$\text{NSPACE}\left(f(n)\right) \subseteq \text{DSPACE}\left(f^2(n)\right).$$

In other words, a DTM can simulate a NDTM without requiring much more space. Thus we also have

$$\textbf{EXPSPACE} = \textbf{NEXPSPACE}.$$

Remark. **2.3.18.** The question whether the complexity class **P** equals the complexity class **NP**, written $\mathbf{P} \overset{?}{=} \mathbf{NP}$, is considered to be the most important open problem in theoretical computer science. (We shall have more to say on this.)

Remark. **2.3.19.** As seen above (Table 2.3.2), there are standard functions whose rate of growth is sublinear, i.e. far slower compared to polynomial or exponential growth rates. These functions can also be associated to tape space (but not moves[43]) of Turing machines, so that we have the following additional complexity classes:

1. The complexity class **L** contains all the languages that are decidable by a DTM with logarithmic-space complexity, i.e.

$$\mathbf{L} = \mathrm{DSPACE}\,(\log n)\,.$$

2. The complexity class **NL** contains all the languages that are decidable by a NDTM with logarithmic-space complexity, i.e.

$$\mathbf{NL} = \mathrm{NSPACE}\,(\log n)\,.$$

Remark. **2.3.20.** Just as in the case of the **P** and **NP** complexity classes, we have the open problem $\mathbf{L} \overset{?}{=} \mathbf{NL}$. As a matter of fact, we have the hierarchy of complexity classes shown in Figure 2.3.3, which depicts not only the relative sizes of each class, but also the open problems in the theory of complexity. Additionally, this Figure shows the tractability status of the computational problems of the different complexity classes.

Definition. 2.3.21. A computational problem is said to be *tractable* if it can be solved in practice. Otherwise, it is said to be *intractable*.

This is a rather vague definition, but the topic of computational tractability is a field of open questions. Nevertheless, some basic "guidelines" can be given.

Because in computational logic time is the resource most often considered we concentrate now on the classes **P** and **NP**. Note that these two classes are directly associated with tractable and intractable problems, respectively (see Fig. 2.3.3), as we can say that a tractable problem is one that can be solved in *polynomial time*, hence a problem in class **P**;

---

[43]Because a sublinear bound does not allow a Turing machine to read the input. This poses problems for space resources, too, but these can be remediated by considering two tapes, a read-only input tape and another read-and-write tape, with only the latter contributing to the space complexity of the machine.

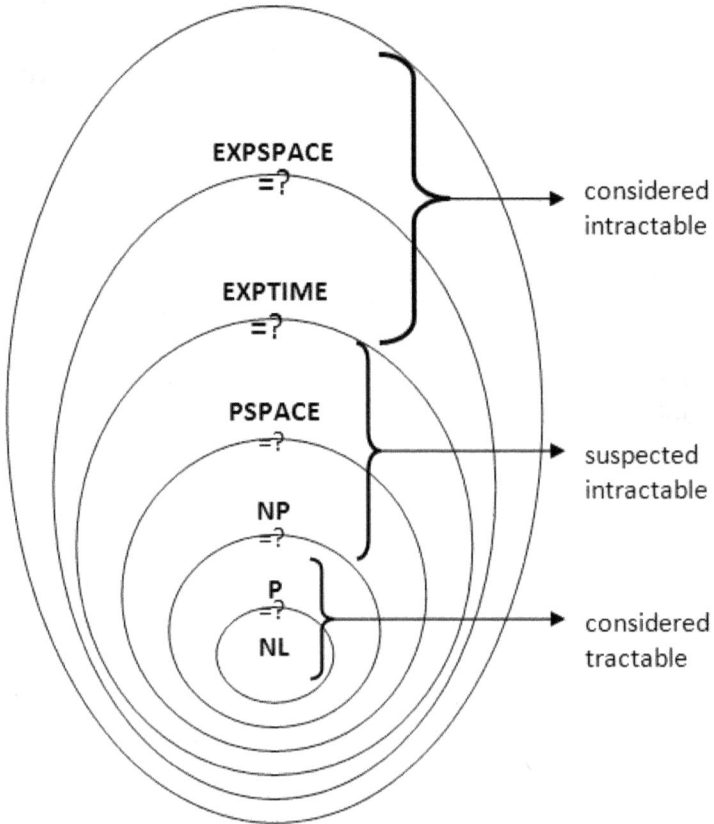

Figure 2.3.3.: The hierarchy of complexity classes with corresponding tractability status.

but also in **NP**, which actually stands for *non-deterministic polynomial time*, as this class corresponds to the problems that can be decided by a NDTM in polynomial time, *if* **P** = **NP**. An intractable problem is one that requires super-polynomial time to be solved. This distinction between tractable = polynomial-time and intractable =super-polynomial-time is known as the *Cook-Karp Thesis*. But if **P** ≠ **NP**, then **NP**-*hard* problems are intractable. This introduces some further fundamental specifications to the classes **P** and **NP**, of which we begin by giving notable examples.

Example 2.3.9. The following are (instances of) problems that are in the class **P**:

The 2-SAT Problem – Given a formula ϕ that is a conjunction of disjunctions of exactly two variables, is ϕ satisfiable? (See Section 7.3.1)

The Path Problem – Given a digraph $\vec{\mathfrak{G}}$, is there a directed path in $\vec{\mathfrak{G}}$ from vertices u to v?

The Shortest Path Problem – Given a graph $\mathfrak{G}$, is there a path in $\mathfrak{G}$ from vertices u and v that uses the fewest edges?

The Relative Primes Problem – We say that two numbers a and b are relative primes if $\gcd(a,b) = 1$. Given two numbers x and y, are they relative primes?

As seen above (cf. Section 2.3.2), once we have determined that a decision problem is in a certain class, we can reduce other problems to this one, so that they are also in this class. This is particularly relevant when we are concerned only with time complexity, as we are then most often faced with problems that are either in **P** or in **NP**. Recall that, in general terms, if a problem A is reducible to a problem B, we write "$A \preceq B$" and read "A is no harder than B" or "B is at least as hard as A." We can actually further specify this "hardness reduction" as polynomial-time reducibility as follows:

Definition. 2.3.22. *Polynomial-time reducibility* – Let A and B be two decision problems and let function f be defined as in Definition 2.3.7. To the condition f_{red} stated in Definition 2.3.7, add the condition

(f_P) f can be computed in polynomial time.

Then, if conditions f_{red} and f_P are both satisfied, we say that A can be *polynomial-time reducible* to B and we write "$A \preceq_P B$".

We state the following theorem in terms of languages for the sake of generality.

Theorem 2.3.7. *Let L_1 and L_2 be languages. If $L_1 \preceq_P L_2$ and $L_2 \in \mathbf{P}$, then $L_1 \in \mathbf{P}$.*

Proof: Let f be the polynomial-time reduction from L_1 to L_2, so that we have $L_1 \preceq_P L_2$. Then, there is a polynomial-time algorithm M_1 deciding L_1 such that, given a polynomial-time algorithm M_2 deciding L_2, on a given input w M_1 (i) computes $f(w)$, and (ii) runs M_2 on input $f(w)$ and outputs the same as M_2. Hence, by condition f_{red} in Definition 2.3.7, M_2 accepts $f(w)$ whenever $w \in L_1$. Because M_2 computes $f(w)$ in polynomial time, $|f(w)|$ is bounded by a polynomial function of $|w|$, and $L_2 \in \mathbf{P}$. Moreover, stages (i) and (ii) constitute the polynomial-time algorithm M_1 that requires the number of steps to compute $f(w)$ plus the number of steps needed by M_2 on the input $f(w)$, so that in fact

M_1 is a composition of polynomials, and a composition of polynomials is also a polynomial, and we have $L_1 \in \mathbf{P}$. **QED**

Definition. 2.3.23. A language L is said to be **NP**-*complete*, or a member of the class **NPC** if (i) $L \in \mathbf{NP}$, and (ii) for every $L_i \in \mathbf{NP}$ it is the case that $L_i \preceq_P L$.

Informally, a problem is **NP**-complete if it is one of the hardest **NP** problems.

Example 2.3.10. The following are (instances of) problems in the class **NPC**:[44]

The Satisfiability Problem (SAT) – An important decision problem in computational logic. See Section 7.1 for a definition and discussion of the SAT.[45]

The k-SAT problem – Given a formula ϕ that is a conjunction of disjunctions of $k \geq 3$ variables, is ϕ satisfiable? (See Section 7.3.1)

The Circuit Satisfiability Problem – Given a Boolean combinational circuit C, is C satisfiable, i.e. is there an assignment of values 0 and 1 to the variables such that the output of C is 1? (see Exercise 7.1.4)

The Hamiltonian Path Problem – See Example 2.3.5 above.

The Subgraph Isomorphism Problem – Given two graphs $\mathfrak{G}_1$ and $\mathfrak{G}_2$, is $\mathfrak{G}_1$ isomorphic to a subgraph of $\mathfrak{G}_2$?

The Vertex Cover Problem – Given an undirected graph $\mathfrak{G} = (V, E)$, a vertex cover is a set $V' \subseteq V$ such that if $(u, v) \in E$, then $u \in V'$ or $v \in V'$, or both. The size of a vertex cover is the number of vertices that are its members. We want to find a vertex cover of (minimum) size k in $\mathfrak{G}$.

The Graph Colorability Problem – Given an undirected graph $\mathfrak{G}$, what is the minimum number k of colors required to color it in a way that no adjacent vertices have the same color?

The Hamiltonian Cycle Problem – Given a graph $\mathfrak{G} = (V, E)$, is there a Hamiltonian cycle–i.e. a simple cycle that contains each $v_i \in V$–in $\mathfrak{G}$?

The Clique Problem – Given an undirected graph $\mathfrak{G}$, is there a clique (i.e. a complete subgraph) of $\mathfrak{G}$ of size k?

---

[44]See Garey & Johnson (1979 or later editions) for an extensive listing and account of problems in the class **NPC**.

[45]This was actually the first problem found to be **NP**-complete, being as such the one problem to which all **NP** problems can be reduced. Because this problem is at the very core of computational logic, we elaborate on it in detail in Chapter 7, where we also approach the reductions CIRCUIT-SAT$\preceq_P$SAT$\preceq_P$3-SAT (see Fig. 2.3.4).

The Subset-Sum Problem – Given a set $S \subseteq \mathbb{N}$, we wish to know whether there is a subset S' whose elements sum to $t \in \mathbb{N}$ (t is called a *target*).

The Traveling Salesman Problem (TSP) – Is there a route of length at most k passing through n places?

The Graph Isomorphism Problem – Are two given graphs $\mathfrak{G}_1$ and $\mathfrak{G}_2$ isomorphic?

Theorem 2.3.8. *If $L_1 \in$ **NPC** and $L_1 \preceq_P L_2$, for $L_2 \in$ **NP**, then $L_2 \in$ **NPC**.*

Proof: If L_1 is **NP**-complete, then every other language $L \in$ **NP** is polynomial-time reducible to L_1. In turn, L_1 is polynomial-time reducible to L_2. Hence, by composition of polynomial-time reductions, every language in **NP** is polynomial-time reducible to L_2. **QED**

Theorem 2.3.8 entails that once we have found a **NP**-complete problem, then we can prove that other problems belong to the **NPC** class by means of polynomial-time reduction. In fact, we have one such problem, the Satisfiability Problem (or SAT):

Theorem 2.3.9. *(Cook-Levin theorem) SAT is **NP**-complete.*

Proof: We leave the proof of this theorem for Chapter 7.

Figure 2.3.4 shows some fundamental polynomial-time reductions. Notice the central position of SAT, but also the transitivity of polynomial-time reducibility, i.e. if $L_1 \preceq_P L_2$ and $L_2 \preceq_P L_3$, then $L_1 \preceq_P L_3$.

Example 2.3.11. We show in general traits how to reduce the Vertex Cover Problem to the Clique Problem. Let $(\mathfrak{G}, k)$ be an instance of the former, for $\mathfrak{G} = (V, E)$. We construct the complement of $\mathfrak{G}$, denoted by $\overline{\mathfrak{G}}$, such that $V_{\overline{\mathfrak{G}}} = V_{\mathfrak{G}}$, but the edge $(u, v) \in \overline{\mathfrak{G}}$ iff $(u, v) \notin \mathfrak{G}$. Where k is the integer parameter for the Vertex Cover Problem, we now define the integer parameter for the Clique Problem as $n - k$. Then–and the reduction consists in this–, $\overline{\mathfrak{G}}$ has a clique of size at least $n - k$ iff $\mathfrak{G}$ has a vertex cover of size at most k. Moreover, the construction for the Clique problem runs in polynomial time, so that this is a polynomial-time reduction.

Theorem 2.3.10. *If L is a **NP**-complete language and $L \in$ **P**, then **P** = **NP**.*

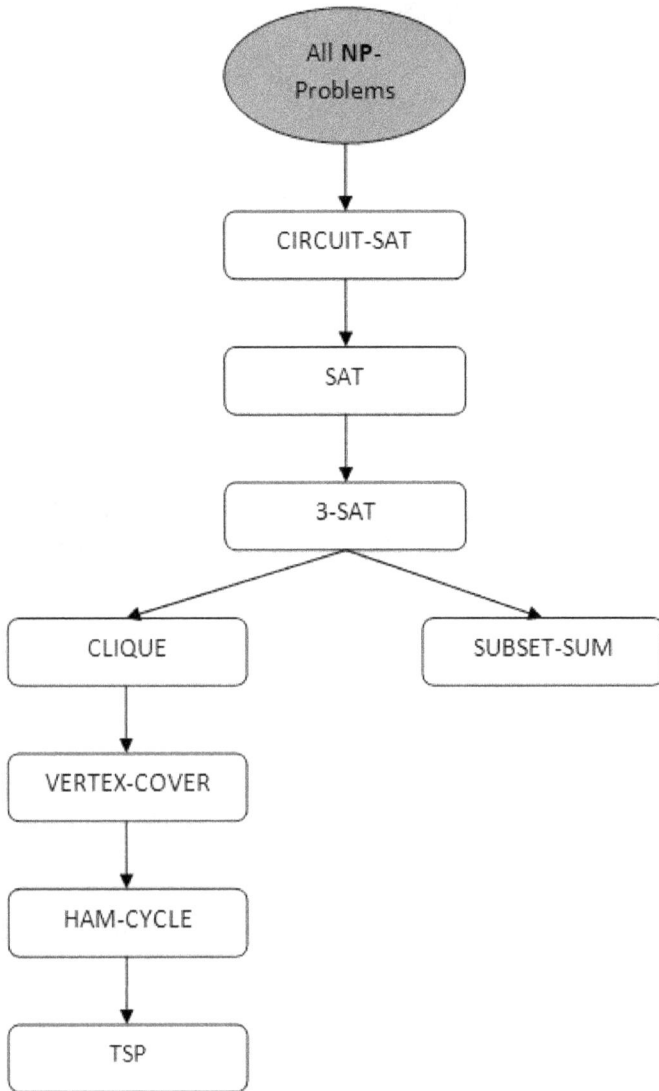

Figure 2.3.4.: Typical structure of **NP**-completeness proofs by polynomial-time reductions.

Proof: Follows directly from the definition of polynomial-time reducibility. **QED**

Definition. 2.3.24. A language L is said to be **NP**-*hard* if for every $L_i \in \textbf{NP}$ we have $L_i \preceq_P L$, but L need not be a member of the class **NP**.

Informally, a problem is **NP**-hard if it is at least as hard as all the **NP** problems.

Example 2.3.12. The following are **NP**-hard problems:
The Halting Problem (Not **NP**-complete; see Example 2.3.3.)
The Circuit Satisfiability Problem (**NP**-complete; cf. Exercise 7.1.4)
The Clique Problem (**NP**-complete; see Example 2.3.10)
The Vertex Cover Problem (**NP**-complete; see Example 2.3.10.)
The Traveling Salesman Problem (**NP**-complete; see Example 2.3.10.)
The Subset-Sum Problem – (**NP**-complete; see Example 2.3.10.)

Above we remarked that if $\textbf{P} \neq \textbf{NP}$, then **NP**-*hard* problems are intractable. Indeed, if $\textbf{P} \neq \textbf{NP}$, we actually have

$$\textbf{NPC} = \textbf{NP} \cap \textbf{NP}\text{-hard}$$

and

$$\textbf{P} \cap \textbf{NPC} = \emptyset.$$

On the other hand, if $\textbf{P} = \textbf{NP}$, then

$$(\textbf{NPC} = \textbf{NP} = \textbf{P}) \subset \textbf{NP}\text{-hard}.$$

The real impact of $\textbf{P} \overset{?}{=} \textbf{NP}$ becomes clearer when we define **P** as the class of problems that can be *solved* in polynomial time (i.e. quickly), and **NP** as the class of problems for which a solution, if one has been found, can be *verified* in polynomial time.[46] The crux posed by problems

---

[46]For which we need a *verifier*, an algorithm Ψ that uses additional information–a *certificate*, or *proof*, of membership, denoted by, say, c–so that a language L is defined as

$$L = \{w | \Psi \text{ accepts } z = e\,(wc) = \langle w, c \rangle \text{ for some string } c\}.$$

A verifier Ψ is a *polynomial-time verifier* if there is a polynomial $p\,(n)$ such that the number of moves of Ψ is no greater than $p\,(|n|)$, where $n = |w|$. For instance, suppose that we have the following sequence $c = \langle v_1, v_2, ..., v_n \rangle$ of $|V| = n$ vertices that has been given as a solution to an instance of the Hamiltonian Cycle Problem. Then, one can verify in polynomial time that $(v_i, v_{i+1}) \in E$ for $i = 1, 2, ..., n-1$ and also that $(v_n, v_1) \in E$.

in **NP** is that there is no known efficient way to find a solution for them in the first place.

However, polynomial time other than $\mathcal{O}(n^2)$, $\mathcal{O}(n^3)$, and $\mathcal{O}(n^4)$ is not necessarily synonymous with *quickly*, or even efficiently computable at all. For instance, an input as low as $n = 100$ will be uncomputable in practice if raised to the power of, say, 15. On the other hand, exponential time algorithms are efficient for very low inputs.

Exercises

Exercise 2.3.4.1. Show that DTIME, DSPACE, NTIME, NSPACE all satisfy the Blum axioms.

Exercise 2.3.4.2. Show that $p(n) = \mathcal{O}(n^k)$ for $p(n) = a_0 + a_1 n + a_2 n^2 + ... + a_k n^k$.

Exercise 2.3.4.3. With respect to Example 2.3.8:

1. Find a Turing machine with time complexity lower than $\mathcal{O}(n^2)$.

2. Show that no single-tape Turing machine can decide L more quickly than in time $\mathcal{O}(n \log n)$.

Exercise 2.3.4.4. Prove the following theorem:

Theorem 2.3.11. *For $T(n) \geq n$, every $T(n)$ time multi-tape Turing machine has an equivalent $\mathcal{O}(T^2(n))$ time single-tape Turing machine.*

Exercise 2.3.4.5. Prove the following theorem:

Theorem 2.3.12. *For $T(n) \geq n$, every $T(n)$ time non-deterministic single-tape Turing machine has an equivalent $2^{\mathcal{O}(T(n))}$ time deterministic single-tape Turing machine.*

Exercise 2.3.4.6. Formulate the problems of Examples 2.3.9 and 2.3.10 as languages. Explain in a detailed way why these problems are in the classes **P**, **NP**, or **NPC**.

Exercise 2.3.4.7. SAT can be solved with a linear-space algorithm, but it cannot even be solved by a polynomial-time algorithm, as it is a **NP**-complete problem. Provide an account of this by appealing to the Turing machine.

Exercise 2.3.4.8. Give an informal account of the following facts:

1. **P⊆PSPACE**

2. **NP⊆NPSPACE**

3. **NP⊆PSPACE**

Exercise 2.3.4.9. Show that if $L \in \mathbf{P}$, L is a language, then $\overline{L} \in \mathbf{P}$.

Exercise 2.3.4.10. Show that the relation $\preceq_P$ on the set of languages is transitive.

Exercise 2.3.4.11. Devise an algorithm to solve the Clique Problem (cf. Example 2.3.10)

Exercise 2.3.4.12. Show that **NP** is closed under union, intersection, concatenation, and Kleene star.

Exercise 2.3.4.13. Show that the Vertex Cover Problem is **NP**-hard.

Exercise 2.3.4.14. Show, by giving an example, that the Vertex Cover Problem can be reduced to the Subset-Sum Problem.

Exercise 2.3.4.15. Give the complete proof of Theorem 2.3.10.

2.3.5. The Chomsky hierarchy (IV)

Above in this Section 2.3, we concentrated our discussion mostly on recursive languages and RELs. We now revisit the Chomsky hierarchy and provide complexity results for the other languages in this hierarchy. Note, however, that these results are not yet well understood, not the least because the complexity classes are themselves far from clearly grasped. For instance, when we consider the family of classes DTIME (n^k), $k = 1, 2, ...$, we are actually considering an infinite number of properly nested complexity classes.

Theorem 2.3.13. *Every regular language is a member of DTIME(n).*

Proof: Left as an exercise.

Theorem 2.3.14. *Every regular language is a member of* **P**.

Proof: Left as an exercise.

Theorem 2.3.15. *Every CFL is a member of* **P**.

Proof: (Sketch) Let a CFL L be generated by a CFG G in Chomsky normal form. Given a string w, in order to test membership of w to $L(G)$ we consider whether every variable in G generates a substring of w. We apply the *CYK algorithm*, which consists in filling in a triangular table with ij entries such that only the cells for $i \leq j$ are filled in with the collection of variables generating the substrings. Given a string $|w| = n$ such that $w = w_1w_2...w_n$, we begin at the bottom of the table for substrings of length $\ell = 1$, next move one row up for substrings of length $\ell = 2$, and so on up to $\ell = n$. For example, for a string $|w| = 5$, we have the following table:

$V_{1,5}$				
$V_{1,4}$	$V_{2,5}$			
$V_{1,3}$	$V_{2,4}$	$V_{3,5}$		
$V_{1,2}$	$V_{2,3}$	$V_{3,4}$	$V_{4,5}$	
$V_{1,1}$	$V_{2,2}$	$V_{3,3}$	$V_{4,4}$	$V_{5,5}$
w_1	w_2	w_3	w_4	w_5

Let us now denote by $V_{i,i}$ the set of variables $V = \{S, A, B, ...\}$ such that $S, A, B, ... \to w_i$ is a production rule of G. Then, the algorithm consists in comparing at most n pairs of previously computed sets

$$(V_{i,i}, V_{i+1,j}), (V_{i,i+1}, V_{i+2,j}), ..., (V_{i,j-1}, V_{j,j}).$$

The algorithm terminates in an accepting state iff $S \in V_{1,n}$; otherwise, the algorithm ends in a rejecting state. It can be shown that the running time by a Turing machine simulation of the CYK algorithm is $\mathcal{O}(n^3)$.[47]
QED

Example 2.3.13. The CFG of Example 2.1.12 can be converted into a CFG in ChNF with the following production set:

$$P = \left\{ \begin{array}{c} S \to CF \\ A \to BE \\ B \to DB \,|\, c \\ C \to a \\ D \to b \\ E \to DC \\ F \to AB \end{array} \right\}$$

For string *acbabc*, we have the following table for its substrings:

---

[47]See, e.g., Younger (1967). D. Younger is one of the three inventors of this algorithm, the other two being J. Cocke and T. Kasami.

acbabc					
acbab	cbabc				
acba	cbab	babc			
acb	cba	bab	abc		
ac	cb	ba	ab	bc	
a	c	b	a	b	c

Application of the CYK algorithm gives us the following table, where empty cells should be considered as containing ∅:

{S}					
	{F}				
	{A}				
	{E}		{B}		
{C}	{B}	{D}	{C}	{D}	{B}
a	c	b	a	b	c

It can be easily verified that the algorithm terminates in an accepting state.

Exercises

Exercise 2.3.5.1. Check the acceptance status of the given strings by means of the CYK algorithm. (Note: Conversion of the given grammars into ChNF is required first.)

1. String *aaabbb* (CFG G of Example 2.1.11).

2. String *abab* (CFG G of Example 2.1.11).

3. String *aaabc* (CFG G of Example 2.2.10).

Exercise 2.3.5.2. Prove (Complete the proof of) the theorems in this Section that were left as an exercise (with a sketchy proof, respectively).

Part IV.

Classical Logic and Classical Deduction

3. Preliminaries: Formal logic, deduction, and deductive computation

In the field of logic, it is important to distinguish an *object language*, i.e. a logical language within which logical objects such as formulae and arguments are proved and/or interpreted, from a *metalanguage*, the language in which the study of an object language is conducted. In this book, the main object language is the first-order language of classical logic, and the metalanguage is English supplemented with logical jargon (e.g., operator, quantifier) and symbols that are not part of the object languages (e.g., $\vdash$, $\Rightarrow$). Given an object language F, μ^{F} denotes the metalanguage of F. In the metalanguage here adopted, arbitrary formulae are denoted by the Greek lowercase letters $\phi, \chi, \psi, ...$, with or without subscripts, and arbitrary sets of formulae are denoted by Greek uppercase letters (e.g., Φ, X, Ψ). This convention notwithstanding, where appropriate we shall also denote formulae in μ^{L1} by the Roman letters $A, B, ..., P, Q, R$, mostly. This practice has the pedagogical advantage of bridging in a more clear way the object language and its corresponding metalanguage.

Given an object language, two aspects are of major importance: *form* and *meaning*. By *logical form*, we mean both what makes a logical expression *well-formed* (e.g., a well-formed formula) and what constitutes a *formal proof*, i.e. a proof that well-formed expressions follow logically from other well-formed expressions by virtue of form alone. Thus, logical form features as the central object of two components of logic related to *syntax*, to wit, the *grammar* of a logical language, and *proof theory*, the study of formal derivations and proofs, and of the diverse proof systems.

This tells us that logical statements and reasoning can be approached from a purely formal perspective, but in fact we more often than not also care for *meaning* in a logical language. This is typically provided by an *interpretation*, i.e. an assignment of meaning to the symbols and expressions of a logical language. This assignment of meaning is essentially an attribution of a distinguished element known as a *truth value* (i.e. a valuation) to a formula: we speak of a formula ϕ being true or false under an interpretation $\mathcal{I}$. This is the fundamental notion of *model theory*, the study of the interpretation of logical languages by means of set-theoretical structures.

Given the two aforementioned components related to form, below we approach logical form firstly from the viewpoint of what makes a formal language a logical language, and then introduce some basic aspects of proof theory, namely arguments and argument form.[1]

3.1. Logical form I: Logical languages

3.1.1. Alphabets, expressions, and formulae logical

Above, in Section 2.1, we studied several formal languages. A *logical language* is a formal language that has the following specificity:

Definition. 3.1.1. A formal language L whose alphabet Σ, denoted by Σ_L, contains logical constants is a logical language.

We shall henceforth denote a logical language by using this font.

Definition. 3.1.2. For F a logical language, Σ_F consists of:

1. an infinite supply of symbols for *variables*, as well as possibly of symbols for *predicates* and *functions* of arity $n \geq 0$ (i.e. n-place arguments), and

2. a finite number of (symbols for) *logical constants*, namely *operators* $\heartsuit_1, ..., \heartsuit_r$ and *quantifiers* $\blacklozenge_1, ..., \blacklozenge_m$.

Furthermore, 0-place function symbols are called *(individual) constants*. *Brackets* are punctuation marks; in fact, in this text they are used mostly for readability. Typically, the variables and the *non-logical constants* (i.e. predicates and functions) of Definition 3.1.2.1 are symbols from the Latin alphabet, with or without subscripts (see below), and the logical constants are special conventional symbols.

---

[1] We remark that this is not a standard way to approach logical languages and formal proofs, as for many authors the expression "logical form" is to be used solely for the latter. We account for our discrepancy with respect to this practice by the explicit emphasis taken in this book that a logical language is first and foremost a formal language.

Example 3.1.1. Typically, arbitrary propositional variables are denoted by $p, q, r, s, ...$, arbitrary individual variables are so by $x, y, z, ...$, and arbitrary constants by $a, b, c, d, ...$; the symbols for arbitrary functions are $f, g, h, ...$, and typical symbols for arbitrary predicates are $P, Q, R, S, ...$ The symbols $\top$ and $\bot$, denoting truth and falsity or absurdity, respectively, are logical constants. The symbol $\neg$, denoting negation, is a logical constant. The symbols $\wedge, \vee, \rightarrow$, denoting the binary operators for conjunction, disjunction, and the conditional, respectively, are logical constants. The symbols $\forall$ and $\exists$, known as universal and existential quantifiers, respectively, are logical constants.

The following definitions frame in a most general way the technical study of the object languages to be approached in this text. Because we are concerned with computing with a logical language, we introduce straightaway the fundamental notion of *ground expression*. Recall Definition 2.1.5.

Definition. 3.1.3. For an alphabet Σ_F and a finite set of rules G_F of a logical language F, we define inductively the *expressions* of F as follows:

1. A *term* t is an expression built up from the *variables* and the n-ary *function*–and therefore *constant* (when $n = 0$)–*symbols* of Σ.

   a) A term that is not a variable or does not contain a variable is called a *ground term*.

2. An *atom* P is an expression of the form $P(t_0, t_1, ..., t_n)$, where P is a n-ary *predicate* symbol for $n \geq 0$ and $t_0, t_1, ..., t_n$, the *arguments* of P (denoted by $arg\,(P)$), are terms.

   a) An atom that contains only ground terms is a *ground atom*.

3. A *formula* A is an atom or a composition of atoms:

   a) An atom is an *(atomic) formula*.
   b) If $\heartsuit_i^n$ is a n-ary operator and $P_1, ..., P_n$ atoms, then the expression $\heartsuit_i\,(P_1, ..., P_n)$ is a *(compound) formula*.

191

c) $\blacklozenge_1 \otimes_1 ... \blacklozenge_n \otimes_n (A)$, for $1 \leq i \leq n$, where $\otimes_i$ is an individual variable, a function symbol, or a predicate symbol, and $\blacklozenge_i$ is a quantifier, is a *(quantified) formula*.

d) A quantifier-free formula whose atoms are all ground atoms is a *ground formula*.

From Example 3.1.1 and this definition, it should be clear that predicates with arity ≥ 1 are always written as uppercase letters. Propositional variables, i.e. 0-ary predicates, may be written indifferently as lower- or uppercase letters; in the first case, their characterization as constants is emphasized, in the second case we emphasize their characterization as atoms.

Some further fundamental specifications for formulae of a logical language F are given now.

Proposition. 3.1.4. Unicity of decomposition – *A formula can be formed in only one way, i.e. for each and every formula $\phi \in$ F one and only one of the following conditions holds:*

1. ϕ is an atom.

2. There is a *unique* formula ψ and a *unique* unary connective $\heartsuit_i^1$ such that $\phi = \heartsuit_i (\psi)$.

3. There is a *unique* pair of formulae (ψ, χ) and a *unique* binary connective $\heartsuit_j^2$ such that $\phi = \heartsuit_j (\psi, \chi)$.

4. There is a *unique* quantifier $\blacklozenge_i$, a *unique* symbol $\otimes$, and a *unique* formula χ such that $\phi = \blacklozenge_i \otimes (\chi)$.

With respect to Proposition 3.1.4.3, we remark that we shall write $\psi \heartsuit \chi$ instead of $\heartsuit (\psi, \chi)$. The former is known as *infix notation*.

Definition. 3.1.5. Given formulae $\phi, \psi, \chi \in$ F, some unary connective $\heartsuit_i^1$, and some binary connective $\heartsuit_j^2$, an *immediate sub-formula* is defined as follows:

1. Atomic formulae have no immediate sub-formulae.

2. $\phi = \heartsuit_i (\psi)$ has as immediate sub-formula only ψ.

3. The formula $\phi = \psi \heartsuit_j \chi$ has as immediate sub-formulae only ψ and χ.

4. $\phi = \blacklozenge_i \otimes (\chi)$ has as immediate subformula only χ.

Exercises

Exercise 3.1.1.1. Let there be given a set $O = \{\neg^1, \wedge^2, \vee^2, \rightarrow^2\}$ of logical connectives with superscripts indicating arity and let there be given the set of quantifiers $Q = \{\forall, \exists\}$. Identify which of the following expressions are terms, atoms, formulae, or none of these:

1. b

2. $R(x, y)$

3. Q

4. $\neg x$

5. $P \vee f(x)$

6. $\exists(P(z))$

7. $g(a, b)$

8. $\forall Q (Q(P))$

9. f

10. $h(a, f(a))$

11. $\forall z \forall w (S(z) \rightarrow T(w))$

12. $A \wedge \neg (B \vee C)$

Exercise 3.1.1.2. Let the sets O and Q be as in Exercise 3.1.1.1 above. Identify (1) the immediate sub-formulae and (2) all the other sub-formulae of the following formulae:

1. $\neg (P \wedge \neg Q)$

2. $R(x) \rightarrow (P(x) \vee (Q(a) \wedge S(b)))$

3. $\neg S \vee \neg (T \rightarrow \neg R)$

4. $P(y) \vee (\neg R(x) \vee R(x))$

5. $(R \vee \neg (P \rightarrow R)) \wedge (Q \wedge (S \rightarrow \neg T))$

6. $\exists P (P(a, b) \rightarrow \forall Q (Q(b, a)))$

7. $\forall x \exists y \, (P \, (x, y) \wedge \forall z \, (R \, (z, a)))$

8. $\forall f \exists x \, (S \, (f \, (x)))$

3.1.2. Orders

A logical language *has order n*, or *is of order n*, according to the following definitions.

Definition. 3.1.6. The *order of a formula* is the highest order of any of its predicates and quantifiers:

1. A predicate has order 1 if all its arguments are terms; otherwise, it has order $n + 1$, for n the highest order of its argument that is not a term.

2. A quantifier has order 1 if it quantifies an individual variable; otherwise, it has order $n + 1$, for n the order of the predicate (or function) quantified.

Predicates and functions, rather than terms, can be considered as *sets*: for instance, $P \, (x) = P = \{x | x \in P\}$, and $f = \{(x, y) \, | x, y \in \mathbb{N}\}$ if f is the function $f \, (x) = y$ for all $x, y \in \mathbb{N}$. By writing the latter as $f \, (x, y)$, we are considering the function f as a predicate, and thus we have $f \, (x, y) = \{(x, y) \, | x, y \in f\}$. This is an important note, as a function seen as a predicate takes only terms for arguments, and thus any function name (e.g., f) is always of order 1.

Example 3.1.2. The predicates $P \, (a)$ and $R \, (f \, (x))$ are of order 1, as both a and $f \, (x)$ are terms. $Q \, (P)$ and $S \, (g)$ are of order 2, as P is a predicate and g is a function. The formula $P \, (Q) \wedge Q \, (S)$ is of order 3, because of the nesting $S \in Q, Q \in P$, i.e. $Q \, (S)$ is of order 2 and $P \, (Q)$ is of order 3. The quantified formulae $\forall x \forall y \, (P \, (x) \to Q \, (y))$ and $\forall x \exists y \exists z \, (R \, (x, y, z))$ are of order 1. $\exists f \forall x \, (P \, (f \, (x)))$ and $\forall y \exists Q \, (Q \, (y))$ are of order 2: in the first case, though $\forall x$ and $P \, (f \, (x))$ are of order 1, $\exists f$ is of order 2; with respect to the second, $\forall y$ and $Q \, (y)$ are of order 1 but $\exists Q$ is of order 2. $\forall x \exists Q \exists S \, (Q \, (x) \to S \, (Q))$ is of order 3, as $\forall x \, (Q \, (x))$ is of order 1, $\exists Q \, (Q \, (x))$ is of order 2, and $\exists S \, (S \, (Q))$ is of order 3.

Definition. 3.1.7. A *n-th order logical language* is a logical language whose formulae have order n or less.

We speak of a zeroth-order logical language when $n = 0$, of a first-order (second-order) logic for $n = 1$ ($n = 2$, respectively), and of a higher-order logic for $n > 2$. When the order is 0, i.e. when there are only 0-ary predicates (i.e. propositional variables), we more commonly speak of a *propositional language*; when the order is 1, we have a *first-order* (abbreviated: *FO*) *predicate language* (often just *FO language*).

We begin by specifying a propositional language F0, and then augment it in order to obtain a FO language F1. For convenience, we do not consider distinct elements for languages of second or higher order, as our object-languages will be essentially of zeroth or first order.

Definition. 3.1.8. A *propositional language* is a pair $\mathsf{F0} = (V, O)$ where $V = \{p, q, r, p_1, q_1, r_1, ...\}$ is a denumerable set of *propositional variables* and $O = \{\heartsuit_1^n, ..., \heartsuit_r^n\}$ is a finite set of r operators, called *connectives*, with arity $n \geq 1$ for finite n.[2]

We can also define a propositional language as the pair $\mathsf{F} = (F_{\mathsf{F0}}, O)$, where F_{F0} is the set of formulae of the propositional language F0 (often simply F if F0 is understood). In very general terms, a language F can be identified with its set of formulae F_{F}, because $F_{\mathsf{F}} \subseteq \mathsf{F}$.

Definition. 3.1.9. A *well-formed formula* of $F \subseteq \mathsf{F0}$ is defined recursively as follows:

1. Every propositional atom is a well-formed formula, i.e., $V \subseteq F$;

2. A formula ϕ is a well-formed formula iff it has a finite number k of propositional variables (or atoms), i.e. iff $\phi = \phi(p_1, ..., p_k)$;

3. If $\phi_1, ..., \phi_n$ are well-formed formulae and $\heartsuit_i^n$ is a logical connective with arity n, then $\heartsuit_i(\phi_1, ..., \phi_n)$ is a well-formed formula.

4. Given a well-formed formula $\phi(p_1, ..., p_k)$ and a substitution σ, the formula $\sigma\phi = \phi(p_1/\sigma p_1, ..., p_k/\sigma p_k)$ obtained from ϕ by simultaneously substituting $p_1, ..., p_k$ by the propositional variables $\sigma p_1, ..., \sigma p_k$, is a well-formed formula.

---

[2] Actually, we often have $n = 0$, for example when we consider $\top$ and $\bot$ as operators rather than propositional constants.

We shall use the term *formula* to abbreviate "well-formed formula," as nothing else is here a formula. Definition 3.1.9.1 defines an *atomic propositional formula*; Definition 3.1.9.2-3 defines a *complex*, or *compound*, propositional *formula*. Definition 3.1.9.4 defines a propositional formula as *invariant under variable substitutions*. This property can be extended to *formula substitutions*.

Definition. 3.1.10. A *FO language* F1 is the language F0 augmented with the following countable sets:

1. $Q = \{\blacklozenge_1, ..., \blacklozenge_m\}$ of quantifiers.

2. $Vi = \{x, y, z, x_1, y_1, z_1, ...\}$ of nominal or individual variables.

3. $Cons = \{a, b, c, a_1, b_1, c_1, ...\}$ of individual constants.

4. $Pred = \{P, Q, R, P_1, Q_1, R_1, ...\}$ of predicates.

5. $Fun = \{f, g, h, f_1, g_1, h_1, ...\}$ of functions.

Definition. 3.1.11. Given the FO language F1, the triple

$$\Upsilon = (Pred_{F1}, Fun_{F1}, ar)$$

where $ar : (Pred_{F1} \cup Fun_{F1}) \longrightarrow \mathbb{N}$ is the arity of the predicate and function symbols of F1, is called a *signature (for* F1).[3]

Definition. 3.1.12. The set $F \subseteq$ F1 is defined as above for F0; atomic formulae are treated similarly as propositional variables and the following additional conditions hold:

1. A formula $\phi \in F$ is well-formed iff it has a finite number of terms, i.e. iff $\phi = \phi(t_0, t_1, ..., t_n)$;

2. If ϕ belongs to F and x is a variable, then $\blacklozenge_i x (\phi) \in F$;

3. Given a substitution σ, we have

   a) $\sigma c = c$, for $c \in Cons$;

---

[3] A 0-ary function symbol is considered a constant. As a matter of fact, an alternative equivalent definition of a signature for F1 is $\Upsilon_{F1} := Pred_{F1} \cup Fun_{F1} \cup Cons_{F1}$.

b) $\sigma\left(\diamond\left(t_1, ..., t_n\right)\right) = \diamond\left(\sigma t_1, ..., \sigma t_n\right)$, for $\diamond \in \left(\Upsilon - Cons\right)$.

The following definition synthesizes the above in Backus-Naur form (in which "::=" denotes left-to-right replacement):

Definition. 3.1.13. Over the signature $\Upsilon = (Pred_{\mathsf{F1}}, Fun_{\mathsf{F1}}, Cons_{\mathsf{F1}})$ for **F1** , the expressions of **F1** are inductively defined as follows:

Terms	t	::=	$x \mid a \mid f\left(t_1, ..., t_n\right)$
Atoms	$P\left(Q, ...\right)$	::=	$p \mid P\left(t_1, ..., t_n\right)$
Formulae	$A\left(B, ...\right)$	::=	$P \mid \heartsuit_i^1 A \mid A \heartsuit_i^2 B \mid \blacklozenge_i x\left(A\right)$

We consider here no higher arity than 2 for the connectives of a logical language. This is solely for the sake of simplicity, given that we shall be working in this text with no higher arity than 2 for connectives.

Definition. 3.1.14. In the formula $\psi = \blacklozenge_i x\left(\phi\right)$, $\left(\phi\right)$ is the *scope* of the quantifier $\blacklozenge_i$ and we say that $\blacklozenge_i$ *binds* (or *quantifies*) a variable x if x is in the scope of $\blacklozenge_i$. If $x \notin Vi\left(\phi\right)$, then ϕ is said to be *trivially quantified*.

1. Every occurrence of a variable x in the scope of a quantifier $\blacklozenge_i$ is said to be *bound*; *free* otherwise. A variable x may be both bound and free in ψ, denoted by $\psi' = \blacklozenge_i x \phi\left(x\right)$.

2. A formula is said to be *closed*, or a *sentence*, if it has no free variables; otherwise, it is *open*.

3. Suppose now that ψ is the formula with free variables $\phi\left(x_1, ..., x_n\right)$. Then:[4]

   a) the *universal closure* of ψ is the formula $\psi' = \forall x_1 ... \forall x_n\left(\phi\right)$;

   b) the *existential closure* of ψ is the formula $\psi' = \exists x_1 ... \exists x_n\left(\phi\right)$.

---

[4]We shall also often write $\psi' = \blacklozenge_i x_1 ... \blacklozenge_i x_n \phi\left(x_1, ..., x_n\right)$, namely to remove trivially quantified formulas from consideration. This does not clash with item 1, as every formula $\psi = \phi\left(x\right)$ can be turned into a formula $\psi' = \blacklozenge_i \phi\left(x\right)$, and vice-versa, as we shall see.

Example 3.1.3. The following are examples of binding and substituting in quantified formulae:

- The formula $\forall x\,(R\,(x,y))$ is open, as the variable y is not bound by any quantifier, but the formula $\forall x\,(R\,(x,a))$ is closed, as the only variable x in $R\,(x,a)$ is bound by $\forall$.

- The existential closure of $\forall x\,(R\,(x,y))$ is $\exists y\forall x\,(R\,(x,y))$.

- The universal closure of the formula $\exists z\,((S\,(x,z)\wedge R\,(y,z))\to Q\,(z))$ is $\forall x\forall y\exists z\,((S\,(x,z)\wedge R\,(y,z))\to Q\,(z))$.

- In the closed formula $\forall x\,(P\,(x)\wedge\exists x\,(Q\,(x)))$, the scope of $\forall$ is $P\,(x)\wedge\exists x\,(Q\,(x))$, but $\forall$ only binds the variable x in $P\,(x)$, as the variable x in $\exists x\,(Q\,(x))$ is already bound by $\exists$. In order to disambiguate it suffices to rename x, resulting in the formula $\forall y\,(P\,(y)\wedge\exists x\,(Q\,(x)))$. Another way of *variable renaming* is by subscripting, producing the formula $\forall x_1\,(P\,(x_1)\wedge\exists x_2\,(Q\,(x_2)))$.

Proposition. 3.1.15. *The following rewritings (denoted by $\Longrightarrow$ or $\Longleftarrow$) are permissible for the quantifiers $\blacklozenge_{i,j}=\forall,\exists$:*

1. *Quantifier reversal* $-\ \blacklozenge_i x\blacklozenge_j y\,(P\,(x,y))\ \overset{\Longrightarrow}{\underset{\Longleftarrow}{}}\ \blacklozenge_j y\blacklozenge_i x\,(P\,(x,y))$ if $i=j$.

2. *Negation distribution* $-\ \neg\blacklozenge_i x\,(P\,(x))\ \overset{\Longrightarrow}{\underset{\Longleftarrow}{}}\ \blacklozenge_j x\,(\neg\,(P\,(x)))$ if $i\neq j$.

3. *Existential distribution* $-\ \exists y\forall x\,(P\,(x,y))\Longrightarrow\forall x\exists y\,(P\,(x,y)).$[5]

Proof: Left as an exercise.

Exercises

Exercise 3.1.2.1. Determine the order of the following formulae:

---

[5]But the reverse rewriting is not permissible.

1. $\exists g \forall x \left(P \left(g \left(x \right) \right) \right)$

2. $\exists g \forall x \left(P \left(g \left(x \right), f \left(x \right) \right) \right)$

3. $\exists P \exists Q \forall y \left(P \left(y \right) \wedge \neg Q \left(y \right) \right)$

4. $R \left(f \right) \rightarrow Q \left(f \left(x \right) \right)$

Exercise 3.1.2.2. True or false: A logical language is of higher order if it allows sets as elements of other sets.

Exercise 3.1.2.3. Identify the closed and the open FO formulae. For the open formulae, give their universal closure.

1. $\forall x \exists y P \left(x, y \right)$

2. $\forall z \exists x \left(R \left(z \right) \rightarrow S \left(z, x \right) \right) \wedge T \left(z \right)$

3. $\exists z \left(P \left(x, y \right) \right) \vee \neg R \left(z \right)$

4. $\forall x \forall y \left(P \left(x \right) \rightarrow \neg Q \left(y \right) \right) \wedge \forall z R \left(z, x \right)$

Exercise 3.1.2.4. Prove Proposition 3.1.15.

3.1.3. Formalization

Although a natural language like English or Sioux is a formal language in a certain sense (see Section 2.1 above), its grammar and lexicon are too complex to allow an unequivocal interpretation of its utterances. This is especially so in the context of scientific discourse, and *very* especially so in mathematics, a field in which ambiguity is to be avoided at all costs. Mathematics, just like any other field of knowledge, uses a natural language to convey linguistically most of its facts and conjectures. But any terms in a natural language come with both a long history and regional, often strictly local, usages. For these reasons, it is often the case that students have to revise their vocabulary when they are introduced to mathematical terms such as *continuous*, *discrete*, *natural*, *real*, etc.

One of the main objectives in conceiving a formal language is that of *formalizing* natural language, i.e. giving an unequivocal form to its utterances. For instance, context-free languages (cf. Section 2.1.2) allow the rigorous analysis of English utterances by means of parsing trees. In the case of a logical language, there is more than this involved. In effect, we formalize utterances in a natural language by means of a logical language when we are interested in unequivocally analyzing some *reasoning*, such as a mathematical proof (see next Section). But we have to start by formalizing individual utterances.

In an important sense, by formalizing an utterance in a natural language via a logical language, we are abstracting it by means of its logical form. In particular, we typically only care to formalize *propositions* (also *assertions*, or *statements*) i.e. utterances that convey factual information;[6] we are commonly not interested in formalizing questions, orders, commands, interjections, or ejaculations.[7]

Technically, formalizing consists in identifying propositions in some natural language, identifying their simpler components (propositions or predicates), and assigning to each of these a propositional variable or a predicate/function of a formal language Fn, for $n \geq 0$. In the latter case, we very likely have to quantify the (sub-)formulae if there are individual variables involved.[8] This done, we decide whether the corresponding formulae are atomic or complex.

This task of formalizing assertions in a natural language may be facilitated by constructing the respective formula trees. From the above Sections 3.1.1-2, it should be evident that every logical formula can be read in a unique way. This *unique readability* can be represented by means of a *formula tree*, a down-growing labeled tree whose root is labeled with the main connective or the main quantifier and the leaves are labeled with propositional atoms or with atomic terms (see Example 3.1.4 below).

A propositional language, albeit *computationally advantageous*, as will be seen, is typically not *expressive* enough to formalize most interesting instances of reasoning, reason why one must know how to formalize statements in a FO language, or even in a language of higher order. For instance, any reasoning involving equality requires at least a first-order language to be formalized properly (see Figure 3.1.2).

The expressiveness of a FO or higher-order language resides to a large extent in its allowing for complex statements about both specific and

---

[6]Below we shall specify that these are utterances that can be true or false.

[7]Note that in discussing logic programming below (Chapter 9) we shall be interested in formalizing questions. In similar ways, utterances such as orders, commands, and interjections can be adequately formalized. On the other hand, some assertions are not–in principle or at all–formalizable. For instance, in classical logic an assertion such as "I took the umbrella to work because it was raining" is, to say the least, problematic: one can argue that it can first be rephrased as "It was raining, so I took the umbrella to work," formalizable as a conditional (cf. Figure 3.1.1), but it is not certain that the two assertions have the same meaning. Playing on the safe side, we do not formalize the English connector *because*.

[8]Constants are typically not quantifiable, at least not by means of $Q = \{\forall, \exists\}$.

unspecific entities, often in the same atom (see Example 3.1.4.d)). In particular, in a FO language a *domain*, or *universe U*, whose members are classes of entities, is required, so that we know what (class) it is specifically that we are reasoning about; furthermore, a *domain of discourse* $\mathscr{D} \subseteq U$ with individual entities is needed, in order that we can reason about particular entities in that domain. For instance, the universe may be *human beings*, and the universe of discourse may be *female students*, or even just *Peter*. For practical ends, however, we may identify both the universe and the domain of discourse, and then specify a function assigning to variables in the domain particular entities in it; this is actually the only thing to do when the domain is infinite and we want to reason about an individual entity in the domain (see next Section).

Example 3.1.4. Let there be given the propositions (a) *Fritz is a cat.*; (b) *Fritz likes fish but it doesn't like pasta.*; (c) *All cats are finicky.*; (d) *Fritz is finicky about everything.* In the case of propositions (a) and (b), we may remain in the terrain of a propositional language; propositions (c) and (d) require a full FO language.[9]

• We can assign to proposition (a) the propositional variable c, and we thus formalize (a) simply as the atomic formula

$$C$$

with the formula tree

$$C$$
$$\bullet$$

with only the root node, which coincides with the atom C.

• Proposition (b) contains two segregable propositions, to wit, *Fritz likes fish* and *Fritz doesn't like pasta*. Assign to the first the propositional variable f. The second proposition requires more care, as it is a negation; in this case, we assign a propositional variable to the affirmative version, i.e. *Fritz likes pasta* $= p$, so that *Fritz doesn't like pasta* $= \neg p$. The presence of the English connector *but* indicates a complex formula, and indeed we have the formula

$$F \wedge \neg P$$

---

[9]We use italics here for convenience only, namely to segregate these propositions with respect to the main text. We shall keep to this usage whenever it proves convenient to do so.

with the formula tree

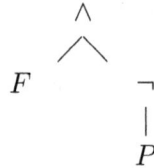

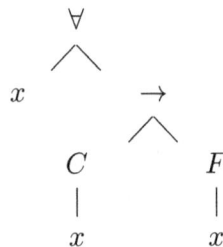

$$\wedge$$

$$F \qquad \neg$$

$$|$$

$$P$$

• Proposition (c) contains a quantifier (*all*), and we thus enter the terrain of a FO language *proper*. This means that we no longer assign propositional variables to atomic assertions, but we assign predicate variables to atomic assertions stating relations. Importantly, we have to identify whether these relations are stated for unspecified individuals or for specific individuals: if the first, then we are employing individual variables; if the second, we employ individual constants. In the case of proposition (c), we have two unary predicates stated for unspecified individuals, to wit, *x is a cat* and *x is finicky*, which we formalize as $C(x)$ and $F(x)$, respectively. Then, note that this is a *universal* assertion, i.e. it asserts something about *all* cats. Thus, we are employing the universal quantifier, denoted by $\forall$. We now have to reformulate proposition (c) in a way that allows for formalization. In fact, proposition (c) means "For all x, if x is a cat, then x is finicky," which abstracts further into the FO formula

$$\forall x \, (C(x) \to F(x))$$

with the formula tree

$$\forall$$

$$x \qquad \to$$

$$C \qquad F$$

$$| \qquad |$$

$$x \qquad x$$

• Clearly, given a domain of discourse $\mathscr{D} = \{fritz\}$, we can now write $C(fritz)$ and $F(fritz)$ to denote that Fritz is a cat and Fritz is finicky. Note how both $C(fritz)$ and $F(fritz)$ are quantifier-free atoms. In effect, in a FO language quantifiers bind solely individual variables. The FO formalization of proposition (d) is the formula

$$\forall x \, (F(fritz, x))$$

with the formula tree

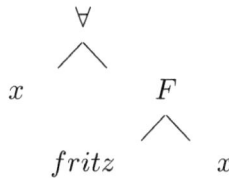

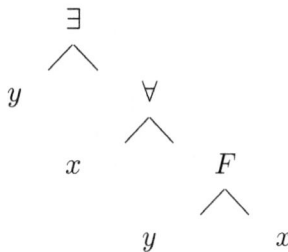

$$\begin{array}{c} \forall \\ \diagdown \\ x \qquad F \\ \diagdown \\ fritz \qquad x \end{array}$$

and where we now have the binary predicate $F(y, x)$ denoting "y is finicky about x." Note that if we do not specify an individual y, then we can have the formula

$$\exists y \forall x \, (F(y, x))$$

with formula tree

$$\begin{array}{c} \exists \\ \diagdown \\ y \qquad \forall \\ \diagdown \\ x \qquad F \\ \diagdown \\ y \qquad x \end{array}$$

formalizing the assertion that there is someone who is finicky about everything. Note also that if we want to specify both individuals denoted by the variables y and x in this formula, we then need to specify two universes, namely $\mathscr{D}_1$ and $\mathscr{D}_2$, as, respectively, the domains of people and, say, food items.

Figure 3.1.1 in Example 3.1.5 provides some common formalizations for English by means of the logical constants of a propositional language. Figure 3.1.2 in Example 3.1.6 does the same for English statements involving quantification and equality. As a matter of fact, the logical language employed in both Examples is that of classical logic (see Section 4.1 below).

Example 3.1.5. Figure 3.1.1 shows some common formalizations for English in a propositional language. We consider a set of connectives $O = \{\neg^1, \wedge^2, \vee^2, \rightarrow^2, \leftrightarrow^2\}$. The connectives $\wedge^2$ and $\vee^2$ are commutative.

Example 3.1.6. Figure 3.1.2 shows some formalizations of English in the FO language of classical logic. We consider a set of quantifiers $Q = \{\forall, \exists\}$ and we augment F1 with a symbol for equality $(=)$.[10]

---

[10] This introduction of the symbol "=" is not without issues in classical FO logic. See Section 4.5 below.

English	Formalization
not p – it is not the case that p	$\neg P$
p and q – both p and q – p but q; q	$P \wedge Q$
neither p nor q	$\neg P \wedge \neg Q$
p or else q	$(P \vee Q) \wedge (\neg P \vee \neg Q)$
p or q – either p or q	$P \vee Q$
there is only one option: p or q	$(P \wedge \neg Q) \vee (\neg P \wedge Q)$
if p then q – if p, q – q only if p – q provided that p	$P \rightarrow Q$
p, so q – not p unless q – p is sufficient for q	
q is necessary for p	
p unless q	$\neg Q \rightarrow P$
p if and only if (iff) q	$P \leftrightarrow Q$
p is necessary and sufficient for q	

Figure 3.1.1.: Formalizations for English by means of the language of classical propositional logic.

Exercises

Exercise 3.1.3.1. Construct the formula trees of the formulae in Exercise 3.1.1.2.

Exercise 3.1.3.2. Formalize the following statements if they are formalizable. If so, decide whether they can be formalized in a propositional language, or require a FO or higher-order language. If the latter, clearly define the universe U and the domain of discourse $\mathscr{D}$.

1. Give them a bonus and they'll be happy.

2. Give me that book, will you?

3. Any formula is a sub-formula of itself.

4. Cats are finicky.

5. Invariably, kids are truants, but occasionally kids are studious.

6. All Vikings, both male and female, were warriors, but for the Greeks they were barbarians.

7. If something is cheap, then it is not of good quality.

8. No cat is both finicky and serendipitous.

9. There is a function that grows faster than the exponential function.

10. Radioactivity is lethal, because it messes with your genes.

11. Karim is a camel if it's not a dromedary.

12. Bob loves Samantha but she doesn't love him back.

13. Gosh, you're such a liar.

14. A superstar told Mary and John that they were Oscar winners.

15. Only warm clothes make good clothes.

16. There is something that Ron likes that everybody else with some commonsense does not like.

17. Not everything that shines is made of gold.

English	Formalization
all/every x – everything	$\forall x$
not all/every x – not everything	$\neg \forall x$
all R are S	$\forall x\,(R(x) \to S(x))$
no R are S	$\forall x\,(R(x) \to \neg S(x))$
there is a x – some x – something	$\exists x$
there isn't any x – nothing	$\neg \exists x$
some R is/are S	$\exists x\,(R(x) \wedge S(x))$
some R is/are not S	$\exists x\,(R(x) \wedge \neg S(x))$
only a is R	$R(a) \wedge \forall x\,(R(x) \to (x = a))$
no R except a is S	$R(a) \wedge S(a) \wedge \forall x\,[(R(x) \wedge S(x)) \to (x = a)]$
all R except a is S	$R(a) \wedge \neg S(a) \wedge \forall x\,[(R(x) \wedge (x \neq a)) \to S(x)]$
there is at most one R	$\forall x \forall y\,[(R(x) \wedge R(y)) \to (x = y)]$
there is exactly one R	$\exists x \exists y\,[(R(x) \wedge R(y)) \to (x = y)]$

Figure 3.1.2.: Formalizations for English by means of the language of classical FO logic.

18. There is at least a white raven and some swans are not white.

19. Anytime, anyone can fool a superb logician but a superb logician can never fool anyone.

20. Alas, the task was too difficult.

21. Some teacher introduced a nurse who was also a philosopher to an astronaut who was also a farmer.

22. Every integer is even or odd.

23. A sequence converges iff it is bounded.

24. There is a set of natural numbers that does not contain the number 7.

25. Are you going out today?

26. Come on!

27. Hurry up, or we'll be late.

28. For every non-empty set there is a set that is a proper subset thereof and that is the empty set.

29. A necessary condition for hypothermia is cold weather. However, it is not a sufficient condition.

30. There are only so many happy marriages.

Exercise 3.1.3.3. Translate the following sentences in F1 into English by paying attention to the universes and the defined predicates:

1. The universe is the set of all the people and $L = \{(x, y) \,|\, x \text{ loves } y\}$:

$$(\forall x \exists y L\,(x, y) \land \neg \exists x \forall y L\,(x, y)) \lor (\exists x \forall y L\,(x, y) \land \exists x \forall y \neg L\,(x, y))$$

2. The universe is the set of all the people and of all the times, and $F = \{(x, y, t) \,|\, x \text{ can fool } y \text{ at time } t\}$:

$$\forall x\,(\exists y \forall t F\,(x, y, t) \land \forall y \exists t F\,(x, y, t) \land \exists y \exists t \neg F\,(x, y, t))$$

Exercise 3.1.3.4. Formalize in F1, specifying the respective domains, the following:

1. The propositions of Definition 1.1.8.

2. The propositions of Definition 1.1.9.

3. The propositions of Definition 1.1.10.

4. Proposition 1.2.15.

5. The propositions of Definition 1.2.16.

6. The propositions of Definition 1.2.19.

7. The propositions of Definition 1.2.21.

8. The Blum axioms (cf. Def. 2.3.9).

3.2. Logical form II: Argument form

Although there may be diverse perspectives on logic and correspondingly different applications thereof, *formal logic* is essentially a means for *reasoning formally*. Indeed, reasoning can be, and is more often than not, carried out *informally*, i.e. without the aid of a logical language. Above we saw how form is fundamental for the definition of both logical languages and their constituents; now we concentrate on form from the viewpoint of reasoning.

Example 3.2.1. Let there be given the following statements in English:

Susan doesn't like dogs but she likes cats and rabbits. She also likes birds, and she has a problem if she also keeps cats. She only keeps dogs and rabbits. Hence, she doesn't have a problem.

This constitutes a reasoning, as attested by the presence of the English connector *hence* in the final statement. That is, anyone uttering or writing the above is just not saying or writing a collection of statements in English, but is concluding something from the previously uttered or written statements. Instead of *hence*, we could have any of the following English connectors: *therefore, thus, so, as a result, consequently, accordingly, in conclusion*, etc. This clearly distinguishes an *argument* from other statements such as explanations, warnings, suggestions and pieces of advice, opinions and beliefs, and illustrations.

Whether in the reasoning of Example 3.2.1 the conclusion is correct or not, that is not easy to determine by analyzing the statements in English.

We require a means to decide *formally* whether our reasoning is correct. The best way we have to do it is precisely by making use of a logical language: We first *formalize* the reasoning above, and then *formally* analyze it in the sense that we determine whether the conclusion follows logically from the previous statements *by virtue of form alone*.

We begin by defining formally *argument*.

Definition. 3.2.1. In very general terms, given a logical language F and formulae $A_1, ..., A_n, B \in F_{\mathsf{F}}$, a *(logical) argument* is a construction of the form

$$(\alpha) \qquad \frac{A_1 \wedge ... \wedge A_n}{B}$$

where $A_1, ..., A_n$ are *premises* and B is the *conclusion*. The following are alternative ways to represent α:

1.
$$A_1, ..., A_n / B$$

2.
$$A_1$$
$$\vdots$$
$$\frac{A_n}{B}$$

3.
$$A_1$$
$$\vdots$$
$$A_n$$
$$\therefore B$$

If for α we have $n = 1$, then α is *not* an argument.[11] Nevertheless, we can still speak of a reasoning of the form α with $n = 1$ that is formally correct–say, p/p–as a *(derived) rule*.

Example 3.2.2. Let there be given a logical language $\mathsf{F0} = (F, O)$ where $O = \{\neg^1, \wedge^2, \vee^2, \rightarrow^2\}$. In order to formalize the reasoning of Example 3.2.1 we begin by assigning a propositional variable to the different atomic statements to be found in it. For example:

Susan likes dogs $= d$

---

[11]This prescribes that a simple conditional statement, i.e. a statement of the form "if p then q" (cf. Figure 3.1.1), is *not* an argument. This is an important note, as the connective $\rightarrow$ plays a rather special role in formalized arguments, as will be seen. In any case, this condition can be relaxed without (too) much ado.

Susan likes cats $= c$
Susan likes rabbits $= r$
Susan likes birds $= b$
Susan has/keeps cats $= k$
Susan has/keeps birds $= a$
Susan has/keeps dogs $= e$
Susan has/keeps rabbits $= f$
Susan has a problem $= p$

According to Definition 3.1.13, propositional atoms are formulae, and we can use either the atoms above or the corresponding formulae in uppercase letters. We choose the latter. In a reasoning, each statement segregated by means of a period is a single premise. We opt for using the argument form specified in Definition 3.2.1.2, and thus every premise is to be written in a different line. By using the connectives in O and the propositional atomic formulae above, we have the following formal argument:

$$
\begin{array}{ll}
1. & \neg D \wedge C \wedge R \\
2. & B \wedge ((K \wedge A) \to P) \\
3. & E \wedge F \wedge \neg K \wedge \neg A \\
\hline
4. & \neg P
\end{array}
$$

We more often than not require formal analysis for more complex kinds of reasoning, such as for theorem proving or software verification, but the principle is the same as applied to Example 3.2.1. However, as stated above, a propositional language is typically not expressive enough for more complex instances of reasoning. In particular, mathematical proofs are typically carried out in a FO language, sometimes in a language even of higher order. Example 3.2.3 shows a very simple mathematical proof that, however simple, cannot be formulated in a propositional language. Below, more sophisticated examples are provided.

Example 3.2.3. We reason that alternate interior angles formed by a diagonal of a trapezoid are equal. In order to formalize this reasoning we require a FO language. We let the alternate angles be abd and cdb. We further let $T(x, y, u, v)$ mean that $xyuv$ is a trapezoid with upper-left vertex x, upper-right vertex y, lower-right vertex u, and lower-left vertex v. We denote the property that the line segment xy is parallel to the line segment uv by $P(x, y, u, v)$. We represent the equality of the angles xyz and uvw as $E(x, y, z, u, v, w)$. Given the specific trapezoid $T(a, b, c, d)$, we reason that $E(a, b, d, c, d, b)$ will follow as a conclusion. We have the following premises and conclusion:

1. $\forall x \forall y \forall u \forall v (T(x,y,u,v) \to P(x,y,u,v))$
2. $\forall x \forall y \forall u \forall v (P(x,y,u,v) \to E(x,y,v,u,v,y))$
3. $T(a,b,c,d)$
---
4. $E(a,b,d,c,d,b)$

In this short Section, we elaborated on argument form solely, omitting any discussion on how to verify the correctness or validity of the reasoning. This requires the fundamental notion of *logical consequence*, to be approached below. However, we can already provide a few classical arguments that are correct by virtue of form alone (see Example 3.2.4). Below, the reader can verify both their correctness and validity. For convenience, in Example 3.2.4 we apply Definition 3.2.1.1. Note the use of Greek lowercase letters to convey the fact that these argument forms are schemata.

Definition. 3.2.2. Given an object-language F, a *schema* ς (plural: *schemata*) is a formula in which one or more variables of μ^{F}, known as *metalinguistic variables*, occur. These metalinguistic variables can be replaced by an expression of F, in order to form an *instance* of the schema ς.

As is evident, a schema ς represents *infinitely* many formulae of an object-language.

Example 3.2.4. The arguments in Figure 3.2.1 are formally correct in classical logic. For convenience, we use Definition 3.2.1.1 and we replace the symbol / by the symbol $\vdash \in \mu^{F}$.[12] The order of the premises is irrelevant.

The above summary treatment of arguments and argument form suffices for the objectives of this text. However, there is much more to be said about this topic, especially from the viewpoint of argumentation, a field in which computational logic is increasingly relevant, to a large extent because of its interest for artificial intelligence (e.g., Rahwan & Simari, 2009). The reader interested in furthering this topic can benefit from, for example, Hurley (2012).

---

[12]See Section 3.4.2 below for this replacement. Note that $\phi, \psi \vdash \chi$ abbreviates $\{\phi, \psi\} \vdash \chi$ or, equivalently, $\phi \wedge \psi \vdash \chi$.

Argument form	Name	Abbr.
$\phi \to \psi, \phi \vdash \psi$	Modus ponens	MP
$\phi \to \psi, \neg\psi \vdash \neg\phi$	Modus tollens	MT
$\phi \to \psi, \psi \to \chi \vdash \phi \to \chi$	Hypothetical syllogism	HS
$\phi \lor \psi, \neg\phi \vdash \psi$	Modus tollendo ponens	TP
$\phi \to \psi, \phi \to \chi \vdash \phi \to (\psi \land \chi)$	Conditional product	CP
$(\phi \to \psi) \land (\chi \to \xi), \phi \lor \chi \vdash \psi \lor \xi$	Constructive dilemma	CD
$(\phi \to \psi) \land (\chi \to \xi), \neg\psi \lor \neg\xi \vdash \neg\phi \lor \neg\chi$	Destructive dilemma	DD
$\phi, \neg\phi \vdash \psi$	Ex contradictione quodlibet	ECQ

Figure 3.2.1.: Some classical formally correct arguments.

Exercises

Exercise 3.2.1. Identify the argument form of the following reasoning instances:

1. $\neg P \rightarrow \neg Q, \neg Q \rightarrow R \vdash \neg P \rightarrow R$

2. $(P \rightarrow \neg Q) \wedge (\neg R \rightarrow S), P \vee \neg R \vdash \neg Q \vee S$

3. $\neg (P \wedge R) \rightarrow \neg Q, Q \vdash P \wedge R$

4. $((T \vee P) \rightarrow (Q \wedge \neg O)) \wedge (S \rightarrow (R \leftrightarrow U)), (T \vee P) \vee S \vdash (Q \wedge \neg O) \vee (R \leftrightarrow U)$

5. $P \vee Q, (P \vee Q) \rightarrow (R \wedge \neg S) \vdash R \wedge \neg S$

6. $P \vee Q, (P \rightarrow R) \wedge (Q \rightarrow S) \vdash R \vee S$

7. $\neg P, ((R \vee S) \wedge (T \rightarrow O)) \rightarrow P \vdash \neg ((R \vee S) \wedge (T \rightarrow O))$

8. $\neg (Q \wedge P) \vee ((R \vee S) \wedge (T \rightarrow O)), Q \wedge P \vdash ((R \vee S) \wedge (T \rightarrow O))$

Exercise 3.2.2. Formalize the following reasoning instances.

1. There is a set $\mathbb{N}$ of natural numbers that does not include the number 6. Therefore, for any numbers $x, y \in \mathbb{N}$, $x, y \neq 6$, it is never the case that $f(x, y) = 6$ for any function $f : \mathbb{N}^2 \longrightarrow \mathbb{N}$.

2. Let $x \in \mathbb{Z}$. If x is even, then it is divisible by 2, and if x is even, then it is not odd. Therefore, if x is even then it is divisible by 2 and it is not odd.

3. All cats are finicky about some food. Fritz is a cat. Hence, Fritz if finicky about fish sticks.

4. A group $\mathcal{G}$ is a set G with an operation $\star$ such that (1) for all elements x, y, z of G we have $(x \star y) \star z$ is equal to $x \star (y \star z)$, (2) for every x of G there is an element e in G such that $x \star e$, $e \star x$, x are all equal, and (3) for every x of G there is an element x^{-1} in G such that $x \star x^{-1}$, $x^{-1} \star x$, e are all equal. If $\mathcal{G} = (G, \star)$ is a group, then for every x, y, z of G the equality of $x \star z$ and $y \star z$ implies that x and y are identical.

5. The same as above, but considering a group $\mathcal{G} = (G, f)$ where $f : G^2 \longrightarrow G$.

6. An idempotent semi-ring is a structure $\mathcal{S} = (S, \star, +, 1, 0)$ such that $(S, +, 0)$ is a commutative monoid with idempotent addition, $(S, \star, 1)$ is a monoid, multiplication distributes over addition from the left and right, and 0 is a right and left annihilator of multiplication. If the relation $\leq$ defined as $x \leq y \leftrightarrow x + y = x$ for all x, y in S is a partial order, then for all $x, y, z \in S$ we have

$$x + y \leq z \leftrightarrow (x < z \wedge y < z)$$

and every idempotent semi-ring is also a semi-lattice $(S, \leq)$ with addition as join.

7. If everybody is a lover (i.e. loves somebody) and everybody loves every lover, then everybody loves everybody.

8. *Russell's paradox*: A barber shaves those men who do not shave themselves. If he shaves himself, then he does not shave himself. (Hint: Make the premise a bi-conditional proposition.)

Exercise 3.2.3. What change(s) would it entail adding the following statement to the reasoning of Examples 3.2.1-2: *But cats are not best friends either with birds or with rabbits.*

Exercise 3.2.4. For convenience reasons, and especially in logic programming, predicates can be formalized in a more "denotational" mode. Reconstruct the original argument from the following formalization:

$(Native\_of\_c(x) \wedge Weapon(y) \wedge Sells(x,y,z) \wedge Hostile(z)) \rightarrow Criminal(x)$
$Missile(m1)$
$Owns(daffy, m1)$
$Enemy(daffy, c)$
$(Missile(x) \wedge Owns(daffy, x)) \rightarrow Sells(west, x, daffy)$
$Native\_of\_c(west)$
$Missile(x) \rightarrow Weapon(x)$
$Enemy(x, c) \rightarrow Hostile(x)$
$\therefore Criminal(west)$

Exercise 3.2.5. Let the following premises be given:

$$\forall x \forall y \, (P(x, y) \rightarrow Q(x, y))$$

$$\forall x \forall y \, (R(x, y) \rightarrow Q(x, y))$$

$$\forall x \forall y \left(P\left(x,y \right) \vee R\left(x,y \right) \right)$$

Without carrying out a formal proof, indicate which of the following two formulae is a conclusion of these premises: (i) $\forall x \exists y \left(Q\left(x,y \right) \right)$; (ii) $\exists y \forall x \left(Q\left(x,y \right) \right)$. Explain your reasoning.

3.3. Logical meaning: Valuations and interpretations

Meaning with respect to a logical language can be approached from different perspectives. In non-mathematical approaches, natural language usually plays a fundamental role, and meaning is object of philosophical discussions within the fields of the philosophy of logic and logical philosophy. We provide here a mathematical, largely algebraic, treatment of meaning in a logical language F1. In any case, *valuations* and *interpretations* are two components of logical meaning that are central both in mathematical and philosophical approaches. In the present Section, we give some basic algebraic foundations for mostly classical valuations and interpretations; this introductory discussion is complemented below in Section 6.3 with a study of Boolean algebras in relation to classical logic. The importance of this algebraic approach notwithstanding, it does not capture all the nuances of the subject of meaning in formal logic; both Tarskian and Herbrand semantics, in Sections 6.1-2 below, contribute with fundamental insights to a fuller grasp of this central topic. Of these, Herbrand semantics is of particular interest to classical deductive computing with classical logic, and it is accordingly to be found in this book in several other Sections.

Definition. 3.3.1. Given a set of formulae $F \subseteq$ F1 and a set $W_n = \{v_0, v_1, ..., v_{n-1}\}$ of *distinguished elements* known as *(logical) truth values*, a *valuation* is a function $val : F \longrightarrow W_n$. Let $\phi \in F$ be an atomic formula; then, we write $val\left(\phi \right) = v_i$.

In other words, a valuation is an assignment of a truth value to the formulae of F1. Given a truth-value set $W_n = \{v_0, v_1, ..., v_{n-1}\}$, typically, $v_0 = 0 = \text{f}$ (false) and $v_{n-1} = 1 = \text{t}$ (true).

Definition. 3.3.2. In a logical language F1 with a truth-value set W_n, with each operator $\heartsuit_i^k$ with arity k and each quantifier $\blacklozenge_i$ there can be associated

1. a *truth table* $\widetilde{\heartsuit}_i^k : W_n^k \longrightarrow W_n$, and

2. a *truth (or distribution) function* $\widetilde{\blacklozenge}_i : \left(2^{W_n} - \emptyset\right) \longrightarrow W_n$.

In this text, we consider mostly the set of *classical* truth values, to wit, $W_2 = \{0, 1\}$. As such, we henceforth omit the subscript in W when we consider W_2.

Clearly, truth tables and truth functions are extensions of valuations; in particular, by means of truth tables the function *val* is extended to all propositional formulae, and by means of truth functions *val* is extended to all quantified formulae.

Example 3.3.1. For the sake of simplicity, we begin by considering the atomic propositions $P, Q \in F_{F0}$ and the set of truth values W. An important feature of propositional atoms is that they have a fixed truth value, in the sense that their valuation, contrarily to predicates and functions (see below), does not depend on truth-value assignments to their arguments (because they have none). Thus, given two atoms $P, Q \in F_{F0}$, there can be four cases: (i) both P and Q are valuated to **true**, (ii) P is valuated to **true** and Q is valuated to **false**; (iii) P is valuated to **false** and Q is valuated to **true**; (iv) both P and Q are valuated to **false**. We represent these four cases in tabular form as follows:

P	Q
1	1
1	0
0	1
0	0

Let now there be given the connective $\rightarrow^2 \in O_{F0}$. Recall Proposition 3.1.4 and Definition 3.1.5: This is a binary connective, so that we have formulae $\rightarrow^2 (\phi_1, \phi_2)$, here written $\phi_1 \rightarrow \phi_2$, where $\phi_1, \phi_2 \in \mu^{F0}$ are either atomic or complex formulae. We now give the truth table for the formula $P \rightarrow Q$:

P	Q	$P \rightarrow Q$
1	1	1
1	0	0
0	1	1
0	0	1

As said above, non-mathematical introductions to logic often provide lengthy explanations on the "meaning" of these four valuations of Example 3.3.1, namely by appealing to reasoning in natural language; in particular, it is often claimed that it is not acceptable that falsity can follow from truth (second line of the table), and from falsity anything can follow (lines three and four), a principle known as *explosion*. But for *meaning* in a strictly formal sense we refer to Proposition 3.3.3 below.

In any case, there is an algebraic basis for classical valuations, Boolean algebras providing a mathematical foundation for the truth tables of the primitive connectives (see Sections 1.2 and 6.3). This said, one can see valuations as formal conventions. This holds for every truth table for every single connective.[13] This follows from the strictly formal sense of truth values as *distinguished elements* in a logical language. By this, it is meant that they are the elements 0 and 1 in a Boolean algebra $\mathfrak{B} = (A,', \wedge, \vee, 0, 1)$ (cf. Def. 1.2.5). In particular, in any complete Boolean lattice, 0 is the least element (the meet of all the atoms) and 1 is the greatest element (the join of all the atoms). For example, in the lattice of Figure 1.2.3, we have $0 = \emptyset$ and $\{a, b, c\} = 1$.

Why 1 and 0 can be, and often are, interchangeable with **true** and **false**, respectively, that seems to be a philosophical hard nut to crack, but we take the latter as distinguished elements simpliciter in the above mathematical sense. By this, we mean that they are not to be confused with the English words *true* and *false*, and we emphasize this by writing them in a distinct **font** from that used for English words.[14]

Proposition. 3.3.3. *Let ϕ be any formula such that $\phi \in F_{F0}$; in order to interpret* FO, *ϕ is provided with a meaning. The meaning of ϕ is its semantical correlate. Let G be the range of all the semantical correlates; a mapping $s : F_{F0} \rightarrow G$ requires two conditions (Frege, 1892):*

1. *With each $\phi \in F_{F0}$, exactly one semantical correlate is associated (i.e., s is a function).*

---

[13]The conventional character of truth tables for single connectives is clearly easier to see in the many-valued logics, there being different truth tables for the same connectives in the different logics (cf. Fig. 3.3.1 below).

[14]This can be seen as a more informal sense of "distinguished." This said, we do not wish to remove completely from **true** and **false** the meanings they have in English, namely for the sake of an *intuitive semantics*, often a desideratum for a logical system. Tarski (1935), without dissolving the philosophical problem (much on the contrary!), provides an interesting definition of *truth* as a property of a formal*ized* language. See Section 6.1 below.

2. *Any two formulae χ, ψ such that $\chi, \psi \in F_{F0}$ are interchangeable in any propositional context $\phi \in F_{F0}$ whenever $s(\chi) = s(\psi)$ or, in other words, when for any $\phi \in F_{F0}$, $p \in V_{F0}$ we have it that*

$$s(\phi[p/\chi]) = s(\phi[p/\psi]) \quad \text{iff} \quad s(\chi) = s(\psi)$$

where $\phi[p/\varsigma]$ stands for the formula that results from ϕ after the substitution ς instead of p.

Proposition 3.3.3 is known as the *Fregean axiom* and 3.3.3.2 is often referred to as the *principle of extensionality*. Generalization of Proposition 3.3.3 to a FO language follows, though not without restrictions (e.g., Wójcicki, 1988).

In particular, Proposition 3.3.3.2, also referred to as *principle of the compositionality of meaning*, states that the meaning of a proposition is a function of the meaning of its components. This is known as *truth-functionality*, and this can find a mathematical foundation in an algebraic interpretation of Proposition 3.3.3.

Definition. 3.3.4. Let $n \geq 2 \in \mathbb{N}$; we denote by G_n the set $\{0, 1, ..., n - 1\}$ and by $\mathfrak{A}_n$ any algebra of the form $(G_n, f_1, ..., f_m)$. In particular, we denote the set of all m-ary mappings defined on G_n with values in the same set by

$$Z_n^m = \{f | f : G_n^m \longrightarrow G_n\}, \quad m \geq 0, m \text{ finite}$$

and we denote the set of all mappings defined on G_n with values in Z_n^m by

$$Z_n = \bigcup_{m \in \omega} Z_n^m$$

for ω the set of all natural numbers.

Example 3.3.2. Let $G_2 = \{0, 1\}$. Then, we have the functions $f_1 : G_2^1 \longrightarrow G_2$ and $f_2 : G_2^2 \longrightarrow G_2$ for unary and binary mappings respectively defined on G_2. The functions f_1 and f_2 are *Boolean functions*, i.e. functions defined on $G_2 = W$. Given the set $O = \{\neg^1, \wedge^2, \vee^2, \rightarrow^2, \leftrightarrow^2\}$, it is evident that $\neg \in f_1$ and $\wedge, \vee, \rightarrow, \leftrightarrow \in f_2$. We may also consider $\top$ and $\bot$ as 0-ary functions, i.e. as degenerate operations. This allows a characterization of classical logic as the set

$$Z_2 = \bigcup_{i=0}^{2} Z_2^i = Z_2^0 \cup Z_2^1 \cup Z_2^2.$$

Definition. 3.3.5. Let now $\mathfrak{L} = (F, o_1, ..., o_m)$ be an algebra of formulae freely generated by the set of generators $V \subseteq F_{\mathsf{F0}}$. Then $\mathfrak{L}$ is a propositional language. Given an algebra $\mathfrak{A} = (G_n, f_1, ..., f_m)$ similar to $\mathfrak{L}$, it is easy to see that a homomorphism $hom\,(\mathfrak{L}, \mathfrak{A})$ can give rise to a function $h : V_{\mathfrak{L}} \longrightarrow G_{\mathfrak{A}}$, which in turn can be extended to the function $h : T_{\mathfrak{L}} \longrightarrow G_{\mathfrak{A}}$ such that for each operation o_i with arity m and given terms $t_i \in T$ we have

$$h\left(o_{i_{\mathfrak{L}}}\left(t_1, ..., t_m\right)\right) = f_{i_{\mathfrak{A}}}\left(h\left(t_1\right), ..., h\left(t_m\right)\right)$$

Then, h is an assignment of truth values, or a valuation, into $\mathfrak{L}$.

More specifically, the algebra $\mathfrak{A}$ is similar to the set of all formulae of $\mathfrak{L}$ formed by means of the m operations of $\mathfrak{L}$ (i.e., the connectives), in other words, the algebra of $F_{\mathfrak{L}}$. From the semantical viewpoint, the fundamental importance of the homomorphism $hom\,(\mathfrak{L}, \mathfrak{A})$ is that it is an embedding of $F_{\mathfrak{L}}$ in $\mathfrak{A}$ that is in fact an *interpretation* for the formulae of $\mathfrak{L}$ (in $\mathfrak{A}$).

Definition. 3.3.6. A set of functions $X \subseteq Z_n$ is said to be *functionally complete* iff every function $f \in Z_n$ can be defined by means of functions in the set X.

In other words, the finite mapping $f : G_n^m \longrightarrow G_n$ can be represented as a composition of the operations $f_1, ..., f_m$. Then, we say that an algebra $\mathfrak{A}_n = (G_n, f_1, ..., f_m)$ is functionally complete. This reduces the problem of functional completeness of logical systems to that of the definability of all unary and binary connectives:

3.3.7. (Prop.) Given $n \in \mathbb{N}$, in a n-valued logic any given place in the truth table can be occupied by n truth values. For a k-place connective, the truth table has room for entries

$$\underbrace{n \times n \times ... \times n}_{k \text{ times}} = n^k$$

and for each there will be any of n possibilities, so that we can have

$$\underbrace{n \times n \times ... \times n}_{n^k \text{ times}} = n^{n^k}$$

possible k-place truth tables for a given n-valued logic.

We thus have it that the number of unary and binary connectives equals n^n and n^{n^2}, respectively (see Example 3.3.3). This explains why

3. Preliminaries: Formal logic, deduction, and deductive computation

truth tables, though they are decision algorithms, find only limited application in logic, classical or non-classical. In effect, given the truth-values set W, we have 2^k for k atomic formulae, which means that a truth-table for classical logic has exponential growth. Below, it should be clear why truth tables are essentially restricted to propositional languages.

Example 3.3.3. Figure 3.3.1 shows the truth table for the connective $\to$ in three different 3-valued logics, i.e. logics with $W_3 = \{\mathtt{f}, \mathtt{i}, \mathtt{t}\}$, where $\mathtt{i}$ denotes indeterminacy or even nonsense. These logics are Łukasiewicz's 3-valued logic (denoted by Ł3), Kleene's 3-valued logic known as "weak" (K_3^W), and Piróg-Rzepecka's 3-valued logic (Rn3). Augusto (2020a) provides many further examples.

$\to_{\text{Ł3}}$	t	i	f
t	t	i	f
i	t	t	i
f	t	t	t

$\to_{\text{KW3}}$	t	i	f
t	t	i	f
i	i	i	i
f	t	i	t

$\to_{\text{Rn3}}$	t	i	f
t	t	f	f
i	t	t	t
f	t	t	t

Figure 3.3.1.: Truth table for the connective $\to$ in the 3-valued logics Ł3, K_3^W, and Rn3.

Definition. 3.3.8. *Logical equivalence 1* – Two formulae $A, B \in$ F0 are said to be (logically) equivalent, denoted by $A \equiv_{\text{(F0)}} B$, iff both are valuated to the same truth value under every assignment of truth values, i.e., if their truth tables for the main logical connective are identical.

Example 3.3.4. We show that $P \to Q \equiv \neg P \vee Q$ by comparing the truth table for the right side of this equivalence with the truth table of Example 3.3.1. As can be seen, the valuations for the main connectives are exactly the same in the four lines of both truth tables. Note that classically a disjunction is true when either of its disjuncts is true; otherwise it is false (cf. Prop. 3.3.12.4 below).

P	Q	$\neg P$	$\vee$	Q
1	1	0	1	1
1	0	0	0	0
0	1	1	1	1
0	0	1	1	0

In terms of functional completeness, it is obvious from Example 3.3.4 that the connective $\to$ can be defined by means of the subset $O' =$

$\{\neg, \vee\}$. Figure 3.1.1 provides a clue for the fact that $\leftrightarrow$ can be defined in terms of $\rightarrow$ and $\wedge$; in fact, we have $P \leftrightarrow Q \equiv (P \rightarrow Q) \wedge (Q \rightarrow P)$. In Section 4.1 we elaborate further on this topic.

We now define some important semantical notions for a logical language F1 that follow from the algebraic exposition above.

Definition. 3.3.9. A *frame* for an object language F1 with an alphabet Σ and truth-value set W is a pair $(\mathscr{D}, \Theta)$ where $\mathscr{D}$ is a non-empty *domain of discourse* and Θ is a *signature interpretation*, i.e., a mapping assigning the functions $\mathscr{D}^n \longrightarrow \mathscr{D}$ and $\mathscr{D}^n \longrightarrow W$ to each n-place function symbol and to each n-place predicate symbol of Σ_{F1}, respectively.

Definition. 3.3.10. An *interpretation* $\mathcal{I}$ for F1 is a triple $(\mathscr{D}, \Theta, \varpi)$ where $(\mathscr{D}, \Theta)$ is a frame and ϖ is a *variable assignment* $\varpi : Vi \longrightarrow \mathscr{D}$. We say that $\mathcal{I}$ is based on the frame $(\mathscr{D}, \Theta)$.

Definition. 3.3.11. For $\mathcal{I} = (\mathscr{D}, \Theta, \varpi)$, $\mathcal{I}$ is an interpretation for F1, and W, there is a corresponding *(e)valuation function* $val_{\mathcal{I}} : F_{\text{L1}}, \Sigma_{\text{L1}} \longrightarrow W$ defined inductively as follows:

1. $val_{\mathcal{I}}(x) = \varpi(x)$ for all x in Σ.

2. $val_{\mathcal{I}}(f(t_1, ..., t_n)) = \Theta(f)(val_{\mathcal{I}}(t_1), ..., val_{\mathcal{I}}(t_n))$ for all n-place function symbols f, $n \geq 0$, in Σ.

3. $val_{\mathcal{I}}(P(t_1, ..., t_n)) = \Theta(P)(val_{\mathcal{I}}(t_1), ..., val_{\mathcal{I}}(t_n))$ for all n-place predicate symbols P, $n \geq 0$, in Σ.

4. $val_{\mathcal{I}}(\heartsuit_i(\phi_1, ..., \phi_n)) = \tilde{\heartsuit}_i(val_{\mathcal{I}}(\phi_1), ..., val_{\mathcal{I}}(\phi_n))$ for all logical connectives $\heartsuit_i$, $n \geq 0$, in Σ.

5. $val_{\mathcal{I}}(\blacklozenge_i x(\phi)) = \tilde{\blacklozenge}_i(\Delta_{\mathcal{I},x}(\phi))$ for all quantifiers $\blacklozenge_i$ in Σ, where $\Delta_{\mathcal{I},x}(\phi)$ is the distribution of ϕ in $\mathcal{I}$ with respect to x, denoted by the set $\Delta_{\mathcal{I},x}(\phi) = \{val_{\mathcal{I}_d^x}(\phi) | d \in \mathscr{D}\}$, and $\mathcal{I}_d^x$ is the interpretation identical to $\mathcal{I}$ when setting $\varpi(x) = d$.

Despite the complexity above, there are in fact some fixed valuations for FO *atomic* formulae.

Proposition. 3.3.12. *The following classical valuations for atomic formulae are fixed for any interpretation* $\mathcal{I} = (\mathscr{D}, \Theta, \varpi)$:

1. $val_{\mathcal{I}}(\top) = 1$ and $val_{\mathcal{I}}(\bot) = 0$

2. $val_{\mathcal{I}}(\neg\phi) = \begin{cases} 1 & \text{if } val_{\mathcal{I}}(\phi) = 0 \\ 0 & \text{otherwise} \end{cases}$

3. $val_{\mathcal{I}}(\phi \wedge \psi) = \begin{cases} 1 & \text{if } val_{\mathcal{I}}(\phi) = 1 \text{ and } val_{\mathcal{I}}(\psi) = 1 \\ 0 & \text{otherwise} \end{cases}$

4. $val_{\mathcal{I}}(\phi \vee \psi) = \begin{cases} 1 & \text{if } val_{\mathcal{I}}(\phi) = 1 \text{ or } val_{\mathcal{I}}(\psi) = 1 \\ 0 & \text{otherwise} \end{cases}$

5. $val_{\mathcal{I}}(\phi \to \psi) = \begin{cases} 1 & \text{if } val_{\mathcal{I}}(\phi) = 0 \text{ or } val_{\mathcal{I}}(\psi) = 1 \\ 0 & \text{if } val_{\mathcal{I}}(\phi) = 1 \text{ and } val_{\mathcal{I}}(\psi) = 0 \end{cases}$

6. $val_{\mathcal{I}}(\phi \leftrightarrow \psi) = \begin{cases} 1 & \text{if } val_{\mathcal{I}}(\phi) = val_{\mathcal{I}}(\psi) \\ 0 & \text{otherwise} \end{cases}$

7. $val_{\mathcal{I}}(\forall x\,(\phi)) = \begin{cases} 1 & \text{if } val_{\mathcal{I}_d^x}(\phi) = 1 \text{ for all } d \in \mathcal{D} \\ 0 & \text{otherwise} \end{cases}$

8. $val_{\mathcal{I}}(\exists x\,(\phi)) = \begin{cases} 1 & \text{if } val_{\mathcal{I}_d^x}(\phi) = 1 \text{ for at least one } d \in \mathcal{D} \\ 0 & \text{otherwise} \end{cases}$

Example 3.3.5. Let there be given the FO formula

$$\chi = \forall x \forall y \exists z\, (L\,(f\,(x,y)\,,z) \to L\,(x,y))\,.$$

We consider the interpretation $\mathcal{I} = (\mathcal{D},\Theta,\varpi)$ such that $\mathcal{D} = \mathbb{Z}$, $\varpi\,(z) = 0 \in \mathbb{Z}$, $f\,(x,y) = x - y$, and $L\,(x,y) = \{(x,y) \in \mathbb{Z} \times \mathbb{Z}\,|\,x < y\}$. Then, we have the formula

$$\chi^{\mu} = \forall x \forall y\,((x - y < 0) \to x < y)$$

and

$$val_{\mathcal{I}}(\chi) = 1$$

as we have

$$val_{\mathcal{I}}(\forall x, y\,(\chi)) = \widetilde{\forall}\,(\Delta_{\mathcal{I},x,y}\,(\chi)) = 1 \text{ for } \Delta_{\mathcal{I},x,y}\,(\chi) = \left\{ val_{\mathcal{I}_d^{x,y}}\,(\chi)\,|\,d \in \mathbb{Z} \right\}$$

and

$$val_{\mathcal{I}}(\exists z\,(\chi)) = \widetilde{\exists}\,(\Delta_{\mathcal{I},z}\,(\chi)) = 1 \text{ for } \Delta_{\mathcal{I},z}\,(\chi) = \left\{ val_{\mathcal{I}_0^z}\,(\chi)\,|\,0 \in \mathbb{Z} \right\}.$$

In particular, we can rewrite χ as

$$\chi' = \underbrace{\forall x \forall y \exists z \left(L \left(f \left(x, y\right), z\right)\right)}_{\phi} \rightarrow \underbrace{\forall x \forall y \left(L \left(x, y\right)\right)}_{\psi}$$

and apply Proposition 3.3.12.5 and 3.3.12.7-8: if $val_{\mathcal{I}_d^{x,y}} \left(val_{\mathcal{I}_0^z} \left(\phi\right)\right) = 1$ and $val_{\mathcal{I}_d^{x,y}} \left(\psi\right) = 1$, then $val_{\mathcal{I}} \left(\chi'\right) = 1$ (check the truth table for the connective $\rightarrow$ in Example 3.3.1). In other words, given any two integers x and y, if $x - y < 0$, then $x < y$.

Note that it is impossible to build a truth table for χ in Example 3.3.5, as the domain $\mathcal{D} = \mathbb{Z}$ is infinite. Given a finite, small, domain, it may be possible to build a truth table for a FO formula, but in general other means are preferred to "decide" on its truth value.[15]

3.3.13. With respect to Definition 3.3.11, we have the following observations: (i) Free variables must be assigned a value in $\mathcal{D}$ and all free occurrences of a variable x must be assigned the same value in $\mathcal{D}$, and (ii) every constant must be assigned a value in $\mathcal{D}$ and all occurrences of the same constant a must be assigned the same value in $\mathcal{D}$.

Definition. 3.3.14. *Logical equivalence 2* – Two formulae $A, B \in$ **F1** are said to be (logically) equivalent, denoted by $A \equiv_{(F1)} B$, iff both are valuated to the same truth value under every assignment of truth values, i.e., if their truth values are identical with respect to every interpretation $\mathcal{I}$.

In other words, two formulae are equivalent iff they have the same meaning.

Proposition. 3.3.15. Quantifier duality – *The following formulae are equivalent:*

1.

$$(\text{QN}_\exists) \qquad \neg \exists x \left(A\right) \equiv \forall x \left(\neg A\right)$$

2.

$$(\text{QN}_\forall) \qquad \neg \forall x \left(A\right) \equiv \exists x \left(\neg A\right)$$

Proof: Left as an exercise.

---

[15] We recall that a FO language is essentially undecidable, being at best semi-decidable. This holds for the standard classical FO language.

Proposition. 3.3.16. *The following formulae are equivalent when x does not appear as a free variable in B:*

1.
$$\forall x \left(A\left(x\right)\right) \wedge B \equiv \forall x \left(A\left(x\right) \wedge B\right)$$

2.
$$\forall x \left(A\left(x\right)\right) \vee B \equiv \forall x \left(A\left(x\right) \vee B\right)$$

3.
$$\exists x \left(A\left(x\right)\right) \wedge B \equiv \exists x \left(A\left(x\right) \wedge B\right)$$

4.
$$\exists x \left(A\left(x\right)\right) \vee B \equiv \exists x \left(A\left(x\right) \vee B\right)$$

5.
$$\forall x \left(A\left(x\right)\right) \rightarrow B \equiv \exists x \left(A\left(x\right) \rightarrow B\right)$$

6.
$$\exists x \left(A\left(x\right)\right) \rightarrow B \equiv \forall x \left(A\left(x\right) \rightarrow B\right)$$

7.
$$B \rightarrow \forall x \left(A\left(x\right)\right) \equiv \forall x \left(B \rightarrow A\left(x\right)\right)$$

8.
$$B \rightarrow \exists x \left(A\left(x\right)\right) \equiv \exists x \left(B \rightarrow A\left(x\right)\right)$$

Proof: Left as an exercise.

Remark. **3.3.17.** With respect to Proposition 3.3.16, if x appears as a free variable in B, then rename the bound variable x in $\blacklozenge x \left(A\left(x\right)\right)$ as y, obtaining the equivalent $\blacklozenge y \left(A\left[x/y\right]\right) = \blacklozenge y \left(A\left(y\right)\right)$.

Exercises

Exercise 3.3.1. Let there be given the formula $\phi = \exists x \left(S\left(g\left(x\right), a\right)\right)$ and the interpretation $\mathcal{I} = \left(\mathscr{D}, \Theta, \varpi\right)$ such that $\mathscr{D} = \mathbb{Z}$, $a \in \mathscr{D} = 5$, $g\left(x\right) = x^2$, and $S\left(x, y\right) = \{(x, y) \in \mathbb{Z} \times \mathbb{Z} | x = y\}$.

1. Rewrite ϕ as ϕ^μ.

2. Evaluate ϕ in the given interpretation.

Exercise 3.3.2. Let there be given the formula $\phi = \exists x \forall y \, (Q \, (f \, (x) \,, y))$ and the interpretation $\mathcal{I} = (\mathscr{D}, \Theta, \varpi)$ such that $\mathscr{D} = \mathbb{N}^+$, $\varpi \, (x) = 2$, $f \, (x) = x - 1$, and $Q \, (x, y) = \{(x, y) \in \mathbb{Z} \times \mathbb{Z} | x \leq y\}$.

1. Rewrite ϕ as ϕ^μ.

2. Evaluate ϕ in the given interpretation.

3. Let now ϕ be the formula $\phi = \forall y \exists x \, (Q \, (f \, (x) \,, y))$. Evaluate ϕ in the given interpretation.

Exercise 3.3.3. Let there be given the interpretation $\mathcal{I} = (\mathscr{D}, \Theta, \varpi)$ such that $\mathscr{D} = \mathbb{R}$ and $P \, (x, y) = \{(x, y) \, | f \, (x) = y\}$ for $f \, (x) = x^2 - 1$. Evaluate the following formulae:

1. $\forall x \forall y \, (P \, (f \, (x) \,, y))$

2. $\forall x \exists y \, (P \, (f \, (x) \,, y))$

3. $\forall y \exists x \, (P \, (f \, (x) \,, y))$

4. $\exists x \forall y \, (P \, (f \, (x) \,, y))$

5. $\exists x \exists y \, (P \, (f \, (x) \,, y))$

Exercise 3.3.4. Let there be given the interpretation $\mathcal{I} = (\mathscr{D}, \Theta, \varpi)$ such that $\mathscr{D} =$ all the people and $F \, (x, y) = \{(x, y) \, | x$ is the father of $y\}$. Evaluate the following formulae:

1. $\forall x \exists y \, (F \, (x, y))$

2. $\forall y \exists x \, (F \, (x, y))$

3. $\exists x \forall y \, (F \, (x, y))$

4. $\exists y \forall x \, (F \, (x, y))$

5. $\exists x \exists y \, (F \, (x, y))$

6. $\exists y \exists x \, (F \, (x, y))$

7. $\forall x \forall y \, (F \, (x, y))$

8. $\forall y \forall x \left(F\left(x,y\right)\right)$

Exercise 3.3.5. With respect to the interpretation above for Exercise 3.3.4, let us assume that John is the father of Sara. We write the individual constants for John and Sara as *john* and *sara*, respectively. Evaluate the following formulae:

1. $F\left(john, sara\right)$

2. $\forall x \left(F\left(x, sara\right)\right)$

3. $\exists x \left(F\left(x, sara\right)\right)$

4. $\exists y \left(F\left(john, y\right)\right)$

5. $\forall y \left(F\left(john, y\right)\right)$

Exercise 3.3.6. For each of the following formulae find an interpretation that makes it evaluate to **true**:

1. $\forall x \left(P\left(x\right) \rightarrow \neg Q\left(x\right)\right)$

2. $\forall x \forall y \left(R\left(x,y\right) \vee R\left(y,x\right)\right)$

3. $\forall x \forall y \left(\neg R\left(x,y\right) \vee R\left(y,x\right)\right)$

4. $\exists x \exists y \left(R\left(x,y\right) \wedge R\left(y,x\right)\right)$

5. $\forall x \exists y \left(P\left(x\right) \rightarrow R\left(x,y\right)\right)$

6. $\forall x \forall y \left(\left(Q\left(x\right) \wedge \neg Q\left(y\right)\right) \rightarrow R\left(x,y\right)\right)$

Exercise 3.3.7. Let there be given an interpretation $\mathcal{I}^{\varnothing} = (\emptyset, \Theta, \varpi)$.

1. Give examples of $\mathscr{D} = \emptyset$. (Hint: empty denotations.)

2. Given $\mathcal{I}^{\varnothing}$, then for any formula ϕ and any variable x, we have $val_{\mathcal{I}^{\varnothing}}\left(\exists x\left(\phi\right)\right) = \mathtt{f}$ and $val_{\mathcal{I}^{\varnothing}}\left(\forall x\left(\phi\right)\right) = \mathtt{t}$. Give the rationale for this.

Exercise 3.3.8. Recall from Definition 3.1.14.3 the notions of universal and existential closure. Let now $\phi = \phi(x_1, ..., x_n)$ be a formula with all $x_1, ..., x_n$ free variables. Then, the universal closure of ϕ preserves its validity, and the existential closure of ϕ preserves solely its (un)satisfiability, i.e.

$$\forall x_1, ..., \forall x_n (\phi) \equiv_\top \phi$$

and

$$\exists x_1, ..., \exists x_n (\phi) \equiv_{sat} \phi$$

Give the rationale for this.

3.4. Logical systems, logics, and logical theories

Above, we saw how formulae in a logical language F are well formed by means of formation rules (Sections 3.1-2) and how they are given logical meaning (Section 3.3). This allows us to formalize utterances in natural language. This formalization has more often than not in view the verification whether some reasoning is valid or correct, i.e. whether in an argument the conclusion does indeed follow from, or is entailed by, the premises. This, derivation or entailment, is an operation that is specific to a *logical system*, which in turn requires a notion of *logical consequence*. This comes in two versions, syntactical and semantical, and when a logical system has both–a common desideratum–it is characterized as adequate.

In this Section, we touch upon general aspects of logical consequence and its connection to the central notions of *inference* and *deduction*, as well as to *logics* and *logical theories*; below (Section 4.2), we elaborate on classical consequence. For an extensive elaboration on the central topic of logical consequence, see Augusto (2020b).

3.4.1. Logical consequence, inference, and deduction

Remark. **3.4.1.** Recall Definition 3.2.1; α is a construct of the object-language F, but there are several other ways to represent α in the metalanguage μ^F, of which we introduce the following two to express formally the fact that $B \in F_F$ is a *logical consequence* of the set of premises $\{A_1, ..., A_n\} \subseteq F_F$:

$$(Cn) \qquad B \in Cn(\{A_1, ..., A_n\})$$

$$(\Vdash) \qquad \{A_1, ..., A_n\} \Vdash B$$

From the viewpoint of the notion of logical consequence, denoted above by Cn or $\Vdash$, a logical system is the theory of what formulae (more generally: assertions or sentences) *follow from* or *are entailed by* (i.e. *are consequences of*) which other formulae.

Definition. 3.4.2. A *logical system* is a pair $L = (F, Cn)$, where F is a logical language and Cn is a *consequence operation*. Equivalently, $L = (F, \Vdash)$, where $\Vdash$ is a (syntactical or semantical) *consequence relation*.

Definition. 3.4.3. Given a logical system $L = (F, Cn)$, for every $X \subseteq F_F$ the set $Cn_L(X)$ is the set of all the consequences of X in L. For a formula $\phi \in F_F$, $\phi \in Cn_L(X)$ denotes the fact that ϕ is a consequence of X in L, or, which is the same, that ϕ can be inferred from X in L.

We shall often omit the subscript in Cn, in particular when the logical system is left unspecified, or we know what the logical system being considered is.

Definition. 3.4.4. For a logical system $L = (F, Cn)$, we define a *(logical) consequence operation* Cn as a mapping $Cn : 2^F \longrightarrow 2^F$ satisfying the following conditions for any $X, Y \subseteq F_F$:

(C1)	$X \subseteq Cn(X)$	*Inclusion*
(C2)	$Cn(Cn(X)) = Cn(X)$	*Idempotency*
(C3)	$Cn(X) \subseteq Cn(Y)$ whenever $X \subseteq Y$	*Monotonicity*

The following are basic properties of a consequence operation Cn:

Definition. 3.4.5. Let Cn be a consequence operation on a set of formulae F_F. Let $X \subseteq F_F$ be given.

1. If for all substitutions σ it is the case that

$$\sigma Cn(X) \subseteq Cn(\sigma X)$$

then Cn is said to be *structural*.

2. Cn is *finitary* if it is the case that

$$Cn\left(X\right) = \bigcup \left\{Cn\left(X'\right) \mid X' \text{ is a finite subset of } X\right\}.$$

Otherwise, Cn is *infinitary*.

3. Cn is said to be *standard* if it is both finitary and structural.

4. The strongest consequence operation on F is the operation Cn such that $Cn\left(X\right) = F = \mathsf{F}$. This is called the *inconsistent* or *trivial consequence operation* on F.

5. The weakest consequence operation on F is the operation Cn defined by $Cn\left(X\right) = X$. This is called the *idle consequence operation* on F.

6. We say that a set $X \subseteq F_{\mathsf{F}}$ is *closed* under a consequence operation Cn if $X = Cn\left(X\right)$. If $X = Cn\left(X\right)$ or, equivalently, $X = Cn\left(Y\right)$ for some Y, then X is a *theory of Cn*, denoted Θ_{Cn}.

7. Two consequence operations can be ordered according to their *strength* (denoted by $\preccurlyeq$): we say that $Cn_1 \preccurlyeq Cn_2$ (Cn_2 is stronger than Cn_1) iff, for all $X \subseteq F_{\mathsf{F}}$, $Cn_1\left(X\right) \subseteq Cn_2\left(X\right)$. In fact, the following conditions are equivalent:

    a) $Cn_1 \preccurlyeq Cn_2$

    b) $\Theta_{Cn_2} \subseteq \Theta_{Cn_1}$

    c) $\Vdash_{Cn_1} \subseteq \Vdash_{Cn_2}$

8. For each set Cq of consequences on F, the consequence

$$Cn_{Cq}\left(X\right) = \bigcap \left\{Cn\left(X\right) \mid Cn \in Cq\right\}$$

is the greatest lower bound of Cq.

9. Let $\mathscr{F}$ be a family of sets. For each consequence operation Cn, Θ_{Cn} is a *closure system*, i.e. for each $\mathscr{F} \subseteq \Theta_{Cn}$, $\bigcap \mathscr{F} \in \Theta_{Cn}$, with $Cn\left(\emptyset\right)$ and F as the least and greatest elements thereof. The former is called the *base theory* and the latter the *trivial theory* of Cn.

10. Let $\mathscr{F} = 2^F$. Then $\mathscr{F}$ is a *closure base* for a consequence operation Cn iff for each $X \subseteq F_\mathsf{F}$,

$$Cn(X) = \bigcap \{Y \in \mathscr{F} \,|\, X \subseteq Y\}.$$

Note that Θ_{Cn} is the greatest closure base for Cn.

11. If Θ_{Cn} is a closure system, then the pair $(\mathscr{F}, \subseteq)$ is a *complete lattice* such that for every $\mathscr{F} \subseteq \Theta_{Cn}$,

a) $inf(\mathscr{F}) = \bigcap \mathscr{F}$,

b) $sup(\mathscr{F}) = inf\{X \in \Theta_{Cn} | \bigcup \mathscr{F} \subseteq X\}$.

Recall from Section 1.1.2 the partial order relation $\leq$ and the definition of a lattice (Def. 1.2.13).

Theorem 3.4.1. *All consequences in a given logical language* F *form a complete lattice under* $\leq$.

Proof: Left as an exercise.

With the consequence operation Cn in a logical system L there is naturally associated a consequence relation $\Vdash$:

Proposition. 3.4.6. *The consequence operation* Cn *induces a consequence relation*

$$(\Vdash_{Cn}) \qquad (X, \phi) \in \Vdash \text{ iff } \phi \in Cn(X)$$

and in turn $\Vdash$ *induces the consequence operation*

$$(Cn_\Vdash) \qquad Cn(X) = \{\phi \in F_\mathsf{L} \,|\, (X, \phi) \in \Vdash\}.$$

Proof: Left as an exercise.

Definition. 3.4.7. For a logical system $\mathsf{L} = (\mathsf{F}, \Vdash)$, we define a *(logical) consequence relation* $\Vdash$ as a relation $\Vdash \subseteq 2^F \times F$ satisfying the following conditions for any sets $X, Y \subseteq F_\mathsf{F}$ and for arbitrary formulae $\phi, \chi \in F_\mathsf{F}$:

(R1)   If $\phi \in X$, then $(X, \phi) \in \Vdash$
(R2)   If $(X, \phi) \in \Vdash$ and $X \subseteq Y$, then $(Y, \phi) \in \Vdash$
(R3)   If $(X, \phi) \in \Vdash$ and $(Y, \chi) \in \Vdash$ for every $\chi \in X$, then $(Y, \phi) \in \Vdash$

R1-3 can be reformulated so that they more clearly convey three important properties of a logical system, to wit, reflexivity (R), monotonicity (M), and transitivity (T).

Proposition. 3.4.8. *We rewrite R1-R3 as follows:*[16]

(R) $\phi \Vdash \phi$
(M)   If $X \Vdash \phi$, then $X, Y \Vdash \phi$
(T)   If $X \Vdash \phi$ and $X, \phi \Vdash \psi$, then $X \Vdash \psi$

Proof: Left as an exercise.

Reflexivity is the case when $X = \{\phi\}$, and it expresses the fact that every formula is a logical consequence of itself. *Monotonicity* is an important property of some logical systems, prominently so of classical logic, and it conveys the fact that the addition of formulae to a theory does not change its set of consequences. *Transitivity* expresses the fact that if ϕ is a lemma in the proof of a theorem, then we are allowed to "cut" ϕ, i.e. substitute it by its proof, reason why transitivity is also spoken of as *cut*.[17]

Given a logical consequence operation or relation, we can also, or alternatively, speak of *logical inference* and of an *inference system*.

Definition. 3.4.9. Let F be a logical language. An *inference* is a couple (X, ψ) such that $X \subseteq F \subseteq \mathsf{F}$ and $\psi \in F$. An alternative notation is $\psi \in Cn(X)$ or $X \Vdash \psi$. In effect,

---

[16] Note the following abbreviation: we write $X, \phi \Vdash \psi$ for $X \cup \{\phi\} \Vdash \{\psi\}$.

[17] In particular, R3 expresses the fact that if $X \Vdash \phi$ and $Y \Vdash \chi$ for every $\chi \in X$, then χ can be used (as a lemma) to obtain ϕ from Y. Other, not necessarily equivalent, formalizations of T/R3 are

(T′)   If $X \Vdash \phi$ and $\phi \Vdash \psi$, then $X \Vdash \psi$
(T″)   If $X \Vdash \phi$ and $Y, \phi \Vdash \psi$, then $X \cup Y \Vdash \psi$
(T‴)   If $X \cup \{\phi_1, ..., \phi_n\} \Vdash \psi$ and $X \Vdash \phi_i$ for $i = 1, ..., n$, then $X \Vdash \psi$

1. a consequence operation Cn (consequence relation $\Vdash$) on F_F is called an *inference operation* (*inference relation*) on F if it satisfies C1 and C2 (R1 and R2, respectively) for every $X \subseteq F \subseteq F$ and $\psi \in F$.

2. an *inference system* on F is a pair (F, Cn) (a pair $(F, \Vdash)$) where Cn ($\Vdash$, respectively) is an inference operation (relation) on F.

Clearly, a logical system is an inference system.[18]

Definition. 3.4.10. A *deductive system* is a system of Cn. In other words, a deductive system of $Cn(X)$ is the least theory of Cn containing X.

As a matter of fact, any set of sentences X of a logical language (a theory) that contains all its consequences (i.e. $X \supseteq Cn(X)$) can be seen as a deductive system (Tarski, 1930). In particular:

Definition. 3.4.11. A logical system whose consequence relation (operation) satisfies R1-3 (C1-3, respectively) and in addition satisfies the following conditions for finite $X' \subseteq X \subseteq F_L$ and $\psi \in F_L$,

$$\left(\Vdash'\right) \qquad \text{If } X \Vdash \psi, \text{ then } X' \Vdash \psi$$

and

$$\left(Cn'\right) \qquad Cn(X) \subseteq \bigcup \{Cn(X') \,|\, X' \subseteq X\}$$

is a deductive system.

Definition. 3.4.12. Conditions $\Vdash'$ and Cn' define the property of *compactness*.

Exercises

Exercise 3.4.1.1. Consider Definition 3.4.5.7 as a Proposition and prove it.

Exercise 3.4.1.2. Show that conditions C1-3 together are equivalent to the following condition for all $X, Y \subseteq F_F$:

$$(C0) \qquad X \subseteq Cn(Cn(X)) \subseteq Cn(X) \subseteq Cn(X \cup Y)$$

---

[18] We are here taking *inference* in a broad sense that allows for either a syntactical or a semantical characterization. Nevertheless, the expression *rule of inference* (or *inference rule*) should be taken strictly in the syntactical sense (cf. Section 3.4.2 below).

Exercise 3.4.1.3. Let us define a *closure operation* on a set A as a function $\mathsf{C} : 2^A \longrightarrow 2^A$ satisfying the following conditions for all $X, Y \subseteq A$:

C1	$X \subseteq \mathsf{C}(X)$	*Inclusion*
C2	$\mathsf{C}(\mathsf{C}(X)) = \mathsf{C}(X)$	*Idempotency*
C3	If $X \subseteq Y$, then $\mathsf{C}(X) \subseteq \mathsf{C}(Y)$	*Monotonicity*

Show that if C0 (cf. Exercise 3.4.1.2) is satisfied by all $X, Y \in F_\mathsf{F}$, then Cn is a closure operation on F_F.

Exercise 3.4.1.4. The following are the properties of a *(Hausdorff) topological closure* for some sets $X, Y \subseteq A$:

HC1	$X \subseteq \mathsf{C}(X)$
HC2	$\mathsf{C}(\mathsf{C}(X)) = \mathsf{C}(X)$
HC3	$\mathsf{C}(X \cup Y) = \mathsf{C}(X) \cup \mathsf{C}(Y)$
HC4	$\mathsf{C}(\emptyset) = \emptyset$

A mapping $\mathsf{C} : 2^A \longrightarrow 2^A$ on a set A verifying conditions HC1-4 is called a *hull operator*. Is Cn a hull operator? Why (not)?

Exercise 3.4.1.5. Reflect on the impact of compactness (cf. Def. 3.4.12) in a logical system.

Exercise 3.4.1.6. Prove Theorem 3.4.1 and Propositions 3.4.6 and 3.4.8.

Exercise 3.4.1.7. The topic of logical consequence, though clearly formulable in a mathematical language, remains for many a problematic topic in philosophical logic and also in applied logic. Give some thought to the following statements. (You might want to read Sections 3.4.2-5 before.)

1. The [consequence] problem is that consequence relations are specifiable by truth conditions, or by proof-theoretic constraints, independently of anything that might be true of any actual agent. (Gabbay & Woods, 2003)

2. [O]ver a large body of logic, the closure structure is basically its essence and its syntactic and linguistic features are secondary, useful for purposes of understanding though they may be. (Martin & Pollard, 1996)

3. A deductive system provides only a way to study a language's consequence relation, to prove results about it, perhaps even mechanize it. But it does not determine or give rise to that relation. This is why the question of whether a particular deductive system for a particular language is sound and complete is always a sensible, and indeed important, one to ask. (Etchemendy, 1999)

4. The crux of the matter is ... the definition of the term "logical consequence." Until this term has been explained, one does not have an opinion as to the nature of mathematics at all. (Curry, 1963)

5. I do not mean to suggest that "thinking" can proceed very far without something like "reasoning." We certainly need (and use) something like syllogistic deduction, but I expect mechanisms for doing such things to emerge in any case from processes for "matching" and "instantiation" required for other functions. Traditional formal logic is a technical tool for discussing either *everything that can be deduced from some data* or *whether a certain consequence can be so deduced*; it cannot discuss at all what ought to be deduced under ordinary circumstances. (Minsky, 1974)

6. [W]hat distinguishes a logic from an algebra is a concept of logical consequence. (Cleave, 1991).

3.4.2. Syntactical consequence and proof theory

In order to fully understand the notion of *syntactical consequence* we require such fundamental notions as *proof* and *proof system*. These are the objects of the subfield of logic known as *proof theory*. The most central notion of proof theory is that of *inference rule*. We provide the most important aspects of these notions in the definitions that follow. We focus on the syntactical consequence relation, but all the statements below can be reformulated equivalently for the syntactical consequence operation.

Definition. 3.4.13. Given an inference system (F, Cn) that is also a logical system, an *inference rule* $\mathbf{r}$ on a set of formulae $F \subseteq \mathsf{F}$ is a mapping assigning to some finite sequence $\chi_1, ..., \chi_n \in X$, $n \geq 0$, of formulae (the *premises*) a formula ψ (the *conclusion*), i.e. $\mathbf{r} : X \longrightarrow F$ where $X \subseteq F^m$ for some $m = 1, 2, ...$ We write $\mathbf{r}(\chi_1, ..., \chi_n) = \psi$, $\mathbf{r}(\{\chi_1, ..., \chi_n\}, \psi)$, or more commonly,

$$(\mathbf{r}) \qquad \frac{\chi_1, ..., \chi_n}{\psi}$$

or

$$(\mathbf{r}) \qquad \chi_1, ..., \chi_n / \psi$$

We denote by RI the set of inference rules $\{\mathbf{r}_1, ..., \mathbf{r}_n\}$.

1. Given a substitution σ, $\mathbf{r}$ is called a *structural inference rule* of F if it has the form

$$(\mathbf{r}) \qquad \sigma\chi_1, ..., \sigma\chi_n / \sigma\psi$$

2. A rule $\mathbf{r}$ is said to *preserve* a set of formulae X, and X is said to *be closed under* $\mathbf{r}$ iff for all $X' \subseteq X$ and for all formulae ψ, if $\mathbf{r}(X', \psi)$, then $\psi \in X$.

3. A subset $F_0 \subseteq F$ is said to be *closed under a rule of inference* $\mathbf{r}$ provided that $(\chi_1, ..., \chi_n) \in F_0^m \cap X$ implies that $\mathbf{r}(\chi_1, ..., \chi_n) \in F_0$.

Definition. 3.4.14. Given an inference of the form X/ψ:

1. We say that ψ is *an axiom* when $X = \emptyset$ in (the application of) an inference rule $\mathbf{r}(X, \psi)$.

2. X can comprise a set AX of *axiom schemata*, formulae in the set $F^\mu \subseteq \mu^{\mathsf{F}}$ that can be replaced by formulae of F.

Definition. 3.4.15. A *proof system* (or *proof calculus*) $\mathcal{P}$ for a logical language F is a pair (RI, AX) where either RI or AX–but not both–can be empty, i.e. a system of rules of inference and/or axiom schemata. The triple (F, RI, AX) is an instance of a *formal system*, namely an *inference system*.[19]

---

[19]Insofar as the pair (RI, AX) gives rise to a consequence relation (see Def. 3.4.21 below).

Definition. 3.4.16. A *proof* of ψ in $\mathcal{P}$ is a finite collection of rules of inference and/or axioms of $\mathcal{P}$ that leads to concluding that ψ is a member of F_{F}.

We make Definition 3.4.16 formally more precise by means of the notion of *provability* or *derivability*.

Definition. 3.4.17. A formula ψ is *provable*, or *derivable*, *from* a (possibly empty) set of formulae X by means of axioms in the set AX and rules in the set RI iff there is a finite sequence of formulae $\psi_1, ..., \psi_n$ that is a *proof* or a *derivation of ψ from X*, i.e. there is a finite sequence $\psi_1, ..., \psi_n$ such that

1. $\psi_1 \in (X \cup AX \cup RI)$;

2. for every $1 < i \leq n$, either $\psi_i \in (X \cup AX \cup RI)$ or ψ_i is the conclusion of one of the rules of inference $\mathbf{r}_j$, $j = 1, ..., k$ of which the premises are some of the $\psi_1, ..., \psi_{i-1}$;

3. $\psi_n = \psi$.

Definition. 3.4.18. Let ψ be a conclusion in a proof (a derivation). When proven (derived) by means of a rule of inference $\mathbf{r}$ from $X = \emptyset$, then ψ is a *theorem*. If proven (derived) by means of a rule of inference $\mathbf{r}$ from the axioms of a theory, then ψ is called a *theorem of the theory*. In any case, theorems are always provable (i.e. derivable) formulae.

Recall now Definition 3.4.14. Obviously, every axiom is a theorem, i.e. is derivable or provable (from $\emptyset$).

Definition. 3.4.19. Two inference rules are particularly important:

1. *Modus ponens*, i.e.
$$(\text{MP}) \qquad \frac{\phi, \phi \to \psi}{\psi}$$

2. The purely syntactical version of the principle of extensionality (cf. Prop. 3.3.3.2), known as *substitution rule*
$$(\text{SUB}) \qquad \frac{\phi}{\sigma\phi}$$

which states that if a formula ϕ is a theorem, then any of its substitution instances (i.e., extensionally equivalent formulae) is also a theorem.

Given the definitions above, it is now an easy matter to define a syntactical consequence relation. We first provide a very general definition (cf. Def. 3.4.20) and then specify it in terms of a binary and a ternary relation.

Definition. 3.4.20. For a logical system L, a *syntactical consequence relation*, denoted by $\vdash_L$, specifies what conclusions are *derivable* or *provable* in L. Let us denote a proof by the symbol ∎. Let X be a (possibly empty) set of formulae and ψ a formula. We say that ψ is *derivable* from (i.e. is a conclusion of) the set of premises (or assumptions) X iff there is a *proof* ∎ such that we have the relation $X \vdash_∎ \psi$. Otherwise, we write $X \nvdash_∎ \psi$ and call ∎ a *counter-proof* or *refutation*.

We often abbreviate $X \vdash_∎ \psi$ as $X \vdash \psi$, especially when ϕ is an axiom, i.e. when we have $\vdash \phi$. We likewise abbreviate $X \nvdash_∎ \psi$ as $X \nvdash \psi$.

Definition. 3.4.21. Let the proof system $\mathcal{P}$ be given for F, and let $F \subseteq$ F. Then, we can define the *syntactical consequence relation* $\vdash$ as the binary relation $\vdash \subseteq \mathcal{P} \times$ F such that for an arbitrary formula $\psi \in F$ and a (possibly empty) set of formulae $X \subseteq F$ we have

$$(\vdash) \qquad X \vdash \psi \quad \text{iff} \quad X \vdash_{\mathcal{P}} \psi.$$

This is more precisely the *syntactical consequence relation induced by $\mathcal{P}$* as, given $\mathcal{P}$, there is some proof ∎ and a ternary relation $\vdash \subseteq \mathcal{P} \times ∎ \times$ F such that we have

$$X \vdash_{\mathcal{P}} \psi \quad \text{iff} \quad X \vdash_{\mathcal{P},∎} \psi.$$

Unless otherwise stated, we shall take the syntactical consequence relation as coinciding with derivability or provability. Clearly, a rule of inference is an instance of the syntactical consequence relation (cf. Definition 3.4.13.1), and we can now write

$$(\mathbf{r}) \qquad \{\chi_1, ..., \chi_n\} \vdash \psi.$$

actually abbreviating $\chi_1 \wedge ... \wedge \chi_n \vdash \psi$. We can also write MP and SUB as

$$(\text{MP}) \qquad \{\phi, \phi \to \psi\} \vdash \psi$$

and

$$(\text{SUB}) \qquad \phi \vdash \sigma\phi.$$

Obviously, α in Definition 3.2.1 can be rewritten as

$$(\alpha) \qquad \{A_1, ..., A_n\} \vdash B$$

This reformulation should give the reader a clearer notion of argument form and arguments whose conclusions follow logically from the premises by virtue of form alone (cf. Figure 3.2.1).

Consistency is another property of relevance in proof theory. We first define it and then show its importance for the notion of proof-theoretical deduction.

Definition. 3.4.22. We say that a set $X = \{\chi_1, ..., \chi_n\}$ (a theory Θ) is *inconsistent* iff we can derive from it both a formula ψ and its negation, i.e.

$$X \vdash (\psi \wedge \neg\psi)$$

Otherwise, the set X (the theory Θ) is *consistent*.[20]

Important results concerning (in)consistency are expressed in the following two theorems, whose proofs are left as exercises.

Theorem 3.4.2. *If X is inconsistent, then $X \vdash \psi$ for any formula ψ.*

From Theorem 3.4.2, a theorem of *reductio ad absurdum* (RA) follows:

Theorem 3.4.3. *If $X \cup \{\neg\psi\}$ is inconsistent, then $X \vdash \psi$.*

The following theorems, dependent on Theorems 3.4.2-3, are central in proof theory, in particular in the proof theory of classical logic.

Theorem 3.4.4. *($\vdash$-Deduction theorem 1) $X \vdash \psi$ iff $X \cup \{\neg\psi\}$ is inconsistent.*

Theorem 3.4.5. *($\vdash$-Deduction theorem 2) Given a set of formulae $X = \{\chi_1, ..., \chi_n\}$ and a formula ψ, ψ is a syntactical consequence of X iff the formula $((\chi_1 \wedge ... \wedge \chi_n) \to \psi)$ is a theorem, i.e. iff*

$$\text{if } (\chi_1 \wedge ... \wedge \chi_n) \vdash \psi, \text{ then } \vdash (\chi_1 \wedge ... \wedge \chi_n) \to \psi.$$

Example 3.4.1. Recall Definition 3.2.1. Theorem 3.4.5 allows us to rewrite any argument α in the form

$$(\alpha') \qquad (A_1 \wedge ... \wedge A_n) \to B$$

---

[20] For simplicity, we make $\Theta = X$. Note that X can be a singleton, in which case we speak of a formula χ being (in)consistent.

so that we can test whether an argument α is a theorem by applying some proof method (cf. Chapter 5) to

$$\vdash (A_1 \wedge ... \wedge A_n) \to B.$$

In fact, Theorem 3.4.5 generalizes in the following way:

Proposition. 3.4.23. *Given a set of formulae* $X = \{\chi_1, ..., \chi_n\}$ *and a formula* ψ:

$$\text{if } (\chi_1 \wedge ... \wedge \chi_n) \vdash \psi, \text{ then } \vdash (\chi_1 \wedge ... \wedge \chi_n) \to \psi,$$

$$\text{then } (\chi_1 \wedge ... \wedge \chi_{n-1}) \vdash (\chi_n \to \psi),$$

$$\text{then } (\chi_1 \wedge ... \wedge \chi_{n-2}) \vdash (\chi_{n-1} \to (\chi_n \to \psi)),$$

$$\vdots$$

$$\text{then } \vdash \chi_1 \to (... \to (\chi_{n-2} \to (\chi_{n-1} \to (\chi_n \to \psi)))).$$

Theorem 3.4.6. *A formula* ψ *is a logical consequence of a set of formulae* $X = \{\chi_1, ..., \chi_n\}$ *iff the formula* $(\chi_1 \wedge ... \wedge \chi_n \wedge \neg\psi)$ *is inconsistent.*

The proofs of the above theorems are left as exercises.

Theorems are of fundamental importance for the definition, given a logical system L, of the logic **L**:

Definition. 3.4.24. Given a logical system $\mathsf{L} = (\mathsf{F}, \vdash)$, where F is a logical language and $\vdash$ is a syntactical consequence relation, the *logic of* L, denoted by **L**, is defined as

$$\mathbf{L} := \{\phi \in \mathsf{F} | \vdash_\mathsf{L} \phi\}$$

where $\vdash \phi$ abbreviates $\emptyset \vdash \phi$.

In other words, the logic **L** is the set of theorems generated by the logical system L. However, we shall often relax this definition and consider a logic **L** to be a pair $(F, \vdash_\mathsf{L})$ where $F \subseteq \mathbf{L}$ is a set of formulae of a logical language L and $\vdash_\mathsf{L}$ denotes the syntactical consequence relation specified by the logical system L.[21]

---

[21]This relaxation is not without grounds; see Augusto (2020b), pp. 78-9, for some critical remarks.

Exercises

Exercise 3.4.2.1. Rewrite $X \nvdash \psi$ in terms of

1. consistency.

2. inconsistency.

Exercise 3.4.2.2. Prove the following statement: Any $\nvdash$-relation rests on a refutation.

Exercise 3.4.2.3. Prove the theorems in this Section.

Exercise 3.4.2.4. Give the rationale for Proposition 3.4.23.

3.4.3. Semantical consequence and model theory

For a logical system L, the *semantical consequence relation*, denoted by $\models_L$, specifies the class of *valid inferences* in L, or, in other terms, which inferences in L *preserve truth*, or are *deductively valid*. We next provide definitions and theorems that specify the consequence relation from the semantical viewpoint. As above, we concentrate on the semantical consequence relation; equivalent formulations can be given for the semantical consequence operation.

The fundamental concepts of semantics *validity* and *satisfiability* depend on the notion of *interpretation* (cf. Def. 3.3.10). This, in turn, is naturally associated to the most central concept of *model theory*, to wit, *model*.

In what follows, let $X \subseteq F_F$, $\psi \in F_F$, and $val_I : F_F \longrightarrow W$.

Definition. 3.4.25. *Satisfiability* – An interpretation I is said to *satisfy*

1. a formula ψ, and ψ is said to be *satisfiable*, iff there is some val_I such that $val_I(\psi) = t$ and we denote this satisfiability relation by $\models_I \psi$. Otherwise, we write $\nvDash_I \psi$. If there is no interpretation I that satisfies ψ, then ψ is *unsatisfiable*, and we write $\nvDash \psi$.

2. a set of formulae (a theory) $X = \{\chi_1, ..., \chi_n\}$, and X is said to be *satisfiable*, iff there is some val_I such that $val_I(\chi_i) = t$ for all $\chi_i \in X$, and we denote this satisfiability relation by $\models_I X$.[22]

---

[22]This implies that X can be viewed as a conjunction of the formulae χ_i, i.e., $X = \bigwedge_{i=1}^{n} \chi_i$.

Otherwise, we write $\not\models_{\mathcal{I}} X$. If there is no interpretation $\mathcal{I}$ that satisfies X, then X is *unsatisfiable*, and we write $\not\models X$.

Definition. 3.4.26. We say that an interpretation $\mathcal{I}$ is a *model* of

1. ψ, and write $\models_{\mathcal{M}} \psi$, iff $\models_{\mathcal{I}} \psi$; otherwise, we write $\not\models_{\mathcal{I}} \psi$ to denote that there is a *counter-model*.

2. X, and write $\models_{\mathcal{M}} X$, iff $\models_{\mathcal{I}} X$; otherwise, we write $\not\models_{\mathcal{I}} X$ to denote that there is a *counter-model*.

Thus, a formula ψ is said to be satisfiable iff there is an interpretation that is a model of ψ, and a set of formulae (a theory) X is satisfiable iff all its elements (formulae, respectively) have a common model. A counter-model is the case when $val_{\mathcal{I}}(\psi) = \mathbf{f}$, or $val_{\mathcal{I}}(\chi_i) = \mathbf{f}$ for at least one $\chi_i \in X$.

Example 3.4.2. In Example 3.3.4, the interpretation in which we have $val_{\mathcal{I}}(P) = \mathbf{f}$ and $val_{\mathcal{I}}(Q) = \mathbf{t}$ (i.e. the third row of the truth table) is a model for the formula $\neg P \vee Q$; the interpretation in which $val_{\mathcal{I}}(P) = \mathbf{t}$ and $val_{\mathcal{I}}(Q) = \mathbf{f}$ (i.e. the second row of the truth table) is a counter-model for the formula $\neg P \vee Q$.

Definition. 3.4.27. A *semantics* $\mathfrak{S}$ for a logical language F is an infinite set of (classes of) models.

Definition. 3.4.28. *Validity* – Let X be a (possibly empty) set of formulae and ψ a formula entailed from X. We say that ψ is *valid* iff there is no interpretation assigning the value $\mathbf{t}$ to all the members of X and $\mathbf{f}$ to ψ, and we write $X \models \psi$ (abbreviating $X \models_{\mathfrak{S}} \psi$, or $X \models_{\mathcal{M}_i} \psi$ for all $\mathcal{M}_i \subseteq \mathfrak{S}$). A formula is said to be *invalid*, written $X \not\models \psi$, iff it is not valid.

Given these definitions, it is now an easy matter to define a semantical consequence relation as a binary or ternary relation.

Definition. 3.4.29. Let $\mathfrak{S} = \{\mathcal{M}_1, \mathcal{M}_2, ...\}$ where the $\mathcal{M}_i$ are models. Then, we can define the *semantical consequence relation* $\models$ as the binary relation $\models \subseteq \mathfrak{S} \times \mathsf{F}$ such that for an arbitrary formula $\psi \in F_{\mathsf{F}}$ and a (possibly empty) set of formulae $X \subseteq F_{\mathsf{F}}$ we have, for $\mathfrak{S}$,

$$(\models) \qquad X \models \psi \quad \text{iff} \quad X \models_{\mathfrak{S}} \psi.$$

This is more precisely the *semantical consequence relation induced by the semantics* $\mathfrak{S}$ as, given a model $\mathcal{M}_i \in \mathfrak{S}$, we have a ternary relation $\models \subseteq \mathfrak{S} \times \mathcal{M}_i \times \mathsf{F}$ such that

$$\models_{\mathfrak{S}} X \quad \text{iff} \quad \models_{\mathfrak{S},\mathcal{M}_i} \psi.$$

Obviously, Definition 3.4.29 means that ψ is a semantical consequence of X iff every semantics (model) that satisfies X also satisfies ψ. Unless otherwise stated, we shall consider the semantical consequence relation as coinciding with the satisfiability relation.

Definition. 3.4.30. A formula ϕ is said to be

1. a *tautology* iff every interpretation $\mathcal{I}$ of ϕ is also a model of ϕ. In other words, ϕ is a tautology if it uniformly takes the truth value t for any and every assignment of truth values to its variables. We consequently have the set $Taut\,(\mathrm{L})$ of all the tautologies in a logical system L with a language F:

$$Taut\,(\mathrm{L}) = \{\phi \in F_{\mathsf{F}} | val\,(\phi) = \mathsf{t} \text{ for every } val : F_{\mathsf{F}} \longrightarrow W_n\}$$

2. a *contradiction* iff there is no interpretation $\mathcal{I}$ of ϕ that is a model of ϕ. In other words, ϕ is said to be a contradiction if it uniformly takes the truth value f for any and every assignment of truth values to its variables, i.e. for the same system we have

$$Cont\,(\mathrm{L}) = \{\phi \in F_{\mathsf{F}} | val\,(\phi) = \mathsf{f} \text{ for every } val : F_{\mathsf{F}} \longrightarrow W_n\}$$

3. *contingent* iff it is neither a tautology nor a contradiction.

We shall often denote an arbitrary tautology by $\top$ and an arbitrary contradiction by $\bot$.

Proposition. 3.4.31. *It is now obvious that a formula ϕ is valid iff*

1. *every interpretation of ϕ is also a model of ϕ, and*

2. *ϕ is a tautology.*

The valid formulae of a logical system L are fundamental to the definition of the logic **L**:

Definition. 3.4.32. Given a logical system L = (F, ⊨), where F is a logical language and ⊨ is a semantical consequence relation, the *logic of* L, denoted by **L**, is defined as

$$\mathbf{L} := \{\phi \in \mathsf{F}|\models_{\mathrm{L}} \phi\}$$

where ⊨ ϕ abbreviates $\emptyset \models \phi$.

In other words, **L** is the set of tautologies generated by L. However, just as in the case of the syntactical consequence relation, we shall often relax this definition and consider a logic **L** to be a pair $(F, \models_{\mathrm{L}})$ where $F \subseteq \mathbf{L}$ is a set of formulae of a logical language L and $\models_{\mathrm{L}}$ denotes the semantical consequence relation specified by the logical system L.[23]

Proposition. 3.4.33. *It is also obvious that a formula ϕ is*

1. *valid iff its negation ($\neg\phi$) is unsatisfiable.*

2. *unsatisfiable iff its negation is valid.*

3. *invalid iff there is at least one interpretation that falsifies it.*

4. *satisfiable iff there is at least one interpretation that makes it true.*

Proposition. 3.4.34. *If a formula is valid, then it is satisfiable (but not vice-versa). If a formula is unsatisfiable, then it is invalid (but not vice-versa).*

Proof: Left as an exercise.

Theorem 3.4.7. *(⊨-Deduction theorem 1) $X \models \psi$ iff $X \cup \{\neg\psi\}$ is unsatisfiable.*

Proof: (⇒) For any interpretation $\mathcal{I}$, either $\mathcal{I}(\chi_i) = \mathsf{t}$ for all $\chi_i \in X$ and $\mathcal{I}(\psi) = \mathsf{t}$ (hence, $\mathcal{I}(\neg\psi) = \mathsf{f}$), or $\mathcal{I}(\chi_i) = \mathsf{f}$ for some $\chi_i \in X$. Either way, $\mathcal{I}(X \cup \{\neg\psi\}) = \mathsf{f}$.[24]
  (⇐) For any interpretation $\mathcal{I}$, either $\mathcal{I}(\chi_i) = \mathsf{t}$ for all $\chi_i \in X$ and $\mathcal{I}(\neg\psi) = \mathsf{f}$ (hence, $\mathcal{I}(\psi) = \mathsf{t}$), or $\mathcal{I}(\chi_i) = \mathsf{f}$ for some $\chi_i \in X$. Therefore, $\mathcal{I}(\psi) = \mathsf{t}$ whenever $\mathcal{I}(X) = \mathsf{t}$, and thus $X \models \psi$. **QED**

It will be useful to provide an equivalent formulation of the deduction theorem:

---

[23] See Augusto (2020b), pp. 78-9, for some critical remarks on this definition.
[24] $\mathcal{I}(\phi)$ abbreviates $val_{\mathcal{I}}(\phi)$.

3. Preliminaries: Formal logic, deduction, and deductive computation

Theorem 3.4.8. ($\models$-*Deduction theorem 2*) *Given a set of formulae* $X = \{\chi_1, ..., \chi_n\}$ *and a formula* ψ, ψ *is a logical consequence of* X *iff the formula* $((\chi_1 \wedge ... \wedge \chi_n) \rightarrow \psi)$ *is valid, i.e. iff*

$$\text{if } (\chi_1 \wedge ... \wedge \chi_n) \models \psi, \text{ then } \models (\chi_1 \wedge ... \wedge \chi_n) \rightarrow \psi.$$

Proof: The proof follows immediately from the above. **QED**

Example 3.4.3. Recall Definition 3.2.1. Theorem 3.4.8 allows us to rewrite any argument α in the form

$$(\alpha') \qquad (A_1 \wedge ... \wedge A_n) \rightarrow B$$

so that we can test whether an argument α is valid by applying some validity testing method (cf. Chapter 5) to

$$\models (A_1 \wedge ... \wedge A_n) \rightarrow B.$$

The above allows a reformulation of the definition of *logical equivalence* (cf. Defs. 3.3.8 and 3.3.14):

Definition. 3.4.35. Two formulae ϕ and ψ are said to be *equivalent*, and we write $\phi \equiv \psi$, iff

1. they have exactly the same truth value in every interpretation.

2. $\phi \models \psi$ and $\psi \models \phi$.

3. $\phi \leftrightarrow \psi$ is a tautology, i.e. $\models (\phi \leftrightarrow \psi)$.

Theorem 3.4.9. *A formula* ψ *is a logical consequence of a set of formulae* $X = \{\chi_1, ..., \chi_n\}$ *iff the formula* $(\chi_1 \wedge ... \wedge \chi_n \wedge \neg\psi)$ *is unsatisfiable.*

Proof: It follows from Theorem 3.4.6 that ψ is a logical consequence of $X = \{\chi_1, ..., \chi_n\}$ iff the negation of $((\chi_1 \wedge ... \wedge \chi_n) \rightarrow \psi)$ is unsatisfiable. In effect,

$$\neg((\chi_1 \wedge ... \wedge \chi_n) \rightarrow \psi) \equiv \chi_1 \wedge ... \wedge \chi_n \wedge \neg\psi$$

by the equivalences $\phi \rightarrow \psi \equiv \neg\phi \vee \psi$, $\neg(\phi \vee \psi) \equiv \neg\phi \wedge \neg\psi$ and $\neg\neg\phi \equiv \phi$, and by the property of associativity of $\wedge$. **QED**

Theorem 3.4.10. *A formula ϕ is unsatisfiable iff it entails a contradiction, i.e. iff we have*

$$\phi \models (\psi \wedge \neg\psi).$$

Proof: By the $\models$-deduction theorem, we have (i) $\phi \models (\psi \wedge \neg\psi)$ iff we have (ii) $\models \phi \rightarrow (\psi \wedge \neg\psi)$. In turn, we have (ii) iff, for every interpretation $\mathcal{I}$, (a) $val_{\mathcal{I}}(\phi) = \mathtt{f}$ or (b) $val_{\mathcal{I}}(\phi) = \mathtt{t}$ and $val_{\mathcal{I}}(\psi \wedge \neg\psi) = \mathtt{t}$. But the truth table of $\psi \wedge \neg\psi$ shows that it is a contradiction, and thus for every interpretation $\mathcal{I}$ we have $val_{\mathcal{I}}(\psi \wedge \neg\psi) = \mathtt{f}$. Therefore, we must have (a). Thus, $\models \phi \rightarrow (\psi \wedge \neg\psi)$ iff ϕ is unsatisfiable, and $\phi \models (\psi \wedge \neg\psi)$ iff ϕ is unsatisfiable. **QED**

Exercises

Exercise 3.4.3.1. Generalize Theorem 3.4.8 and give the rationale for this generalization.

Exercise 3.4.3.2. Give all the steps of the proof of Theorem 3.4.9.

Exercise 3.4.3.3. Prove (Complete the proof of) the theorems and propositions of this Section left without a proof (with a sketchy proof, respectively).

3.4.4. Adequateness of a deductive system

Now suppose that our logical system L at hand is equipped with both a syntactical and a semantical consequence relation such that we have the deductive system $\mathsf{L} = (\mathsf{F}, \Vdash)$ satisfying the following theorem:

Theorem 3.4.11. *(Deduction theorem) For a (possibly empty) set $X \subseteq F_\mathsf{F}$ and for any formulae $\phi, \psi \in F_\mathsf{F}$,*

$$\text{(DT)} \qquad \text{If } X, \phi \Vdash \psi, \text{ then } X \Vdash \phi \rightarrow \psi$$

Proof: Proofs for $\vdash$ and $\models$ were given or sketched in above Sections in this Chapter. **QED**

Any logical system $\mathsf{L} = (\mathsf{L}, \Vdash)$ in which DT holds is indeed a deductive system. Consider now that the symbol $\Vdash$, otherwise employed to denote indifferently a syntactical consequence relation $\vdash$ or a semantical consequence relation $\models$, denotes the *coincidence* of both these relations into a single consequence operation. Then, we say that the logical system L is both sound and complete.

We next give the fundamental concepts of *soundness* and *completeness* of a logical system formulated as theorems. We consider DT in the two versions for $\vdash$ and $\models$, denoted by $DT_\vdash$ and $DT_\models$, respectively.

Theorem 3.4.12. L *is* sound *if, if* $X \vdash_L \phi$, *then* $X \models_L \phi$.

Proof: (Sketch) We first sketch a proof when $X = \emptyset$. Assume that $\vdash_L \phi$. Then, there is a sequence $\phi_1, ..., \phi_n$ that is a proof of ϕ in L in which every step is either an axiom or derived from previous steps by means of rules of inference (cf. Def. 3.4.17). The idea is to show, by induction, that every step of the proof is a tautology: assume that all steps up to ϕ_i are tautologies; it is now necessary to show that ϕ_i itself is a tautology. But ϕ_i is either an axiom, in which case it is a tautology (a truth-table verification of the axioms of the deductive system at hand may elucidate this), or is derived from a previous result by means of a rule of inference, which assures us that it is also a tautology. Therefore, $\models_L \phi$ by proof induction up to $\phi_n = \phi$. For $X = \{\chi_1, ..., \chi_n\}$, we have $\vdash (\chi_1 \rightarrow (\chi_2 \rightarrow ... (\chi_n \rightarrow \phi)))$ by multiple applications of $DT_\vdash$ from $\{\chi_1, ..., \chi_n\} \vdash \phi$; by soundness, we conclude $\models (\chi_1 \rightarrow (\chi_2 \rightarrow ... (\chi_n \rightarrow \phi)))$, and hence $\{\chi_1, ..., \chi_n\} \models \phi$. **QED**

As can be seen in its proof, Theorem 3.4.12 is actually a corollary of the case when $X = \emptyset$.

Theorem 3.4.13. L *is* strongly complete *if, if* $X \models_L \phi$, *then* $X \vdash_L \phi$.

Proof: (Sketch) Let us assume that $X \models_L \phi$. Then, $X \cup \{\neg\phi\}$ does not have a model in L. This means that we can have $X \cup \{\neg\phi\} \vdash \bot$ for some contradiction $\bot$, and $X \cup \{\neg\phi\}$ is inconsistent. Hence, by Theorems 3.4.2-3, we have $X \vdash_L \phi$. **QED**

Just as in the case of Theorem 3.4.12, this is actually the corollary to the *completeness theorem*

$$\text{if } \models_L \phi, \text{ then } \vdash_L \phi.$$

In Section 4.4, we prove the completeness of CFOL.

Theorem 3.4.12 expresses the fact that everything that is provable in L (i.e. every theorem of L) is also logically true, i.e., L does not prove falsities. As for Theorem 3.4.13, it expresses the fact that every logical truth in L is provable in L (i.e. is a theorem of L); that is to say that L requires no additional inference rules to prove every logical truth in L, being thus complete in this sense. Hence, with Gödel (1930) we say that a logical system or a theory is complete (incomplete) if every logical truth of the system/theory is (not, respectively) a theorem thereof. Together, soundness and completeness express the fact that in a deductive system L one can derive everything one should (i.e., the system is complete), and nothing one should not (the system is sound). However, it is obvious that if one has to choose, then one should choose soundness; completeness is a nice property of a logical system, but one has to make sure first and foremost that one does not prove falsities.

Given Theorems 3.4.12-3, the following proposition is obvious.

Proposition. 3.4.36. *L is sound if $\vdash_L \subseteq \models_L$ and complete if $\models_L \subseteq \vdash_L$.*

Proof: Left as an exercise.

Definition. 3.4.37. Given the deductive system $L = (F, \Vdash)$, we say that the axiomatization $\mathcal{P}$ of L is *adequate* whenever the set of theorems of L coincides with the set of tautologies of L, i.e.

$$\vdash_{L,\mathcal{P}} \psi \quad \text{iff} \quad \models_L \psi.$$

We say that a semantics $\mathfrak{S}$ is *adequate* for a deductive system L whenever the set of tautologies of L coincides with the set of theorems of L, i.e.

$$\models_{L,\mathfrak{S}} \psi \quad \text{iff} \quad \vdash_L \psi.$$

Proposition. 3.4.38. *A deductive system $L = (F, \Vdash)$ is adequate iff, given a (possibly empty) set of formulae X and a formula ψ, $X \subseteq F_F$ and $\psi \in F_F$, we have*

$$X \vdash_{L,\mathcal{P}} \psi \quad \text{iff} \quad X \models_{L,\mathfrak{S}} \psi.$$

Proof: Trivial. **QED**

The property of *adequateness* is especially desirable when talking about theories, i.e. deductively closed sets of formulae (see Section 3.4.5 below). We thus say that a deductive system L is adequate for, say, group theory. It is also a central property for (automated) theorem proving, as we can answer questions related to validity by ultimately checking for derivability (cf. Fig. 3.4.1). In particular, Herbrand's theorem (see Section 7.3.3) relates validity with satisfiability, and relates the latter with consistency by appealing to computable functions.

Theorem 3.4.14. $\chi_1, ..., \chi_n \Vdash \psi$ *iff* $\{\chi_1, ..., \chi_n, \neg\psi\}$ *is unsatisfiable / inconsistent.*

Proof: Left as an exercise.

Example 3.4.4. Recall from Examples 3.4.1 and 3.4.3 that we can rewrite any argument α in the form

$$(\alpha') \qquad (A_1 \wedge ... \wedge A_n) \to B.$$

Adequateness allows us to test for theoremhood or validity by applying either a proof method or a validity-test method to

$$\Vdash (A_1 \wedge ... \wedge A_n) \to B.$$

In particular, Theorem 3.4.14 allows us to test an argument α for theoremhood or validity by testing whether with respect to $\neg\alpha'$ we have

$$(\neg\alpha_{ref}) \qquad \nVdash A_1 \wedge ... \wedge A_n \wedge \neg B.$$

If $\neg\alpha_{ref}$ holds, then α' is a theorem or valid formula, and the argument α is a correct or valid argument.

Importantly, if a deductive system L for a specific logic is adequate, then it is equivalent to any other adequate deductive system S for the same logic, i.e. any proof in L can be converted to a proof in S.

We now present an important result for the adequateness of CFOL. Recall the deduction theorem (DT) above. We have the following important lemma:

Lemma 3.4.15. *If* $\Vdash \phi(x) \to \psi(x)$, *then* $\Vdash \forall x\,(\phi(x)) \to \forall x\,(\psi(x))$.

Proof: We give the semantical version of the proof. Assume that we have $\models \phi(x) \to \psi(x)$. Let $\mathcal{I}$ be an interpretation and let $\varpi \in \mathcal{I}$ be a variable assignment with $val_\varpi\,(\forall x \phi(x)) = \mathsf{t}$. Then, for all x-variants ϖ' of ϖ we have $val_{\varpi'}\,(\phi(x)) = \mathsf{t}$. By assumption, we have $val_{\varpi'}\,(\psi(x)) = \mathsf{t}$, and consequently $val_\varpi\,(\forall x \psi(x)) = \mathsf{t}$. **QED**

Exercises

Exercise 3.4.4.1. A stronger theorem than DT (Theorem 3.4.11) is the *deduction-detachment theorem*, formalized as

$$(\text{DDT}) \qquad X, \phi \Vdash \psi \quad \text{iff} \quad X \Vdash \phi \to \psi$$

MODEL THEORY	PROOF THEORY
Validity Given a semantics $\mathfrak{S}$, we have $\chi_1, ..., \chi_n \models \psi$ iff, if $\models_\mathfrak{S} \chi_1, ..., \chi_n$, then $\models_\mathfrak{S} \psi$	*Derivability* Given a proof $\blacksquare$, we have $\chi_1, ..., \chi_n \vdash \psi$ iff $\chi_1, ..., \chi_n \vdash_\blacksquare \psi$
Satisfiability Given a model $\mathcal{M}$, $\{\chi_1, ..., \chi_n\}$ is satisfiable iff $\models_\mathcal{M} \chi_1$ ... and ... and $\models_\mathcal{M} \chi_n$	*Consistency* $\{\chi_1, ..., \chi_n\}$ is consistent iff $\{\chi_1, ..., \chi_n\} \nvdash (\psi \wedge \neg\psi)$
$\Downarrow$ $\models$-deduction theorems $\mathrm{DT}_\models$	$\Downarrow$ $\vdash$-deduction theorems $\mathrm{DT}_\vdash$
$\underrightarrow{\textit{Completeness}}$	$\underleftarrow{\textit{Soundness}}$
THEOREMS	**3.4.11** and **3.4.14**

Figure 3.4.1.: Adequateness of a deductive system $\mathsf{L} = (\mathsf{L}, \Vdash)$.

1. Explain informally why DDT is stronger than DT.

2. Prove DDT.

Exercise 3.4.4.2. Our proof of Theorem 3.4.13 above is a summary of the proof first given by L. Henkin (1949) for CFOL. Research into the details of this proof.

3.4.5. Logical theories

As said above, the adequateness of a logical system is so with respect to the theories that are formalizable within it. We begin by providing a few precisions with respect to theories, so far treated largely as unspecified sets of formulae.

Definition. 3.4.39. A *theory* Θ is a deductively closed (sub)set of formulae of some logical language F, i.e. $\Theta = F_{\mathsf{F}}^{(')} \cup AX \cup RI$.

1. Let Θ be a theory. Θ^* is said to be an *extension* of Θ if every theorem of Θ is a theorem of Θ^*, i.e. if $\Theta \subset \Theta^*$, and Θ' is said to be a *subtheory* of Θ if $\Theta' \subset \Theta$.

2. A theory Θ is said to be *consistent* if it is *not* the case that we have *both* $\Theta \vdash \phi$ *and* $\Theta \vdash \neg\phi$ for some formula ϕ. Otherwise, Θ is *inconsistent*.

3. A theory Θ is *complete* if it is the case that *either* $\Theta \vdash \phi$ *or* $\Theta \vdash \neg\phi$ for some formula ϕ. Otherwise, Θ is *incomplete*.

Recall now Definition 3.4.14.

Definition. 3.4.40. Let $\Theta = F_{\mathsf{F}}^{(')} \cup AX \cup RI$ be a FO theory.

1. An axiom $\psi \in AX$ is said to be a *logical axiom* if it is an axiom schema of the utmost generality, i.e. with no domain specified.

2. An axiom $\psi \in AX$ is a *non-logical* or *proper axiom* if it is formulated for a specific domain.

Definition. 3.4.41. Let $\Theta = F_{\mathsf{F}}^{(')} \cup AX \cup RI$ be a FO theory. If all the $\psi \in AX$ are logical axioms, then Θ is a FO *calculus*.

Clearly, a FO theory Θ is a FO calculus Θ' together with the proper axioms $AX\,(\Theta) - AX\,(\Theta')$.

Example 3.4.5. We give an example of a theory and a calculus:

- In Exercise 3.2.2.4, (1)-(3) are the proper axioms of *group theory*. If the conclusion does indeed follow from the premises (it does), then it too is part of the theory, i.e. is a theorem thereof, as are the rules of inference–if any–that were employed to prove it.

- Below, in Section 5.1,[25] £1-3 and Q1-2 are logical axioms; these five axioms constitute–together with the rules MP and GEN–the *Frege-Lukasiewicz calculus*.

We expand now on group theory from the viewpoint of logical theories.

Example 3.4.6. Let group theory be the theory of Exercise 3.2.2.4 and denote it by $\Theta_{\mathcal{G}}$. The model for $\Theta_{\mathcal{G}}$ is (called) a group. Obtain the extension $\Theta_{\mathcal{G}}^{\divideontimes}$ by adding to $\Theta_{\mathcal{G}}$ the following proper axiom:[26]

$$\forall x \forall y\,(x \star y = y \star x)$$

Then, $\Theta_{\mathcal{G}}^{\divideontimes}$ is the theory of Abelian groups, and a model for $\Theta_{\mathcal{G}}^{\divideontimes}$ is (called) an Abelian group. Restrict now $\Theta_{\mathcal{G}}$ to the axiom of associativity; then, the subtheory $\Theta_{\mathcal{G}}'$ obtained is the theory of semigroups and a model for $\Theta_{\mathcal{G}}'$ is (called) a semigroup (cf. Def. 1.2.4). Extend $\Theta_{\mathcal{G}}'$ with the axiom of the identity element; then, $\Theta_{\mathcal{G}}'^{\divideontimes}$ is the theory of monoids and a model for $\Theta_{\mathcal{G}}'^{\divideontimes}$ is (called) a monoid (cf. Def. 1.2.4). Let $\Theta_{\mathcal{G}}'^{\divideontimes} = \Theta_{\mathcal{G}}''$; then,

$$\Theta_{\mathcal{G}}' \subset \Theta_{\mathcal{G}}'' \subset \Theta_{\mathcal{G}} \subset \Theta_{\mathcal{G}}^{\divideontimes}$$

which entails that $\Theta_{\mathcal{G}}^{\divideontimes}$ has at least one model. This, in turn, by the compactness property entails that $\Theta_{\mathcal{G}}^{\divideontimes}$ has a model.

The following theorem on consistent theories will play an important role in the proof of completeness of the classical FO predicate calculus (see Section 4.4).

---

[25]Cf. §§5.5.1-2.

[26]For convenience, we introduce the symbol for equality (=) in F1, *but see Section 4.5 below*.

Theorem 3.4.16. *(Lindenbaum's theorem) If Θ is a consistent theory, then there is a consistent, complete extension Θ^* of Θ.*

Proof: Left as an exercise.

Exercises

Exercise 3.4.5.1. Prove the following statement: *Every FO predicate calculus is consistent.*

Exercise 3.4.5.2. Let $\Theta \nvdash \neg\phi$, ϕ is a ground formula. Prove that if we extend Θ by adding ϕ as a new axiom, then Θ^* is consistent.

Exercise 3.4.5.3. Prove Lindenbaum's theorem.

3.5. Deductive computation

As seen above, what truly distinguishes a logical language from other formal languages is the fact that the former can be used from the perspective of logical consequence. That is to say that with a formal language that is also a logical language we can obtain formulae from (sets of) formulae, not in terms of rewriting, but in terms of consequence or deduction. When this is carried out by means of implementations in a computing device, such as a digital computer, we call it *deductive computation*.

As for other kinds of computation implementable in a computing device, time and space requirements need to be considered in deductive computation; this we do in Section 3.5.3. As for decidability, issues specific to FO logical theories and FO deductive systems need to be addressed; this we do in Section 3.5.2. Before these preliminary discussions on decidability and complexity issues for logical languages and deductive systems, we elaborate on logical problems as computational problems.

3.5.1. Logical problems and computational solutions

Recall now the material above on validity and satisfiability (Section 3.4.3), as well as the discussion on decision problems (Section 2.3).

Definition. 3.5.1. Given some model of computation M, a *logical problem* is defined as

$$LOGP_M = \{\langle \Theta, \phi \rangle \mid \Theta \models \phi\}$$

where Θ is a logical theory (or, more simply, a–possibly empty–set of logical formulae) and ϕ is a logical formula, and $\langle \Theta, \phi \rangle = e(\Theta)\, e(\phi)$ for e an appropriate encoding.

1. If we wish to know whether $\Theta \models \phi$ holds in every interpretation, then we speak of the *validity problem*, and abbreviate it as VAL.

2. If we wish to know whether $\Theta \models \phi$ holds in at least one interpretation, then we speak of the *satisfiability problem*, and abbreviate it as SAT.[27]

Clearly, LOGP is a *computational problem* in the strict sense (cf. Def. 2.3.8) if the pair (Θ, ϕ) (or $(\emptyset, \phi)$) is a solution for (an instance of) the problem LOGP, i.e. if we have $\phi \in Cn(\Theta)$ ($\phi \in Cn(\emptyset)$, respectively). Moreover, it is a *decision problem*, now in the sense that we wish to know whether $\Theta \Vdash \phi$ for ϕ a logical formula and a logical theory Θ (or a possibly empty set Θ). Recast in the original terms of an unspecified formal language, we wish to know whether $\phi \in \Theta$. If we identify the logical theory Θ with the logical language F, then the decision problem for a logical formula ϕ is that of whether $\phi \in$ F (cf. Def. 2.3.1).

This means two things: (i) given some computation model or device M (the human brain or a Turing machine, for instance), there is a computation of ϕ from Θ that is a finite sequence of configurations of M (cf., for example, Def. 2.2.4), and (ii) this computation ends with a "Yes/No" answer.

With respect to the first point, given some computation model or device M the deduction theorems assure us that there is a finite sequence of configurations of M that is a computation from a finite theory $\Theta = \{\chi_1, ..., \chi_n\}$ to a formula ϕ, as we have the configuration

$$(C_0) \qquad \{\chi_1, ..., \chi_n\} \vdash_M \phi$$

iff we have the configurations

$$(C_1) \qquad \{\chi_1, ..., \chi_{n-1}\} \vdash_M (\chi_n \to \phi)$$

$$(C_2) \qquad \{\chi_1, ..., \chi_{n-2}\} \vdash_M (\chi_{n-1} \to (\chi_n \to \phi))$$

$$\vdots$$

---

[27]Note, however, that more properly we use the abbreviation SAT for the *propositional* satisfiability problem. We shall disambiguate, if need occurs. See Section 3.5.3 for the diverse satisfiability problems.

$$(C_f) \qquad \vdash_M \chi_1 \to (\dots \to (\chi_{n-2} \to (\chi_{n-1} \to (\chi_n \to \phi))))$$

We considered the sequence of configurations above as actually sequences of proofs of M, but we could have done so for models of M, as given an interpretation $\mathcal{I} = (\mathcal{D}, \Theta, \varpi)$ we have a sequence of n models $\mathcal{M}_i$ and n configurations C_i, $i = 1, \dots, n$, where $n = f$, such that we have C_i iff we have $\mathcal{M}_i$.

We make this formally precise:

Definition. 3.5.2. Given a finite theory $\Theta = \{\chi_1, \dots, \chi_n\} = \{\chi_j\}_{j=1}^n$ and a formula ϕ, we say that there is a *computational deduction* of ϕ from Θ, written $C_0 = \{\chi_j\}_{j=1}^n \Vdash_M \phi$, if, given some computational model or device M, there is a sequence of $i = 1, 2, \dots, n$ configurations C_i of M, $|\{C_i\}_{i=1}^n| = |\Theta|$, such that we have for some $1 < l < n$,

$$C_i = \{\chi_j\}_{j=1}^{n-i} \Vdash_M \left(\underbrace{\chi_l \to (\cdots (\chi_{n-1} \to (\chi_n \to \phi)))}_{i+1 \text{ formulae}} \right)$$

where we write $\vdash_M$ if M produces a proof or $\models_M$ if M produces a model.

1. If we have the above for all the j up to n such that we end up with no χ_j on the left of $\Vdash_M$, then there is a "Yes" answer. Taking into consideration the fact that Θ may be empty, we actually have it that M reaches configuration C_f in zero or more iterations, so that in fact we have

$$C_0 \vdash_M^* C_f \quad \text{iff} \quad \Theta \Vdash_M^* \phi.$$

2. A "No" answer occurs if for any of the i, $i = 1, \dots, n-1$, we have either $\nvdash_M^*$ or $\nvDash_M^*$, in which case we have

$$\Theta \nVdash_M^* \phi.$$

In either answer given by M, $\Theta \Vdash_M^* \phi$ or $\Theta \nVdash_M^* \phi$, we say that the logical / computational problem LOGP is *decidable*.

Definition. 3.5.3. A theory is *decidable* if there is an algorithm for it that terminates with a "Yes/No" answer.

We know, however, from Chapter 2 that there is yet another possibility: the computation of M may not stop. If the theory Θ is actually or virtually infinite–for instance, it has an infinite domain of discourse–, then it should not be surprising that we have no final answer; that is, the theory is *undecidable*. Otherwise, we should expect an answer, but it may be the case that, if in fact $\Theta \not\vdash \phi$, we end up with no answer, too. This, *semi-decidability*, is accounted for by M entering a loop (cf., for example, Def. 2.2.25.7). In any case, we say that the decidability of a theory depends on there being a procedure for it that is in fact a decision procedure.

Definition. 3.5.4. A procedure is a *decision procedure* for a theory Θ if it is sound *and* complete with respect to Θ.

These two terms, *sound* and *complete*, are already well known to us, but we now specify them with respect to theories and decision procedures therefor:

Definition. 3.5.5. Let Ψ be a procedure. Given an input (Θ, ϕ),

1. Ψ is said to be *sound* if when it returns $\Theta \vdash \phi$, it is the case that $\Theta \models \phi$.

2. Ψ is said to be *complete* if

   a) it always terminates, and

   b) it returns $\Theta \models \phi$ when it is the case that $\Theta \vdash \phi$.

Although there are undecidable propositional theories, propositional logic is generally decidable. That is to say that there is a decision procedure for it. Given a formula $\phi \in$ F0, we can always say whether $\Vdash \phi$ or $\not\Vdash \phi$ by means of a truth table, for instance. We also know that FOL is generally undecidable, being at best semi-decidable. This feature of FOL is directly connected to its *expressiveness*, a property it owes to quantification associated with infinite domains of discourse. But, in fact, semi-decidability is a significant improvement with relation to undecidability, because some procedure Ψ *always* terminates on input (Θ, ϕ) if indeed we have $\Theta \models \phi$. Thus, all we require is a complete deductive system, i.e. a system in which there is a complete procedure for the computational problem LOGP.

Exercises

Exercise 3.5.1.1. Formulate the Hamiltonian Path Problem HAM-PATH (cf. Example 2.3.5) as a logical problem.

Exercise 3.5.1.2. Let Θ be an axiomatizable theory. Show that if Θ is complete, then Θ is decidable.

Exercise 3.5.1.3. Show that the theory of groups $\Theta_{\mathcal{G}}$ is decidable by outlining a decision procedure Ψ for it.

Exercise 3.5.1.4. Give examples of undecidable theories.

3.5.2. Taming FOL undecidability

3.5.2.1. Finite satisfiability and ground extensions

In Chapter 4.4, we shall prove the completeness of the classical FO deductive system, but we begin now by proving the completeness of an arbitrary FO deductive system with the logical language F1.[28]

But we want to do more than prove FO completeness; we want to do this while at the same time reducing this property to propositional completeness, as this guarantees decidability, rather than only semi-decidability. With this objective in mind, if we present completeness as a consequence of compactness,[29] then we want to reduce FO-compactness to propositional-compactness, too. In order to do this some further elaboration on contents already discussed is required, namely on satisfiability; some anticipations are also needed, namely with respect to material from Section 6.2 and, especially, Section 7.3.3. Internal references aid the reader in these recallings and anticipations.

Recall from Definition 3.4.25 that a set of formulae X is satisfiable iff there is an interpretation $\mathcal{I}$ that satisfies all the elements of X; in other words, X is satisfiable iff every element thereof has a (common) model.

Definition. 3.5.6. *Finite and infinite satisfiability* – Let $X \subseteq F_{\mathsf{F1}}$ be a (possibly infinite) satisfiable set of FO formulae. We say that X is *finitely satisfiable* iff every finite subset $X'_i \subseteq X$ has a model; otherwise, X is said to be *infinitely satisfiable*.

---

[28] Clearly, once the latter is done, the former is also proven, but we will introduce some aspects in the former proof that are, if not of computational interest, of import for a general theory of classical theories and CFOL.

[29] For convenience, we choose to support our proof of FO completeness on an important property already mentioned: *compactness* (cf. Def. 3.4.12).

Obviously, a satisfiable finite set of formulae is trivially finitely satisfiable, and a finite set of propositional formulae is always finitely satisfiable if it is satisfiable. The importance of finite satisfiability resides in the well-known property of the infinity of models for FO formulae. This infinity is, in turn, accounted for by the infinity of domains for a FO language.[30] We know that a valuation $val_{\mathcal{I}}(\phi)$ for any FO formula $\phi(x_1, ..., x_n)$ given any interpretation $\mathcal{I} = (\mathscr{D}, \Theta, \varpi)$ depends on the variable assignment ϖ that replaces all bound or free variables $\vec{x} = x_1, ..., x_n$ of ϕ by some constants $\vec{a} = a_1, ..., a_n$ in the domain $\mathscr{D}$; thus, this valuation is more precisely a valuation $val_{\mathcal{I}_{\vec{a}}^{\vec{x}}}(\phi)$, so that a FO formula $\phi(\vec{x})$ is said to be *(un)satisfiable in a domain* $\mathscr{D}$. If the domain is infinite, the interpretations $\mathcal{I}_{\vec{a}}^{\vec{x}}$ are so, too, and so are the models of ϕ.

So, we have it that a satisfiable FO formula ϕ can be either finitely or infinitely satisfiable. If we can determine that ϕ is finitely satisfiable, this is an improvement over tackling infinite satisfiability, which is of little to no practical use (see Section 7.3.3). Is there a way, given some signature Υ_ϕ, to fix the interpretations of ϕ such that ϕ has a *fixed* interpretation that is independent of $\mathscr{D}$ and thus of ϖ, too? In other words, is there a way to make a FO formula ϕ behave like a propositional formula?

Indeed there is. All we need to do is, given a FO formula, firstly remove all the quantifiers and substitute all individual variables by ground terms, and secondly, substitute all obtained ground atoms by propositional variables. That is, we first need a ground extension, and then a transfer function mapping ground atoms to propositional variables. We next explain this jargon.[31]

---

[30]This infinity with respect to the domains is double: for any FO language, there may be arbitrarily finitely or infinitely many domains, and these domains may themselves be arbitrarily finitely or infinitely large.

[31]We apply here somehow loosely what is known as the *transfer principle*, which allows FO assertions in a (algebraic) structure $\mathcal{A}$ to be transferred or translated to another (algebraic) structure $\mathcal{B}$, called its *(elementary) extension*, by means of a transfer or translation function $\tau : \mathcal{A} \longrightarrow \mathcal{B}$ in such a way that

$$\mathcal{B} = \tau(\mathcal{A}).$$

This is a principle with important applications, namely in non-standard analysis, where it is applied to guarantee that there is a non-standard model $\mathcal{Q} = \tau(\mathcal{R})$ of the standard model $\mathcal{R}$ of the reals. An elaboration on this topic is outside the scope of this book, but two aspects need specification: firstly, because we often identify the structure $\mathcal{A}$ with its universe A, the transfer principle can be applied to (infinite) sets; secondly, the function τ is more rigorously applied to *superstructures*, i.e. $\tau : \mathcal{S}_\infty(A) \longrightarrow \mathcal{S}_\infty(B)$, given $\mathcal{S}_0(A) = A$, $\mathcal{S}_n(A) = \mathscr{P}\left(\bigcup_{k=0}^{n-1} \mathcal{S}_k(A)\right)$ for $\mathscr{P}(A) = 2^A$, and $\mathcal{S}_\infty(A) = \bigcup_{k=0}^\infty \mathcal{S}_k(A)$, $\mathcal{S}_\infty(A)$ is the superstructure over A. In more logical terms, the identity $\mathcal{B} = \tau(\mathcal{A})$ entails that $\mathcal{B}$ inherits all the FO properties of $\mathcal{A}$ in such a way that, given a *FO model* $\mathcal{A}$, we have $X \models_\mathcal{A} \phi$ iff $X \models_\mathcal{B} \phi$. The

Definition. 3.5.7. Let us consider the FO formula ϕ in the form

$$\forall x_1, ..., \forall x_n \, (\phi_M)$$

where ϕ_M is quantifier-free, only atoms are negated in ϕ_M, and the only connectives in ϕ_M are $\wedge$ and $\vee$.[32] Let us consider also the substitution set $\sigma = \{x_1 \mapsto t_1, ..., x_n \mapsto t_n\}$, where every $t_i \in T_g \, (\phi)$ and $T_g \, (\phi)$ is the set of ground terms in ϕ. Then, the *ground extension* of ϕ is defined as

$$GE \, (\phi) = \{\phi_M \, [x_1/t_1, ..., x_n/t_n] \, | t_i \in T_g \, (\phi)\} \, .$$

By this means, we have now actually a set of quantifier-free formulae $\Phi^g = GE \, (\phi)$ to which the following lemma can be applied:

Lemma 3.5.1. *Let ϕ be a FO formula. Then, ϕ is satisfiable iff $GE \, (\phi)$ is satisfiable.*

Proof: (Idea) We know that satisfiability of a FO formula $\phi \, (x_1, ..., x_n) = \phi \, (\vec{x})$ depends on there being at least one model $\mathcal{M}$ in an interpretation $\mathcal{I}_{\vec{t}}^{\vec{x}}$ in which every variable in $\vec{x}$ is replaced by a ground term in $\vec{t}$. In assigning to every variable in $\vec{x}$ a term in $\vec{t}$, the quantifiers are eliminated and we obtain a quantifier-free formula ϕ^g only with ground terms–i.e. a ground formula–that is equisatisfiable to ϕ, a relation we denote by $\phi \equiv_{sat} \phi^g$ and which means that ϕ is satisfiable iff ϕ^g is. The proof of this equisatisfiability relation is by induction on the single-variable FO formula $\phi \, (x)$, for which we have $\models_{\mathcal{M}} \forall x \, (\phi \, (x))$ iff $\models_{\mathcal{M}} \phi_t^x$ (cf. Prop. 3.3.12.7).[33] **QED**

But if $T_g \, (\phi) = T_g \, (\phi_M)$ is infinite, then $GE \, (\phi)$ is also infinite, and we have ended up with infinite satisfiability. Can we do better than this?

Definition. 3.5.8. Let Φ^g be the set of quantifier-free formulae in $GE \, (\phi)$; then, all the (sub)formulae in Φ^g are atoms $A_i \in At_g \, (\Phi^g)$, where

---

principle can then be stated as: for any FO formula $\phi \, (x_1, ..., x_n)$ with bounded quantifiers and any sequence $a_1, ..., a_n \in S_\infty \, (\mathcal{A})$, the formula $\phi \, (a_1, ..., a_n)$ is *true* in $\mathcal{A}$ iff $\phi \, (\tau \, (a_1), ..., \tau \, (a_n))$ is *true* in $\mathcal{B}$.

[32] This form is known as Skolem normal form (cf. Def. 7.2.10): it has only universal quantifiers, ϕ_M is called the matrix of ϕ, and ϕ_M is said to be in prenex normal form.

[33] We actually leave the proof of this lemma as an exercise in Section 6.2, but the idea provided should allow the reader a working grasp of this lemma for the reading of this Section. Importantly, in order to invoke Proposition 3.3.12.7 we need to consider the implicit identity $T_g \, (\phi) = \{t_1, ..., t_n\} = \mathscr{D} \, (\phi)$ and implicitly take it that there is some interpretation $\mathcal{I}$ such that $\varpi \in \mathcal{I}$ corresponds to some substitution σ. In Section 6.2 below, we shall see how the finite set $T_g \, (\phi) = \{t_1, ..., t_n\}$ corresponds explicitly to the Herbrand universe of a function-free formula $\forall x_1 ... \forall x_n \, (\phi_M)$.

$At_g\left(\Phi^g\right)$ is the set of the ground atoms of Φ^g. For arbitrary (possibly infinite) sets At_g and At_p of ground atoms and propositional variables, respectively, we define a *transfer function* $\tau : At_g \longrightarrow At_p$ by means of which each ground atom A_i is replaced by some propositional atom $P_i \in At_p = V$ such that for the finite sets $\{\phi\}_g = \phi^g$ and Φ^g we have

$$\tau\left(\Phi^g\right) = \{\phi\left[A_1/P_{A_1}, ..., A_n/P_{A_n}\right] | \phi \in \Phi, \{A_1, ..., A_n\} \subseteq At_g\left(\phi^g\right)\}$$

and

$$\tau\left(At_g\left(\phi^g\right)\right) = \{P_{A_i} | A_i \in At_g\left(\Phi^g\right)\}$$

where we write P_{A_i} to denote the propositional atom obtained from the ground atom A_i.

Proposition. 3.5.9. *All the elements of $\tau\left(\Phi^g\right)$ are (well-formed) propositional formulae (over $\tau\left(At_g\left(\phi^g\right)\right)$).*

Proof: (Sketch) Apply structural induction on $\tau\left(a\right) = p$ for some $a \in At_g\left(\phi^g\right)$ and $p \in V$. **QED**

Lemma 3.5.2. *Let ϕ be a FO formula. Then ϕ is (FO-)satisfiable iff $\tau\left(\Phi^g\right)$ is (propositional-)satisfiable.*

Proof: (Sketch) By Lemma 3.5.1, ϕ is satisfiable iff Φ^g is satisfiable. So we need only show now that Φ^g is satisfiable iff $\tau\left(\Phi^g\right)$ is. This one can do by showing that

$$\models_{\mathcal{I}} \Phi^g \quad \Leftrightarrow \quad \models_{val} \tau\left(\Phi^g\right)$$

for some interpretation $\mathcal{I}$ and some propositional valuation *val*, i.e. a valuation independent of an interpretation $\mathcal{I} = (\mathscr{D}, \Theta, \varpi)$. **QED**

Definition. 3.5.10. Let us now define the ground extension of Φ as

$$GE\left(\Phi\right) = \bigcup\{GE\left(\phi\right) | \phi \in \Phi\}.$$

We now have the following fundamental result:

Theorem 3.5.3. *Let Φ be a (possibly infinite) set of FO formulae and let $\Phi^g = GE\left(\Phi\right)$ be the corresponding set of quantifier-free formulae. Then the following holds:*

1. *Φ is (FO-)satisfiable iff Φ^g is (FO-)satisfiable.*

2. If Φ is finitely (FO-) satisfiable, then Φ^g is finitely (FO-)satisfiable.

3. Φ^g is (FO-)satisfiable iff $\tau\left(\Phi^g\right)$ is (propositional)-satisfiable.

4. If Φ^g is finitely (FO-)satisfiable, then $\tau\left(\Phi^g\right)$ is (propositional)-satisfiable.

Proof: Easy from the above by induction or generalization on the proofs for Φ where $|\Phi| = 1$. **QED**

We are now ready to give the FO-compactness result that will allow us to prove the completeness of FO languages.

Theorem 3.5.4. *(Compactness of FOL) Let $X \subseteq F_{F1}$ be a (possibly infinite) set of formulae. Then, X is satisfiable iff X is finitely satisfiable.*

Proof: ($\Rightarrow$) Trivial.
($\Leftarrow$) (Hint: Theorem 3.5.3.) **QED**

Lemma 3.5.5. *For X a set of FO formulae and some $\phi \in X$, we have $X \models \phi$ iff $X \cup \{\neg\phi\}$ is unsatisfiable.*

Proof: (Idea) This is Theorem 3.4.7 for FO formulae. Think now in terms of models $\mathcal{M}(X)$, $\mathcal{M}(\phi)$, and $\mathcal{M}(\neg\phi)$. **QED**

Corollary 3.5.6. *For X a set of FO formulae and some $\phi \in X$, we have $X \models \phi$ iff there is a finite set $X' \subseteq X$ such that $X' \models \phi$.*

Proof: Easy, by Lemma 3.5.5 and Theorem 3.5.4. **QED**

Finally:

Theorem 3.5.7. *(Completeness of FOL) For X a (possibly infinite) set of FO formulae and some $\phi \in X$, if $X \models \phi$, then $X \vdash \phi$.*

Proof: (Idea) Firstly, prove that FOL is complete if,

$$\text{if } \chi_1, ..., \chi_n \models \phi, \text{ then } \chi_1, ..., \chi_n \vdash \phi.$$

Make it a lemma to Theorem 3.4.13. Invoke then Corollary 3.5.6. **QED**

We have thus proved the *completeness* of FOL. Although we have tamed the *undecidability* of FO languages, we have not eliminated their intrinsic expressiveness. This would be possible only if we would be ready and willing to remove all function symbols from a FO language.

In other words, the above results are essentially only for some signature $\Upsilon - Fun$. This removal is in fact possible, but, as the reader may deduce from (especially) Section 3.2 above, this is not always feasible or even desirable. In Section 7.3.3, we show how *refutation-completeness* considered in a particular semantics–Herbrand semantics (Section 6.2)–does effectively eliminate this problem.

3.5.2.2. Finite models and prefix classes

However, function symbols are not the only factor that impacts on FO decidability. In fact, the form of the quantifier prefix in a prenex normal form, the arity of predicates, and the inclusion of the symbol "=" for equality in a FO alphabet determine the decidable or undecidable character of a logical theory. And indeed, though FOL theories are generally undecidable, some have been proven to be decidable. We give a complete characterization below and refer the reader to, e.g., Börger, Grädel, & Gurevich (2001) for a comprehensive discussion of FO decidability.

Above, we were concerned with *finite satisfiability*; now, *finite models* are our main concern.[34]

Definition. 3.5.11. Given an interpretation $\mathcal{I} = (\mathscr{D}, \Theta, \varpi)$, a model $\mathcal{M}$ is said to be *finite* if the cardinality of $\mathscr{D}$ is finite; otherwise, it is said to be *infinite*.

Definition. 3.5.12. Let $\mathscr{F}$ be a class of FO formulae. We say that $\mathscr{F}$ has the *finite-model property (FMP)* if, for every formula ϕ in $\mathscr{F}$, either ϕ is unsatisfiable or it has a finite model.

Proposition. 3.5.13. *Let a* F1*-formula be in the form*

$$(\dagger) \qquad \blacklozenge_1 x_1, ..., \blacklozenge_n x_n (M)$$

where every $\blacklozenge_i x_i$, $i = 1, ..., n$ is either $\forall x_i$ or $\exists x_i$, and M is a formula containing no quantifiers. $\blacklozenge_1 x_1, ..., \blacklozenge_n x_n$ is the prefix and (M) is the matrix of A. There are no function symbols in M. Let now $\blacklozenge, \blacklozenge^n, \blacklozenge^*$ denote the single occurrence of a quantifier $\blacklozenge$, the sequence of n occurrences of a quantifier $\blacklozenge$, a sequence of zero or more occurrences of a quantifier $\blacklozenge$, respectively. Then, the following prefix forms determine prefix classes of FO formulae for which there are decision procedures:

---

[34] Don't let the word "finite" trick you into thinking that these are equivalent; in effect, compactness, for instance, fails over finite models, thus making finite model theory a field distinct from model theory. See Libkin (2012) for an elaboration on finite model theory largely from the perspective of computational logic.

1. $\exists^*\forall^*$

2. $\exists^*\forall\exists^*$

3. $\exists^*\forall^2\exists^*$ (if the symbol $=$ does not occur in M)

Proof: (Idea) The proof is based on the fact that these classes have the FMP. See also Exercise 3.5.2.5. **QED**

Proposition. 3.5.14. *The following prefix patterns determine prefix classes of undecidable FO formulae:*

1. $\forall^3\exists$

2. $\forall\exists\forall$

Proof: Left as a research exercise.

Let now $\bar{\blacklozenge}$ denote the dual of the quantifier $\blacklozenge$, i.e. $\bar{\forall} = \exists$ and $\bar{\exists} = \forall$. Then we have the following important result:

Proposition. 3.5.15. *Let $\mathscr{P}$ be a prefix class and $\bar{\mathscr{P}}$ be its dual class obtained by replacing each quantifier $\blacklozenge$ by its dual $\bar{\blacklozenge}$. Then, the following statements are equivalent:*

1. *VAL for $\mathscr{P}$ is decidable.*

2. *SAT for $\bar{\mathscr{P}}$ is decidable.*

Proof: Hint: Proposition 3.4.33.1. **QED**

Exercises

Exercise 3.5.2.1. Consider the argument in Example 3.2.3 as a set of FO formulae C. Apply to this set the notion of ground extension and the transfer function τ in order to obtain the set of propositional formulae $\tau(C^g)$ via $\tau(At_g(C^g)) = \{T, P, E\}$.

Exercise 3.5.2.2. The *Löwenheim-Skolem Theorem* states that if a set X of formulae has a model, then X has an enumerable model.

1. This theorem can also be referred to as the *Löwenheim-Skolem Transfer Theorem*. Explain why.

2. This theorem is a corollary of compactness. Explain.

3. What happens if X is a set of formulae of a non-enumerable language?

4. In spite of item 2 above, this theorem can be given an independent proof. Sketch such a proof.

Exercise 3.5.2.3. Besides the formulae in Proposition 3.5.13, it has also been proven that *monadic FO formulae*, i.e. formulae whose predicates are unary and have no function symbols, are decidable. Research into the proof for this result.

Exercise 3.5.2.4. Propositions 3.5.13-14 give us a complete classification of the prefix classes in terms of decidability for formulae without function symbols. Why is this classification complete?

Exercise 3.5.2.5. Prove that there is a decision procedure for the VAL of all FO formulae with the FMP.

Exercise 3.5.2.6. Complete all the sketchy proofs in this Section.

3.5.3. The complexity of logical problems

Virtually all computational problems can be formulated in a logical language. In particular, most mathematical theories can be formulated in the logical language F1, and thus most computational problems in mathematics can be formulated in this language. As seen in Chapter 2, any decidable computational problem requires an amount of time and space resources to be effectively computed; if the upper bounds on these resources are too high, it may simply be the case that a decidable computational problem is in practice not computable.

Example 3.5.1. Tarski proved in 1949 that the theory $\Theta_{\mathbb{R}}$ of the infinite structure of the reals $\mathfrak{R} = (\mathbb{R}, +, \cdot, 0, 1)$ is decidable, but the complexity of $\Theta_{\mathbb{R}}$ is thought to be at least **PSPACE**-hard and at most **EXPSPACE**; in any case, the computational problem for $\Theta_{\mathbb{R}}$ is thought to be essentially unsolvable in practice (cf. Fig. 2.3.3).

These upper bounds on the resources are determined by both the size and the kind of the input, though the model of computation also plays an important role, with some models returning an output with fewer time and space costs. Because computational problems are today massively solved in the digital computer, we have really to consider only

263

the size and the kind of the input.[35] By the size of a logical input, we obviously mean some n of (atomic) formulae. By the kind of the input, we mean the type of formulae and expressions thereof used to formulate the problem. For instance, the depth of a quantifier (i.e. the number of variables bound by a single quantifier) and the arity of predicates can impact significantly on the complexity of a theory; also, whether a logical language has functions or not does have a significant impact.

It is well known that an input of conjunctions of disjunctions of atomic (propositional) formulae is optimal for solving computational problems in a digital computer. This kind of input is known as the CNF-SAT, an instance of the SAT, or *(propositional) satisfiability problem.*

The SAT is a central logical problem because any logical problem that can be formalized in **F1** can be formulated in its terms, and these are so using a fragment of **F0**, namely the quantifier-free fragment based on the subset of connectives $O' = \{\neg, \wedge, \vee\}$. This, together with the fact that the SAT asks whether a formula (or a set of formulae) is satisfiable, makes it evident why the SAT is also known as the *Boolean propositional satisfiability problem* (cf. Boolean algebras in Section 1.2). In effect, Theorem 7.2.1 assures us that any FO formula $\phi \in F_{\mathsf{F1}}$ can be rewritten in a Boolean logical language;[36] more specifically, ϕ can be expressed in a conjunctive normal form, abbreviated as CNF, or in a disjunctive normal form, abbreviated DNF,[37] and given any algebra of formulae $\mathfrak{F} = (F, \neg, \wedge, \vee, \bot, \top)$ of type $(1, 2, 2, 0, 0)$ there is a homomorphism $h : \mathfrak{F} \longrightarrow 2$ such that every formula of F is valuated exactly as 1 or 0.

Recall from Section 2.3.4 that a computational problem is of time complexity $\mathcal{O}(g(n))$ if, given an input of size n, there is a decision procedure Ψ that runs on time at most $f(n)$ for a function $f(n) \in \mathcal{O}(g(n))$. Revisit Table 2.3.2 and put it into relation with the complexity classes and the corresponding tractability status (cf. Fig. 2.3.3). Given the information that the SAT is of time complexity $\mathcal{O}(n^2)$, we can easily conclude that this is a hard-to-solve problem. As a matter of fact, it was the first decision problem to be proven to be **NP**-complete (cf. Theorem 2.3.9), and although many logical problems are easier to solve, this logical problem remains as the measure for hard-to-solve problems; that is, a problem is said to be in the complexity class **NP** if it is at most as difficult to solve as the SAT.

Although the SAT is first and foremost a classical propositional-logic

---

[35] Of course, the software used may impact on the upper bounds, but never to the point of exceeding those determined for the worst-case scenario (see Section 2.3.4).

[36] A logical language that is isomorphic to the algebra of formulae $\mathfrak{F} = (F, \neg, \wedge, \vee, \bot, \top)$.

[37] See Section 7.2.5 for how to rewrite any **F1**-formula in a CNF/DNF.

Complete Problem	Complexity Class
2-SAT	**NLOGSPACE**
HORN-SAT	**PTIME**
SAT, 3-SAT	**NP**
QBF-SAT	**PSPACE**

Table 3.5.1.: Complete SATs and their complexity classes.

problem, it can be generalized to CFOL, as well as to non-classical logics. In the case of CFOL, the SAT uses only (implicitly) the existential quantifier. When the universal quantifier is allowed instead of the existential one, then we speak of the *validity problem*, abbreviated as VAL, which is known to be a co-**NP**-complete problem. If both quantifiers are allowed, then we have a *quantified Boolean formula (QBF) problem*. Although the QBF-SAT is in the complexity class **PSPACE**, whose problems are believed to be strictly harder to solve than **NP** problems, they can be solved in linear time by some SAT solvers.

Although, as seen, many computational problems are polynomial-time reducible to the SAT, reason why we say it works as a measure of complexity of computational problems, including the logical problems, it is not always easy or even feasible to formulate a logical theory in terms of the CNF-SAT. We now specifically say "CNF-SAT" because, though a computational problem formulated as a DNF-SAT may be computable in linear time, the conversion of the formulae to DNFs may take more than exponential time, thus bringing no benefits in many cases.

This said, the (CNF-)SAT can be further fragmented into subclasses other than the QBF-SAT: we speak of the 2-SAT (3-SAT) when we have conjunctions of exactly two (three, respectively) disjuncts, and of the HORN-SAT when we have conjunctions of disjunctions with at most one positive literal. The complexity classes for all the classes of the SAT have been established (see Table 3.5.1).

We thus have it that **PSPACE** is the highest complexity class for a SAT subclass. As a matter of fact, we can determine the complexity of a logical theory Θ of a specific logic **L** by combining two different kinds of complexity, one related to the structure (the finite model) and the other to the formulae of **L**. However, when one of the parameters–the structure or the formula–is fixed, we measure the complexity only in terms of the other.

Definition. 3.5.16. Let $\Theta \subseteq \mathbf{L}$, $\mathbf{L}$ is a logic.

1. The *data complexity of* Θ is the family of the decision problems

$$DATACOMP = \{\mathcal{M}| \models_{\mathcal{M}} \phi\}$$

for every *fixed* formula $\phi \in \Theta$ given a finite model $\mathcal{M}$.

2. The *expression complexity of* Θ is the family of the decision problems

$$EXPCOMP = \{\phi| \models_{\mathcal{M}} \phi\}$$

for every *fixed* finite model $\mathcal{M}$ given a formula $\phi \in \Theta$.

3. The *combined complexity of* Θ for a formula $\phi \in \Theta$ given a finite model $\mathcal{M}$ is

$$MODELCHECK = \{\mathcal{M}, \phi| \models_{\mathcal{M}} \phi\}$$

i.e. the decision problem whether we have $\models_{\mathcal{M}} \phi$.[38]

Definition 3.5.16.3 is more commonly known as the *model checking problem* for Θ (or for $\mathbf{L}$).

We abbreviate the above decision problems as DCP, ECP, and MCP, respectively. We now provide a way to measure the data complexity and the expression complexity of a theory Θ (or of a logic $\mathbf{L}$).

Definition. 3.5.17. Let $\Theta \subseteq \mathbf{L}$. Let further $\mathscr{C}$ be a complexity class.

1. The data complexity of $\Theta \subseteq \mathbf{L}$ is

    a) in $\mathscr{C}$ if for every formula $\phi \in \Theta \subseteq \mathbf{L}$ the decision problem DCP is in $\mathscr{C}$.

    b) $\mathscr{C}$-complete if it is in $\mathscr{C}$ and there is at least one formula $\phi \in \Theta \subseteq \mathbf{L}$ such that the decision problem DCP is $\mathscr{C}$-complete.

---

[38]More rigorously, we should have some computation model M and encodings $e(\mathcal{M}) = \langle\mathcal{M}\rangle$, $e(\phi) = \langle\phi\rangle$ such that we we define, say, the DATACOMP problem as

$$DATACOMP_M = \{\langle\mathcal{M}\rangle| \models_{\mathcal{M}} \phi\}.$$

However, we are now committed to the digital computer and skip this formal precision for the sake of simplification. Also, consistent with our notation, we should write $\mathscr{DATACOMP}$, etc., as these are families of decision problems, but again we opt for simplifying.

2. The expression complexity of $\Theta \subseteq \mathbf{L}$ is

    a) in $\mathscr{C}$ if for every finite model $\mathcal{M}$ the decision problem ECP is in $\mathscr{C}$.

    b) $\mathscr{C}$-complete if it is in $\mathscr{C}$ and there is at least one finite model $\mathcal{M}$ such that the decision problem ECP is $\mathscr{C}$-complete.

From the above, the following result follows for a FOL with F1 with the symbol "=" added and without function symbols. Note the discrepancy between DCP and both ECP and MCP as far as their complexity classes is concerned.

Theorem 3.5.8. *For FOL, the following statements hold:*

1. *The data complexity of FOL is in the complexity class* $\mathbf{L}$.

2. *The expression complexity of FOL is* **PSPACE**-*complete.*

3. *The combined complexity of FOL is* **PSPACE**-*complete.*

Proof: (Sketch) It should be obvious that logarithmic space is required to check, given a finite model $\mathcal{M}$, all the instantiations $\vec{t}$ of the quantifiers of a FO formula $\phi(\vec{x})$ one at a time, and to keep track of them as well. That is, given a model $\mathcal{M}$ of finite domain $\mathscr{D}$, this computation (whose details the reader is expected to provide) requires $\mathcal{O}(\log(|\mathscr{D}|))$. Given now ϕ as part of the input, this computation can be carried out in a space bounded by a polynomial in $|\mathscr{D}|$, and we thus have MCP∈**PSPACE**. If $\mathcal{M}$ has at least two distinct elements, then the QBF-SAT is reducible to the ECP. **QED**

Theorem 3.5.8 accounts for our saying that FOL *captures* the complexity class **PSPACE** given a domain of finite structures, it being meant by this that the properties of these structures that are definable in FOL are precisely those that are definable in the complexity class **PSPACE**.

We then also say that FOL is a *logic for* **PSPACE**. Clearly, we should aim at finding a FOL **L** for a class of lower complexity, and this is a problem of ongoing research. Our interest falls on a FOL **L**, because logics of higher order than 1 are not complete.

Exercises

Exercise 3.5.3.1. Give examples of theories not expressible in FOL, i.e. not formalizable in F1.

Exercise 3.5.3.2. How is the discrepancy between the complexity classes of DCP and of both ECP and MCP (cf. Theorem 3.5.8) to be accounted for?

Exercise 3.5.3.3. Research into the proof of incompleteness of second order logic.

Exercise 3.5.3.4. Complete the proof of Theorem 3.5.8.

---

4. The system CL and the logic CL

The logical system known as *classical logic* (abbreviated: CL) has a well-known and well-studied characterization from diverse perspectives. From the algebraic viewpoint, CL is *Boolean*, i.e. a Boolean algebra provides it with a mathematical foundation for both its syntax and its semantics. To say that CL is Boolean in terms of syntax means that all formulae of CL can be rewritten using only the subset of Boolean operators (cf. Section 7.2). With respect to its semantics, to say the same means that CL has a bivalent semantics (see Section 6.3). This, in turn, has an impact on the set-theoretical foundations: the sets of CL are all *crisp*, with crispiness–a term arising in the context of fuzzy and/or rough sets–being the property of a set membership to which is decidable by a "Yes/No" answer.

A large part of the orthodox characterization of CL, of both its language and consequence operation/relation, has already been carried out in Chapter 3. Indeed, most of the material in this Chapter relates directly to it. In the present Chapter, we mostly give specifications to complete this characterization. Although striving for self-containment, we cannot here provide as comprehensive a characterization of CL, namely in terms of mathematical logic, as is given in, for example, Mendelson (2015) or Enderton (2001).

4.1. The language of classical logic

4.1.1. The language L1

We shall consider the language of classical logic to be F1 (cf. Chapter 3) with the specifications that follow.

Definition. 4.1.1. Let there be given the sets of connectives $O_L = \left\{ \neg^1, \wedge^2, \vee^2, \rightarrow^2, \leftrightarrow^2 \right\}$ and of quantifiers $Q_{L1} = \{\forall, \exists\}$. If we define terms, atoms and formulae (whose corresponding sets are T, At, and F, respectively) over the signature $\Upsilon_{L1} = \Upsilon_{F1}$ inductively in the Backus-Naur notation as follows

Terms	t	$::=$	$x \mid a \mid f\,(t_1, ..., t_n)$
Atoms	$P\;(Q, ...)$	$::=$	$p \mid P\,(t_1, ..., t_n)$
Formulae	$A\;(B, ...)$	$::=$	$P \mid \neg A \mid A \wedge B \mid A \vee B \mid A \to B \mid A \leftrightarrow B \mid$
			$\forall x\,(A) \mid \exists x\,(A)$

then we have the syntax of the logical connectives $\neg$ (*negation*), $\wedge$ (*conjunction*), $\vee$ (*disjunction*), $\to$ (*material implication*, or *conditional*), and $\leftrightarrow$ (*material equivalence*, or *biconditional*), and of the quantifiers $\forall$ (*universal quantifier*) and $\exists$ (*existential quantifier*) that we shall refer to as L1.

We henceforth omit the superscripts indicating arity for the connectives of O_L.

In terms of orders, L0 provides the language for the *classical propositional logic* (CPL), and L1 does so for the *classical first-order logic* (CFOL). We denote by L the language of classical logic when it is not necessary to specify the order; by this, we usually mean L1, though higher orders might also be meant.

Definition. 4.1.2. The language L is functionally complete (cf. Def. 3.3.6), as it has at least one set of connectives that is functionally complete.

Example 4.1.1. The subset $O'_{\neg,\to} = \{\neg, \to\}$ of L is functionally complete. In effect, we have the following inductive definitions of the remaining connectives of O_L:

1.
$$(\vee_{df}) \qquad A \vee B \quad := \quad \neg A \to B$$

2.
$$(\wedge_{df}) \qquad A \wedge B \quad := \quad \neg(A \to \neg B)$$

3.
$$(\leftrightarrow_{df}) \quad A \leftrightarrow B \quad := \quad (A \to B) \wedge (B \to A)\,;\,(A \wedge B) \vee (\neg A \wedge \neg B)$$

As a matter of fact, the subsets $O'_{\neg,\vee}$ and $O'_{\neg,\wedge}$ of L are also functionally complete. Obviously, $O' = \{\neg, \wedge, \vee\}$ is also functionally complete.

4.1.2. Substitutions and unification for L1

The existence of individual variables in the alphabet of L1 requires special methods that allow deduction over it. These are mostly *substitutions* and *unification*.

Definition. 4.1.3. Given a set $Vi = \{x_1, ..., x_n\}$ of variables and a set $T = \{t_1, ..., t_n\}$ of terms, a *substitution* is a mapping $\sigma : Vi \longrightarrow T$ such that $\sigma x_i = t_i$ almost everywhere, and where most often $t_i \neq x_i$. We represent a substitution σ as a finite set of expressions of the form $x_i \mapsto t_i$, where no two different terms substitute the same variable, i.e.

$$\sigma = \{x_1 \mapsto t_1, ..., x_n \mapsto t_n\}.$$

We define the domain of a substitution σ as $dom(\sigma) = \{x | \sigma x = t\}$ and its range as $rg(\sigma) = \{\sigma x | x \in dom(\sigma)\}$.

1. We say that σ is a *ground substitution* when $Vi(rg(\sigma)) = \emptyset$.

2. For $\sigma = \emptyset$, we speak of the *empty substitution*, denoted by ϵ.

3. A substitution σ such that $rg(\sigma) \subseteq Vi$ is a *(variable) renaming*.

Definition. 4.1.4. For the unary connective $\neg$ and the binary connectives $\rightarrow, \leftrightarrow$ and $\heartsuit = \wedge, \vee$, as well as for the quantifiers $\blacklozenge = \forall, \exists$, the following facts concerning a substitution σ hold in CL (cf. also Def. 3.1.12):

1. $\sigma(\neg A) = \neg(\sigma A)$.

2. $\sigma\left(A \underset{\leftrightarrow}{\overset{\rightarrow}{}} B\right) = \sigma A \underset{\leftrightarrow}{\overset{\rightarrow}{}} \sigma B.$

3. $\sigma(A_1 \heartsuit ... \heartsuit A_n) = \sigma A_1 \heartsuit ... \heartsuit \sigma A_n.$

4. Let $A = \blacklozenge x A(x)$. Then, t is said to be *substitutable* (or *free*) *for x* in A, and we write $\sigma(A) = A_t^x$ or $\sigma(A) = A[x/t]$, if (i) $Vi(t) = \emptyset$, (ii) $t \neq y$ or $y \notin Vi(t)$ if y is bound in A, or (iii) $t = x$.

Note with respect to Def. 4.1.4.4 that x must be free in A. For instance, $[\forall x(P(x)) \wedge Q(x)]_a^x = \forall x(P(x)) \wedge Q(a)$. But $[\forall x(R(x))]_x^x = R(x)$.

Definition. 4.1.5. For a substitution $\theta = \{x_1 \mapsto t_1, ..., x_n \mapsto t_n\}$ and an expression E, $E\theta = \theta E$ is an expression obtained from E by replacing simultaneously each occurrence of the variable x_i, $1 \leq i \leq n$, in E by the term t_i. We say that $E\theta$ is an *instance* of E. If $Vi(E\theta) = \emptyset$, then $E\theta$ is called a *ground instance*.

Example 4.1.2. Let $\theta = \{x \mapsto a, y \mapsto f(b), z \mapsto c\}$, $E = P(x, y, z)$. Then $E\theta = P(a, f(b), c)$.

Definition. 4.1.6. Consider substitutions $\theta = \{x_1 \mapsto t_1, ..., x_n \mapsto t_n\}$ and $\lambda = \{y_1 \mapsto u_1, ..., y_n \mapsto u_n\}$. Their *composition*, denoted by $\theta \circ \lambda$, is obtained from the set $\{x_1 \mapsto t_1\lambda, ..., x_n \mapsto t_n\lambda, y_1 \mapsto u_1, ..., y_n \mapsto u_n\}$ by deleting any element $x_j \mapsto t_j\lambda$ for which $t_j\lambda = x_j$, and any element $y_i \mapsto u_i$ such that y_i is among $\{x_1, ..., x_n\}$.

Example 4.1.3. Given the substitution sets $\theta = \{x \mapsto f(y), y \mapsto z\}$ and $\lambda = \{x \mapsto a, y \mapsto b, z \mapsto y\}$, then $\theta \circ \lambda = \{x \mapsto f(b), z \mapsto y\}$.

Definition. 4.1.7. We say that a set of expressions $E = \{E_1, ..., E_n\}$ is *unifiable* by a substitution σ (called the *unifier* for E) if $E_i\sigma = E_j\sigma$ for all $E_i, E_j \in E$. A unifier σ for the set E is a *most general unifier* (*MGU*) iff for each unifier θ for the set there is a substitution λ such that $\theta = \sigma \circ \lambda$. (Equivalently, let us say that λ is more general than θ, denoted $\lambda \leq_s \theta$, if there is a substitution σ such that $\sigma \circ \lambda = \theta$. Then, σ is the MGU of the set E if for every other unifier λ it is the case that $\sigma \leq_s \lambda$.)

Example 4.1.4. For substitutions $\theta = \{x \mapsto z, y \mapsto z, u \mapsto f(z, z)\}$, $\lambda = \{y \mapsto z\}$, and $\sigma = \{x \mapsto y, u \mapsto f(y, z)\}$, σ is the MGU, because $\theta = \sigma \circ \lambda$.

Example 4.1.5. $\theta = \{x \mapsto f(b), y \mapsto b, z \mapsto u\}$ is a unifier of the expressions $E_1 = f(x, b, g(z))$ and $E_2 = f(f(y), y, g(u))$, because $E_1\theta = E_2\theta = f(f(b), b, g(u))$.

Definition. 4.1.8. Let E_1 and E_2 be two expressions. Then, $E_1 \leq_{ss} E_2$ if there is a substitution σ such that $E_1\sigma \subseteq E_2$, and we say that E_1 *subsumes* E_2.

Definition. 4.1.9. For a non-empty set of expressions E, the *disagreement set* $D(E)$ of E is the set of sub-expressions of E obtained by locating the leftmost symbol at which not all expressions in E have exactly the same symbol, and then extracting from each expression in E the sub-expression that begins with the symbol occupying that position.

Example 4.1.6. The disagreement set of

$$E = \{P\left(x, f\left(y, z\right)\right), P(x, a), P\left(x, g\left(h\left(k\left(x\right)\right)\right)\right)\}$$

is

$$D\left(E\right) = \{f\left(y, z\right), a, g\left(h\left(k\left(x\right)\right)\right)\}.$$

Definition. 4.1.10. The *unification problem* is that of finding a MGU of two given terms.

This is a purely mechanical procedure for which there is more than one algorithm. In Example 4.1.7, we give a formulation of the Robinson algorithm (Robinson, 1965).

Algorithm 4.1 The Robinson algorithm.

Input: A set *Lit* of literals
Output: A substitution set σ or "Not unifiable"

```
Let σ := ε
while |σ(Lit)| > 1 {
  select a disagreement pair d in σ(Lit);
  if d does not contain any variable then {
    stop and return "not unifiable";
  } else {
    let d = (x, t), x is a variable;
    if x occurs in t, then "occurs check" {
      stop and return "not unifiable";
    } else {
      let σ := σ ∘ (x ↦ t);
    }
  }
}return σ;
```

Example 4.1.7. A possible way to formulate the Robinson algorithm is given in Algorithm 4.1.

4. The system CL and the logic CL

The application of *unification rules* is an alternative procedure more efficient than the Robinson algorithm.

Definition. 4.1.11. Let σ and θ stand for substitutions and denote failure in the application of a rule by $\bot$. For P and Q pairs of expressions $\{\langle E_1, F_1 \rangle, ..., \langle E_n, F_n \rangle\}$, the unification rules for P and Q have the general form

$$P; \sigma \Longrightarrow Q; \theta$$

or

$$P; \sigma \Longrightarrow \bot.$$

The (successive) application of the unification rules ends either in success, denoted by $\emptyset$, or in failure. The order of application of the rules is non-deterministic.

Definition. 4.1.12. The *unification (inference) rules* are as follows:

1. *Trivial:*
$$\{\langle s, s \rangle\} \cup P'; \sigma \Longrightarrow P'; \sigma$$

2. *Decomposition:*
$$\{\langle f(s_1, ..., s_n), f(t_1, ..., t_n) \rangle\} \cup P'; \sigma \Longrightarrow \{\langle s_1, t_1 \rangle, ..., \langle s_n, t_n \rangle\} \cup P'; \sigma$$

   if $f(s_1, ..., s_n) \neq f(t_1, ..., t_n)$.

3. *Orient:*
$$\{\langle t, x \rangle\} \cup P'; \sigma \Longrightarrow \{\langle x, t \rangle\} \cup P'; \sigma$$

   if t is not a variable.

4. *Variable elimination:*
$$\{\langle x, t \rangle\} \cup P'; \sigma \Longrightarrow P'\theta; \sigma\theta$$

   given that x does not occur in t and $\theta = \{x \to t\}$.

5. *Symbol clash:*
$$\{\langle f(s_1, ..., s_n), g(t_1, ..., t_m) \rangle\} \cup P'; \sigma \Longrightarrow \bot$$

   if $f \neq g$.

6. *Occurs check:*

$$\{\langle x, t \rangle\} \cup P'; \sigma \Longrightarrow \bot$$

if x occurs in t but $x \neq t$.

Example 4.1.8. We provide examples of the application of each of the above rules:

1. For the pair of expressions $\langle P(a), P(a) \rangle$ we have

$$\{\langle P(a), P(a) \rangle\}; \epsilon \Longrightarrow_{Tr}$$
$$\emptyset; \epsilon$$

2. Given the pair of expressions $P(f(a), g(x))$ and $P(y, z)$ we have

$$\{\langle P(f(a), g(x)), P(y, z) \rangle\}; \epsilon \Longrightarrow_{Dec}$$
$$\{\langle f(a), y \rangle, \langle g(x), z \rangle\}; \epsilon$$

3. Applying this rule to the example immediately above, we have

$$\{\langle f(a), y \rangle, \langle g(x), z \rangle\}; \epsilon \Longrightarrow_{Or}$$
$$\left\{ \left\langle \underline{y, f(a)} \right\rangle, \langle g(x), z \rangle \right\}; \epsilon$$

4. Given the pair of expressions $P(x, f(a))$ and $P(g(y), x)$ we have

$$\{\langle P(x, f(a)), P(g(y), x) \rangle\}; \epsilon \Longrightarrow_{VE}$$
$$\{\langle g(y), f(a) \rangle\}; \{x \to f(a)\}$$

5. Applying this rule to the result obtained immediately above, we have

$$\{\langle g(y), f(a) \rangle\}; \epsilon \Longrightarrow_{SCl}$$
$$\bot$$

6. For the pair of expressions $P(x)$ and $P(f(x))$ we have

$$\{\langle x, f(x) \rangle\}; \epsilon \Longrightarrow_{OCh}$$
$$\bot$$

Example 4.1.9. The pair $P(a, x, h(g(z)))$ and $P(z, h(y), h(y))$ is unifiable (see Fig. 4.1.1).

$$\{\langle P(a, x, h\,(g\,(z))), P\,(z, h\,(y), h\,(y))\rangle\}\,; \epsilon \Longrightarrow_{Dec}$$

$$\{\langle a, z\rangle, \langle x, h\,(y)\rangle, \langle h\,(g\,(z)), h\,(y)\rangle\}; \epsilon \Longrightarrow_{Or}$$

$$\{\langle z, a\rangle, \langle x, h\,(y)\rangle, \langle h\,(g\,(z)), h\,(y)\rangle\}\,; \epsilon \Longrightarrow_{VE}$$

$$\{\langle x, h\,(y)\rangle, \langle h\,(g\,(a)), h\,(y)\rangle\}\,; \{z \mapsto a\} \Longrightarrow_{VE}$$

$$\{\langle h\,(g\,(a)), h\,(y)\rangle\}\,; \{z \mapsto a, x \mapsto h\,(y)\} \Longrightarrow_{Dec}$$

$$\{\langle g\,(a), y\rangle\}\,; \{z \mapsto a, x \mapsto h\,(y)\} \Longrightarrow_{Or}$$

$$\{\langle y, g\,(a)\rangle\}\,; \{z \mapsto a, x \mapsto h\,(y)\} \Longrightarrow_{VE}$$

$$\emptyset; \{z \mapsto a, x \mapsto h\,(y), y \mapsto g\,(a)\}$$

Figure 4.1.1.: Unifying the pair $\langle P\,(a, x, h\,(g\,(z))), P\,(z, h\,(y), h\,(y))\rangle$.

Exercises

Exercise 4.1.1. For the pairs of expressions E and F, determine whether σ is (i) a unifier and (ii) a MGU of E and F:

1. $E = P\,(a, f\,(y), z)\,; F = S\,(x, f\,(f\,(b)), b)$

   $\sigma = \{x \mapsto a, y \mapsto f\,(b), z \to b\}$

2. $E = P\,(x, h\,(a, z), f\,(x))\,; F = P\,(g\,(g\,(v)), y, f\,(w))$

   $\sigma = \{x \mapsto g\,(g\,(v)), y \mapsto h\,(a, z), w \mapsto x\}$

3. $E = Q\,(f\,(x), g\,(y))\,; F = Q\,(z, g\,(v))$

   $\sigma = \{x \mapsto a, z \mapsto f\,(a), y \mapsto v\}$

4. $E = Knows\,(john, x)\,; F = Knows\,(john, jane)$

   $\sigma = \{x \mapsto jane\}$

Exercise 4.1.2. Determine whether you can unify the following pairs of atoms:

1. $Q(x, y, z)$ and $Q(u, h(v, v), u)$

2. $P(x, f(x))$ and $P(y, y)$

3. $P(f(x))$ and $P(a)$

4. $P(x, y)$ and $P(y, f(y))$

5. $P(a, y)$ and $P(x, f(x), b)$

6. $S(f(a), g(x))$ and $S(u, u)$

7. $R(a, x, h(g(z)))$ and $R(z, h(y), h(y))$

8. $P(f(a), g(x))$ and $P(x, x)$

9. $R(a, x, f(x))$ and $R(a, y, y)$

4.2. Classical logical consequence

4.2.1. Classical $\heartsuit$-consequences

To define the k_i-ary operations $\heartsuit_i^k$ on a set of formulae F_L, $0 \leq k \leq 2$, equates with imposing specific conditions on the consequence relation/operation in presence of the $\heartsuit_i^k$. We thus speak of $\heartsuit$-*consequences*. By this means we define not only negation, conjunction, disjunction, etc., but also their *behavior* (e.g., classical, modal, paraconsistent, etc.). In this particular case, we focus on the classical $\heartsuit$-consequences and we do it with generality in mind in this Section. Below, in Chapter 5, we specify this behavior for the calculi $\mathcal{LK}$ (cf. the logical rules) and $\mathcal{NK}$ having inference rules in mind.

We consider the language L with $O_L = \{\top, \bot, \neg, \wedge, \vee, \rightarrow\}$ the set of connectives; $\top$ and $\bot$ are taken as 0-ary connectives and the connective $\leftrightarrow$ can be defined in terms of the other members of O_L in the usual ways (cf. Example 4.1.1.3).

With generality in mind, we first introduce the $\heartsuit$-consequence operations, and then separately approach the syntactical and semantical $\heartsuit$-relations. With respect to the $\heartsuit$-consequence operations Cn, they are spoken of as *Tarski-style conditions for Cn* if they involve exactly one connective of the language at hand.

Definition. 4.2.1. The following are the Tarski-style conditions for the classical consequence operation Cn_{L} for a (possibly empty) consistent set $\Gamma \subseteq F_{\mathsf{L}}$ and any formulae $\phi, \psi \in F_{\mathsf{L}}$:[1]

1. $(\top_{Cn})$ $\top \in Cn(\Gamma)$ and $\phi \in Cn(\Gamma, \top)$

2. $(\bot_{Cn})$ $\bot \notin Cn(\Gamma)$ but $\phi \in Cn(\Gamma, \bot)$

3. $(\neg_{Cn})$ $\phi \in Cn(\Gamma)$ iff $Cn(\Gamma, \neg\phi) = F_{\mathsf{L}}$

4. $(\wedge_{Cn})$ $Cn(\Gamma, \phi \wedge \psi) = Cn(\Gamma, \phi, \psi)$

5. $(\vee_{Cn})$ $Cn(\Gamma, \phi \vee \psi) = Cn(\Gamma, \phi) \cap Cn(\Gamma, \psi)$

6. $(\rightarrow_{Cn})$ $\phi \rightarrow \psi \in Cn(\Gamma)$ iff $\psi \in Cn(\Gamma, \phi)$

Informally, Definition 4.2.1.1 tells us that $\top$ is a consequence of any empty or consistent set Γ and any formula ϕ is a consequence of $\Gamma \cup \{\top\}$. Definition 4.2.1.2 formulates the fact that, if Γ is a consistent set, then $\bot$ is not one of its consequences, but any formula whatsoever is a consequence of $\Gamma \cup \{\bot\}$, a principle known as *ex falso* (or *contradictione*) *quodlibet*, abbreviated as EFQ (ECQ, respectively). In 4.2.1.3, Cn is defined with relation to $\neg$ in terms of consistency: an arbitrary formula ϕ is a consequence of Γ iff $\Gamma \cup \{\neg\phi\}$ is inconsistent. Definition 4.2.1.4 tells us that the consequences of a set Γ and of a formula $\phi \wedge \psi$ are exactly the consequences of $\Gamma \cup \{\phi\} \cup \{\psi\}$, whereas Definition 4.2.1.5 tells us that the consequences of a set Γ and of a formula $\phi \vee \psi$ are the intersection of the consequences of $\Gamma \cup \{\phi\}$ and of $\Gamma \cup \{\psi\}$. Finally, Definition 4.2.1.6 defines the $\rightarrow$-consequence in terms of DDT (cf. Exercise 3.4.4.1), which is made clearer in Proposition 4.2.5.6 below.

Because the Tarski-style conditions for the connectives are in fact rules (of inference), we can also determine Cn with respect to the rules MP and DT:[2]

Proposition. 4.2.2. MP_{Cn} *holds iff MP is a rule of* Cn, *and* DT_{Cn} *holds if DT is a rule of* Cn.

---

[1] Not all of them were considered by Tarski, despite this label; in fact, Tarski (1930) considered solely $\neg_{Cn}$ and $\rightarrow_{Cn}$. We obtain $\wedge_{Cn}$ and $\vee_{Cn}$ as derivations from these after defining the connectives $\wedge$ and $\vee$ in terms of $O' = \{\neg, \rightarrow\}$ (cf. Example 4.1.1.1-2).

[2] Clearly, DT can be seen as a rule of inference.

4.2.1.1. Classical syntactical ♡-consequences

Given the inter-definition with respect to consequence relations and operations (cf. Prop. 3.4.6), we can now obtain the classical syntactical ♡-consequence relations from the Tarski-style definitions above of the classical ♡-consequence operations. We use the symbols $\Rightarrow, \Leftrightarrow \in O^\mu$ for conditional and iff-sentences, respectively, at the metalanguage level.

Definition. 4.2.3. A consequence relation $\vdash$ over the language L, for $\Gamma \subseteq F$ and all $\phi, \psi, \chi \in F$, is

1. ⊤-*classical* iff

    a) $\vdash \top$

    b) $\Gamma, \top \vdash \phi \Rightarrow \Gamma \vdash \phi$

2. ⊥-*classical* iff $\bot \vdash \phi$

3. ¬-*classical* iff

    a) $(\Gamma, \phi \vdash \psi \text{ and } \Gamma, \neg\phi \vdash \psi) \Rightarrow \Gamma \vdash \psi$ and

    b) $\phi, \neg\phi \vdash \psi$

4. ∧-*classical* iff $\Gamma, \phi \wedge \psi \vdash \chi \Leftrightarrow \Gamma, \phi, \psi \vdash \chi$

5. ∨-*classical* iff $\Gamma, \phi \vee \psi \vdash \chi \Leftrightarrow \Gamma, \phi \vdash \chi$ and $\Gamma, \psi \vdash \chi$

6. →-*classical* iff $\Gamma, \phi \vdash \psi \Leftrightarrow \Gamma \vdash \phi \rightarrow \psi$

Although essentially the same as the Definitions in 4.2.1, the above Definitions 4.2.3.1-6 allow us to specify a few more properties of the classical logic system. With respect to Definition 4.2.3.3.a), this is known as *redundancy* (abbr.: RD) and 4.2.3.3.b) can also be called the *principle of explosion* (PE). Definition 4.2.3.1.b) expresses *monotonicity* (M; a weak version thereof, actually). Further characterizations of classicality for the connectives, some involving more than one connective, are as follows.[3]

Definition. 4.2.4. For a set of formulae $\Gamma \subseteq F_L$ and for all $\phi, \psi, \chi \in F_L$, if $\vdash$ is

1. ¬-classical, then $\Gamma, \neg\phi \vdash \phi \Rightarrow \Gamma \vdash \phi$ (*reductio ad absurdum*, strong version; RAstr) and $\phi \dashv\vdash \neg\neg\phi$ (*involution*, or *double negation, law*, DN).

---

[3]Whenever it is adequate to do so we omit Γ in the premises and abbreviate (e.g., written in full, AD (Def. 4.2.4.2.a) is: if $\Gamma \vdash \phi$ and $\Gamma \vdash \psi$, then $\Gamma \vdash \phi \wedge \psi$).

2. ∧-classical, then

    a) $\phi, \psi \vdash \phi \wedge \psi$ (*adjunction*, AD) and

    b) $\begin{cases} \phi \wedge \psi \vdash \phi \\ \phi \wedge \psi \vdash \psi \end{cases}$ (*simplification*, SP).

3. ∨-classical, then

    a) $\begin{cases} \phi \vdash \phi \vee \psi \\ \phi \vdash \psi \vee \phi \end{cases}$ (*addition*, AT) and

    b) $(\Gamma, \phi \vdash \chi$ and $\Gamma, \psi \vdash \chi \Rightarrow \Gamma, \phi \vee \psi \vdash \chi)$ (*summation*, SM).

4. ∧- *and* ¬-classical, then $\Gamma, \phi \vdash \psi \wedge \neg\psi$ implies $\Gamma \vdash \neg\phi$ (*principle of non-contradiction*, PNC).[4]

5. ∨- *and* ¬-classical, then $\vdash \phi \vee \neg\phi$ (*tertium non datur* or *principle of excluded middle*, PEM).

The above Definitions can be added to the argument forms of Figure 3.2.1 whenever there is more than one premise. These, as well as some of the Definitions above, can be formulated as classical tautologies (see below Prop. 4.3.1).

We can now characterize the classical syntactical ♡-consequence operations in terms of their associated relations as follows:

Proposition. 4.2.5. Classical Cn *satisfies*

1. $\top_{Cn}$ *iff it satisfies* M.

2. $\bot_{Cn}$ *iff it satisfies* EFQ.

3. $\neg_{Cn}$ *iff it satisfies* RD, PE, RAstr, PNC, PEM, *and* DN.

4. $\wedge_{Cn}$ *iff it satisfies* AD, SP, *and* PNC.

5. $\vee_{Cn}$ *iff it satisfies* AT, SM, *and* PEM.

6. $\to_{Cn}$ *iff it satisfies both* MP$_{Cn}$ *and* DT$_{Cn}$.

---

[4]Or, more correctly perhaps, *proof by contradiction* or *reductio ad impossibilem*.

4.2.1.2. Classical semantical $\heartsuit$-consequences

Recall from above that classical logic is a deductive system. We can define deduction from the semantical viewpoint as *preservation of truth* (or *truth-preservation*), so that we say that an argument is deductively valid iff, if the premises are true, then the conclusion must be true, too. For CPL, it suffices to define the semantical $\heartsuit$-consequence relations in Boolean terms, i.e. in terms of their truth tables (cf. Defs. 3.3.1 and 3.3.2.1; see also Section 6.3).

Proposition. *4.2.6. For all $\phi, \psi \in F_L$, a valuation* val *is*

1. $\top$-Boolean *iff* $val\,(\top) = \mathbf{t}$

2. $\bot$-Boolean *iff* $val\,(\bot) = \mathbf{f}$

3. $\neg$-Boolean *iff* $val\,(\neg\phi) = \mathbf{t} \Leftrightarrow val\,(\phi) = \mathbf{f}$

4. $\wedge$- Boolean *iff* $val\,(\phi \wedge \psi) = \mathbf{t} \Leftrightarrow val\,(\phi) = \mathbf{t}$ *and* $val\,(\psi) = \mathbf{t}$

5. $\vee$- Boolean *iff* $val\,(\phi \vee \psi) = \mathbf{t} \Leftrightarrow val\,(\phi) = \mathbf{t}$ *or* $val\,(\psi) = \mathbf{t}$

6. $\rightarrow$-Boolean *iff* $val\,(\phi \rightarrow \psi) = \mathbf{t} \Leftrightarrow val\,(\phi) = \mathbf{f}$ *or* $val\,(\psi) = \mathbf{t}$

Proposition. *4.2.7. For $A, B \in F_L$, the following are the truth tables for the $\heartsuit_i^k$ connectives of CPL with $k = 1, 2$:*

A	$\neg A$		A	B	$A \wedge B$	$A \vee B$	$A \rightarrow B$	$A \leftrightarrow B$
t	f		t	t	t	t	t	t
f	t		t	f	f	t	f	f
			f	t	f	t	t	f
			f	f	f	f	t	t

In order to extend the above results to CFOL an interpretation $\mathcal{I} = (\mathcal{D}, \Theta, \varpi)$ (cf. Definition 3.3.10) is required. In particular, the Boolean valuations are now constrained by val_ϖ, i.e. a valuation under the variable assignment ϖ, which assigns to a variable x a constant a in the domain of discourse $\mathcal{D}$ (denoted by $\varpi\,(x/a)$).

Definition. *4.2.8. For all $\phi, \psi \in F_{L1}$, given an interpretation $\mathcal{I} = (\mathcal{D}, \Theta, \varpi)$ a valuation $val_\mathcal{I}$ is*[5]

---

[5]Compare with Proposition 3.3.12.

1. $\top$-Boolean iff $val_{\mathcal{I}} = \mathbf{t}$

2. $\bot$-Boolean iff $val_{\mathcal{I}} = \mathbf{f}$

3. $\neg$-Boolean iff $val_{\mathcal{I},\varpi}(\neg\phi) = \mathbf{t} \Leftrightarrow val_{\mathcal{I},\varpi}(\phi) = \mathbf{f}$

4. $\wedge$- Boolean iff $val_{\mathcal{I},\varpi}(\phi \wedge \psi) = \mathbf{t} \Leftrightarrow val_{\mathcal{I},\varpi}(\phi) = \mathbf{t}$ and $val_{\mathcal{I},\varpi}(\psi) = \mathbf{t}$

5. $\vee$- Boolean iff $val_{\mathcal{I},\varpi}(\phi \vee \psi) = \mathbf{t} \Leftrightarrow val_{\mathcal{I},\varpi}(\phi) = \mathbf{t}$ or $val_{\mathcal{I},\varpi}(\psi) = \mathbf{t}$

6. $\rightarrow$-Boolean iff $val_{\mathcal{I},\varpi}(\phi \rightarrow \psi) = \mathbf{t} \Leftrightarrow val_{\mathcal{I},\varpi}(\phi) = \mathbf{f}$ or $val_{\mathcal{I},\varpi}(\psi) = \mathbf{t}$

We can now define the classical conditions for the semantical $\heartsuit$-consequence relations by means of the connectives of **L1** (cf. Prop. 3.3.12).

Definition. 4.2.9. Let $\mathcal{I} = (\mathscr{D}, \Theta, \varpi)$ be an interpretation for **L1**. We have

1. $\models_{\mathcal{I},\varpi} \top$ is classical

2. $\nvDash_{\mathcal{I},\varpi} \bot$ is classical

3. $\models_{\mathcal{I},\varpi} \neg\phi$ is classical iff $\nvDash_{\mathcal{I},\varpi} \phi$

4. $\models_{\mathcal{I},\varpi} \phi \wedge \psi$ is classical iff $\models_{\mathcal{I},\varpi} \phi$ and $\models_{\mathcal{I},\varpi} \psi$

5. $\models_{\mathcal{I},\varpi} \phi \vee \psi$ is classical iff $\models_{\mathcal{I},\varpi} \phi$ or $\models_{\mathcal{I},\varpi} \psi$

6. $\models_{\mathcal{I},\varpi} \phi \rightarrow \psi$ is classical iff $\nvDash_{\mathcal{I},\varpi} \phi$ or $\models_{\mathcal{I},\varpi} \psi$

Given the adequateness of classical logic, the classical semantical $\heartsuit$-consequence relations can be *mutatis mutandis* further specified as in Definitions 4.2.3-4 for $\models$, as in fact the conditions for the classical consequence operation in Proposition 4.2.5 hold also in a semantical perspective.

Exercises

Exercise 4.2.1.1. Define $\leftrightarrow_{Cn}$ and formalize the syntactical and semantical classicality conditions for the $\leftrightarrow$-consequence.

Exercise 4.2.1.2. Could we have

$$(\neg^*_{Cn}) \qquad \neg\phi \in Cn\,(\Gamma) \text{ iff } Cn\,(\Gamma, \phi) = F_{\mathsf{L}}$$

as a Tarski-style condition for classical $\neg_{Cn}$ instead of Definition 4.2.1.3? Why (not)?

Exercise 4.2.1.3. Let $\phi, \psi, \chi \in F_{\mathsf{L0}}$ be arbitrary formulae. Indicate whether the following statements are true or false. Account for your answer, with counter-examples if necessary.

1. $\phi \Vdash \psi$ and $\phi \Vdash \chi \quad \Rightarrow \quad \psi \Vdash \chi$

2. $\phi \Vdash \psi$ and $\psi \Vdash \chi \quad \Rightarrow \quad \phi \Vdash \chi$

3. $\phi \Vdash \psi \vee \chi \quad \Rightarrow \quad \phi \Vdash \psi$

4. $\phi \equiv \psi \vee \chi \quad \Rightarrow \quad \psi \Vdash \phi$

4.2.2. Classical ♦-consequences

The extension from L0 to L1 can be quite problematic in philosophical terms insofar as the introduction of the quantifiers is concerned. In particular, Boolean truth valuations are missing with respect to the quantifiers, and this motivates the suspicion of arbitrariness with respect to the choice of–and restriction to–the quantifiers $\forall$ and $\exists$. Nevertheless, completeness for classical FOL has been proven (see below), and from a mathematical perspective things are (more) clearly defined, namely in regard to the definitions of the consequence relations with respect to the quantifiers.

The syntactical classicality conditions for the $\forall$- and $\exists$-consequence relations can be formulated in terms of rules of inference. More specifically, these classicality conditions are formulated in terms of the rules for the introduction and elimination of the quantifiers (see Chapter 5 below).

Definition. 4.2.10. For the language L1, the consequence relation $\vdash$ is classical iff

1. for $\forall$ it is the case that

 a) $\phi(a) \vdash \forall x (\phi(x))$, provided a is arbitrary;

 b) $\forall x (\phi(x)) \vdash \phi(a)$.

2. for $\exists$ it is the case that

 a) $\phi(a) \vdash \exists x (\phi(x))$;

 b) $\exists x (\phi(x)) \vdash \phi(a)$, provided a has a merely auxiliary role.

The semantical counterpart of the classical $\forall$- and $\exists$-consequence relations originated in the work of Tarski (1935). The next definition stipulates the *Tarskian truth conditions* for the quantifiers:

Definition. 4.2.11. (Tarski, 1935) Given an assignment $\varpi : Vi \to \mathscr{D}$,

1. $\forall x (\phi(x))$ is true under ϖ iff every a in $\mathscr{D}$ satisfies $\phi(x)$ under ϖ.

2. $\exists x (\phi(x))$ is true under ϖ iff at least one a in $\mathscr{D}$ satisfies $\phi(x)$ under ϖ.

It is easy to see that this corresponds to the following definitions in terms of the semantical $\blacklozenge$-consequence relations:

Definition. 4.2.12. For the language L1, given an interpretation $\mathcal{I} = (\mathscr{D}, \Theta, \varpi)$ the consequence relation $\models$ is classical iff

1. $\models_{\mathcal{I},\varpi} \forall x (\phi(x))$ iff for every $a \in \mathscr{D}$ it is the case that $\models_{\mathcal{I}_a^x} \phi(a)$.

2. $\models_{\mathcal{I},\varpi} \exists x (\phi(x))$ iff for some $a \in \mathscr{D}$ it is the case that $\models_{\mathcal{I}_a^x} \phi(a)$.

And the corresponding semantical $\blacklozenge$-consequence operation is defined as follows:

Definition. 4.2.13. For the language L1, given an interpretation $\mathcal{I} = (\mathscr{D}, \Theta, \varpi)$ the consequence operation Cn is classical iff

1. $\phi[x/a] \in Cn_{\mathcal{I},\varpi} (\forall x (\phi(x)))$ if every $a \in \mathscr{D}$ is substitutable for x in $\phi(x)$ under ϖ.

2. $\phi[x/a] \in Cn_{\mathcal{I},\varpi} (\exists x (\phi(x)))$ if some $a \in \mathscr{D}$ is substitutable for x in $\phi(x)$ under ϖ.

Exercises

Exercise 4.2.2.1. Comment on the following condition for the classical ∀-consequence operation first stated by Tarski (1935):

$$(C\forall) \qquad \forall x \, (\phi \, (x)) \in Cn \, (\phi \, (x)) \, .$$

Exercise 4.2.2.2. Formalize the Tarski-style classicality conditions for the connective ↔.

4.3. The logic of CL

Given the Tarski-style conditions above, the deductive system CL generates tautologies that characterize it uniquely as a logic.

Proposition. 4.3.1. *The following are some of the valid sentences of the logic* **CL***:*

($\Vdash$1)	$A \vee \neg A$	PEM
($\Vdash$2)	$\neg (A \wedge \neg A)$	PNC
($\Vdash$3)	$A \rightarrow A$	Law of identity
($\Vdash$4)	$\neg\neg A \leftrightarrow A$	DN
($\Vdash$5)	$((A \rightarrow B) \wedge A) \rightarrow B$	MP
($\Vdash$6)	$((A \rightarrow B) \wedge \neg B) \rightarrow \neg A$	MT
($\Vdash$7)	$(A \rightarrow B) \leftrightarrow (\neg B \rightarrow \neg A)$	Law of contraposition
($\Vdash$8)	$((A \rightarrow B) \wedge (B \rightarrow C)) \rightarrow (A \rightarrow C)$	HS
($\Vdash$9)	$(\neg A \rightarrow (B \wedge \neg B)) \rightarrow A$	*Reductio ad absurdum* (RA)
($\Vdash$10)	$\neg (A \wedge B) \leftrightarrow (\neg A \vee \neg B)$	De Morgan's law (DeM$_\wedge$)
($\Vdash$11)	$\neg (A \vee B) \leftrightarrow (\neg A \wedge \neg B)$	De Morgan's law (DeM$_\vee$)

Proof: The proof is by examination of the truth tables or by producing the proofs in a calculus. **QED**

Proposition. *4.3.2. The negation of any tautology of CL is a contradiction thereof, and the negation of a contradiction of CL is a tautology thereof.*

Proof: Obvious given the Boolean nature of CL. **QED**

In particular, and denoting a contradiction or falsity by the symbol $\Box$:

Definition. 4.3.3. We have in CL the definition

$$(\Box_{df}) \qquad A \wedge \neg A.$$

Proposition. 4.3.4. *Given any formula A, in CL we have:*

$$A \wedge \Box = \Box$$

Proof: Obvious given $\Box_{df}$ and the classical truth table for $\wedge$. **QED**

We began this Section by speaking of unique characterizations with respect to a deductive system. By *characterizing uniquely*, we mean that given different conditions for the connectives of a given logical language, this typically gives rise to different logics. This unique characterization is particularly important in the case of **CL**, as it stands as the reference logic which other logics either restrict or extend. If a deductive system S does not generate at least one of the tautologies of the logic **CL**, or does not derive at least one of the theorems of **CL**, then we say that S is *non*-classical and we speak of the logic **S** as being a *non*-classical logic. For instance, intuitionistic logic does not generate or derive ⊩ 1 above, and with respect to ⊩ 4 only the ←-direction is a sentence of this logic. If a deductive system S has a many-valued semantics, then $A \wedge \neg A$ may not be a contradiction, and thus ⊨2 does not belong to **S**. These non-classical logics are thus subsets of **CL**, which makes them–arguably–less expressive logics, but the opposite is true of logics whose sets of tautologies/theorems are actually bigger than **CL**'s; examples of these are the modal logics.[6]

In some cases, there is only partial acceptance or rejection of some of the formulae in Proposition 4.3.1. For instance, PEM (⊩1) is not eliminated *tout court* in intuitionistic logic, but rather in (the FO formalization of) mathematical theories involving infinity. For instance, let $A = \forall x \exists y \, (P\,(y,x))$; if $x, y \in \mathbb{Z}^+$ and $P\,(y,x)$ stands for the ordering relation $y > x$, then we should have A or $\neg A$, but this is rejected in intuitionistic logic on the grounds that it does not appear possible to

---

[6] For this, usually extensions of L1 are required. For instance, in the case of the modal logics, we have the extended set of connectives $O_M = \{\neg^1, \Box^1, \Diamond^1, \wedge^2, \vee^2, \to^2\}$ ($\leftrightarrow$ is defined as usually by means of the other connectives).

provide a proof of either.[7] Take now ⊩ 2; clearly, this is a tautology iff $A \wedge \neg A$ is a contradiction. The reason to reject contradictions from CL is that, if accepted, then CL would be a trivial deductive system, as anything whatsoever follows from a contradiction (the principle of explosion; cf. Def. 4.2.3.3.b). However, some logics–the dialetheic logics–are founded on the stance that there are indeed true contradictions, and some other–the paraconsistent logics–"tolerate" them in view of the fact that they do not necessarily lead to trivial theories.[8]

We concentrated the discussion in the last paragraphs on ⊩1-3.[9] In effect, a logic in which any or all of these do not hold is indeed a non-classical logic. In fact, ⊩1-3 are often spoken of as the pillars of CL. Historically, they can all be traced back to Aristotle, not without reason called the father of CL. We give here the Aristotelian passages where these classical logical *laws* were first conceived, not as historical curiosities, but as food for thought on the very core of CL.[10]

> [⊩ 1]   But on the other hand there cannot be an inter-
> mediate between contradictories, but of one subject we must
> either affirm or deny any one predicate. This is clear, in the
> first place, if we define what the true and the false are. To
> say of what is that it is not, or of what is not that it is, is
> false, while to say of what is that it is, and of what is not
> that it is not, is true; so that he who says of anything that
> it is, or that it is not, will say either what is true or what is
> false; but neither what is nor what is not is said to be or not
> to be. (Aristotle, *Metaphysics*, Book IV, Part 7)

---

[7]This requirement of intuitionistic logic is known as the *disjunction property*, according to which, given some theory Θ,

$$\vdash_{\Theta} (\phi \vee \psi) \quad \Rightarrow \quad \vdash_{\Theta} \phi \text{ or } \vdash_{\Theta} \psi$$

for two formulae $\phi, \psi \in F_{L1}$, where $\vdash_{(\Theta)} \phi$ actually denotes that there is a proof (in Θ) of ϕ. Note how this is different from the same formalization in CL, where $\vdash_{(\Theta)} \phi$ denotes that ϕ is a theorem (of Θ).

[8]See Chapter 4 of Augusto (2020b) for a discussion on non-classical deductive consequences.

[9]Note that ⊩ 1 ≡ ⊩ 3, as we have

$$A \rightarrow A \equiv \neg A \vee A.$$

This formal identity, however, does not necessarily entail equivalent linguistic accounts of the two laws. In fact, ⊩ 3 is often formalized by employing the symbol for equality, i.e. $A = A$. But the employment of the symbol for equality in CL is not without issues (cf. Section 4.5 below), and so we prefer the formalization above.

[10]Traditionally, the first law to be presented is ⊩ 3. However, the order is irrelevant given ⊩ 1 ≡ ⊩ 3.

[⊪ 2]　　[I]t is impossible, then, that "being a man" should mean precisely "not being a man", if "man" not only signifies something about one subject but also has one significance ... And it will not be possible to be and not to be the same thing, except in virtue of an ambiguity, just as if one whom we call "man", others were to call "not-man"; but the point in question is not this, whether the same thing can at the same time be and not be a man in name, but whether it can be in fact. (Aristotle, *Metaphysics*, Book IV, Part 4)

[⊪ 3]　　First then this at least is obviously true, that the word "be" or "not be" has a definite meaning, so that not everything will be "so and not so". Again, if "man" has one meaning, let this be "two-footed animal"; by having one meaning I understand this:—if "man" means "X", then if A is a man "X" will be what "being a man" means for him. (It makes no difference even if one were to say a word has several meanings, if only they are limited in number; for to each definition there might be assigned a different word. For instance, we might say that "man" has not one meaning but several, one of which would have one definition, viz. "two-footed animal", while there might be also several other definitions if only they were limited in number; for a peculiar name might be assigned to each of the definitions. If, however, they were not limited but one were to say that the word has an infinite number of meanings, obviously reasoning would be impossible; for not to have one meaning is to have no meaning, and if words have no meaning our reasoning with one another, and indeed with ourselves, has been annihilated; for it is impossible to think of anything if we do not think of one thing; but if this is possible, one name might be assigned to this thing.) (Aristotle, *Metaphysics*, Book IV, Part 4)

Exercises

Exercise 4.3.1. Research into further rationales to expand or restrict the set of sentences of **CL**.

4.4. Classical FO theories and the adequateness of CFOL

The reader might have noticed that in the Section immediately above we approached the logic of CL solely from the propositional perspective. In effect, in a strict sense we may say that classical FO logic (CFOL) has no tautologies proper, as any open or closed formula $\phi(x_1, ..., x_n) = \phi(\vec{x})$ requires an interpretation of the form $\mathcal{I}_{\vec{a}}^{\vec{x}}$ for $\vec{x} \in Vi(\phi)$ and $\vec{a} \in \mathscr{D}$, which can entail arbitrarily finitely or infinitely many models. Thus, rather than tautologies, in CFOL we must speak of *instances* of FO tautologies. On the syntactical side, we may consider $\neg\forall x(\phi) \leftrightarrow \exists x(\neg\phi)$ and $\neg\exists x(\phi) \leftrightarrow \forall x(\neg\phi)$ (cf. Prop. 3.1.15.2), as well as $\forall x(\phi \wedge \psi) \leftrightarrow (\forall x(\phi) \wedge \forall x(\psi))$ and $\exists x(\phi \vee \psi) \leftrightarrow (\exists x(\phi) \vee \exists x(\psi))$, as theorems, but this is quite unnecessary, as these–and other examples–can be taken as equivalence-based rules of inference proper. Nevertheless, in a broader sense, we may say that there are FO tautologies and theorems, which allows us to say, for instance, that the set of (instances of) classical FO tautologies/theorems is recursively enumerable. Hence, semi-decidability is assured for CFOL theories.

Besides (semi-)decidability, we also often require adequateness of the deductive systems in which we treat our theories, i.e. we want to be assured that we only prove tautologies or valid formulae, and that all the valid formulae in the theories are provable (see Section 3.4.4). While adequateness of the deductive systems has to be evaluated with specific theories in mind,[11] we can give some general results. In this perspective, CFOL is an adequate deductive system for many theories, in particular so for many mathematical theories. The proof of the soundness of CFOL is essentially the same as that given for Theorem 3.4.12, but the reader is referred to, for instance, Enderton (2001) for a complete proof. The proof of CFOL completeness, firstly given by Gödel (1930), is not such a simple matter, being, in mathematical lingo, a *deeper result*.

In Section 3.5.2, we discussed the topic of FO completeness with two main aspects in mind: generality and decidability. That is, we were concerned with FO languages in general and how to tame their expressiveness so as to allow for increased decidability in logical systems based on them. Given these two aspects, all the results (including the restrictions) in the mentioned Section hold for CFOL. These results are actually fundamental for us, as we are here concerned with computing with CFOL, and without them this would be inefficient at best. Never-

---

[11]For instance, any FO theory containing elementary arithmetic is incomplete. This is the famous incompleteness result first published in Gödel (1931).

theless, we think it only appropriate to provide the reader with the more traditional results on CFOL that are typically invoked in mathematical and philosophical logic, and we think it appropriate also because they are often invoked in computational logic as well. This we do now, based on the proof given by L. Henkin (1949), in the present Section. However, the proof of completeness of CFOL–Gödel's or anyone else's[12]–is convoluted, as befits a deep result, and we provide only some of the main aspects thereof. This we do also with the objective of leaving the details as exercises for the reader, who is referred to Henkin's original proof, as well as to Enderton (2001) and Mendelson (2015) for different presentations of this celebrated result.

As said above, though we can prove the adequateness of a deductive system, this is often done with specific theories in mind. We shall now write $\vdash_\Theta \phi$ instead of $\Theta \vdash \phi$ to emphasize that Θ is not just a set of formulae, actually including axioms and/or rules of inference; in other words, Θ may be simply a calculus.[13] The whole point of the completeness theorem for CFOL is that, given some theory Θ, we have $\vdash_\Theta \phi$ if we have $\models_\Theta \phi$, so a theory Θ can also be defined semantically as a closed set of formulae for which, given an interpretation $\mathcal{I}$, there is some model $\mathcal{M}$, i.e. there is an interpretation in which all the axioms of Θ are valuated to **true**. In other words, the validity of the formulae of Θ is preserved by the rules of inference of Θ, and Θ is said to be *complete* in the sense that no more rules of inference are necessary to prove all the tautologies of Θ.

Although somehow unorthodoxly coined, the following theories will play an important role in the proof of completeness of the classical FO predicate calculus.

Definition. 4.4.1. Let Θ be a theory in a FO language. Then, Θ is a *scapegoat theory* if for any formula $\phi(x)$ that has x as its only free variable there is a ground term t such that

$$(\Upsilon) \qquad \vdash_\Theta \exists x \, (\neg\phi(x)) \to \neg\phi(t).$$

Note that Υ is actually an application of DT to $\exists x \, (\neg\phi(x)) \vdash_\Theta \neg\phi(t)$, and this is equivalent to $\neg\forall x \, (\phi(x)) \vdash_\Theta \neg\phi(t)$. Clearly, only a language with the required resources–such as L1–can produce a scapegoat theory.

---

[12]E.g., Beth (1960). As it is easy to guess, completeness of CPL is a far less complex affair to prove; see, e.g., Quine (1938).
[13]See Chapter 5 below for an elaboration on axiomatic systems. However, we may wish to distinguish the proper axioms from the calculus proper, in which case we may write, say, $\Theta \vdash_\mathcal{E} \phi$.

In the following fundamental theorem, the symbol $\Vdash$ without any subscripts denotes the classical FO consequence relation for a FO language with countably many symbols and formulae.

Theorem 4.4.1. (*The completeness theorem; Gödel, 1930*) *The following statements are equivalent:*

1. *In any FO theory Θ, the theorems of Θ are precisely the valid formulae of Θ, i.e.*

$$if \ \models_\Theta \phi, \ then \ \vdash_\Theta \phi.$$

2. *Any consistent FO theory has a model.*

Proof: First, we have to prove that statements 1 and 2 are equivalent. This done, it suffices to prove either of the statements. We leave the first part as an exercise, and choose to prove statement 2. We begin by preparing the "setting" (step I) and then proceed to (the sketch of) the proof proper (step II).

I. Let Θ be a consistent set of FO formulae of the countable language L1. Henkin (1949) considers a theory Θ,[14] with MP and GEN[15] for inference rules, comprising the following axioms 1-5 (with restrictions indicated by *) and the classical FO theorems 6-10:

1. $C \rightarrow (B \rightarrow C)$
2. $(A \rightarrow B) \rightarrow ((A \rightarrow (B \rightarrow C)) \rightarrow (A \rightarrow C))$
3. $\neg\neg A \rightarrow A$
4. $\forall x (A \rightarrow B) \rightarrow (A \rightarrow \forall x (B))$ (*)
5. $\forall x (A) \rightarrow B [x/y]$ (*)
6. $DT_\vdash$ (Theorem 3.4.11; version for $\vdash$)
7. $\vdash \neg B \rightarrow (B \rightarrow C)$
8. $\vdash B \rightarrow (\neg C \rightarrow \neg (B \rightarrow C))$
9. $\vdash \forall x (\neg A) \rightarrow \neg\exists x (A)$
10. $\vdash \neg\forall x (B) \rightarrow \exists x (\neg B)$

The restrictions are as follows: In 4, x does not occur freely in A; in 5, y is any symbol replacing each free occurrence of x in A and no free occurrence of x in A occurs in a sub-formula of A of the form $\forall y (C)$.

---

[14] Actually, Henkin considers a *class* of deductive systems. This is obviously the class of classical FO logics, or CFOL for short. We work with a theory for reasons that should become clear in the proof below.

[15] This rule, called *generalization rule*, states that one can infer $\forall x (A)$ from A. See below § 5.1.2.

Axiom 1 is so also in the axiom system $\mathcal{L}q$ of Section 5.1. Axiom 3 is our DN, i.e. involution or double negation law (cf. Def. 4.2.4.1). The theorems have already been given in some form or can easily be verified to be theorems of CFOL. We give here Henkin's formal system, even though we do not strictly follow his original proof, and the reader is free to add axioms and/or theorems at will as long as they are well established as classical and can actually be invoked in this proof of FOL completeness. In particular, we shall invoke the following axiom in our theory Θ:

$$(\dagger) \qquad \forall x_i \, (\phi \, (x_i)) \to \phi \, (t)$$

which holds if $\phi \, (x_i) \in F_{\mathsf{L1}}$ and t is a term free for x_i in $\phi \, (x_i)$. In fact, it may be the case that $x_i = t$, so that we have the axiom

$$(\dagger') \qquad \forall x_i \, (\phi) \to \phi.$$

We shall also invoke a particularization of $\dagger$, denoted by $\ddagger$, which states that if t is free for x in $\phi \, (x)$, then

$$(\ddagger) \qquad \forall x \, (\phi \, (x)) \vdash \phi \, (t) \, .$$

As x is free for x in $\phi \, (x)$, a special form of $\ddagger$ is

$$(\ddagger') \qquad \forall x \, (\phi) \vdash \phi.$$

Besides the above rules of inference, Henkin considers an additional rule:

11. Let X be a set of formulae no one of which contains a free occurrence of the individual symbol c. Let further a formula B be obtained from a formula A by replacing each free occurrence of c by the individual symbol x, it being the case that none of these occurrences of x is bound in B. Then, if $X \vdash A$, also $X \vdash B$.

With respect to the rule GEN, we have the following specification: Given $X \Vdash \phi$, then we can obtain $X \Vdash \forall x \, (\phi \, [c/x])$ if $c \notin X$ is a constant symbol and x does not occur in ϕ; furthermore, there is a deduction $X \Vdash \forall x \, (\phi \, [c/x])$ in which c does not occur. We call this specification *constant generalization* (abbr.: CGEN).

II. Expand $\mathsf{L1}$ with a countably infinite set of new constants $Cons^{*}$, so that the extension Θ^{*} remains consistent in the new language $\mathsf{L1}^{*}$ (1. How?). For every $\phi \in \mathsf{L1}^{*}$, for any variable x, and any constant $c \in Cons^{*}$, we let $A \in \Theta^{*}$,

$$A = \neg \forall x \, (\phi \, (x)) \to \neg \phi \, [x/c]$$

(2. Why?). Let now

$$\theta_1 = \neg\forall x_1 \left(\phi\left(x_1\right)\right) \to \neg\phi_1\left[x_1/c_1\right]$$

$$\vdots$$

$$\theta_n = \neg\forall x_n \left(\phi\left(x_n\right)\right) \to \neg\phi_n\left[x_n/c_n\right]$$

where c_i is the i-th new constant not occurring in ϕ_i or in θ_j for $j < i$.
Let

$$\Theta^* - \Theta = \bigcup_{i=1}^{n} \theta_i.$$

Then, $\Theta \cup \Theta^*$ is consistent. (3. Proof?) We extend (4. How?) $\Theta \cup \Theta^*$
to a consistent set Θ^∞; Θ^∞ is maximal, i.e. for any closed formula
$\psi \in \mathsf{L1}^*$, either $\psi \in \Theta^\infty$ or $\neg\psi \in \Theta^\infty$, and Θ^∞ is deductively closed. In
effect, we have:

$$\vdash_{\Theta^\infty} \psi \quad \Rightarrow \quad \nvdash_{\Theta^\infty} \neg\psi$$

$$\Rightarrow \neg\psi \notin \Theta^\infty$$

$$\Rightarrow \psi \in \Theta^\infty$$

From Θ^∞, we now construct (5. How?) an interpretation $\mathcal{I} = (\mathscr{D}, \Theta, \varpi)$
for $\mathsf{L1}^*$ where $\mathscr{D}$ is the set of ground terms of Θ^∞ and for every predicate
symbol P of $\mathsf{L1}^*$ we have

$$val_{\mathcal{I}}\left(P\left(t_1, ..., t_n\right)\right) = \{(t_1, ..., t_n) \mid \vdash_{\Theta^\infty} P\left(t_1, ..., t_n\right)\}.$$

In other words, the closed formula $P\left(t_1, ..., t_n\right)$ is true in $\mathcal{I}$ iff it is provable in Θ^∞, so that, given the consistency of Θ^∞, we must have

$$(\clubsuit) \qquad \vdash_{\Theta^\infty} P\left(t_1, ..., t_n\right) \quad \Rightarrow \quad (t_1, ..., t_n) \in \mathcal{I}\left(P\right)$$

$$(\spadesuit) \qquad \vdash_{\Theta^\infty} \neg P\left(t_1, ..., t_n\right) \quad \Rightarrow \quad (t_1, ..., t_n) \notin \mathcal{I}\left(P\right)$$

So, we need to prove that, given $\mathcal{I}$, there is a model $\mathcal{M}$ such that if
$\vdash_{\Theta^\infty} \psi$, then $\models_{\mathcal{M}} \psi$ for a formula ψ. This we do by induction on atomic
ψ.

(i) Let $\psi = P\left(t_1, ..., t_n\right)$. By $\clubsuit$, we have $\models_{\mathcal{M}} P\left(t_1, ..., t_n\right)$, and so
$\models_{\mathcal{M}} \psi$.

(ii) Let now $\psi = \neg\chi$, so that we have $\models_{\mathcal{M}} \neg\chi$.

(iii) Let now $\psi = \chi \to \omega$ and let $val_{\mathcal{I}}\left(\chi \to \omega\right) = \mathsf{f}$ to apply *reductio ad
absurdum*, so that we have $\vdash_{\Theta^\infty} \neg\psi$ (6. How?) and Θ^∞ is inconsistent.
But this is a contradiction.

(iv) Make now $\psi = \forall x_i \left(\phi \left(x_i \right) \right)$. If x_i does not occur free in ϕ, ϕ is closed, and by the induction hypothesis, if $\vdash_{\Theta\infty} \phi$, then $\models_{\mathcal{M}} \phi$. If $\vdash_{\Theta\infty} \phi$, then $\vdash_{\Theta\infty} \forall x_i \left(\phi \right)$, and $\models_{\mathcal{M}} \forall x_i \left(\phi \right)$ iff $\models_{\mathcal{M}} \phi$ (7. How?). Hence, we have $\models_{\mathcal{M}} \psi$ iff $\vdash_{\Theta\infty} \psi$. On the other hand, if x_i occurs free in $\phi \left(x_i \right)$, then it must be the only free variable, given that $\psi = \forall x_i \left(\phi \left(x_i \right) \right)$ is closed. Thus, it must be the case that $\phi \left(x_i \right) = \theta_i \in \Theta^{*}$. Let it be θ_k. Then, $\psi = \forall x_k \left(\phi \left(x_k \right) \right)$. Suppose now $\vdash_{\Theta\infty} \psi$ but $\not\models_{\mathcal{M}} \psi$. Then, there is a sequence that does not satisfy $\forall x_k \left(\phi \left(x_k \right) \right)$ and so neither does it satisfy $\phi \left(x_k \right)$. But then neither does it satisfy $\phi \left[x_k / c_k \right]$ and we have $\models_{\mathcal{M}} \neg \phi \left(c_k \right)$ (8. How?). But $\vdash_{\Theta\infty} \forall x_k \left(\phi \left(x_k \right) \right) \to \phi \left(c_k \right)$, from which by MP we have $\vdash_{\Theta\infty} \phi \left(c_k \right)$; by the induction hypothesis, we have $\models_{\mathcal{M}} \phi \left(c_k \right)$, and we reach a contradiction.

Thus, all the formulae in Θ^{*} have a common model $\mathcal{M}$ in $\mathcal{I}$ iff they are derivable in Θ^{∞}, and we proved that any consistent FOL theory has a model. **QED**

Exercises

Exercise 4.4.1. Give an intuitive account of a scapegoat theory. (Hint: Find the "scapegoat" in Υ.)

Exercise 4.4.2. For Theorem 4.4.1, prove the equivalence of statements 1 and 2.

Exercise 4.4.3. In the proof of Theorem 4.4.1, give answers to the questions posed in it. The following are hints to some of these questions:

1. Invoke constant generalization.

2. Focus on $\neg$.

3. The proof is by RA. Assume that $\Theta \cup \left(\Theta^{*} = \{ \theta_1, ..., \theta_k, \theta_{k+1} \} \right)$, $k \geq 0$, is inconsistent. Let $\theta_{k+1} = A$ and take the least k to obtain

$$\vdash_{\Theta \cup \{ \theta_1, ..., \theta_k \}} \begin{cases} \neg \forall x \left(\phi \right) \\ \phi \left[x / c \right] \end{cases}$$

so that we have both $\neg \forall x \left(\phi \right)$ and $\forall x \left(\phi \right)$, the latter by $\forall y \left(\phi \left[y / x \right] \right)$.

4. Let $\Theta^{\infty} = \Theta \cup \Theta^{*} \cup \Xi$, Ξ is the set of axioms for $\mathsf{L1}^{*}$, and let $\Theta^{\infty} = \{ \psi | val \left(\psi \right) = \mathsf{t} \}$. What are the *additional* axioms of Θ^{∞}?

5. Define $\mathcal{I}$ for every constant symbol c, every function symbol f, and every predicate symbol P of $\mathsf{L1}^{*}$.

Exercise 4.4.4. The proof of Theorem 4.4.1 is given for countable languages. Indicate, in general terms, what changes might be required to this proof to accommodate languages with a higher cardinality.

Exercise 4.4.5. Let the following lemma be given:

Lemma 4.4.2. *Every consistent theory Θ has a consistent extension Θ^* that contains denumerably many ground terms and is a scapegoat theory.*

1. Where was this lemma employed in the proof of Theorem 4.4.1?

2. Prove it in detail.

Exercise 4.4.6. Let the following lemma be given:

Lemma 4.4.3. *Let Θ be a complete scapegoat theory. Then Θ has a model $\mathcal{M}$ whose domain is the set of ground terms of Θ.*

1. Where was this lemma employed in the proof of Theorem 4.4.1?

2. Prove it in detail.

Exercise 4.4.7. Where in the proof of Theorem 4.4.1 is Lindenbaum's theorem (Theorem 3.4.16) implicitly invoked?

Exercise 4.4.8. Which of the following statements, given any FOL theory Θ, can be a (multiple-statement) corollary to the completeness theorem?

1. For $\phi \in \Theta$ and $X \subseteq \Theta$, if $\phi \in Cn(X)$, then we have $X \vdash_\Theta \phi$.

2. ϕ is **true** in every denumerable model of Θ iff $\vdash_\Theta \phi$.

3. The set of terms of a language L1 is denumerable.

4. If, in every model of Θ, every sequence that satisfies all $\chi_i \in X$ also satisfies ϕ, then $X \vdash_\Theta \phi$.

5. For $\phi, \chi \in \Theta$, if $\phi \in Cn(\chi)$, then $\chi \vdash_\Theta \phi$.

6. If Θ is consistent, then it has a denumerable model.

Exercise 4.4.9. In Section 3.5.2.1, we made the completeness theorem for FOL follow from the FO-compactness theorem. Given the completeness proof in this Section, how is the reverse obtained, i.e. how do we obtain the FO-compactness theorem from the completeness proof?

4.5. The extension CL$^=$: CL with equality

So far, other than some brief treatments in Chapter 3, we have been mostly silent on the topic of *equality* in CL. The reason for this silence is that this relation happens to have more to it than first meets the eye. Put simply, it "complicates" CFOL if we introduce it in the alphabet of L1 by means of a new symbol, typically "=". However, equality is a property that is quite important, especially for the formalization of mathematical theories, and many authors see it as actually indispensable. As it is often put intuitively, equality is a predicate that is meaningful regardless of the domain of discourse.

In Example 3.2.3, we avoided the introduction of the symbol for equality by means of a predicate E of arity n, but equality in a logical language is a *logical* symbol rather than a parameter. In particular, it allows the expression of two fundamental laws:

$$(\text{SubP}) \qquad \forall x \forall y \left[(x = y) \to \forall P \left(P(x) \leftrightarrow P(y) \right) \right]$$

and

$$(\text{IdI}) \qquad \forall x \forall y \left[\forall P \left(P(x) \leftrightarrow P(y) \right) \to (x = y) \right]$$

known as *substitution principle* and (Leibniz's principle of) *identity of indiscernibles*, respectively. These two laws, in turn, allow us to express *Leibniz's law*:

$$(\text{LL}) \qquad \forall x \forall y \left[\forall P \left(P(x) \leftrightarrow P(y) \right) \leftrightarrow (x = y) \right].$$

It is obvious, however, that the three laws above are *not* expressible in FOL. In the above paragraph we wrote *complicate* between inverted commas to express the fact that we use this term somehow reluctantly. In effect, to be rigorous, from the formal point of view, a FO language cannot fully express equality. This can easily be verified in the following axiomatization in the language L1$^=$, i.e. L1 augmented with the equality symbol.

Proposition. 4.5.1. *The following is an axiomatization of equality as a logical symbol:*

$$(\mathcal{E}1) \quad \forall x \, (x = x)$$
$$(\mathcal{E}2) \quad \forall x \forall y \left((x = y) \to (y = x) \right)$$
$$(\mathcal{E}3) \quad \forall x \forall y \forall z \left[((x = y) \wedge (y = z)) \to (x = z) \right]$$
$$(\mathcal{E}4) \quad \forall f \forall x \forall y \left[(x = y) \to (f(x) = f(y)) \right]$$
$$(\mathcal{E}5) \quad \forall P \forall x \forall y \left[(x = y) \to (P(x) = P(y)) \right]$$

Proof: Left as an exercise (Hint: Use SubP, IdI, and $\mathcal{E}1$ to prove $\mathcal{E}2$-5).

To be more precise, in axioms $\mathcal{E}4$-5 we must introduce sequences of variables $\vec{u} = u_1, ..., u_n$ and $\vec{z} = z_1, ..., z_m$, for $n, m \geq 0$, such that we have

$$(\mathcal{E}4) \qquad \forall f \forall x \forall y \left[(x = y) \rightarrow (f(\vec{u}, x, \vec{z}) = f(\vec{u}, y, \vec{z})) \right]$$

and

$$(\mathcal{E}5) \qquad \forall P \forall x \forall y \left[(x = y) \rightarrow (P(\vec{u}, x, \vec{z}) = P(\vec{u}, y, \vec{z})) \right].$$

Again, it will be easily verified that this axiomatization of equality is actually a *second-order* axiomatization (cf. axioms $\mathcal{E}4$-5). This means that the quantifiers are allowed to quantify over functions and predicates, and this entails that we are no longer using L1, but rather L2$^=$, i.e. a second-order extension of L1 with the equality symbol. It is thus obvious that the above theory

$$\Theta_{Id} = \{\text{SubP}\} \cup \{\text{IdI}\} \cup \{\text{LL}\} \cup \{AX_{Eq}\}$$

where AX_{Eq} is the set of the five axioms of Proposition 4.5.1 and which is sometimes called *theory of logical identity* (e.g., Tarski, 1994), is not a FO theory.

Is there some way to introduce *de facto* the symbol "=" in CFOL? Just as in the case of all other logical symbols in a truth-preserving logical system, we require what can be called *a semantics of the symbol* "=". In effect, the fact that CFOL is not capable of expressing the theory of equality Θ_{Id} does not entail that we cannot extend it by including "=" in its set of logical symbols.

Definition. 4.5.2. Given an interpretation $\mathcal{I} = (\mathcal{D}, \Theta, \varpi)$, we say that a model $\mathcal{M}$ is *normal* if = denotes the equality relation on $\mathcal{D}$.

By this simple means, we have in fact obtained an extension of CL that we shall denote by CL$^=$ and speak of as *classical FO logic with equality*. Similarly, the language L1 is now L1$^=$.

Definition. 4.5.3. Let us denote the semantical consequence relation in CL$^=$ by $\models_=$. Then, given a set of formulae $X \subseteq$ L1$^=$ and a formula $\phi \in$ L1$^=$, we write

$$X \models_= \phi$$

provided that $val_{\mathcal{I}}(\phi) = \mathbf{t}$ in every normal model in which $val_{\mathcal{I}}(\chi_i) = \mathbf{t}$ for all the $\chi_i \in X$.

We then have the following trivial proposition, where we denote the classical semantical consequence relation simply by $\models$:

Proposition. 4.5.4. *If $X \models \phi$, then $X \models_= \phi$.*

Proof: Left as an exercise.

Example 4.5.1. We formalize the argument of Exercise 3.2.2.4 in L1$^=$. Axioms 1-3 are the proper axioms of the group theory Θ_G for a group a pair $\mathcal{G} = (G, \star)$. Note that we did not add the logical axioms of equality, nor did we include any of SubP, IdI, or LL. Note also that we chose to write the predicate $\star$ in infix notation, as is more common in mathematics.

$$
\begin{array}{ll}
(1) & \forall x \forall y \forall z\, ((x \star y) \star z = x \star (y \star z)) \\
(2) & \forall x \exists w\, (x \star w = x \wedge w \star x = x) \\
(3) & \forall x \exists y \exists w\, (x \star y = w \wedge y \star x = w) \\
\hline
(4) & \forall x \forall y \forall z\, ((x \star z = y \star z) \rightarrow x = y)
\end{array}
$$

We can prove the correctness of the argument in Example 4.5.1 by applying directly some proof calculus, say, resolution. Does this mean that we can do without Θ_{Id} when working in L1$^=$? Not really, as a rigorous formal FO theory of equality cannot dispense with Θ_{Id}, totally or in part. However, we might not need it if we use a proof calculus that has its own way to tackle equality. This is the case of the resolution calculus, which has a proof technique called paramodulation (see Section 8.1.5). In other cases, there might be a specific equality predicate; for instance, Prolog has a built-in predicate for this relation (see Section 9.2.1).

Of course, one may wonder whether these correspond to the equality defined in Θ_{Id}. Fitting (1996) gives a formal means to integrate the axioms of AX_{Eq} in CFOL via function and predicate *replacement axioms*.

Definition. 4.5.5. Let $f \in Fun$ and $P \in Pred$. Then,

1. a replacement axiom $\mathcal{F}_{repl} \in Fun_{repl}$ for f is

$$\forall v_1...\forall v_n \forall w_1...\forall w_n[((v_1 = w_1) \wedge ... \wedge (v_n = w_n)) \rightarrow$$

$$(f(v_1, ..., v_n) = f(w_1, ..., w_n))].$$

2. a replacement axiom $\mathcal{P}_{repl} \in Pred_{repl}$ for P is

$$\forall v_1...\forall v_n \forall w_1...\forall w_n [((v_1 = w_1) \wedge ... \wedge (v_n = w_n)) \rightarrow$$

$$(P(v_1, ..., v_n) = P(w_1, ..., w_n))].$$

Example 4.5.2. Let us consider $\star$ as a binary function symbol. Let there be given the replacement axiom

$$(\mathcal{F}_{repl}1) \quad \forall x \forall y \forall z \forall w [((x = z) \wedge (y = w)) \rightarrow (x \star y = z \star w)].$$

Let now $c \in Cons$. Then we have

$$\{\mathcal{F}_{repl}1\} \models \forall x \forall z [((x = z) \wedge (c = c)) \rightarrow (x \star c = z \star c)]$$

and

$$\{\mathcal{F}_{repl}1\} \cup \{\mathcal{E}1\} \models \forall x \forall z ((x = z) \rightarrow (x \star c = z \star c)).$$

Let $AX_{repl} = Fun_{repl} \cup Pred_{repl}$ be the set of the replacement axioms. Then we have the following result:

Theorem 4.5.1. *For a set of FOL formulae X and a FO formula ϕ,*

$$X \models_= \phi \quad iff \quad X \cup \{\mathcal{E}1\} \cup \{AX_{repl}\} \models \phi.$$

Proof: Left as an exercise.

Rigorously, we now have

$$\mathbf{CL}^= = \mathbf{CL} \cup \{\mathcal{E}1\} \cup \{AX_{repl}\}.$$

However, just as we may dispense with Θ_{Id} in some cases, we may also relax this and simply consider $\mathbf{CL}^=$ as classical FO logic with equality without going all the formal way above. Were our main concern mathematical proofs, we would not so easily get away with this relaxation; but we are concerned here with a general notion of deductive computation, and thus feel this relaxation is permissible. The reader not willing to condescend with this relaxation can benefit from Fitting (1996), which integrates equality in a rigorous way in both the resolution and the analytic tableaux calculi.

In the Chapters that follow, we consider mostly CL *without* equality. Whenever this is considered, namely by means of the extension CL$^=$, we shall make it clear; otherwise, in presence of the symbol "=" the reader may safely assume that we are at the metalanguage level.[16]

Exercises

Exercise 4.5.1. Show that $\mathcal{E}$1-3 define an equivalence relation.

Exercise 4.5.2. Formalize the argument in Example 3.2.3 in FOL with equality.

Exercise 4.5.3. The substitution principle (SubP) is often also called *principle of extensionality*. Account for this synonymy.

Exercise 4.5.4. Sometimes the following axioms are given for CFOL instead of $\mathcal{E}$4 and $\mathcal{E}$5, respectively:

$$\forall x \forall y \forall z \left[((f(x) = z) \wedge (x = y)) \rightarrow (f(y) = z) \right]$$

$$\forall x \forall y \left[(P(x) \wedge (x = y)) \rightarrow P(y) \right].$$

Are these equivalent to $\mathcal{E}$4 and $\mathcal{E}$5, respectively?

Exercise 4.5.5. Produce an example showing that the converse of Proposition 4.5.4 is not true.

Exercise 4.5.6. Consider $<$ as a binary predicate symbol and give an example of a replacement axiom for it.

Exercise 4.5.7. Explain formally why $\mathbf{CL}^=$ is an extension of $\mathbf{CL}$.

Exercise 4.5.8. Show that the compactness theorem carries over to $\mathbf{CL}^=$.

Exercise 4.5.9. Prove Theorem 4.4.1 for $\mathbf{CL}^=$ (Hint: It suffices to make some changes to the proof given in Section 4.4).

Exercise 4.5.10. Prove Propositions 4.5.1 and 4.5.4.

Exercise 4.5.11. Prove Theorem 4.5.1.

---

[16] Given our restricted and localized use of the symbol "=" at the object-language level, we see no need to distinguish this from equality at the metalanguage level by means of a different symbol. In particular, we see no need to distinguish *here* mathematical *equality* from logical *identity*, as, for instance, Tarski (1994) advocates.

5. Classical proofs

Proof systems are essentially systems of proof for *theoremhood*, i.e. they are used to prove theorems. However, they can be used to prove validity if the given logic is sound. This is the case of CL (see Chapter 4).

As stated and accounted for in the Introduction (cf. Section 0.3), in this book we are mostly concerned with satisfiability (vs. validity) testing, namely with refutation in view, both for automated theorem proving and logic programming; this justifies our choice of the resolution and the analytic tableaux calculi (see Chapter 8) as our main proof calculi. Nevertheless, we believe that knowledge of the direct classical proof methods is, if not mandatory, useful, and we provide in this Section brief expositions and examples of the main standard direct proof calculi for classical logic, to wit, the axiom system $\mathcal{L}$, the natural deduction calculus $\mathcal{NK}$, and the sequent calculus $\mathcal{LK}$. With respect to this last calculus, we give it a somehow more extensive treatment having in mind its kinship with the analytic tableaux calculus. Augusto (2019) thoroughly focuses on classical proofs and classical proof systems. The reader seeking a more comprehensive treatment of classical proof theory can benefit from, for example, Troelstra & Schwichtenberg (2000).

For what follows, recall the material in Section 3.4.2. An *axiom system* or *calculus* for the language L is a triple $\mathcal{A} = (F_\mathsf{L}, AX, RI)$, where F_L is a set of formulae, AX is a set of axiom schemata, and RI is a set of rules of inference. Clearly, this triple can be a proof calculus $\mathcal{P} = (F_\mathsf{L}, AX, RI)$, so we have to state where the distinction lies: in an axiom system $\mathcal{A}$, the set RI is either empty or minimal,[1] whereas in a *rule-based* proof calculus $\mathcal{P}$, the reverse is true, i.e. the set AX is either empty or minimal. Nevertheless, it is often the case that proof calculi, regardless of whether rule-based or axiomatic proper, are referred to as axiom systems, and the term "axiomatization" is often employed to mean that a logical language F is provided with a pair (AX, RI) where either AX or RI may be empty.

It should be obvious from the examples below that, while axiomatic proofs proper allow for a characterization of a logic as a whole, they do not allow for a characterization of the behavior of the individual

---

[1] Typically, in an axiom system we have $RI = \{\text{MP,SUB}\}$ or a subset thereof (cf. Def. 3.4.19).

connectives of the logic at hand. This shortcoming is overcome in the rule-based proof systems, such as natural deduction and the sequent calculi (see below). This helps one to decide which proof system to use.

5.1. The axiom system $\mathcal{L}$

When checking for theoremhood in an axiom system, one has to observe carefully that every line of the proof is a theorem. In an axiom system $\mathcal{A} = (F_{\mathsf{L}}, AX, RI)$, the set AX plays a more important role than the set RI. The cardinality of the set AX can vary greatly from system to system, with some systems having as little as three axioms and others more than ten, but it is important to check the *independence* of the axioms from each other; if one can derive an axiom from other axioms of the same system, then one has a superfluous axiom. Other than independence, *consistency* and *sufficiency* are also important properties of an axiom system: no inconsistent formula should be derivable from the set of axioms, and these should suffice to prove every theorem of the logic in consideration.

We elaborate here on the axiom system known as Frege-Łukasiewicz's, which is a simplification of the original six-axiom system conceived by Frege (see Exercise 5.1.3). We denote it generally by $\mathcal{L}$ and specify its FO extension as $\mathcal{L}q$ (where q stands for "quantification" or "quantified").

Proposition. 5.1.1. *The following axiom schemata $\mathcal{L}$1-3, together with the rule MP, constitute $\mathcal{L} = (F_{\mathsf{L}}, \{\mathcal{L}1, \mathcal{L}2, \mathcal{L}3\}, \{MP\})$, a proof system for classical propositional logic known as the* Frege-Łukasiewicz *axiom system:*

$$(\mathcal{L}1) \quad \phi \to (\psi \to \phi)$$
$$(\mathcal{L}2) \quad (\phi \to (\psi \to \chi)) \to ((\phi \to \psi) \to (\phi \to \chi))$$
$$(\mathcal{L}3) \quad (\neg\phi \to \neg\psi) \to (\psi \to \phi)$$

Proof: Left as a (research) exercise.

Example 5.1.1. Figure 5.1.1 shows a proof of the theorem $\vdash \phi \to \phi$ in $\mathcal{L}$. On the right, the axiom schemata and/or rule used in each step of the derivation are indicated. For instance, "MP (1, 2)" denotes the application of the rule MP on the steps 1 and 2. The symbol $\vdash$ features on every step of the proof solely with the aim of making it clear that each step of a proof in this axiom system is a theorem; in fact, this is a symbol of $\mu^{\mathcal{L}}$, the metalanguage of $\mathcal{L}$, whereas a proof typically belongs to $\mathcal{L}$ as an object-language.

1.	$\vdash (\phi \to ((\phi \to \phi) \to \phi)) \to ((\phi \to (\phi \to \phi)) \to (\phi \to \phi))$	$\mathcal{L}2$
2.	$\vdash \phi \to ((\phi \to \phi) \to \phi)$	$\mathcal{L}1$
3.	$\vdash (\phi \to (\phi \to \phi)) \to (\phi \to \phi)$	MP (1,2)
4.	$\vdash \phi \to (\phi \to \phi)$	$\mathcal{L}1$
5.	$\vdash \phi \to \phi$	MP (3,4)

Figure 5.1.1.: Proof of $\vdash_{\mathcal{L}} \phi \to \phi$.

Remark. **5.1.2.** In order to obtain the axiom system $\mathcal{L}q$ for L1 we add to $\mathcal{L}$1-3 further axiom schemata–the *quantifier axioms* Q1-2, which assume ϕ_t^x–and an additional inference rule, known as *generalization rule* (GEN):[2]

$$(\text{Q1}) \qquad \forall x\,(\phi\,(x)) \to \phi\,(t)$$
$$(\text{Q2}) \qquad \phi\,(t) \to \exists x\,(\phi\,(x))$$

$$(\text{GEN}) \qquad \frac{\phi\,(x)}{\forall y\,(\phi\,(y))}; \quad \frac{\phi}{\forall x\,(\phi)}$$

Example 5.1.2. We prove in $\mathcal{L}q$ that $\forall x\,(\chi)$ follows from the set of premises $\{\phi, \forall x\,(\phi) \to \chi\}$ in CFOL, i.e. $\{\phi, \forall x\,(\phi) \to \chi\} \vdash_{\mathcal{L}q} \forall x\,(\chi)$. Figure 5.1.2 shows the proof. (Instead of premises, *assumptions* may be used in proofs of the kind $\vdash \phi$.)

Recall that CFOL is an adequate system (cf. Section 4.4 above). Thus, although $\mathcal{L}q$ is a proof system, and thus a system for proving theoremhood, a feature denoted by $\vdash_{\mathcal{L}q}$, it is also a system for proving validity, and we denote this by $\models_{\mathcal{L}q}$.

Further relevant axiom systems for classical logic are those of Hilbert & Ackermann (1928) and Kleene (1952).

---

[2]We give two versions of GEN.

5. Classical proofs

1.	$\vdash \phi$	Premise
2.	$\vdash \forall x\,(\phi) \to \chi$	Premise
3.	$\vdash \forall x\,(\phi)$	GEN (1)
4.	$\vdash \chi$	MP (2, 3)
5.	$\vdash \forall x\,(\chi)$	GEN (4)

Figure 5.1.2.: Proof of $\{\phi, \forall x\,(\phi) \to \chi\} \vdash_{\mathcal{L}q} \forall x\,(\chi)$.

Exercises

Exercise 5.1.1. Prove in $\mathcal{L}q$ the classical *validity* of the following formulae, derived rules, or arguments:

1. $P \to P$

2. $\neg P \to (P \to Q)$

3. $\{P \to Q, \neg P \to Q, Q\}$

4. $\{P \to R, R \to Q, \neg(\neg Q \wedge P)\}$

5. $\neg(P \wedge R) \to (R \to \neg P)$

6. $\{\neg P \to \neg Q, Q \to P\}$

7. $\{P \to R, (Q \wedge P) \to (Q \wedge R)\}$

8. $(\neg P \to P) \to P$

9. $\neg P \to (\neg R \to \neg(P \vee R))$

10. $\neg\neg\neg P \to \neg P$

11. $\forall x\,(\phi \to \chi) \to (\exists x\,(\phi) \to \exists x\,(\chi))$

12. $\neg\forall x\,(\phi) \to \exists x\,(\neg\phi)$

13. $\forall x\,(\phi \to \chi) \to (\forall x\,(\phi) \to \forall x\,(\chi))$

14. $\forall x\,(\phi) \to \forall x\,(\phi \vee \chi)$

15. $(\phi \to \forall x\,(\chi)) \leftrightarrow \forall x\,(\phi \to \chi)$

Exercise 5.1.2. Prove the following arguments in the axiom system $\mathcal{L}q$ over the language L1:

1. All men are mortal. All philosophers are men. Hence, all philosophers are mortal.

2. Some cats have no tail. All cats are mammals. Therefore, some mammals have no tail.

3. No logic exercise is fun. Some thoughts are logic exercises. Thus, some thoughts are not fun.

Exercise 5.1.3. The original Frege axiom system, denoted by $\mathcal{F}$, had the following six axiom schemata:

$$
\begin{array}{ll}
(\mathcal{F}1) & \phi \to (\psi \to \phi) \\
(\mathcal{F}2) & (\phi \to (\psi \to \chi)) \to ((\phi \to \psi) \to (\phi \to \chi)) \\
(\mathcal{F}3) & (\phi \to (\psi \to \chi)) \to (\psi \to (\phi \to \chi)) \\
(\mathcal{F}4) & (\phi \to \psi) \to (\neg\psi \to \neg\phi) \\
(\mathcal{F}5) & \neg\neg\phi \to \phi \\
(\mathcal{F}6) & \phi \to \neg\neg\phi
\end{array}
$$

1. Łukasiewicz showed that axiom $\mathcal{F}3$ could be derived from $\mathcal{F}1$-2 (= $\mathcal{L}1$-2). Give the proof(s).

2. Łukasiewicz further substituted $\mathcal{F}4$-6 by $\mathcal{L}3$. How do you account for this substitution?

5.2. The natural deduction calculus $\mathcal{NK}$

Around 1930, there was in Europe a concern with *natural* ways of reasoning in mathematical proofs; more specifically, axiom systems were seen as quite remote from the reasoning instances typically carried out by mathematicians, and other proof methods closer to these were actively searched for. In this quest, natural deduction, denoted here by $\mathcal{NK}$, was invented independently by Gentzen (1934-5) and Jaśkowski (1934) as the proof system for classical logic $\mathcal{NK} = (RI, \emptyset)$ over L1, where the

$\mathbf{r}_i \in RI$ are rules for the *introduction* (denoted by I) and the *elimination* (E) of the corresponding connectives and quantifiers.

In what follows, $[\![\phi]\!]$ denotes that *assumption* ϕ is *discharged* (see Example 5.2.1 for meanings), and $\bot$ denotes absurdity. The rules for the quantifiers implicitly apply operations such as ϕ_x^a (ϕ_t^x), denoting the substitution in ϕ of the variable x for the constant a (the term t free for the variable x, respectively).[3]

Proposition. 5.2.1. *The following are the inference (or formation) rules of* $\mathcal{NK}$.[4]

1.
$$(\wedge I) \quad \frac{A \quad B}{A \wedge B}$$

2.
$$(\wedge E) \quad \frac{A \wedge B}{A}; \frac{A \wedge B}{B}$$

3.
$$(\vee I) \quad \frac{A}{A \vee B}; \frac{B}{A \vee B}$$

4.
$$(\vee E) \quad \frac{A \vee B \quad \overset{[\![A]\!]}{C} \quad \overset{[\![B]\!]}{C}}{C}$$

5.
$$(\to I) \quad \frac{\overset{[\![A]\!]}{B}}{A \to B}$$

6.
$$(\to E) \quad \frac{A \quad A \to B}{B}$$

7.
$$(\neg I) \quad \frac{\overset{[\![A]\!]}{\bot}}{\neg A}$$

8.
$$(\neg E) \quad \frac{A \quad \neg A}{\bot}$$

---

[3]More strictly, a is a *parameter*, i.e. a free variable. But as $[\forall x\,(\phi\,(x))]_x^x = \phi\,(x)$, for some $\mathcal{I}_a^x$ we can apply "directly" $\phi_a^x = \phi\,(a)$ where $a \in \mathcal{D}$. More formally, take a parameter to be a new constant symbol in L1 (cf. Section 4.4).

[4]Concerning negation, rules 7 and 8 are considered by Gentzen, but Prawitz (1965), who considers instead rule 13, sees them as special cases of 5 and 6, respectively.

9.

$$(\forall I) \qquad \frac{A\,(a)}{\forall x A\,(x)}$$

Restriction: a must not occur in any assumption on which $A\,(a)$ depends.

10.

$$(\forall E) \qquad \frac{\forall x A\,(x)}{A\,(t)}$$

11.

$$(\exists I) \qquad \frac{A\,(t)}{\exists x A\,(x)}$$

12.

$$(\exists E) \qquad \frac{\exists x A\,(x) \qquad \overset{\displaystyle [\![A\,(a)]\!]}{B}}{B}$$

Restriction: a must not occur in either $\exists x A$ or B, or in any assumption on which the upper occurrence of B depends other than $A\,(a)$.

13.

$$(\perp) \qquad \frac{\overset{\displaystyle [\![\neg A]\!]}{\perp}}{A}$$

Restrictions: A should be different from $\perp$ and it should not have the form $B \rightarrow \perp$.

Proof: Left as a (research) exercise.

Example 5.2.1. Figure 5.2.1 shows a $\mathcal{NK}$ proof of $A \rightarrow (B \wedge C)$ from $(A \rightarrow B) \wedge (A \rightarrow C)$, i.e. $\vdash ((A \rightarrow B) \wedge (A \rightarrow C)) \rightarrow (A \rightarrow (B \wedge C))$. This is the proof of a conditional statement (i.e., the conditional is the main connective), so the best strategy is to apply $\rightarrow I$. We begin by *assuming* the antecedent (step 1). This antecedent has itself two conditional propositions in which A is an antecedent, and we actually want to prove first that from this antecedent $(B \wedge C)$ can be proved. We thus assume A (step 2) and enter a new level of the proof (a *subproof*), which is graphically represented by an indentation. Steps 3-7 represent the application of the rules of inference of $\mathcal{NK}$ with respect to the former steps to obtain a proof that $A \rightarrow (B \wedge C)$ (step 8). Note in 8 that the indentation started in 2 is closed; this represents graphically the fact that the assumption in 2 was *discharged*. Step 9 concludes the proof of

the main conditional, and resumes the indentation in 1, i.e. it discharges
the first assumption in 1. Note thus that a (sub-)formula is proved iff it
is proved under certain assumptions, unless it is a theorem.

1.	$(A \to B) \wedge (A \to C)$	Assumption
2.	A	Assumption
3.	$A \to B$	$\wedge E1\ (1)$
4.	$A \to C$	$\wedge E2\ (1)$
5.	B	$\to E\ (3, 2)$
6.	C	$\to E\ (4, 2)$
7.	$B \wedge C$	$\wedge I\ (5, 6)$
8.	$A \to (B \wedge C)$	$\to I\ (2, 7)$
9.	$((A \to B) \wedge (A \to C)) \to (A \to (B \wedge C))$	$\to I\ (1, 8)$

Figure 5.2.1.: Proof of $\vdash_{\mathcal{NK}} ((A \to B) \wedge (A \to C)) \to (A \to (B \wedge C))$.

Although the proof in Example 5.2.1 was completed solely by means
of the introduction and elimination rules of $\mathcal{NK}$, this is actually not the
common practice, especially in the case of complex arguments. Clearly,
any classical theorem can be proven from these and only these rules (otherwise, it would not be a classical theorem), but restricting the proofs
to these rules can be cumbersome and impractical. In order to avoid
this we may apply any *derived rule*, i.e. any rule that has been derived
from these main rules, and thus any proven theorem, as well as *inder-
definitions*, *equivalences*, and *properties* (e.g., distributivity of $\wedge$ and $\vee$)
can be used in proving theoremhood in the calculus $\mathcal{NK}$. This holds
also for any argument form of Figure 3.2.1, as well as for all the classical
$\heartsuit$- and $\blacklozenge$-consequences of Chapter 4. If one so wishes, this calculus can
be referred to as *extended $\mathcal{NK}$*.

Example 5.2.2. Figure 5.2.2 shows a proof in (extended) $\mathcal{NK}$ of the
argument

$$\{(P \vee R) \to \neg (Q \wedge T), \neg S \to Q, P \wedge T\} \vdash S$$

In this proof, besides the $\mathcal{NK}$ rules proper, De Morgan's law for conjunction (DeM$_\wedge$; cf. Prop. 4.3.1, $\models$ 10), the argument form known as
modus tollendo ponens (TP; cf. Fig. 3.2.1), and the double-negation
law (DN; cf. Def. 4.2.4.1) were employed.

1.	$(P \vee R) \rightarrow \neg (Q \wedge T)$	Premise
2.	$\neg S \rightarrow Q$	Premise
3.	$P \wedge T$	Premise
4.	P	$\wedge E$ (3)
5.	T	$\wedge E$ (3)
6.	$P \vee R$	$\vee I$ (4)
7.	$\neg (Q \wedge T)$	$\rightarrow E$ (6, 1)
8.	$\neg Q \vee \neg T$	$DM_\wedge$ (7)
9.	$\neg S$	Assumption
10.	Q	$\rightarrow E$ (9, 2)
11.	$\neg T$	TP (8, 10)
12.	$T \wedge \neg T$	$\wedge I$ (5, 11)
13.	$\neg \neg S$	$\neg I$ (9, 12)
14.	S	DN (13)

Figure 5.2.2.: Proof of an argument in (extended) $\mathcal{NK}$.

Example 5.2.3. Figure 5.2.3 shows a $\mathcal{NK}$ proof of the FO argument

$$\frac{\forall x \, (R \, (x)) \vee \forall y \, (\neg P \, (y))}{\forall y \, (\neg P \, (y))}$$

Finally, (extended) $\mathcal{NK}$ can also be used as a refutation proof system: if one negates the formula ϕ (the conclusion of argument α) to be proven and derives a contradiction, this is a *reductio-ad-absurdum* proof of ϕ (α, respectively).

Exercises

Exercise 5.2.1. Prove in (extended) $\mathcal{NK}$ the validity of the formulae, rules, and arguments of Exercise 5.1.1.

Exercise 5.2.2. Prove in (extended) $\mathcal{NK}$ the validity of the arguments of Exercise 5.1.2.

Exercise 5.2.3. Explain Step 5 in the proof of Figure 5.2.3.

1.	$\forall x\,(R\,(x)) \vee \forall y\,(\neg P\,(y))$	Premise
2.	$\neg \exists z\,(P\,(z) \wedge R\,(z))$	Premise
3.	$\forall z \neg\,(P\,(z) \wedge R\,(z))$	QN$_\exists$ (2)
4.	$\forall z\,(\neg P\,(z) \vee \neg R\,(z))$	DeM$_\wedge$ (3)
5.	$\quad y_1$	
6.	$\quad \neg P\,(y_1) \vee \neg R\,(y_1)$	$\forall E$ (4)
7.	$\qquad \forall x\,(R\,(x))$	Assumption
8.	$\qquad R\,(y_1)$	$\forall E$ (7)
9.	$\qquad \neg\neg R\,(y_1)$	DN (8)
10.	$\qquad P\,(y_1)$	TP (6, 9)
11.	$\qquad \forall y\,(\neg P\,(y_1))$	Assumption
12.	$\qquad \neg P\,(y_1)$	$\forall E$ (11)
13.	$\quad \neg P\,(y_1)$	$\vee E$ (1, (7, 12))
14.	$\forall y\,(\neg P\,(y))$	$\forall I$ (5, 13)

Figure 5.2.3.: A FO $\mathcal{N K}$ proof.

5.3. The sequent calculus $\mathcal{L K}$

Sequent calculi are rather more interesting for the analysis of proofs than for theorem proving, for which natural deduction is more adequate. In any case, and also because they are related to the analytic tableaux calculus to be approached in depth below, we provide a more developed exposition of this proof calculus.

The sequent calculus for classical logic $\mathcal{L K}$ (abbreviating the German expression **k***lassische Prädikaten***l***ogik*, "classical predicate logic" in English) was firstly conceived by Gentzen (1934-5). As the name of this calculus indicates, sequents are the central object of $\mathcal{L K}$, and we begin our short exposition of this calculus by providing the corresponding definition.

Definition. 5.3.1. A *sequent s* is an expression of the form

$$(s) \qquad A_1, ..., A_n \Rightarrow B_1, ..., B_k$$

where A_i, $0 \leq i \leq n$, and B_j, $0 \leq j \leq k$, are formulae, and $\vdash$ is a new symbol denoting that the disjunction of the B_j *follows from* the conjunction of the A_i, i.e.

$$(s) \qquad \left(\bigwedge_{i=0}^{n} A_i\right) \Rightarrow \left(\bigvee_{j=0}^{k} B_j\right)$$

which, in turn, is equivalent to

$$(s') \qquad \Rightarrow \left(\bigwedge_{i=0}^{n} A_i\right) \rightarrow \left(\bigvee_{j=0}^{k} B_j\right)$$

it being the case that a sequent of the form "$\Rightarrow A$" denotes a theorem. This is the same as saying that a sequent is actually a structure asserting that whenever all the A_i are true at least one of the B_j will also be true, and we can replace $\Rightarrow$ by $\vdash$ (see Exercise 5.3.3). We call $\bigwedge_{i=0}^{n} A_i$ the *antecedent* and $\bigvee_{j=0}^{k} B_j$ the *succedent* (or *consequent*).

Definition. 5.3.2. A *context* (or *side formula*), denoted by $\Gamma, \Lambda, \Sigma, \Pi$, is a finite, possibly empty sequence of formulae. In the conclusion of each rule, the formula not in the context is the *principal* (or *main*) formula.

Definition. 5.3.3. A *sequent rule* **s** is of the general form

$$(\mathbf{s}) \qquad \frac{s_n}{s_{n-1}}$$

and a *sequent axiom* $\mathbf{a}_s$ has the form

$$(\mathbf{a}_s) \qquad \frac{}{s_n}.$$

Note the "upward direction" of the sequent rules and axioms.

Definition 5.3.1. 5.3.4. A *proof* in a sequent calculus is a (branching) sequence

$$s_0, s_{11}, ..., s_{ij}..., s_{nm}$$

where s_0 is the proved sequent / formula, and each of the sequents s_{ij} in the proof is inferred (or derived) from earlier sequents in the sequence by means of rules of inference (or derivation). This sequence is an (upwards growing) *ordered finite tree*, with the *root* s_0, the sequents $s_{ij}, 0 < i < n$ are the *i*-th *node* in the *j*-th *branch*, $0 < j \le m$, and s_{nj} is the *leaf* of the *j*-th branch.[5]

---

[5]There are variations to this structure; see, e.g., Troelstra & Schwichtenberg (2000).

Importantly, this proof structure without the rule CUT (see below) guarantees the following property of the sequent calculus:

Proposition. 5.3.5. Sub-formula property: *all formulae occurring in a proof of a sequent s_0 are sub-formulae of the formulae in s_0.*

Proof: Trivial. **QED**

In effect, the possibility of applying rules along each j-th branch stops when we have obtained atomic sub-formulae, the above s_{nj} sequents. The semantical explanation for this proof structure is as follows for material implication (for example): if all branches terminate in sequents of the form $\Gamma', L \vdash L, \Lambda'$, then there is *no* interpretation $\mathcal{I}$ for which $val_{\mathcal{I}}(\Gamma) = \mathtt{t}$ *and* $val_{\mathcal{I}}(\Lambda) = \mathtt{f}$.

Definition. 5.3.6. A *sequent calculus* is thus a pair (RI_s, AX_s) of rules of inference and axioms for sequents.

$\mathcal{LK}$ has several rules of inference (see below) and a single axiom:

Proposition. 5.3.7. Axiom of identity:

$$(Ax) \qquad \frac{}{A \vdash A}$$

The inference rules are of two kinds, *structural* and *logical* (or *operational*), the latter being rules for the use of the logical operators (connectives, as well as quantifiers). Each rule, structural or logical, has a right and left version, denoted by R and L, respectively.

Proposition. 5.3.8. *The following are the structural rules of $\mathcal{LK}$:*

1. *Weakening* (W):

$$(\text{WL}) \quad \frac{\Gamma \vdash \Lambda}{\Gamma, A \vdash \Lambda} \qquad\qquad (\text{WR}) \quad \frac{\Gamma \vdash \Lambda}{\Gamma \vdash A, \Lambda}$$

2. *Contraction* (C):

$$(\text{CL}) \quad \frac{\Gamma, A, A \vdash \Lambda}{\Gamma, A \vdash \Lambda} \qquad\qquad (\text{CR}) \quad \frac{\Gamma \vdash A, A, \Lambda}{\Gamma \vdash A, \Lambda}$$

3. *Permutation* (P):

$$(\text{PL}) \quad \frac{\Gamma, A, B, \Sigma \vdash \Lambda}{\Gamma, B, A, \Sigma \vdash \Lambda} \qquad\qquad (\text{PR}) \quad \frac{\Gamma \vdash \Lambda, A, B, \Pi}{\Gamma \vdash \Lambda, B, A, \Pi}$$

Proof: Left as a (research) exercise.

Proposition. 5.3.9. *The following are the logical rules for the connectives of* $\mathcal{LK}$:

1. $\wedge$:

$$(\wedge L_1) \quad \frac{\Gamma, A \vdash \Lambda}{\Gamma, A \wedge B \vdash \Lambda} \qquad (\wedge L_2) \quad \frac{\Gamma, B \vdash \Lambda}{\Gamma, A \wedge B \vdash \Lambda}$$

$$(\wedge R) \quad \frac{\Gamma \vdash A, \Lambda \quad \Sigma \vdash B, \Pi}{\Gamma, \Sigma \vdash A \wedge B, \Lambda, \Pi}$$

2. $\vee$:

$$(\vee L) \quad \frac{\Gamma, A \vdash \Lambda \quad \Sigma, B \vdash \Pi}{\Gamma, \Sigma, A \vee B \vdash \Lambda, \Pi}$$

$$(\vee R_1) \quad \frac{\Gamma \vdash A, \Lambda}{\Gamma \vdash A \vee B, \Lambda} \qquad (\vee R_2) \quad \frac{\Gamma \vdash B, \Lambda}{\Gamma \vdash A \vee B, \Lambda}$$

3. $\rightarrow$:

$$(\rightarrow L) \quad \frac{\Gamma \vdash A, \Lambda \quad \Sigma, B \vdash \Pi}{\Gamma, \Sigma, A \rightarrow B \vdash \Lambda, \Pi}$$

$$(\rightarrow R) \quad \frac{\Gamma, A \vdash B, \Lambda}{\Gamma \vdash A \rightarrow B, \Lambda}$$

4. $\neg$:

$$(\neg L) \quad \frac{\Gamma \vdash A, \Lambda}{\Gamma, \neg A \vdash \Lambda}$$

$$(\neg R) \quad \frac{\Gamma, A \vdash \Lambda}{\Gamma \vdash \neg A, \Lambda}$$

Proof: Left as a (research) exercise.

Note that, contrarily to natural deduction calculi, in which there are rules for the introduction and the elimination of the connectives, in the sequent calculi there are only rules for the introduction of the connectives in the antecedent or in the succedent of a sequent; the eliminations in the latter calculi take the form of introductions in the antecedent.

The sequent calculus $\mathcal{LK}$ can be extended to classical FOL, sufficing to that end to add two more logical rules for the quantifiers.

Proposition. 5.3.10. *Let* t *be a term free for* x *and let* a *be a constant (more strictly: a parameter). The following are the logical rules of* $\mathcal{LK}$ *for the quantifiers:*

1. ($\forall$):

$$(\forall L) \quad \frac{\Gamma, A\,(t) \vdash \Lambda}{\Gamma, \forall x\,(A\,(x)) \vdash \Lambda}$$

$$(\forall R) \quad \frac{\Gamma \vdash A\,(a), \Lambda}{\Gamma \vdash \forall x\,(A\,(x)), \Lambda}$$

2. ($\exists$):

$$(\exists L) \quad \frac{\Gamma, A\,(a) \vdash \Lambda}{\Gamma, \exists x\,(A\,(x)) \vdash \Lambda}$$

$$(\exists R) \quad \frac{\Gamma \vdash A\,(t), \Lambda}{\Gamma \vdash \exists x\,(A\,(x)), \Lambda}$$

Proof: Left as a (research) exercise.

Note that restrictions apply to the rules $\forall$R and $\exists$L: a must not occur within Γ and Λ, *or* it must not appear anywhere in the respective lower sequents.

Example 5.3.1. Figure 5.3.1 shows a proof in $\mathcal{LK}$ of the CFOL theorem

$$\vdash \forall x\,(A\,(x) \to B) \to \exists x\,(A\,(x) \to B)$$

where $t = a$. Figure 5.3.2 shows a proof in $\mathcal{LK}$ of axiom $\mathscr{L}2$.

$$
\frac{
\dfrac{
\dfrac{
\dfrac{\overline{A\,(a) \vdash A\,(a), B}}{\vdash A\,(a), A\,(a) \to B}}{\vdash A\,(a), \exists x A\,(x) \to B}
\quad
\dfrac{
\dfrac{\overline{A\,(a), B \vdash B}}{B \vdash A\,(a) \to B}}{B \vdash \exists x A\,(x) \to B}
}{
\dfrac{A\,(a) \to B \vdash \exists x A\,(x) \to B}{\forall x A\,(x) \to B \vdash \exists x A\,(x) \to B}
}
}{\vdash \forall x\,(A\,(x) \to B) \to \exists x\,(A\,(x) \to B)}
$$

Figure 5.3.1.: Proof in $\mathcal{LK}$ of a FO validity.

Finally, an additional structural rule of $\mathcal{LK}$ is the following:

Proposition. 5.3.11. Cut:

$$(\text{CUT}) \quad \frac{\Gamma \vdash \Lambda, A \quad A, \Sigma \vdash \Pi}{\Gamma, \Sigma \vdash \Lambda, \Pi}$$

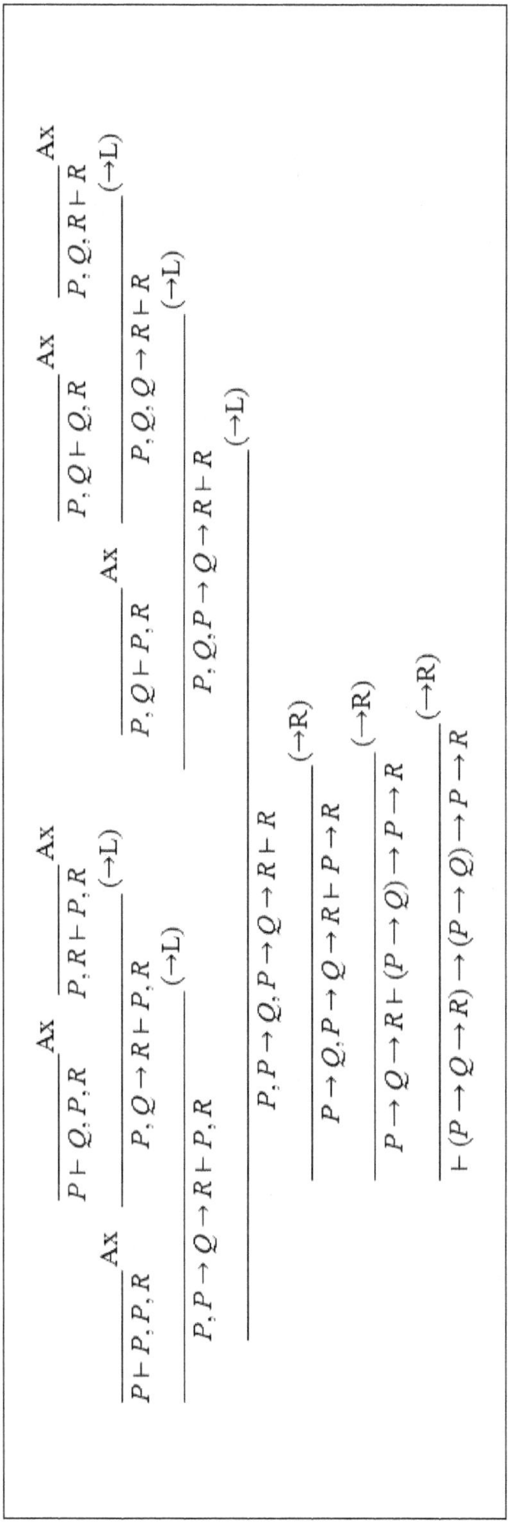

$$\dfrac{\dfrac{}{P \vdash P,P,R}\text{ Ax} \qquad \dfrac{\dfrac{}{P \vdash Q,P,R}\text{ Ax} \quad \dfrac{}{P,R \vdash P,R}\text{ Ax}}{P,Q \to R \vdash P,R}(\to\text{L})}{P,P \to Q \to R \vdash R}(\to\text{L})$$

$$\dfrac{\dfrac{}{P,Q \vdash P,R}\text{ Ax} \qquad \dfrac{\dfrac{}{P,Q \vdash Q,R}\text{ Ax} \quad \dfrac{}{P,Q,R \vdash R}\text{ Ax}}{P,Q,Q \to R \vdash R}(\to\text{L})}{P,Q,P \to Q \to R \vdash R}(\to\text{L})$$

$$P,P \to Q,P \to Q \to R \vdash R$$

$$\dfrac{}{P \to Q,P \to Q \to R \vdash P \to R}(\to\text{R})$$

$$\dfrac{}{P \to Q \to R \vdash (P \to Q) \to P \to R}(\to\text{R})$$

$$\dfrac{}{\vdash (P \to Q \to R) \to (P \to Q) \to P \to R}(\to\text{R})$$

Figure 5.3.2.: Proof in $\mathcal{LK}$ of axiom $\mathcal{L}2$ of the axiom system $\mathcal{L}$.

Proof: Left as a (research) exercise.

CUT has motivated much work–if not furore–in modern logic, in particular in the field of automated deduction. Despite this, CUT plays no important role in this text, and we leave it at that. For its "special" character and status in the sequent calculus above, see, e.g., Fitting (1999).

Exercises

Exercise 5.3.1. Label the steps of the proof in Figure 5.3.1.

Exercise 5.3.2. Prove in the sequent calculus $\mathcal{LK}$ the formulae, derived rules, or arguments of Exercise 5.1.1.

Exercise 5.3.3. Account formally for the permissible substitution of the symbol $\vdash$ for the symbol $\Rightarrow$ in the sequent calculus $\mathcal{LK}$.

---

6. Classical models

In Chapter 2 and in Sections 3.1-2, we concentrated on form, or structure, alone with respect to formal languages. That is to say, we learned how to arrange and combine symbols of an alphabet into *legal strings* (*(well-formed) formulae*, in logic; *statements*, in programming languages; *sentences*, in natural language), and we learned how to change legal strings into also legal strings. We were not concerned with *meaning* at all, but it is obvious that structure alone does not suffice if we want to move from *legal* strings to *true* formulae, statements, or sentences. This little step that is actually a giant leap is made possible, as seen in Section 3.3, by the introduction of a *semantics*. This is said to be *classical* if it is based on bivalent valuations, i.e. if the semantical correlates of the formulae are restricted to the set $G = \{0, 1\}$ or their denotations are limited to *truth* and *falsity*.

Above, in Chapters 3 and 4, we introduced some fundamental semantical notions, and we did so especially in relation with the classical logical consequence relation / operation. We now elaborate on the semantics that are commonly invoked for classical logic–here referred to as *classical semantics*–, both generally and in a computational context. With respect to the former, Tarski semantics and Boolean algebras are essential for an understanding of classical logic. Both, and especially Boolean algebras, also play an important role in the computational applications of classical logic, but Herbrand semantics is, so to say, the jewel of the crown, as without it we would be at a loss as far as much of deductive computing with classical logic is concerned. This is particularly true of automated theorem proving, but this semantics is also important for logic programming. This importance accounts for our extended elaboration on this semantics (cf. Sections 6.2 and 7.3.3, mostly).

6.1. Tarskian semantics

Most of the semantical notions discussed above were so from the viewpoint of what can be called *Tarskian semantics*. This despite the fact that this semantics can be captured by the term *denotational semantics*, which, in turn, is actually a Fregean achievement. Recall the principle of extensionality or of the compositionality of meaning, formalized in

Proposition 3.3.3, in particular in item 2 of this proposition; this principle states that the meaning of a proposition (its semantical correlate or, more generally, its truth value) is a function of the meaning of its components. We already know from Section 3.3 that this lies at the root of the concept of truth-functionality, but it can be further specified in terms of *denotation* as follows: as seen in the aforementioned Section, when we formalize natural language assertions by means of the alphabet of a logical language, we do so by assigning denotations to its symbols; these denotations are in fact the predicates, functions, and constants (or objects) of the logical language at hand.[1] Then, the meaning of an atomic formula is considered in terms of what the expressions thereof denote, and the meaning of a compound formula is the rule-based composition of the meaning of its atomic formulae.

But by *denotation* of an expression at a metalogical level we mean actually the truth value assigned to it by means of an interpretation, more precisely so a valuation (cf. Def. 3.3.11), so that in fact meaning and denotation coincide with the semantical correlates, which can be *logical truth values*, such as **true** and **false**, but also *algebraic truth values*, as seen in Definition 3.3.4. In either case, we are not speaking here of the natural-language adjectives "true" and "false," and it is this specification of "truth" in a formal(ized) language that earned A. Tarski the honor of having this denotational semantics named after him (cf. Tarski, 1935).

It is also this specification that makes Tarskian semantics relevant for computing with classical logic. Above, we spoke of rule-based composition of meaning. This immediately suggests adequateness of a logical system, in particular in the sense that one can speak of the *truth* or *falsity* of a rule of inference. For instance, in Chapter 9 below, we give an intuitive explanation for a logic programming rule of the characteristic form

$$(\mathbf{r}_{LP}) \qquad A \leftarrow B_1, ..., B_n$$

as follows: if every B_i is true, then A is true, or A can be proved by proving all the B_i. This truth- or provability-preservation actually implicitly entails that rule $\mathbf{r}_{LP}$, concretized in the connective $\leftarrow$ (essentially a reversed $\rightarrow$), is itself *true*. Without this *truth* of the inference rule at hand there simply is no way to assure us that the rule "holds," but this *truth* is a function of $\leftarrow$ with respect to the meanings of A and the B_i.

---

[1] More properly, the denotations are so of predicates and functions when their terms are replaced or specified by constants, i.e. objects of the domain of discourse, so that Tarskian semantics can be seen as an interpretation of constants.

As it is, $\mathbf{r}_{LP}$ coincides with a Horn clause, a formula of the form

$$A \vee \neg B_1 \vee ... \vee \neg B_n$$

which is valuated as **true** if at least one of its sub-formulae is so valuated. But this valuation depends on the denotation of the expressions constituting each of the sub-formulae. For instance, the atomic logic-programming formula **father (john, sara)** is valuated to **true** iff there is an interpretation $\mathcal{I} = (\mathscr{D}, \Theta, \varpi)$ such that, given the predicate **father (X, Y)** and the variable assignments $\varpi\,(\mathtt{X}) = \mathtt{john}$ and $\varpi\,(\mathtt{Y}) = \mathtt{sara}$, we have[2]

$$val_{\mathcal{I}}\,(\texttt{father (john, sara)}) = \Theta\,(\texttt{father})\,(val_{\mathcal{I}}\,(\texttt{john})\,, val_{\mathcal{I}}\,(\texttt{sara})) = \mathtt{t}.$$

Although this might be terrain for philosophical debates, from a formal perspective we talk here of the soundness and completeness of a rule of inference, it being the case that this adequateness holds in virtue of the denotational semantics that determines that the meaning of a classical FO formula is the meaning, or the denotation, of its constituting parts, which meaning in turn is dependent on a rule of inference concretized by some logical operation. Even though a deductive system might be given such a Tarskian semantics only implicitly–as is the case in logic programming–, this provides it with a warranty that whatever the logic generated by it is, it is a *rule-based truth-preserving logic*, and this is what classical logic is, after all, (mostly, if not entirely) about (cf. especially Chapter 4).

In effect, although resolution and analytic tableaux are essentially proof calculi, their ability to prove a set of formulae inconsistent is determined against a Tarskian semantics, making unsatisfiability out of this inconsistency. To be sure, especially in the case of the resolution calculus–but also not negligibly so in the analytic tableaux calculus–, it is of Herbrand (un)satisfiability that we speak, mainly due to Herbrand's theorem (cf. Section 7.3.3), but Herbrand semantics is employed at a for some very high cost: the *propositionalization* of CFOL. And this is precisely so because, as elaborated on in the next Section, Herbrand semantics is not denotational, treating predicates, functions, and constants at a fixed-interpretation level.

---

[2]The logic-programming language is somehow different from L1, though not essentially so: predicate names are written with an initial lowercase letter and individual variables are written with (initial) uppercase letters. Because this language is employed mainly in programming, it helps to make the "denotations" clear by writing, for example, **father (john, sara)** instead of $F\,(j, s)$ or $P\,(a, b)$, more typical of L1.

Exercises

Exercise 6.1.1. Reflect on the following passages by A. Tarski on the concept of truth.

1. We regard the truth of a sentence as its "correspondence with reality". This rather vague phrase, which can certainly lead to various misunderstandings and has often done so in the past, is interpreted as follows. We shall regard as valid all such statements as:

the sentence "it is snowing" is true if and only if it is snowing; the sentence "the world war will begin in the year 1963" is true if and only if the world war will begin in 1963.

Quite generally we shall accept as valid any sentence of the form

the sentence x is true if and only if p

where "*p*" is to be replaced by any sentence of the language under investigation and "*x*" by any individual name of that sentence provided this name occurs in the metalanguage. (In colloquial language such names are usually formed by means of quotation marks.) Statements of this form can be regarded as partial definitions of the concept of truth. They explain in a precise way, and in conformity with common usage, the sense of all special expressions of the type: *the sentence x is true.* Now, if we succeed in introducing the term "true" into the metalanguage in such a way that every statement of the form discussed can be proved on the basis of the axioms and rules of inference of the metalanguage, then we shall say that the way of using the concept of truth which has thus been established is *materially adequate.* In particular, if we succeed in introducing such a concept of truth by means of a definition, then we shall also say that the corresponding definition is materially adequate. We can apply an analogous method to any other semantical concepts as well. (Tarski, 1935/1956)

2. Let us consider the following sentence:

if 1 *is a positive number and* 1 < 2 *, then* 1 *is a positive number.*

This sentence is obviously true, it contains exclusively constants belonging to the domain of logic and arithmetic, and yet the idea of listing this sentence as a special theorem in a textbook of mathematics would not occur to anybody. If one reflects why this is so, one comes to the conclusion that this sentence is completely uninteresting from the standpoint of arithmetic; it fails to enrich our knowledge about numbers in any way, since its truth does not depend at all upon the content of the arithmetical terms occurring within it, but only upon the meaning of the words *"and"*, *"if"*, *"then"*. (Tarski, 1994)[3]

3. [W]hen dealing with the problem of truth, we are concerned with relating expressions (in this case, sentences) and the objects to which the expressions refer, or, which they "talk about". (Tarski, 1994)

6.2. Herbrand semantics

The semantics for CL that we approach now was conceived by the French mathematician J. Herbrand in his doctoral dissertation (Herbrand, 1930). As a matter of fact, Herbrand's dissertation was first and foremost in the proof theory of CFOL, but it so happens that from this work a semantics for CFOL can be extracted. We call it *Herbrand semantics*.

From a computational viewpoint, Herbrand semantics exhibits many advantages over the Tarskian semantics elaborated on in (especially) the Section immediately above, and in particular so with respect to automated deduction. This is so mostly because Herbrand semantics provides a fixed interpretation for CFOL, thus allowing for a purely syntactical manipulation of symbols in FO formulae while at the same time providing these symbols with a bivalent semantics. We thank this feature to Herbrand's concern with finitistic approaches to CFOL, in line with Hilbert's formalist-finitist program, and in particular to his dislike–or outright rejection–of infinite models.

The use of Herbrand semantics in automated deduction is, however, sanctioned mostly by the fact that in this we are mainly concerned with satisfiability. Indeed, Herbrand semantics is solely satisfiability-preserving, being incomplete and not validity-preserving.

---

[3]This is an English translation of a text originally published in Polish.

In fact, we shall be more precisely concerned with unsatisfiability-preservation. Contrary to Tarskian semantics, which is diffused in different parts of this book, we concentrate our discussion of Herbrand semantics in this and in Section 7.3.3 below. In the present Section, we introduce the basic elements of this semantics, as well as the main results concerning it, and below we elaborate on it from the viewpoint of unsatisfiability.

Although our discussion of Herbrand semantics serves mainly our treatment of resolution, which is clause-based, this semantics is equally relevant to the analytic tableaux calculus; the reason for our concentration in resolution lies in the fact that discussing this semantics in a clause-based formalism is particularly adequate (see Section 7.3.3 below). In effect, although we restrict our discussion here to sets of clauses, i.e. finite disjunctions of literals, Herbrand's results hold for any set X of FOL formulae (see Def. 6.2.5 below).

Let C be a set of clauses. Then the following definitions hold.

Definition. 6.2.1. The *Herbrand universe* of C, denoted by H_C, is the set of all ground terms built up from the constants and functions of C in the following way:[4]

1. if C *contains no function*, then $H_C = H_0$ is the set of constants occurring in C; if no constant occurs in C, then $H_C = H_0$ consists of a single arbitrary constant, say, $H_C = \{a\}$, i.e.

$$H_0 = \begin{cases} Cons\,(C) & \text{if } Cons\,(C) \neq \emptyset \\ \{a\} & \text{if } Cons\,(C) = \emptyset \end{cases} \quad ;$$

2. if C *contains a function*, then $H_C = \bigcup_{i=0}^{\infty} H_i$, $H_i = H_{i-1} \cup Fun$,

$$Fun = \{f\,(t_1, ..., t_n)\,|\,f \in Fun\,(C)\,,(t_1, ..., t_n) \in H_{i-1}, n \in \mathbb{N}\}\,.$$

H_i, for $1 \leq i \leq \infty$, is called the *i-level constant set of* C. Clearly, H_C is infinite iff $Fun\,(C) \neq \emptyset$.

Example 6.2.1. We exemplify the above:

- Let $C_1 = \{P(b), \neg P(x) \vee Q(y)\}$. Then $H_{C_1} = H_0 = \{b\}$.

---

[4]In other words, the Herbrand universe of C is the set of all ground terms definable over $\Upsilon_C - Pred\,(C)$. See Definition 3.1.11.

- Let $C_2 = \{P(x) \lor Q(x), R(z), T(y) \lor \neg S(y)\}$. Then we let $H_{C_2} = H_0 = \{a\}$.

- Let $C_3 = \{P(f(x)), Q(a), R(g(y), b)\}$. Then

$$H_0 = \{a, b\}$$
$$H_1 = \{a, b, f(a), f(b), g(a), g(b)\}$$
$$H_2 = \{a, b, f(a), f(b), g(a), g(b), f(f(a)), f(f(b)), f(g(a)), f(g(b)),$$
$$g(f(a)), g(f(b)), g(g(a)), g(g(b))\}$$
$$\vdots$$
$$H_{C_3} = \{a, b, f(a), f(b), g(a), g(b), f(f(a)), f(f(b)), f(g(a)), f(g(b)),$$
$$g(f(a)), g(f(b)), g(g(a)), g(g(b)), ...\}$$

Definition. 6.2.2. A *ground instance* of a clause $\mathcal{C}$ of C is a clause obtained by replacing variables in $\mathcal{C}$ by members of H_C. A *Herbrand instance (H-instance)* of $\mathcal{C}$ is a ground instance $\mathcal{C}\theta$ of $\mathcal{C}$ such that θ is based on C. The *Herbrand base* of C, denoted by $H(C)$, is the set of all Herbrand instances of atoms occurring in clauses of C.

Example 6.2.2. The Herbrand base of the clauses in Example 6.2.1 is as follows:

- $H(C_1) = \{P(b), Q(b)\}$

- $H(C_2) = \{P(a), Q(a), R(a), T(a), S(a)\}$

- $H(C_3) = \{P(a), Q(a), R(a,a), R(a,b), P(b), Q(b), R(b,a), ...\}$

Obviously, $H(C)$ is finite iff H_C is finite.

Definition. 6.2.3. A *Herbrand interpretation (H-interpretation)* for C, denoted by $H\mathcal{I}_C$, is a triple (H_C, Θ, ϖ) (cf. Defs. 3.3.10-11) such that

1. $\Theta(c) = c$ for every $c \in Cons(C)$;

2. $\Theta(f)(t_1, ..., t_n) = f(t_1, ..., t_n)$ for all $t_1, ..., t_n \in H_C$, if $f \in Fun(C)$.

$H\mathcal{I}_C$ provides a *fixed* interpretation, as every constant symbol is interpreted as itself (i.e., $H\mathcal{I}_C$ maps every constant to itself), and every function symbol is interpreted as a term builder over H_C, or, in other words, as the function that applies it (i.e. $H\mathcal{I}_C$ maps every function symbol $f \in Fun\,(C)$ with arity > 0 to the n-ary function that maps every n-tuple $(t_1, ..., t_n)$ of terms $t_1, ..., t_n \in H_C$ to the term $f\,(t_1, ..., t_n)$). Moreover, because clauses are interpreted as closed formulae, ϖ is irrelevant in $H\mathcal{I}_C$.

All this entails that we end up with a purely syntactical interpretation, being meant by this that the symbols in a set of clauses are interpreted independently of any domain.

Definition. 6.2.4. A H-interpretation $H\mathcal{I}_C$ is a subset $H'\,(C)$ of $H\,(C)$ such that the truth value t is assigned to all elements of $H\mathcal{I}_C$ and the truth value f is assigned to all atoms in $H(C) - H\mathcal{I}_C$. The subset $H'\,(C)$ is in fact a *Herbrand model (H-model)* $H\mathcal{M}_C$ of C, because for an interpretation $H\mathcal{I}_C$ and some $P \in Pred\,(C)$ we have

$$\Theta\,(P)\,(t_1, ..., t_n) = val_{HI}\,(P\,(t_1, ..., t_n))$$

so that for $t_i \in H\,(C)$ we have

$$H\mathcal{M}_C = \{P\,(t_1, ..., t_n)\,|\Theta\,(P)\,(t_1, ..., t_n) = \mathsf{t}\}.$$

Example 6.2.2. Let $C = \{P\,(x) \vee Q\,(x), R\,(f\,(y))\}$. Then,

$$H_C = \{a, f\,(a), f\,(f\,(a)), ...\}$$

and

$$H(C) = \{P\,(a), Q\,(a), R\,(a), P\,(f\,(a)), Q\,(f\,(a)), R\,(f\,(a)), ...\}$$

The following are H-interpretations for C:

$H\mathcal{I}_{C_1} = \{P\,(a), Q\,(a), R\,(a), P\,(f\,(a)), Q\,(f\,(a)), R\,(f\,(a)), ...\}$
$H\mathcal{I}_{C_2} = \{\neg P\,(a), \neg Q\,(a), \neg R\,(a), \neg P\,(f\,(a)), \neg Q\,(f\,(a)), \neg R\,(f\,(a)), ...\}$
$H\mathcal{I}_{C_3} = \{P\,(a), Q\,(a), \neg R\,(a), P\,(f\,(a)), Q\,(f\,(a)), \neg R\,(f\,(a)), ...\}$

It is easy to see that C is satisfied by $H\mathcal{I}_{C_1}$, but falsified by $H\mathcal{I}_{C_2}$ and $H\mathcal{I}_{C_3}$. Thus, only $H\mathcal{I}_{C_1}$ is a H-model $H\mathcal{M}_C$.

As said above, Herbrand semantics applies to any set of FOL formulae, and not only to sets of clauses. The following shows this by means of the inductive definition of the satisfiability relation in Herbrand semantics.

Definition. 6.2.5. Let $P(t_1, ..., t_n)$, $n \geq 1$, be an atomic ground formula and let ϕ, ψ be (atomic) ground formulae. Then, given a Herbrand interpretation $H\mathcal{I}$, the *Herbrand satisfiability (H-satisfiability) relation* is defined inductively as follows:

1. $\models_{H\mathcal{I}} \top$

2. $\not\models_{H\mathcal{I}} \bot$

3. $\models_{H\mathcal{I}} P(t_1, ..., t_n)$ iff $P(t_1, ..., t_n) \in H\mathcal{M}$

4. $\models_{H\mathcal{I}} \neg\phi$ iff $\not\models_{H\mathcal{I}} \phi$

5. $\models_{H\mathcal{I}} \phi \wedge \psi$ iff $\models_{H\mathcal{I}} \phi$ and $\models_{H\mathcal{I}} \psi$

6. $\models_{H\mathcal{I}} \phi \vee \psi$ iff $\models_{H\mathcal{I}} \phi$ or $\models_{H\mathcal{I}} \psi$

7. $\models_{H\mathcal{I}} \phi \to \psi$ iff $\not\models_{H\mathcal{I}} \phi$ or $\models_{H\mathcal{I}} \psi$

8. $\models_{H\mathcal{I}} \forall x (\phi(x))$ iff $\models_{H\mathcal{I}} \phi(t)$ for every ground term $t \in H_\phi$

9. $\models_{H\mathcal{I}} \exists x (\phi(x))$ iff $\models_{H\mathcal{I}} \phi(t)$ for some ground term $t \in H_\phi$

A comparison of Definition 6.2.5 with the Tarski-style semantical classicality conditions in Section 4.2 reveals the essentially classical character of Herbrand semantics. However, the condition above that an atomic formula be of the form $P(x_1, ..., x_n)$ for $n \geq 1$ makes this semantics uninteresting to CPL itself. In effect, the main pay-off of employing the less-studied Herbrand semantics is that of "simulating" a classical propositional calculus for a given FO language, which has attached advantages such as finiteness of models and thus the existence of algorithmic decision procedures. In fact, the "simplification" that Herbrand semantics entails with respect to Tarskian semantics is to be found in Definition 6.2.5.3; in the latter semantics, and according to Definition 3.3.11.3, this condition would be formulated as

$$\models_\mathcal{I} P(t_1, ..., t_n) \text{ iff } \Theta(P)(val_\mathcal{I}(t_1), ..., val_\mathcal{I}(t_n)) = \mathbf{t}$$

for every $t_1, ..., t_n$ in the alphabet Σ of some FO language. This means that in Tarskian semantics every term t_i in P is valuated in an interpretation $\mathcal{I} = (\mathscr{D}, \Theta, \varpi)$ and we have $val_\mathcal{I}(P(t_1, ..., t_n)) = \mathbf{t}$ iff every t_i is valuated to $\mathbf{t}$. Given an infinite domain, we have infinite models, i.e. interpretations in which $val_\mathcal{I}(P(t_1, ..., t_n)) = \mathbf{t}$. Compare with Definition

6.2.4: a H-model of $P(t_1, ..., t_n)$ just is a subset of the Herbrand base in which for every t_i in this base it is the case that $val_{HI}(P(t_1, ..., t_n)) = \mathsf{t}$, it being the case that Definition 6.2.3 eliminates the domain of discourse, replacing it by the Herbrand universe, and thus renders the variable assignment ϖ wholly superfluous. If we reduce the alphabet of a FO language to $\Sigma = \{P, a\}$ then there is only one model for the atomic formula $P(a)$, and that is $\{P(a)\}$. Increase finitely the alphabet with the constants $b, c, d, ...$ and one still has a finite number of models for $P(a)$: $\{P(a), P(b)\}$, $\{P(a), P(b), P(c)\}$, etc.

In terms of the relation between satisfiability and validity, the results obtained in Tarskian semantics (cf. Prop. 3.4.31) hold in Herbrand semantics. But this entails an advantage of the latter over the former in CFOL: under this, CFOL is generally undecidable (at best semi-decidable), but it becomes decidable under Herbrand semantics, even if only given some restrictions to a classical signature.

Theorem 6.2.1. *Given the language* L1, *let* $\Upsilon_{L1} = (Pred, Fun, Cons)$ *where* $Fun = \emptyset$ *and* $|Cons|$ *is finite. Then, the problem of validity in* L1 *is decidable under Herbrand semantics.*

Proof: For the given Υ_{L1}, the Herbrand universe H_F for a finite set of formulae $F \subseteq$ L1 is finite by Definition 6.2.1.1. Hence, $H(F)$, the set of ground instances of atoms of F, is also finite, and the models HM_F are necessarily in finite number, too. Assign to each ground atom in $H(F)$ a propositional symbol: you have a propositional rewriting or transformation of the FO set of formulae F. (Cf. Section 3.5.2.1 for the details of this transformation.) Thus, the validity problem in L1 can be reduced by means of grounding to the validity problem in L0, which is known to be decidable. **QED**

Let us denote the above *function-free* fragment of L1 by $L1_{ff}$ and let us denote the Herbrand semantics above by $H\mathfrak{S}$. Then we have the following obvious result for some set $X \subseteq L1_{ff}$ and some formula $\phi \in L1_{ff}$.

Corollary 6.2.2. $X \models_{H\mathfrak{S}} \phi$ *is decidable.*

Proof: $L1_{ff}$ generates solely (sets of) ground formulae that are substitutable by propositional formulae of L0, for which there is a semantical decision procedure, namely the truth-table construction. **QED**

In the following two theorems, we consider some *Herbrand axiomatization*, and we denote it by $H\mathcal{P}$.[5]

---

[5]Such an axiomatization may well be the (extended) $\mathcal{NK}$ calculus (see Section 5.2

Theorem 6.2.3. *Given a set of formulae $X \subseteq F_{L1}$ and a formula $\phi \in F_{L1}$, if $X \vdash_{HP} \phi$, then $X \models_{H\mathfrak{S}} \phi$.*

Proof: (Idea) A Herbrand axiomatization of L1 only proves (sets of) formulae of L1 that have a model in Herbrand semantics, both proofs and models being finite. (Of course, you have to show that finite models are finitely axiomatizable.) **QED**

Theorem 6.2.4. *There exist a set of formulae $X \subseteq F_{L1}$ and a formula $\phi \in F_{L1}$ such that $X \models_{H\mathfrak{S}} \phi$ but $X \nvdash_{HP} \phi$.*

Proof: (Idea) Some formulae derivable in $H\mathfrak{S}$ require infinite proofs, but by definition proofs are finite objects. (Hint: Compactness is missing in $H\mathfrak{S}$. See Exercise 6.2.3.) **QED**

Exercises

Exercise 6.2.1. For each of the following sets of clauses C, find the Herbrand universe H_C and the Herbrand base $H(C)$:

1. $C = \{P(x, y) \vee \neg Q(b), \neg P(a, x) \vee Q(b)\}$

2. $C = \{P(x, f(y)), P(z, g(z))\}$

3. $C = \{P(a, f(x, y)), P(b, f(x, y))\}$

4. $C = \{(R(g(a, f(b))))\}$

Exercise 6.2.2. Interpret formula χ of Example 3.3.5 under Herbrand semantics.

Exercise 6.2.3. Herbrand semantics is not FO-compact.

1. Show that there are infinite sets of FO formulae that are unsatisfiable while every finite set thereof is satisfiable.

2. Say how this impacts on satisfiability testing.

---

above). Herbrand (1930) does indeed provide his own proof calculus for CFOL, but it does not differ significantly from this calculus, although it considers the elimination of the rule of inference MP.

Exercise 6.2.4. Show by means of an example that Skolemization (cf. Section 7.2.4) does not preserve satisfiability in Herbrand semantics.

Exercise 6.2.5. Herbrand semantics allows for more expressive power of a logic in comparison to Tarskian semantics. Comment on this.

Exercise 6.2.6. CFOL with Herbrand semantics loses semi-decidability. Why?

Exercise 6.2.7. Complete the proof of Lemma 3.5.1. (Hint: Restate the lemma in terms of a model $H\mathcal{M}$.)

Exercise 6.2.8. Prove Theorems 6.2.3-4 by following the given ideas and hint.

6.3. Algebraic semantics: Boolean algebras

By *algebraic semantics*, we mean here the employment of algebraic structures to give meaning to logical formulae. As seen in Section 1.2, models are algebraic structures, and we have been employing them so far in association to interpretations and logical consequence, in both Tarskian and Herbrand semantics. In both these semantics, a FO formula ϕ is said to be satisfiable iff, given an interpretation $\mathcal{I} = (\mathcal{D}, \Theta, \varpi)$ (a H-interpretation $H\mathcal{I} = (H_\phi, \Theta, \varpi)$), there is a model $\mathcal{M}$ (a H-model $H\mathcal{M}$) such that we have $\models_\mathcal{M} \phi$ ($\models_{H\mathcal{M}} \phi$, respectively). But this is not quite what we mean here by *algebraic semantics*; by this label, we intend to capture the fact that logical theories and calculi obey the laws of some (classes of) algebras. In particular, the theories or calculi obey the *logical laws* that are in fact exactly the *algebraic laws* of some given (class of) algebras. In this case, we say that the given (classes of) algebras *characterize* a certain logic. For instance, the Boolean algebras are known to characterize CPL, and Heyting algebras do so for propositional intuitionistic logic.

This characterization, often inessential from not only the philosophical but also the mathematical viewpoint, is essential for computing with logic, as it provides a mathematical description of logics or logical systems. In Chapter 2, we saw that any computational problem can be reduced to the SAT (but see especially Chapter 7), which is in fact a problem formulated in algebraic terms, to wit, in Boolean algebra.

In effect, in an algebraic semantics, a logic or a logical system is provided with models that are in fact order-theoretical structures, such as lattices: the meaning of some proposition $p \in F_\mathsf{L}$, where L is a logical language, is identified with an element of a specific lattice, and the connectives $\heartsuit_i \in O_L$ are interpreted as lattice-operations thereof.[6] We then speak not only of a model $\mathcal{M}$, but of an *algebraic model* $\mathcal{M} = \mathcal{U}_{\mathcal{L}}$, for a given lattice $\mathcal{L}$. If a logic **L** is characterizable by some algebra $\mathfrak{A}$, we then write

$$\models_{\mathcal{M}} p \quad \Leftrightarrow \quad \leq_{\mathcal{U}_{\mathcal{L}}} [p]$$

where $\mathcal{U}_{\mathcal{L}}$ denotes a filter of $\mathcal{L}$ and $[p]$ denotes the equivalence class of p. More precisely, and recalling notions from Section 1.2, we have $p \in hull\,(h)$ for h a Boolean homomorphism $h : \mathfrak{L} \longrightarrow 2$ where $\mathfrak{L} \supseteq \{p\}$ is an algebra of (propositional) formulae and 2 is the two-element Boolean algebra. In other words, $\mathcal{U}_{\mathcal{L}}$ is an ultrafilter.

The symbol $\Leftrightarrow$ denotes here a "bridge" between the logic **L** and an algebra $\mathfrak{A}$ established by a Lindenbaum-Tarski algebra. We give below the basic details of how this "bridge" is obtained between CPL and a Boolean algebra, but this construction carries over to many classes of algebras and many logics.[7]

Definition. 6.3.1. Given the language L0 and the set of all formulae F_{L0}, we define the equivalence relation $\sim$ over F_{L0} as

$$\phi \sim \psi \quad \text{iff} \quad \Theta \Vdash \phi \leftrightarrow \psi$$

for a consistent theory $\Theta \subseteq F_{L0}$. The set B of all equivalence classes defined as $[\phi] = \{\psi |\; \Vdash \phi \leftrightarrow \psi\}$ is a Boolean algebra under the operations $\lor, \land, ', 1$, and 0, in the following ways:

$$[\phi] \lor [\psi] = [\phi \lor \psi]$$
$$[\phi] \land [\psi] = [\phi \land \psi]$$
$$[\phi]' = [\neg\phi]$$
$$1 = [\phi \lor \neg\phi]$$
$$0 = [\phi \land \neg\phi]$$

This algebra is called the *Lindenbaum-Tarski algebra of the language* L0. This algebra can be extended to L1 by considering its set T of all terms.

---

[6]Thus, the syntax of some logical language can be defined as an algebra, too. In particular, a propositional language L0 can be seen as an algebra of formulae $\mathfrak{L}$ freely generated by the set $V \subseteq F_{L0}$ (cf. Def. 3.3.5 and, below, Def. 6.3.2).

[7]This is a topic we elaborate on in the second volume of this work.

Then, it can be shown that

$$\bigvee_{t \in T} [\phi(t)] := [\exists x \, (\phi(x))]$$

and

$$\bigwedge_{t \in T} [\phi(t)] := [\forall x \, (\phi(x))].$$

Definition. 6.3.2. Let the set $B = \mathsf{L0}/\sim$ of propositional formulae of $\mathsf{L0}$ *modulo equivalence* be the set of equivalence classes. Then, the algebra $\mathfrak{B} = (B, \vee, \wedge, ', 0, 1)$ of type $(2, 2, 1, 0, 0)$ is a Boolean algebra. Moreover, this is a free Boolean algebra freely generated by the set of propositional variables V_{L0}.

Note that what we have here is a means of investigating sets of CFOL formulae by investigating their associated Lindenbaum-Tarski algebras instead, and as these can be shown to be Boolean algebras,[8] we may investigate these sets of formulae by investigating Boolean algebras instead.

With the two above-discussed semantics, Tarskian and Herbrand's, we were able to prove the completeness of CFOL (see Sections 3.4.4 and 4.4); in other words, we related the logical language $\mathsf{L1}$ to both model-theoretical and proof-theoretical structures, to wit, models and proofs. Because logical languages, models, and proofs are all objects of logic, we did not leave the terrain of logic proper. But now, having shown that classical logical theories coincide with Boolean algebras, we have a much easier way to prove the completeness of CFOL via a *representation theorem*, i.e. a theorem that puts families or classes of mathematical structures in relation with one of its proper sub-families or sub-classes. In fact, the completeness theorem can be considered a corollary of the representation theorem for Boolean algebras.

Theorem 6.3.1. *(Representation theorem; Stone, 1936) Let A be the set of two-valued homomorphisms on a Boolean algebra $\mathfrak{B}$. Then, $\mathfrak{B}$ is embeddable into 2^A via the mapping defined by*

$$f(p) = \{h \in A | h(p) = 1\}$$

---

[8]More specifically, Lindenbaum-Tarski algebras of propositional languages coincide with free Boolean algebras, and Lindenbaum-Tarski algebras of FO languages coincide with polyadic Boolean algebras. We eschew any discussion of the latter algebras for several reasons, not the least of which is that our interest here in algebraic semantics is mostly connected to the SAT, which can be naturally generalized to CFOL and adequately tackled then with Herbrand semantics.

for every $p \in \mathfrak{B}$.

Proof: Recall that a Boolean algebra $\mathfrak{B}$ has the operations of complementation, meet, and join, denoted by $',\wedge,\vee$, respectively. Recall also from Definition 1.2.3.4 that an embedding, or monomorphism, is a one-to-one homomorphism. Let $p, q \in \mathfrak{B}$. We show that h is a homomorphism:

$$
\begin{aligned}
h\,(p \vee q) \;&= \{h \in A | h\,(p \vee q) = 1\} \\
&= \{h \in A | h\,(p) \vee h\,(q) = 1\} \\
&= \{h \in A | h\,(p) = 1 \text{ or } h\,(q) = 1\} \\
&= \{h \in A | h\,(p) = 1\} \cup \{h \in A | h\,(q) = 1\} \\
&= h\,(p) \cup h\,(q)
\end{aligned}
$$

Similarly, we have

$$
\begin{aligned}
h\,(p') \;&= \{h \in A | h\,(p') = 1\} \\
&= \{h \in A | h\,(p)' = 1\} \\
&= \{h \in A | h\,(p) = 0\} \\
&= \{h \in A | h\,(p) = 1\}' \\
&= h\,(p)'
\end{aligned}
$$

The meet-operation can be defined by means of the join and complementation operations $(p \wedge q = (p' \vee q')')$, and we have shown that h is a homomorphism. Now, we have to show that h is one-to-one, an easy thing to do in algebra (hint: kernel!). **QED**

It should be easy to see that in the above homomorphism $h : \mathfrak{B} \longrightarrow 2^A$, A can be seen as the set of ultrafilters of $\mathfrak{B}$.

Corollary 6.3.2. *Every Boolean algebra is isomorphic to a field of sets.*

Proof: Left as an exercise.

The exercises in this Section are all geared to a fuller grasp of the characterization of CPL by means of Boolean algebras. As a matter of fact, it is quite customary to speak of classical logic as *Boolean logic*, a coinage that honors both the mathematical conception of these algebras and their association with the laws of thought carried out by G. Boole (1847, 1854). The following words by G. Boole fully justify this coinage:

> [A]ny system of propositions may be expressed by equations
> involving symbols x, y, z, which, whenever interpretation is
> possible, are subject to laws identical in form with the laws
> of a system of quantitative symbols, susceptible only of the
> values 0 and 1. But as the formal processes of reasoning
> depend only upon the laws of the symbols, and not upon the
> nature of their interpretation, we are permitted to treat the
> above symbols x, y, z, as if they were quantitative symbols
> of the kind above described. (Boole, 1854)

Although the label "Boolean logic" is actually more often than not used
as a synonym for *Boolean algebra* (e.g., in switching theory), the following notions, building up on contents of Sections 1.2.1-2, show the
natural kinship between this and classical logic. This kinship is particularly explicit in the literature on the SAT, especially so on (automated)
refutation methods therefor, given the essentially exclusive use of the
Boolean operators $\neg$, $\wedge$, and $\vee$ in the conjunctive and disjunctive normal forms commonly used with most SAT solvers.[9]

Exercises

Exercise 6.3.1. Let Θ be a propositional theory and $\mathcal{M}_\Theta$ be a set
of Θ-models such that for every model $\mathcal{M}$ satisfying all the formulae in
Θ there is a model $\mathcal{M}' \in \mathcal{M}_\Theta$ that is elementarily equivalent to $\mathcal{M}$. Let
$\mathcal{M}_\phi$ denote the set of models in $\mathcal{M}_\Theta$ that satisfy a given formula $\phi \in \Theta$.
Recall now the definition of a field of sets (Exercise 1.2.1.8). Show that
the $\mathcal{M}'_\phi$ form a field of sets with unit universe $\mathcal{M}_\Theta$.

Exercise 6.3.2. Denote the above field of sets by $\mathfrak{F}_\Theta$. Then, we say
that the algebraization of CPL is the class $Iso\,(\mathfrak{F}_\Theta)$ of algebras, where
Iso denotes isomorphism and Θ varies over the possible theories on the
possible propositional languages. Show that $Iso\,(\mathfrak{F}_\Theta)$ of CPL coincides
with the isomorphic closure of $\mathfrak{F}_\Theta$.

Exercise 6.3.3. A class of algebras is said to be a *variety* if it can
be axiomatized by a set of equations.

1. Show that $Iso\,(\mathfrak{F}_\Theta)$ is a variety and that it is the class of Boolean
   algebras.

---

[9]See Chapter 7. For obvious reasons, the label "Boolean logic" applied to logic proper
is also very frequent in the many-valued logic literature.

2. Say informally how this result can be seen as a completeness theorem for CPL.

Exercise 6.3.4. Let us denote the Lindenbaum-Tarski algebra for a propositional theory Θ by $\mathfrak{B}_\Theta$. Let $Iso\,(\mathfrak{B}_\Theta)$ denote the algebraization of CPL based on a given proof system.

1. Show that $\mathfrak{B}_\Theta$ and $\mathfrak{F}_\Theta$ are isomorphic.

2. Explain how this result can be considered a completeness theorem for CPL.

Exercise 6.3.5. Match the expressions/symbols in column A (Boolean algebras) with the expressions in column B (propositional logics):

A	B
1. Boolean filter	a. *Taut*
2. $\mathfrak{B}_\emptyset \longrightarrow \mathscr{D}$	b. complete logical theory
3. Boolean ultrafilter	c. CPL is complete
4. $\leq$	d. propositional model
5. Boolean algebras are a variety	e. logical theory
6. 1	f. $\models$

Exercise 6.3.6. The *Ultrafilter theorem* states that every filter is included in an ultrafilter. Establish the relation for propositional logic between this and Lindenbaum's theorem (Theorem 3.4.16).

Exercise 6.3.7. Let $\mathfrak{B}$ be a Lindenbaum-Tarski algebra. Show that the following statements are equivalent for a set of formulae $A \in F_{\mathsf{L}}$:

1. A is consistent.

2. A is satisfiable.

3. The set $\{[x]\,|\,x \in A\}$ has the finite-meet property in $\mathfrak{B}$ (cf. Def. 1.2.28).

Exercise 6.3.8. Show that all the properties of a Boolean lattice (cf. Def. 1.2.21) are logical laws of CPL.

Exercise 6.3.9. Prove Corollary 6.3.2.

Exercise 6.3.10. Complete the proof of Theorem 6.3.1.

Part V.

Classical Deductive Computing with Classical Logic

7. Classical logic and deductive computation

We here consider the Boolean problem of satisfiability, or SAT, to constitute the core of classical deductive computing with classical logic. In effect, any computational problem, which means also any decision problem, that can be formulated in CL and solved in a classical computing device, namely in a Turing machine, can be so in terms of the SAT. Given a logical formula (a logical theory), we ask whether there is some interpretation that makes the formula (all the formulae of the theory, respectively) **true**. We then give the formula (the theory) as input to some classical computing device that computes the characteristic function

$$\chi_{\text{SAT}}\left(\phi\right) = \begin{cases} 1 & \text{if } \phi \in \text{SAT} \\ 0 & \text{otherwise} \end{cases}$$

for a formula ϕ, or for a theory $\Theta = \bigwedge_{i=1}^{n} \phi_i$,[1]

$$\chi_{\text{SAT}}\left(\Theta\right) = \begin{cases} 1 & \text{if } \Theta \in \text{SAT} \\ 0 & \text{otherwise} \end{cases} .$$

We know that CFOL is not decidable for standard FO formulae or theories. Nevertheless, there are techniques to rewrite FO formulae in such a way that equisatisfiable ground formulae are decidable–given an adequate semantics, such as Herbrand semantics. As a matter of fact, the SAT works as a measure of the tractability status of a computational problem: if we can polynomial-time reduce the SAT to a computational problem, then we know that this is a **NP**-complete problem (cf. Theorems 2.3.8-9).

In the following Sections we discuss these topics at length. The reader seeking a comprehensive elaboration on the SAT can benefit from, for example, Biere et al. (2009).

---

[1] Recall the the deduction theorem (DT) allows us to formulate a logical theory as a single formula.

7.1. The computational problem of classical satisfiability, or SAT

Much of classical computing has to do with the decidability of theories. Let us revisit the group theory $\Theta_{\mathcal{G}}$ for a group a pair $\mathcal{G} = (G, \star)$. Let the statements 1-3 of Exercise 3.2.2.4, known as the axioms of associativity, of the identity element, and of the inverse element, respectively, be the axioms $\theta_1 - \theta_3$ of this theory. Given $x, y, z \in G$, we wish to know if the following formula belongs to $\Theta_{\mathcal{G}}$:[2]

$$\phi = \forall x \forall y \forall z \left((x \star z = y \star z) \to x = z \right).$$

In other words, we wish to know whether ϕ is a valid formula, or a theorem, of $\Theta_{\mathcal{G}}$, i.e.

$$\Theta_{\mathcal{G}} \overset{?}{\Vdash} \phi.$$

Equivalently, we wish to know whether $(\theta_1 \wedge \theta_2 \wedge \theta_3) \to \phi$, for $\{\theta_1, \theta_2, \theta_3\} \subseteq \Theta_{\mathcal{G}}$, is a valid formula, which, if so, allows us to decide that $\phi \in \Theta_{\mathcal{G}}$, in which case we extend $\Theta_{\mathcal{G}}$ as $\Theta_{\mathcal{G}}^*$.

Because, as said above, validity testing is less efficient than testing for satisfiability, relying on Propositions 3.4.33-4 we ask instead whether the formula $\theta_1 \wedge \theta_2 \wedge \theta_3 \wedge \neg\phi$ is satisfiable. Theorem 3.4.14 gives us a decision procedure in more than one way. Firstly, it tells us that, for any theory Θ, if $\Theta^* = \{\theta_1, ..., \theta_n, \neg\phi\}$ is an unsatisfiable set of formulae, then the formula $(\theta_1 \wedge ... \wedge \theta_n) \to \phi$ is indeed valid, and we have $\Theta \models \phi$; otherwise, Θ^* is satisfiable and we have $\Theta \nvDash \phi$. Secondly, if we have an adequate deductive system at hand, then we can simply apply some proof calculus to decide algorithmically whether $\Theta \vdash \phi$, as inconsistency is the syntactical counterpart of unsatisfiability. This means that a decision problem that is expressed in a model-theoretical way can find a computational solution by proof-theoretical means. In particular, these proof-theoretical means are refutation calculi, which exhibit many computational advantages over both the direct proof calculi of Chapter 5 and the model-theoretical means such as the DPLL procedure.

The model-theoretical formulation of this decision problem is known as the satisfiability problem.

Definition. 7.1.1. The language

$$SAT = \{\langle\phi\rangle \,|\, \phi \text{ is satisfiable}\}$$

---

[2]The fact that this formula belongs to CL with equality is negligible for our purposes.

for a given formula ϕ and $\langle\phi\rangle$ an adequate encoding thereof, is called the *satisfiability problem*, or, abbreviated, *SAT*.

We now give the adequate encoding. Recall Definitions 1.2.9-10.

Definition. 7.1.2. An instance of SAT is a formula $\phi = \phi(x_1, ..., x_n)$ composed of

1. $x_1, ..., x_n$ Boolean variables;

2. k Boolean connectives, i.e. connectives in the set $O'_{L1} = \{\neg, \wedge, \vee\}$.

In other words, ϕ is a *Boolean formula*. This entails that a FO formula must undergo some rewriting using only $O'_{L1} = \{\neg, \wedge, \vee\}$ that preserves its original satisfiability; we give this rewriting in the next Section, but we can already make more specific the definition of the SAT.

Definition. 7.1.3. The language

$$SAT = \{\langle\phi\rangle \mid \phi \text{ is a satisfiable Boolean formula}\}$$

is called the *Boolean satisfiability problem*, or, abbreviated, *SAT*.

From now on, when we write "SAT" we mean the Boolean satisfiability problem, even if we do not write "Boolean." This is so because we are concerned here with classical logic, and so we can reformulate the SAT in the following way:

Definition. 7.1.4. The language

$$SAT = \{\phi \mid \models_{\mathcal{M}} \phi\}$$

for ϕ a Boolean formula and some model $\mathcal{M}$ is the satisfiability problem, or SAT.

We can now give a strictly logical definition of the satisfiability problem:

Definition. 7.1.5. Given a (propositional) formula $\phi(x_1, ..., x_n)$ with finitely many k connectives in $O'_{L1} = \{\neg, \wedge, \vee\}$, it is asked if ϕ can be evaluated to $\mathbf{t}$ by some assignment of the truth-value set $W = \{\mathbf{f}, \mathbf{t}\}$ to the x_i, $1 \leq i \leq n$. We call this question the (Boolean) satisfiability problem, or SAT.

This entails that a Boolean formula $\phi = \phi(x_1, ..., x_n)$ belongs to SAT iff there is a truth-value assignment to each and every one of its n variables that makes ϕ **true**. Thus, the SAT reduces to finding a model for a Boolean formula ϕ. Moreover, because this problem is formulated in Boolean terms, any formula or theory can be easily and adequately encoded in the binary notation using the alphabet $\Sigma = \{0, 1\}$. We need not here carry out a full encoding, being satisfied with Boolean formulae and their corresponding Boolean truth-value assignments, as illustrated in Example 7.1.1.

Example 7.1.1. Let there be given the following formula:

$$\phi = (x_1 \vee x_2) \wedge ((\neg x_1 \wedge x_2) \vee x_3)$$

The truth-value assignment $x_1 = 1$, $x_2 = 0$, $x_3 = 1$ produces a model for ϕ. In effect, applying the laws of Boolean algebra (cf. Def. 1.2.5.2) we have:

$$(1 \vee 0) \wedge ((0 \wedge 0) \vee 1) =$$

$$= 1 \wedge (0 \vee 1) =$$

$$= 1 \wedge 1 = 1$$

We conclude: $\phi \in SAT$.

The formula in Example 7.1.1 is actually given below (Example 7.2.6) in a propositional form, along with a truth table for it, in which all the models of this formula can be verified. In effect, the variables in Example 7.1.1 are to be considered as propositional variables, and we shall write "SAT" only in the case of CPL. Below we specify the satisfiability problem for quantified Boolean formulae, but we anticipate the information that, by means of the rewritings below together with Herbrand semantics, classical FO formulae can be processed as classical propositional formulae, with the advantages associated with these, namely decidability. (The reader can benefit here from a (re)reading of Sections 3.5.2.1 and 6.2 above, and Section 7.3.3 below.)

This said, it might in fact be very difficult, if not wholly impossible, to solve a computational problem formulated as an instance of the SAT, as this is an intractable problem in the **NPC** class. This was stated as Theorem 2.3.9, and we now focus on this result. This is a fundamental result as it gives us a measure of the difficulty of a computational problem: if SAT (or 3SAT, for that matter) is polynomial-time reducible to some problem, then we know that this is also a **NP**-complete problem. So far, thousands of computational problems have been shown to be in the **NPC** class via this reduction.[3]

However, the proof of Theorem 2.3.9 is quite fastidious, and thus we give here only the most important aspects thereof, leaving as an exercise the research into its details.

Proof of Theorem 2.3.9, Part I – Recall Definition 2.3.23. This dictates that the first step in the proof of Theorem 2.3.9 is to show that SAT is in the class **NP**. Let us consider a Boolean formula in the form

$$\phi = \bigwedge_i \left(\bigvee_j L_{ij} \right).$$

This is an abbreviation of a conjunctive normal form (CNF; see Section 7.2.5 below for the full form), i.e. conjunctions of disjunctions of (negated) atoms. An example of a formula in this form is

$$\phi = (x_1 \vee x_3 \vee \neg x_4) \wedge \neg x_2 \wedge (\neg x_1 \vee x_2).$$

Because these are Boolean variables, it is easy to devise a deterministic algorithm: we take all the possible truth-value assignments of the variables $x_1, ..., x_n$ and evaluate ϕ for each of the assignments. As there are 2^n different assignments, this exhaustive algorithm is of exponential-time complexity, i.e. in the **EXPTIME** class. However, we can improve on this by means of a non-deterministic polynomial-time algorithm that guesses a truth-value assignment to ϕ and returns "accept" if the assignment satisfies ϕ. Hence, SAT is a **NP** problem.

Now, the fastidious part is showing that any **NP** problem is polynomial-time reducible to the SAT, i.e. SAT is a **NP**-complete problem. We give the essentials of this proof, which requires knowledge of the Turing machines (cf. Section 2.2.3).

Proof of Theorem 2.3.9, Part II – Let L be any language in the **NP** class. We aim to show that

$$L \preceq_P SAT.$$

We consider a non-deterministic Turing machine M_T that decides L in time n^k for some constant k; so we have $L = L(M_T)$. (We abbreviate M_T as M.) The alphabet of L is $\Sigma = \{\neg, \wedge, \vee, x, 0, 1\}$ possibly with parentheses. For convenience, we shall write $\bar{x}$ instead of $\neg x$. Recall

---

[3]See Figure 2.3.4 for the most typical reductions.

from Definition 2.3.7 that $L \preceq SAT$ means that $w \in L$ iff $f(w) \in SAT$. We aim to show that we can construct a function

$$f_L : \Sigma_L^* \longrightarrow CNF_L$$

such that (i) for every $w \in \Sigma^*$, w is accepted by M iff $f_L(w) = \phi \in CNF_L$, ϕ is a (Boolean) formula in CNF, is satisfiable, and (ii) the corresponding function $g_M : \Sigma^* \to \{\neg, \wedge, x, 1\}^*$,

$$g_M(w) = \phi \in L$$

is computable in polynomial time.

Part (i) is the fastidious one; (ii) is easily verifiable. The idea of the proof is based on the fact that there will be satisfying assignments of $f_L(w)$ iff there are accepting configurations of M on w, so that in fact the reduction of L to SAT is carried out via what we can call "computation histories." In order to achieve this objective we have to both describe the computations of M by Boolean variables and express accepting states of M on w by Boolean formulae.

But first of all, we need a *tableau* for M, i.e. a $n^k \times n^k$ table each row of which is one of the n^k configurations $C_i = Q \cup \Gamma \cup \{\$\}$, \$ denotes a delimiter, of a branch of the computation of M on w for $|w| = n$ (cf. Fig. 7.1.1).[4] Let C_1 be the start configuration. Let each cell (i, j) of the $(n^k)^2$ cells of the tableau have its contents represented by $|C|$ Boolean variables of ϕ

$$\{x_{i,j,\sigma} | \sigma \in C\}$$

indicating that cell (i, j) contains σ if $x_{i,j,\sigma} = 1$. Given $1 \leq i, j \leq n^k$, there will be $|C| \cdot (n^k)^2$ Boolean variables.

A tableau is said to be an *accepting tableau* if there is a row that is an accepting configuration $C_{p(|w|)=n^k}$, and $f_L(w)$ is satisfiable iff there is a "computation history" (a computation branch) of consecutive configurations of M

$$C_1, C_2, ..., C_{n^k}$$

where two configurations are said to be consecutive if we have $C_{i-1} \vdash_M C_i$ (or $C = C'$). In order to produce an accepting tableau for ϕ we have to produce the corresponding CNF of ϕ

$$CNF(\phi) = \phi_{cell} \wedge \phi_{start} \wedge \phi_{move} \wedge \phi_{accept}.$$

The first thing to consider in this correspondence between the tableau

---

[4]Not to be confused with a *tableau* as in the *tableaux calculus* of Section 8.2.

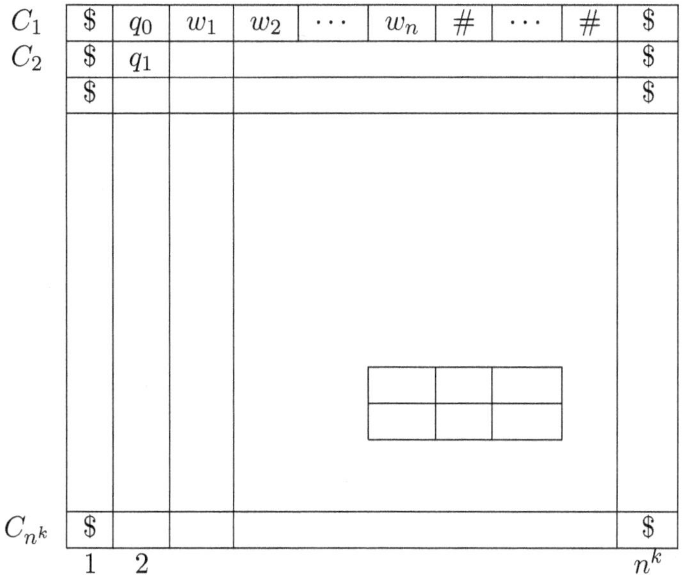

Figure 7.1.1.: A tableau for the Turing machine M.

and $CNF(\phi)$ is to make sure that for each assignment there must be exactly one symbol $\sigma \in C$ such that $x_{i,j,\sigma} = 1$. This is expressible in Boolean terms as follows:

$$\phi_{cell} = \bigwedge_{\substack{1 \le i \\ j \le n^k}} \left[\left(\bigvee_{\sigma \in C} x_{i,j,\sigma} \right) \wedge \left(\bigwedge_{\substack{\sigma, \tau \in C \\ \sigma \ne \tau}} (\overline{x}_{i,j,\sigma} \vee \overline{x}_{i,j,\tau}) \right) \right]$$

where $\left(\bigvee_{\sigma \in C} x_{i,j,\sigma} \right)$ means that at least for one of the variables it is the case that $x_{i,j,\sigma} = 1$ and $\bigwedge_{\sigma, \tau \in C; \sigma \ne \tau} (\overline{x}_{i,j,\sigma} \vee \overline{x}_{i,j,\tau})$ expresses the fact that for no more than one variable it is the case that $x_{i,j,\sigma} = 1$.

The formula ϕ_{start} guarantees that the first row of the tableau is the starting configuration. This it expresses as

$$\phi_{start} = x_{1,1,\$} \wedge x_{1,2,q_0} \wedge x_{1,3,w_1} \wedge x_{1,4,w_2} \wedge \ldots$$

$$\wedge x_{1,n+2,w_n} \wedge x_{1,n+3,\#} \wedge \ldots$$

$$\wedge x_{1,n^k-1,\#} \wedge x_{1,n^k,\$}$$

We now require a formula that guarantees that there occurs an ac-

343

cepting configuration in the tableau. This is the formula

$$\phi_{accept} = \bigvee_{\substack{1 \leq i \\ j \leq n^k}} x_{i,j,q_a}$$

where q_a denotes the accepting state. Obviously, if for some cell (i, j) we have a variable x_{i,j,q_a} such that $x_{i,j,q_a} = 1$, then we have an accepting configuration.

We spoke above of a "computation history." This entails that in the sequence of consecutive configurations of M

$$C_1, C_2, ..., C_{n^k}$$

we have it that C_{i+1} follows *legally* from C_i, $i = 1, 2, ..., n^k - 1$. In terms of the tableau for M, this means that each 2×3 *window* of cells (see center of the tableau in Fig. 7.1.1) strictly follows–or does not violate–the transition function of M. In order to verify that each row of the tableau corresponds to a configuration that preserves this legality we have the formula

$$\phi_{move} = \bigwedge_{\substack{1 \leq i < n^k \\ 1 < j < n^k}} \mathcal{W}_{i,j}$$

where $\mathcal{W}_{i,j}$ denotes the valid 2×3 window with top-middle cell at (i, j) and

$$\mathcal{W}_{i,j} =$$

$$\bigvee_{(a_1,...,a_6) \in Val} \left(x_{i,j',a_1} \wedge x_{i,j,a_2} \wedge x_{i,j'',a_3} \wedge x_{i+1,j',a_4} \wedge x_{i+1,j,a_5} \wedge x_{i+1,j'',a_6} \right)$$

where Val is the set of 6-tuples of valid assignments to the cells of $\mathcal{W}$, and $j' = j - 1, j'' = j + 1$.

This done, we now have to prove that each of the four formulae above can be expressed by a formula of size $\mathcal{O}\left(n^{2k}\right)$ and can be constructed in polynomial time from w. That is, we have to show that the "computation history" above can be abbreviated as

$$C_1 \vdash_M^* C_{n^k}$$

for $n^k = p(|w|)$ for a given formula w. (This is left as an exercise.) **QED**

Exercises

Exercise 7.1.1. Determine whether the following Boolean formulae are satisfiable by applying Boolean algebra as in Example 7.1.1.

1. $(x_1 \lor x_3 \lor \neg x_4) \land \neg x_2 \land (\neg x_1 \lor x_2)$

2. $(p_2 \lor \neg p_2 \lor p_3) \land p_1 \land (\neg p_1 \lor \neg p_3)$

3. $p \lor (q \land \neg r) \lor (\neg p \land \neg q \land r)$

4. $(x_2 \land \neg x_1) \lor (x_1 \lor (\neg x_2 \land x_3))$

5. $x_2 \land x_1 \land (x_3 \lor \neg x_4) \land \neg x_2$

6. $(p \lor \neg q) \land (\neg p \lor \neg q) \land (p \lor q) \land (\neg p \lor q)$

Exercise 7.1.2. With respect to Part II of the proof above of Theorem 2.3.9:

1. Provide the details of the construction and verification of the set *Val*.

2. Prove (ii).

Exercise 7.1.3. Show that the *general SAT*, i.e. the SAT for Boolean formulae not necessarily in CNF, is **NP**-complete.

Exercise 7.1.4. It can be shown that CIRCUIT-SAT$\preceq_P$SAT (cf. Fig. 2.3.4). Research into the CIRCUIT-SAT and

1. give the essentials of the straightforward proof that this is an **NP**-complete problem.

2. give the proof that CIRCUIT-SAT$\preceq_P$SAT.

7.2. Computerizing CFOL

We are here interested in making L1 amenable to classical computing, i.e. we wish to carry out classical deductive computation with sets of formulae F_{L1} that have been submitted to *normalization* processes. In effect, in order to computerize logical theories–for example, with automated theorem proving in mind–certain specific *normal forms* are often required for the formulae of a logical language. These normal forms, in turn, are associated with further important notions such as *literal* and *clause*. These have already occasionally occurred in this text and we now give them a full treatment. The aim of this Section is to enable the reader to make the transformations into normal forms, and thus we skip theoretical elaborations; for these see, for example, Baaz, Egly, & Leitsch (2001).

7.2.1. Literals and clauses

Definition. 7.2.1. We define a *literal*, denoted by L, to be an atom (e.g., P) or the negation of an atom (e.g., $\neg P$). We say that the literals P and $\neg P$ are *complementary*.

Definition. 7.2.2. A *clause* C is a finite disjunction of literals, i.e. $C = L_1 \vee ... \vee L_n = \|L_1, ..., L_n\|$.

1. A one-literal clause is a *unit* clause.

2. C is a *Horn clause* if it contains at most one positive literal.

   a) A Horn clause with exactly one positive literal is a *definite clause*.

   b) C is a *dual-Horn clause* if it has at most one negative literal.

3. The *empty clause* $\|\ \|$, denoted by $\square$, is a clause that contains no literals.

4. A clause is called *ground* if no individual variables occur in it.

We shall further denote a finite set of clauses $\{\|\cdot\|_1, ..., \|\cdot\|_n\}$ by C.[5]

Example 7.2.1. The formula $\forall x \neg (\neg P(x) \vee \exists y Q(x, y))$ has two atoms, to wit, $P(x)$ and $Q(x, y)$, and two literals, to wit, $\neg P(x)$ and $Q(x, y)$. The subformula $(\neg P(x) \vee \exists y Q(x, y))$ is a clause, so that the given formula contains the clause $C = \|\neg P(x), Q(x, y)\|$. C is both a

---

[5]Note that $\|\cdot\|$ just is another way to represent the *set* of literals of a clause C. It is a convenient way, as the common practice of representing a set of clauses as $C = \{\{\cdot\}, ..., \{\cdot\}\}$ can be confusing.

Horn clause and a dual-Horn clause. By means of substitutions $\sigma_1 = \{x \mapsto a\}$ and $\sigma_2 = \{y \to b\}$, $\sigma = \sigma_1 \cup \sigma_2$, we obtain the ground clause $C\sigma = \|\neg P(a), Q(a,b)\|$.

Proposition. 7.2.3. *The empty clause $\square$ is always* **false**.

Proof: The *falsity* of the empty clause derives from the fact that it has no literal that can be satisfied by an interpretation. **QED**

Remark. **7.2.4.** In particular, and denoting a contradiction or *falsity* by $\square$, we have in CL

$$(\square_{df}) \qquad A \wedge \neg A$$

and consequently

$$A \wedge \square = \square.$$

7.2.2. Negation normal form

Definition. 7.2.5. A formula $A \in F_{L1}$ is said to be in *negation normal form* (NNF) iff the negation connective $\neg$ is applied only to atoms and the only other connectives are conjunction and disjunction.

The latter requirement–that the connectives other than $\neg$ be solely conjunction and disjunction–is not a necessary condition for a formula to be in NNF, but it helps if the aim is to have a final formula in a conjunctive or disjunctive normal form (see below). It is indeed the case that these two forms are often essential for computational applications of CFOL.

Proposition. 7.2.6. *Any formula $A \in F_{L1}$ can be transformed into NNF by applying the following rewriting rules:*

1. $(\to_{df})$ $A \to B \implies \neg A \vee B$

2. *De Morgan's laws:*

     a) $(DeM_\vee)$ $\neg (A \vee B) \implies \neg A \wedge \neg B$

     b) $(DeM_\wedge)$ $\neg (A \wedge B) \implies \neg A \vee \neg B$

7. Classical logic and deductive computation

3. *Double negation, or involution, law (DN):* $\neg\neg A \implies A$

4. *Quantifier duality:*

   a) $(QN_\exists)$ $\neg\exists x\,(A) \implies \forall x\,(\neg A)$

   b) $(QN_\forall)$ $\neg\forall x\,(A) \implies \exists x\,(\neg A)$

Proof: Left as an exercise.

The application of $\text{DeM}_{\wedge(\vee)}$ and $QN_{\exists(\forall)}$ has as a result the "pushing inwards" of the connective $\neg$.

Example 7.2.2. We show the transformation of the formula

$$A = \exists x\,(P\,(x) \wedge \forall y Q\,(x,y)) \to \forall x \neg\,(P\,(x) \to \exists y Q\,(x,y))\,.$$

into the NNF formula A' such that $A \equiv A'$. By applying $\to_{df}$ to the consequent of A, we obtain the equivalent formula:

$$\exists x\,(P\,(x) \wedge \forall y Q\,(x,y)) \to \forall x \neg \underbrace{(\neg P\,(x) \vee \exists y Q\,(x,y))}_{\to_{df}}$$

Applying $\text{DeM}_\vee$ to the second disjunct gives the equivalent formula:

$$\exists x\,(P\,(x) \wedge \forall y Q\,(x,y)) \to \forall x \underbrace{(\neg\neg P\,(x) \wedge \neg\exists y Q\,(x,y))}_{\text{DeM}_\vee}$$

An application of DN gives:

$$\exists x\,(P\,(x) \wedge \forall y Q\,(x,y)) \to \forall x \left(\underbrace{P\,(x)}_{\text{DN}} \wedge \neg\exists y Q\,(x,y)\right)$$

We next apply $QN_\exists$ to obtain:

$$\exists x\,(P\,(x) \wedge \forall y Q\,(x,y)) \to \forall x \left(P\,(x) \wedge \underbrace{\forall y\,(\neg Q\,(x,y))}_{QN_\exists}\right)$$

We have:

$$\exists x\,(P\,(x) \wedge \forall y Q\,(x,y)) \to \forall x\,(P\,(x) \wedge \forall y\,(\neg Q\,(x,y)))$$

We now apply $\rightarrow_{df}$ to the whole formula, which outputs:

$$\neg\exists x \left(P\left(x\right) \wedge \forall y Q\left(x,y\right)\right) \vee \forall x \left(P\left(x\right) \wedge \forall y \left(\neg Q\left(x,y\right)\right)\right)$$

An application of $QN_\exists$ gives:

$$\underbrace{\forall x \neg}_{QN_\exists} \left(P\left(x\right) \wedge \forall y Q\left(x,y\right)\right) \vee \forall x \left(P\left(x\right) \wedge \forall y \left(\neg Q\left(x,y\right)\right)\right)$$

And by applying $DeM_\wedge$ we obtain:

$$\forall x \underbrace{\left(\neg P\left(x\right) \vee \neg \forall y Q\left(x,y\right)\right)}_{DeM_\wedge} \vee \forall x \left(P\left(x\right) \wedge \forall y \left(\neg Q\left(x,y\right)\right)\right)$$

Finally, application of $QN_\forall$ produces:

$$\forall x \left(\neg P\left(x\right) \vee \underbrace{\exists y \left(\neg Q\left(x,y\right)\right)}_{QN_\forall} \right) \vee \forall x \left(P\left(x\right) \wedge \forall y \left(\neg Q\left(x,y\right)\right)\right).$$

The application of all the rewriting rules outputs the formula:

$$A' = \forall x \left(\neg P\left(x\right) \vee \exists y \left(\neg Q\left(x,y\right)\right)\right) \vee \forall x \left(P\left(x\right) \wedge \forall y \left(\neg Q\left(x,y\right)\right)\right)$$

A' is in NNF and $A \equiv A'$.

7.2.3. Prenex normal form

Definition. 7.2.7. A formula $A \in F_{L1}$ is said to be in *prenex normal form* (PNF) iff if it is in the form

$$(\dagger) \qquad \blacklozenge_1 x_1, ..., \blacklozenge_n x_n(M)$$

where every $\blacklozenge_i x_i$, $i = 1, ..., n$ is either $\forall x_i$ or $\exists x_i$, and M is a formula containing no quantifiers. $\blacklozenge_1 x_1, ..., \blacklozenge_n x_n$ is the *prefix* and (M) is the *matrix* of A.

Proposition. 7.2.8. *Every formula $A \in L1$ can be transformed into a PNF by an algorithmic process constituted of the recursively applicable steps in Algorithm 7.1.*

Proof: Left as an exercise.

With respect to Algorithm 7.1, Step 2.e-h is to be applied only to formulas in NNF that have the connectives $\rightarrow$and/or $\leftrightarrow$. (Recall that the

---

Algorithm 7.1 PNF transformation.

Input: A FO formula A in NNF
Output: Formula A' in PNF such that $A' \equiv A$

Steps:

1. Rename the bound variables so that each variable occurs only once.

2. Push the quantifiers outwards by means of the following rewriting rules *when x does not appear as a free variable in B*:

   a) $\forall x \left(A\left(x \right) \right) \wedge B \Longrightarrow \forall x \left(A\left(x \right) \wedge B \right)$

   b) $\forall x \left(A\left(x \right) \right) \vee B \Longrightarrow \forall x \left(A\left(x \right) \vee B \right)$

   c) $\exists x \left(A\left(x \right) \right) \wedge B \Longrightarrow \exists x \left(A\left(x \right) \wedge B \right)$

   d) $\exists x \left(A\left(x \right) \right) \vee B \Longrightarrow \exists x \left(A\left(x \right) \vee B \right)$

   e) $\forall x \left(A\left(x \right) \right) \rightarrow B \Longrightarrow \exists x \left(A\left(x \right) \rightarrow B \right)$

   f) $\exists x \left(A\left(x \right) \right) \rightarrow B \Longrightarrow \forall x \left(A\left(x \right) \rightarrow B \right)$

   g) $B \rightarrow \forall x \left(A\left(x \right) \right) \Longrightarrow \forall x \left(B \rightarrow A\left(x \right) \right)$

   h) $B \rightarrow \exists x \left(A\left(x \right) \right) \Longrightarrow \exists x \left(B \rightarrow A\left(x \right) \right)$

3. If x appears as a free variable in B, then rename the bound variable x in $\blacklozenge x \left(A\left(x \right) \right)$ as y, obtaining the equivalent $\blacklozenge y \left(A\left[x/y \right] \right) = \blacklozenge y \left(A\left(y \right) \right)$.

---

absence of these connectives is not a necessary condition for a formula to be in NNF.)

Example 7.2.3. We apply Algorithm 7.1 to the formula obtained in Example 7.2.2, in order to convert it into an equivalent formula in PNF. We begin now by renaming the bound variables (Step 3 of Algorithm 7.1) in the second disjunct of A' by means of the renaming substitutions $\varsigma_1 = \{x \mapsto u\}$ and $\varsigma_2 = \{y \mapsto z\}$, $\sigma = \varsigma_1 \cup \varsigma_2$. This gives the formula:

$$\forall x \left(\neg P\left(x\right) \vee \exists y \left(\neg Q\left(x, y\right)\right)\right) \vee \underbrace{\forall u \left(P\left(u\right) \wedge \forall z \left(\neg Q\left(u, z\right)\right)\right)}_{\sigma}$$

Repeated application of Step 2.a-d produces first

$$\forall x \exists y \left(\neg P\left(x\right) \vee \neg Q\left(x, y\right)\right) \vee \forall u \forall z \left(P\left(u\right) \wedge \neg Q\left(u, z\right)\right)$$

and then

$$A'' = \forall x \exists y \forall u \forall z \left(\left(\neg P\left(x\right) \vee \neg Q\left(x, y\right)\right) \vee \left(P\left(u\right) \wedge \neg Q\left(u, z\right)\right)\right).$$

A'' is in PNF and $A' \equiv A''$.

7.2.4. Skolem normal form

Definition. 7.2.9. The procedure in Algorithm 7.2 is called *Skolemization*.

Example 7.2.4. We give two examples of Skolemization.

- By applying Algorithm 7.2 to the formula in PNF of Example 7.2.3

$$A = \forall x \exists y \forall u \forall z \left(\left(\neg P\left(x\right) \vee \neg Q\left(x, y\right)\right) \vee \left(P\left(u\right) \wedge \neg Q\left(u, z\right)\right)\right)$$

  we obtain the formula

$$A_{Sk} = \forall x \forall u \forall z \left(\left(\neg P\left(x\right) \vee \neg Q\left(x, f\left(x\right)\right)\right) \vee \left(P\left(u\right) \wedge \neg Q\left(u, z\right)\right)\right).$$

- From the formula in PNF

$$B = \exists x \forall y \forall z \forall u \exists v \left(\left(P\left(x\right) \vee Q\left(y\right)\right) \wedge R\left(z, u, v\right)\right)$$

  by applying Algorithm 7.2 we obtain

$$B_{Sk} = \forall y \forall z \forall u \left(\left(P(a) \vee Q\left(y\right)\right) \wedge R\left(z, u, f\left(y, z, u\right)\right)\right)$$

Algorithm 7.2 Skolemization

Input: A formula A be in the PNF † (where M is in CNF/DNF; see below).
Output: A skolemized formula A_{Sk} such that $A_{Sk} \equiv_{sat} A$

Let $\blacklozenge_r$, $1 \leq r \leq n$, be an existential quantifier in the prefix of A. Then,

1. If there is no universal quantifier before $\blacklozenge_r$, we choose a new constant c that does not occur in M, replace all x_r in M with c, and delete $\blacklozenge_r x_r$ from the prefix.

2. If before $\blacklozenge_r$ there are the universal quantifiers $\blacklozenge_{s_1}, ..., \blacklozenge_{s_m}$, i.e., $1 \leq s_1 < s_2 < ... < s_m < r$, we select a new m-place function symbol f which does not occur in M, replace all $\blacklozenge_r$ in M by $f(x_{s_1}, ..., x_{s_m})$ and delete $\blacklozenge_r x_r$ from the prefix.

Definition. 7.2.10. Formula B_{Sk} is in *Skolem normal form* (SNF). The constant a and the function $f(y, z, u)$ used to replace the existential variables of A are called *Skolem constant* and *Skolem function*, respectively.

As is well known, the universal quantifiers can be removed from any formula ϕ_{Sk}. In order to carry out some proof procedures (e.g., resolution), a formula in SNF must be represented as a set of clauses (understood to be universally quantified). For instance, B_{Sk} must be represented by the set

$$C_{B_{Sk}} = \{P(a) \vee Q(y), R(z, u, f(y, z, u))\} =$$
$$= \{\|P(a), Q(y)\|, \|R(z, u, f(y, z, u))\|\}$$

The process of transforming a formula ϕ into a formula in SNF does not necessarily preserve equivalence, but only satisfiability. We say that the formula ϕ_{Sk} in SNF is satisfiable iff ϕ is satisfiable, and we call this relation equisatisfiability.

Definition. 7.2.11. Two formulae ϕ and ψ are *satisfiability-equivalent*, or *equisatisfiable*, denoted by $\phi \equiv_{sat} \psi$, iff they are both satisfiable or they are both unsatisfiable.

Preservation of satisfiability is, however, all that is required in the case of refutation-based proof procedures, such as resolution and analytic tableaux.

7.2.5. Conjunctive and disjunctive normal forms

Definition. 7.2.12. A formula $A \in \mathsf{L1}$ is said to be in a *conjunctive normal form* (CNF) iff A has the form $A = C_1 \wedge \ldots \wedge C_n$, $n \geq 1$, where each of $C_1, \ldots, C_n$ is a clause, i.e.,

$$A = \bigwedge_{i=1}^{n} \left(\bigvee_{j=1}^{m_i} L_{i,j} \right)$$

and is in NNF.

Definition. 7.2.13. A formula $A \in \mathsf{L1}$ is said to be in a *disjunctive normal form* (DNF) iff A has the form $A = \mathcal{E}_1 \vee \ldots \vee \mathcal{E}_n$, $n \geq 1$, where each of $\mathcal{E}_1, \ldots, \mathcal{E}_n$ is a conjunction of literals, i.e.,

$$A = \bigvee_{i=1}^{n} \left(\bigwedge_{j=1}^{m_i} L_{i,j} \right)$$

and is in NNF.

Example 7.2.5. The formula $(P \vee R \vee \neg S) \wedge \neg Q \wedge (\neg P \vee Q)$ is in CNF. An example of a formula in DNF is $(S \wedge P \wedge \neg S) \vee (Q \wedge \neg R) \vee (\neg P \wedge R \wedge \neg Q)$.

Proposition. 7.2.14. *Let* $\{A_1, \ldots, A_n\}$ *be a finite set of formulae. Then*

$$\neg \left(\bigwedge_{i=1}^{n} A_i \right) \equiv \left(\bigvee_{i=1}^{n} \neg A_i \right)$$

and

$$\neg \left(\bigvee_{i=1}^{n} A_i \right) \equiv \left(\bigwedge_{i=1}^{n} \neg A_i \right).$$

Proof: (Sketch) We have it that $\neg(A) \equiv (\neg A)$. Then, obviously the proposition is proved for $n = 1$ by the equivalence $\neg \left(\bigwedge_{i=1}^{1} A_i \right) \equiv \left(\bigvee_{i=1}^{1} \neg A_i \right)$. The proof then follows by induction on n. **QED**

Proposition. 7.2.15. *Let A be a formula in CNF and B be a formula in DNF. Then $\neg A$ is equivalent to a formula in DNF, and $\neg B$ is equivalent to a formula in CNF.*

Proof: If A is in CNF, then A is the formula $\bigwedge_{i=1}^{n}\left(\bigvee_{j=1}^{m_i} L_{i,j}\right)$. By Proposition 7.2.14, we have

$$\neg A = \neg \bigwedge_{i=1}^{n}\left(\bigvee_{j=1}^{m_i} L_{i,j}\right) \equiv \bigvee_{i=1}^{n} \neg\left(\bigvee_{j=1}^{m_i} L_{i,j}\right) \equiv \bigvee_{i=1}^{n}\left(\bigwedge_{j=1}^{m_i} \neg L_{i,j}\right).$$

The proof runs similarly for $\neg B$ being equivalent to a formula in CNF. **QED**

Proposition. 7.2.16. *Any formula can be transformed into a CNF/DNF by applying $\rightarrow_{def}$ and/or $\leftrightarrow_{def}$, together with DN, DeM$_{\wedge(\vee)}$, and the distributive laws (cf. Def. 1.2.21.c).*

Proof: Left as an exercise.

Theorem 7.2.1. *Every formula A is equivalent to any formula A_1 in DNF and some formula A_2 in CNF.*

Proof: Follows immediately from the above. The proof is by induction on the complexity of A. **QED**

Theorem 7.2.1 guarantees the existence of a formula in DNF that is equivalent to a formula A. In order to find this formula, it suffices to compute a truth table for A. By considering the rows in which A is true, we obtain a formula in DNF equivalent to A.

Example 7.2.6. Let $A = (B \vee C)\wedge((\neg B \wedge C) \vee D)$. The truth table for A is as follows:

B	C	D	A	$\neg A$
t	t	t	t	f
t	t	f	f	t
t	f	t	t	f
t	f	f	f	t
f	t	t	t	f
f	t	f	t	f
f	f	t	f	t
f	f	f	f	t

Then, the DNF of A is:

$$(B \wedge C \wedge D) \vee (B \wedge \neg C \wedge D) \vee (\neg B \wedge C \wedge D) \vee (\neg B \wedge C \wedge \neg D)$$

Algorithm 7.3 Tseitin transformation

Input: Some formula A
Output: Formula A' in CNF such that $A' \equiv_{sat} A$

Steps:

1. Introduce a new atom, say p_1, not occurring anywhere else in the formula to abbreviate the innermost sub-formula and conjoin the abbreviated formula with the definition of p_1.

2. Proceed as in 1., introducing new atoms $p_2, p_3, ..., p_n$ as required in the direction of the main connective in A.

3. Put each of the conjuncts in CNF by applying the rules in Figure 7.2.1.

Likewise, by negatively considering the rows in which A is false, i.e., by considering the DNF for $\neg A$ and subsequently negating it, we obtain a formula in CNF equivalent to A. Thus, the DNF of $\neg A$ is:

$$(B \wedge C \wedge \neg D) \vee (B \wedge \neg C \wedge \neg D) \vee (\neg B \wedge \neg C \wedge D) \vee (\neg B \wedge \neg C \wedge \neg D)$$

By Proposition 7.2.15, the negation of this DNF produces the CNF:

$$(\neg B \vee \neg C \vee D) \wedge (\neg B \vee C \vee D) \wedge (B \vee C \vee \neg D) \wedge (B \vee C \vee D)$$

Although the procedure in Example 7.2.6 guarantees logical equivalence of a formula A with its DNF or CNF, it can lead to an exponential growth (i.e. "explosion") of the formula A. If one wishes to restrict this to a linear growth, then one may have to give up on equivalence in favor of equisatisfiability. This can be done by applying the Tseitin algorithm for CNF formulae (Tseitin, 1968). We give this as Algorithm 7.3.

Note in Figure 7.2.1 the following equivalences:

$$(\circledast) \qquad p \to (q \wedge r) \equiv (p \wedge q) \to (p \wedge r)$$

$$(\odot) \qquad ((p \wedge q) \vee (\neg p \wedge \neg q)) \to r \equiv ((p \wedge q) \to r) \wedge ((\neg p \wedge \neg q) \to r)$$

Example 7.2.7. Let $A = (p \vee (q \wedge \neg r)) \to s$. Then we have the

7. Classical logic and deductive computation

(♡)		Rule
$x \leftrightarrow (\neg p)$ $\equiv$	$(x \to \neg p) \wedge (\neg p \to x)$	$\leftrightarrow_{df}$
$x \leftrightarrow (p \wedge q)$ $\equiv$	$(x \to p) \wedge (x \to q) \wedge ((p \wedge q) \to x)$	$\leftrightarrow_{df}, \circledast$
$\equiv$	$(\neg x \vee p) \wedge (\neg x \vee q) \wedge (\neg (p \wedge q) \vee x)$	$\to_{df}$
$\equiv$	$(\neg x \vee p) \wedge (\neg x \vee q) \wedge (\underline{\neg p \vee \neg q} \vee x)$	$DM_{\wedge}$
$x \leftrightarrow (p \vee q)$ $\equiv$	$(p \to x) \wedge (q \to x) \wedge (x \to (p \vee q))$	$\leftrightarrow_{df}, \circledast$
$\equiv$	$(\neg p \vee x) \wedge (\neg q \vee x) \wedge (\neg x \vee p \vee q)$	$\to_{df}$
$x \leftrightarrow (p \leftrightarrow q)$ $\equiv$	$(x \to (p \leftrightarrow q)) \wedge ((p \leftrightarrow q) \to x)$	$\leftrightarrow_{df}$
$\equiv$	$(x \to ((p \to q) \wedge (q \to p))) \wedge$ $((p \leftrightarrow q) \to x)$	$\leftrightarrow_{df}$
$\equiv$	$(\neg x \vee ((\neg p \vee q) \wedge (\neg q \vee p))) \wedge$ $\left(((p \wedge q) \vee (\neg p \wedge \neg q)) \to x \right)$	$\to_{df}$ $\leftrightarrow_{df}$
$\equiv$	$\left(\neg x \vee \underline{(\neg (\neg p \vee q) \vee \neg (\neg q \vee p))} \right) \wedge$ $((p \wedge q) \to x) \wedge ((\neg p \wedge \neg q) \to x)$	$DM_{\wedge}$ $\circledcirc$
$\equiv$	$\left(\neg x \vee \underline{((p \wedge \neg q) \vee (q \wedge \neg p))} \right) \wedge$ $\left((\neg p \vee \neg q) \to x \right) \wedge \left((p \vee q) \to x \right)$	$DM_{\vee}$ $DM_{\wedge}$
$\equiv$	$\left(\neg x \vee \underline{(\neg (p \wedge \neg q) \wedge \neg (q \wedge \neg p))} \right) \wedge$ $(\neg (\neg p \vee \neg q) \vee x) \wedge (\neg (p \vee q) \vee x)$	$DM_{\vee}$ $\to_{df}$
$\equiv$	$\neg x \vee \left((\neg p \vee q) \wedge (\neg q \vee p) \right) \wedge$ $\left(\underline{(p \wedge q)} \vee x \right) \wedge \left(\underline{(\neg p \wedge \neg q)} \vee x \right)$	$DM_{\wedge}$ $DM_{\vee}$
$\equiv$	$(\neg x \vee \neg p \vee q) \wedge (\neg x \vee \neg q \vee p) \wedge$ $\left(\underline{\neg p \vee \neg q} \vee x \right) \wedge \left(q \vee \underline{p} \vee x \right)$	$D_{\vee}$ $DM_{\wedge}$

Figure 7.2.1.: Tseitin transformations for the connectives of L.

following definitions:

$$p_1 \leftrightarrow (q \wedge \neg r)$$
$$p_2 \leftrightarrow (p \vee p_1)$$
$$p_3 \leftrightarrow (p_2 \to s) \equiv p_3 \leftrightarrow (\neg p_2 \vee s) \equiv p_3 \leftrightarrow (p_2 \wedge \neg s)$$

We have the conjuncts:

$$(p_1 \leftrightarrow (q \wedge \neg r)) \wedge (p_2 \leftrightarrow (p \vee p_1)) \wedge (p_3 \leftrightarrow (p_2 \wedge \neg s)) \wedge p_3$$

By putting the conjuncts into CNF (see Fig. 7.2.1), we obtain the formula:

$$(\neg p_1 \vee q) \wedge (\neg p_1 \vee r) \wedge (\neg q \vee \neg r \vee p_1) \wedge$$
$$(\neg p_2 \vee p \vee p_1) \wedge (p_2 \vee \neg p) \wedge (p_2 \vee \neg p_1) \wedge$$
$$(\neg p_3 \vee p_2) \wedge (\neg p_3 \vee \neg s) \wedge (\neg p_2 \vee s \vee p_3) \wedge p_3$$

Proposition. 7.2.17. *The Tseitin transformation algorithm preserves formula equivalence.*

Proof: Left as an exercise.

Exercises

Exercise 7.2.1. Transform each of the following formulae into a set of ground clauses:

1. $(x_1 \wedge x_2) \leftrightarrow \neg (x_2 \vee (\neg x_1 \wedge x_2))$

2. $(p_1 \wedge p_2) \to (\neg (p_3 \wedge p_4) \to p_1)$

3. $\exists y \forall x \, (P(x, y) \to Q(x))$

4. $\exists y \forall x \, (P(x, y)) \leftrightarrow Q(x)$

5. $\forall x \forall y \, (P(x, y) \vee \exists z Q(x, y, z))$

6. $\forall w \exists u \exists v \, ((\neg P(w, u) \wedge Q(w, v)) \to R(w, u, v))$

7. $\forall y \neg \exists z \, (P(y) \to Q(y, z)) \wedge \forall x \forall u \exists w \, (R(x, w) \leftrightarrow S(x, u, w))$

8. $\forall x \, ((P(x) \leftrightarrow Q(x)) \wedge \forall z \neg \exists y \, (R(x, z) \to S(x, y, z)))$

Exercise 7.2.2. Rewrite the following formulae in CNF clearly showing every step of the transformation:

1. $x_1 \lor (x_2 \land (x_1 \lor x_3) \lor (x_2 \land \neg x_1))$

2. $\forall x \neg \exists y \, [(P(x) \rightarrow S(y)) \land \forall z \, (Q(x,y,z) \leftrightarrow R(x,z))]$

3. $\forall x((P(x) \leftrightarrow Q(x)) \land \forall z \neg \exists y(R(x,z) \rightarrow S(x,y,z)))$

4. $\forall y \neg \exists z \, (P(y) \rightarrow Q(y,z)) \land \forall x \forall u \exists w \, (R(x,w) \leftrightarrow S(x,u,w))$

Exercise 7.2.3. How many clauses has the CNF of the formula ϕ by applying the Tseitin transformation?

$$\phi = \neg \, (p_1 \land (p_2 \lor ... \lor p_n))$$

Exercise 7.2.4. Prove (Complete the proof of) the propositions above left as exercises (left with a sketchy proof, respectively).

7.3. Computing SAT

7.3.1. The different forms of SAT

Although the SAT is generally formulated in terms of the set of connectives $O'_{L1} = \{\neg, \land, \lor\}$ (cf. Def. 7.1.5), it is in fact more often than not formulated for formulae in CNF, and this to such an extent that "SAT" is often synonymous with "CNF-SAT." However, we can further specify instances of this problem according to the number of literals per clause, their kind (positive or negative), and/or the order of the formulae. Far from being a mere specification, this actually involves different complexity classes, so that if we can formulate or even convert an instance of SAT into another we may end up with significant computational gains (cf. Table 3.5.1).

Definition. 7.3.1. The language

$$2\text{-}SAT = \{\phi \mid \models_M \phi, \, \phi \text{ is a 2-CNF formula}\}$$

where "2-CNF formula" means a CNF formula whose clauses have at most two literals, is called the *satisfiability problem for 2-CNF formulae*, or, abbreviated, *2-SAT*.

Definition. 7.3.2. The language

$$3\text{-}SAT = \{\phi \mid \models_{\mathcal{M}} \phi, \phi \text{ is a 3-CNF formula}\}$$

where "3-CNF formula" means a CNF formula whose clauses have at most three literals, is called the *satisfiability problem for 3-CNF formulae*, or, abbreviated, *3-SAT*.

For clauses with more than three literals we write "k-CNF formulae" and speak of the *k-SAT*.

Definition. 7.3.3. The language

$$HORN\text{-}SAT = \{\phi \mid \models_{\mathcal{M}} \phi, \phi \text{ is a Horn formula}\}$$

is called the *satisfiability problem for Horn formulae*, or, abbreviated, *HORN-SAT*.

Although the Boolean propositional satisfiability problem is typically formulated in terms of formulae in CNF, it cannot be ignored that it can also be formulated for formulae in DNF. In effect, this is sanctioned by the following theorem:

Theorem 7.3.1. *A formula in CNF is a tautology iff each of its clauses is a tautology. Dually, a formula in DNF is unsatisfiable iff each of its conjunctions of literals is unsatisfiable.*

Proof: (Sketch) Let us abbreviate

$$A = \bigwedge_{i=1}^{n} \left(\bigvee_{j=1}^{m_i} L_{i,j} \right)$$

as

$$A = \bigwedge_{i=1}^{n} \mathcal{C}_i.$$

Clearly, if each $\mathcal{C}_i$, $1 \leq i \leq n$, is a tautology, then A is a tautology, and if some $\mathcal{C}_i$ is not a tautology, then A is not a tautology. Apply now the duality principle to prove the dual result for a formula in DNF. **QED**

Nevertheless, a formulation of the SAT in terms of DNFs is far less usual (cf. Exercise 7.3.1.3), and we shall be contented with a very general definition.

Definition. 7.3.4. The language

$$DNF\text{-}SAT = \{\phi \mid \models_{\mathcal{M}} \phi, \phi \text{ is a DNF formula}\}$$

is called the *satisfiability problem for formulae in DNF*, or, abbreviated, *DNF-SAT*.

Were it not for the fact that truth tables are inefficient SAT solvers for more than a few atomic formulae, we would happily resort to them; but they are not–at all, for FO formulae. In effect, there is no decision procedure for CFOL, which means that SAT is basically undecidable for Boolean FO formulae, or quantified Boolean formulae (abbr.: QBF). Of course, some instances thereof are decidable, some even in linear time, but they are **PSPACE**-complete in terms of complexity classes (cf. Table 3.5.1), which is suspected to be composed of problems harder to solve than those in the **NP**-complete class (see Fig. 2.3.3).

Definition. 7.3.5. The language

$$QBF\text{-}SAT = \{\phi|\models_{\mathcal{M}} \phi, \phi \text{ is a QBF}\}$$

is called the *satisfiability problem for QBFs*, or, abbreviated, *QBF-SAT*.

Exercises

Exercise 7.3.1.1. Prove that 3-SAT is **NP**-complete by the reduction

$$SAT \preceq_P 3\text{-}SAT.$$

Exercise 7.3.1.2. Prove the following statements:

1. 2-SAT$\in$ **P**.

2. HORN-SAT$\in$ **P**.

3. DNF-SAT$\in$ **P**.

4. 3-SAT is **NP**-complete.

5. k-SAT, $k \geq 3$, is **NP**-complete.

6. QBF-SAT is **PSPACE**-complete.

Exercise 7.3.1.3. If DNF-SAT$\in$ **P**, but (CNF-)SAT is **NP**-complete, why is the Boolean satisfiability problem not typically formulated in terms of DNFs?

Exercise 7.3.1.4. Research into the *maximum satisfiability problem*, abbreviated MAX-SAT:

1. Give a formal definition of this problem and its complexity class.

2. Is MAX-SAT a decision problem?

Exercise 7.3.1.5. The DUAL-HORN-SAT is the satisfiability problem for dual-Horn formulae (cf. Def. 7.2.2.2.b). Show that DUAL-HORN-SAT∈**P**.

Exercise 7.3.1.6. Complete the proof of Theorem 7.3.1.

7.3.2. SAT and unsatisfiability I: The DPLL procedure and model finding

The Davis-Putnam-Logemann-Loveland procedure (Davis, Logemann, & Loveland, 1962), abbreviated DPLL, is the basis for many classical propositional satisfiability solvers, or classes of algorithms for large subsets of SAT instances. It is actually a refinement of the resolution-based procedure first implemented by Davis and Putnam for propositional logic formulae (Davis & Putnam, 1960), and it can be straightforwardly applied to the CNF-SAT problem.

We leave the original version of the DPLL proof procedure as an exercise and work with a variation that is adequate for pedagogical reasons. **Definition. 7.3.6.** Let ϕ be a formula in CNF and let C_ϕ be the set of clauses of ϕ. The *DPLL1-8 rules* to apply to C_ϕ that constitute the *DPLL proof procedure* are as follows:

1. *UNSAT:* If C_ϕ contains the empty clause, then ϕ is unsatisfiable.

2. *SAT:* If C_ϕ is empty, then ϕ is satisfiable.

3. *MULT:* If a literal occurs more than once in a clause, then all but one of its occurrences can be deleted.

4. *SUBS:* If a clause C_1 is a superset of a clause C_2, $C_1, C_2 \subseteq C_\phi$, then C_1 can be deleted.

5. *UNIT:* If C_ϕ contains $\|L\|$, then an element $\neg L \in C_i \subset C_\phi$ can be deleted.

6. *TAUT:* If a clause C contains a literal and its complement, then it can be deleted.

7. *PURE:* If a clause $\mathcal{C}$ contains a literal L and $\neg L$ does not occur in ϕ, then $\mathcal{C}$ can be deleted.

8. *SPLIT:* If C_ϕ is semantically equivalent to a formula of the form

$$\{(\mathcal{C}_1 \vee L), ..., (\mathcal{C}_k \vee L), (\mathcal{C}_{k+1} \vee \neg L), ..., (\mathcal{C}_m \vee \neg L), \mathcal{C}_{m+1}, ..., \mathcal{C}_n\}$$

where neither L nor $\neg L$ occur in $\mathcal{C}_i$, $1 < i < n$, then replace C_ϕ by

$$\{\mathcal{C}_1, ..., \mathcal{C}_k, \mathcal{C}_{m+1}, ..., \mathcal{C}_n\} \vee \{\mathcal{C}_{k+1}, ..., \mathcal{C}_m, \mathcal{C}_{m+1}, ..., \mathcal{C}_n\}.$$

Example 7.3.1. We provide examples of the DPLL rules above:

(DPLL1) $C = \{\Box, \mathcal{C}_1, ..., \mathcal{C}_n\} \equiv \Box$; UNSAT.

(DPLL2) $C = \{L_1, L_2, \neg L_1\} \equiv \{\ \}$; SAT.

(DPLL3) $\|L, L, L_1, ..., L_m\| \equiv \|L, L_1, ..., L_m\|$.

(DPLL4) $C = \{\|L_1, ..., L_m\|, \|L_1, ..., L_m, ..., L_k\|, \mathcal{C}_1, ..., \mathcal{C}_n\} \equiv$
$\{\|L_1, ..., L_m\|, \mathcal{C}_1, ..., \mathcal{C}_n\}$.

(DPLL5) $C = \{\|L_1, ..., L_m, \neg L\|, \|L\|\} \equiv \{\|L_1, ..., L_m\|, \|L\|\}$.

(DPLL6) $C = \{\|L_1, ..., L_m, L, \neg L\|, \mathcal{C}_1, ..., \mathcal{C}_n\} \equiv \{\mathcal{C}_1, ..., \mathcal{C}_n\}$.

(DPLL7) $C = \{\|P, \neg Q\|, \|\neg R\|\}$ is *not* equivalent to $C' = \{\|P, \neg Q\|\}$, but $\{\mathcal{C}_1, ..., \mathcal{C}_m, \|L, L_1, ..., L_n\|\}$ is unsatisfiable iff $\{\mathcal{C}_1, ..., \mathcal{C}_m\}$ is unsatisfiable, where $\neg L$ occurs neither in $\mathcal{C}_i$, $1 < i < m$, nor in L_j, $1 < j < n$.

(DPLL8) $C = \{\|P, R\|, \|\neg R\|, \|Q\|\}$ and $C' = \{\|P\|, \|Q\|\} \vee \{\Box, \|Q\|\}$ are *not* equivalent, but the rule preserves unsatisfiability.

It is obvious that while rules DPLL3-6 are equivalence-preserving, rules DPLL7-8 are unsatisfiability-preserving, i.e., for a formula ψ obtained from a formula ϕ by application of any of these two rules, we have it that ϕ is unsatisfiable iff ψ is unsatisfiable. We give the DPLL-algorithm as Algorithm 7.4.

Proposition. 7.3.7. *The DPLL-algorithm constitutes a sound and complete proof procedure.*

Proof: Left as an exercise.

The strategy we mentioned above applies to Step 1 of Algorithm 7.4: Rules DPLL3 to DPLL7 are to be applied eagerly, but wise decisions should be made when applying DPLL8 with regard to the literal to be eliminated.[6]

Example 7.3.2. Let $C = (L \vee M) \wedge (O \vee \neg M \vee \neg N) \wedge (\neg L \vee N) \wedge \neg O$. We apply the DPLL-algorithm to C (see Fig. 7.3.1).

---

[6]The original DPLL algorithm is backtracking-search based, namely based on

---

Algorithm 7.4 DPLL algorithm

Input: A formula *A* in CNF
Output: A proof of (un)satisfiability of *A*

---

1. Apply rules DPLL3 to DPLL8, until either SAT (DPLL2) or UNSAT (DPLL1) can be applied.

2. If SAT is applicable, then terminate with "*A* is satisfiable"; else, UNSAT is applicable and terminate with "*A* is unsatisfiable".

$\{\|L,M\|,\|O,\neg M,\neg N\|,\|\neg L,N\|,\|\neg O\|\}$	RULES
$\{\|L,M\|,\|O,\neg M,\neg N\|,\|\neg L,N\|,\|\neg O\|\}$	Initialization
$\{\|L,M\|,\|\neg M,\neg N\|,\|\neg L,N\|,\|\neg O\|\}$	$\|\neg O\|$, DPLL5
$\{\|L,M\|,\|\neg M,\neg N\|,\|\neg L,N\|\}$	$\neg O$, DPLL 7
$\{\|M,\neg M,\neg N\|,\|N,\neg M,\neg N\|\}$	L, DPLL8
$\{\|N,\neg M,\neg N\|\}$	$\|M,\neg M,\neg N\|$, DPLL6
$\{\}$	$\|N,\neg M,\neg N\|$, DPLL6
"Satisfiable"	$\{\}$, DPLL2

Figure 7.3.1.: A DPLL proof procedure.

Exercises

Exercise 7.3.2.1. Show that the following sets of clauses are unsatisfiable using the DPLL proof procedure (Algorithm 7.4):

1. $\{\|P,Q,\neg R\|,\|P,Q,R\|,\|P,\neg Q\|,\|\neg P\|\}$

2. $\{\|R,S\|,\|\neg R,S\|,\|R,\neg S\|,\|\neg R,\neg S\|\}$

3. $\{\|P,\neg Q,R\|,\|Q,R\|,\|\neg P,R\|,\|Q,\neg R\|,\|\neg Q\|\}$

4. $\{\|P,\neg Q\|,\|P,R\|,\|\neg Q,R\|,\|\neg P,Q\|,\|Q,\neg R\|,\|\neg P,\neg R\|\}$

Exercise 7.3.2.2. Search into the original DPLL algorithm. (Hint: Use the original literature given above.)

---

branching rules (see, e.g., Ouyang, 1998).

1. Give a formulation of the algorithm.

2. Compare it with the variation given in Algorithm 7.4.

7.3.3. SAT and unsatisfiability II: Herbrand's theorem and refutation

As anticipated in Section 6.2 above, Herbrand semantics is of particular interest when testing for the SAT. Recall that the SAT is a decision problem defined in the context of classical computation. As our focus is on refutation procedures, we shall be more concerned with unsatisfiability. This sanctions our application of Herbrand semantics, as it is unsatisfiability-preserving for formulae in SNF.

In effect, by Proposition 3.4.34, Theorem 6.2.1 assures us of finite satisfiability[7] for formulae of $\mathsf{L1}_{ff}$. Although this holds only for this function-free fragment of $\mathsf{L1}$, on which for instance deductive databases are founded (see Section 9.3 below), it is an important result in terms of satisfiability, as infinite satisfiability is of little practical use, if not altogether undesirable. Say we consider models as states of some database; clearly, infinite states simply cannot be stored, let alone manipulated. The importance of unsatisfiability for automated theorem proving resides precisely in this: if a given ground formula ϕ (a set of ground formulae X) is *not* unsatisfiable, then it can be infinitely satisfiable; but if it is unsatisfiable, then we can construct a finite proof–for instance, a closed semantic tree–of its unsatisfiability. This provided we choose to work with Herbrand semantics.

But the employment of Herbrand semantics has further advantages, as we discuss below.

We begin with a very general result on unsatisfiability-preservation, introducing then Herbrand's theorem and briefly discussing its impact in the context of the SAT. This done, we elaborate on versions of this theorem that gives us a semantical foundation for proof-theoretical based unsatisfiability testing.

Theorem 7.3.2. *Let C_ϕ be a set of clauses that represents a SNF of a FOL formula ϕ. Then ϕ is unsatisfiable iff C_ϕ is unsatisfiable.*

Proof: Left as an exercise.

This theorem requires that we specify when C is unsatisfiable. This was done by J. Herbrand (1930) in a theorem that can be rephrased in

---

[7] Cf. Def. 3.5.6.

two equivalent versions for practical purposes. These two versions are actually equivalent formulations of Herbrand's original theorem, which, in an already simplified formulation, runs as follows:

> Let Θ be a theory axiomatized by exclusively universal formulae. Suppose that $\Theta \models \forall x \exists y_1, ..., y_k \, (P\,(x,y))$ where $P\,(x,y)$ is a quantifier-free formula. Then, there is a finite sequence $t_{ij} = t_{ij}\,(x)$ of terms, $1 \leq i \leq r$ and $1 \leq j \leq k$, such that

$$\Theta \vdash \forall x \left(\bigvee_{i=1}^{r} P\,(x, t_{i1}, ..., t_{ik}) \right).$$

In other words, a closed FOL formula ϕ is satisfiable iff all its sets of ground clauses are truth-functionally satisfiable. Reformulated in terms of unsatisfiability, this theorem states that a closed formula ϕ in SNF is unsatisfiable iff there is a finite number of clause instances of ϕ whose conjunction is unsatisfiable in terms of truth-functionality, i.e. propositionally.

In Section 6.2, we treated this "propositionalization" rather informally, but now, in light of Herbrand's theorem, we can give it a more formal treatment.

Definition. 7.3.8. Let C be a set of skolemized FO formulae. We define the *Herbrand expansion* of C as the set:

$$HE_C = \{\phi\,[x_1/t_1, ..., x_n/t_n] \,|\forall x_1, ..., \forall x_n\,(\phi) \in C, t_i \in H_C\}$$

Clearly, HE_C contains solely ground atoms, and thus C_ϕ is amenable to a propositional interpretation via a transformation $\tau\,(C_\Phi)$ of C_ϕ (cf. Def. 3.5.8). Indeed, $HE\,(C)$ *just is* the ground extension $GE\,(C_\phi)$ of Section 3.5.2. Formally, this means that we have now a H-interpretation $HI_C = \tau\,(C_\Phi)$. This guarantees, by Lemma 3.5.2, that C is finitely satisfiable iff $\tau\,(C_\Phi)$ is finitely satisfiable.

But, as we know, HE_C only guarantees finite satisfiability in case of a function-free fragment of L1, as HE_C is finite iff $Fun\,(C) = \emptyset$. What about unsatisfiability? It so happens that if C is unsatisfiable, we can always find a *finite* subset of HE_C that is unsatisfiable, i.e. Herbrand semantics is propositional-compact with respect to unsatisfiability.

With this in mind, we can actually improve on the compactness results of Section 3.5.2 as follows:

Theorem 7.3.3. *(Compactness of propositional logic) A set X of propositional formulae is unsatisfiable iff at least one finite subset X' of X is unsatisfiable.*

Proof: Left as an exercise.

To the obvious question of how to find such a counter-model we have an easy answer: by the application of the rules of inference of a refutation calculus such as resolution or analytic tableaux (see Chapter 8).

Summing up, two fundamental results are obtained from Herbrand's theorem: firstly, and as seen, it gives us a means to treat FOL formulae as propositional formulae, the advantage being that there are decision procedures for the latter; secondly, but equally importantly, it provides us with a proof-theoretical means to test for unsatisfiability, the advantage being that proofs are by definition finite objects.

We now elaborate on the above by considering the two aforementioned equivalent versions of Herbrand's theorem. We chose to present first version 2 of the theorem; we prove it after stating and proving version 1 thereof.

Theorem 7.3.4. *(Herbrand, 1930 - version 2) A set C of clauses is unsatisfiable iff there is a finite unsatisfiable set C' of ground instances of C.*

Proof: Given below.

Although not problem-free, Herbrand's results are fundamental in more than one way. Firstly, this theorem tells us that in order to verify whether C is unsatisfiable we need only focus on the H-interpretations, as a set of clauses C is unsatisfiable iff C is false under all H-interpretations, i.e. iff C has no Herbrand model HM_C. This, in turn, means that we need only consider the Herbrand universe H_C (vs. all possible domains).

We can now provide the following central result:

Theorem 7.3.5. *A set C of clauses is unsatisfiable iff C is false under all H-interpretations.*

Proof: Obvious from the above. **QED**

Given n elements in $H(C)$, there will generally be 2^n H-interpretations. This, however, is where Herbrand's results might be problematic, as there are infinitely many H-interpretations for $n = \infty$. It so happens that a core requirement for a decision procedure to be an algorithm is that the number of steps that constitute it be finite (cf. Section 1.4). In order to organize all H-interpretations in a systematic way we can apply the notion of a semantic tree.

Definition. 7.3.9. A *semantic tree* $\mathcal{T}$ for a set C of clauses, denoted by $\mathcal{T}_C$, is a (downward growing) *labeled binary tree* in which each link is attached with a finite set of (negations of) atoms from $H(C)$ in such a way that:

1. For each node N there are only finitely many immediate links $L_1, ..., L_n$ from N. For $\mathcal{E}_i$ the conjunction of all the literals in the set attached to $L_i, i = 1, ..., n$, $\mathcal{E}_1 \vee \mathcal{E}_2 \vee ... \vee \mathcal{E}_n$ is a valid formula.

2. For each node N, let $I(N)$ be the union of all the sets attached to the links of the branch of $\mathcal{T}_C$ down to and including N. Then $I(N)$ does not contain any complementary pair.

Definition. 7.3.10. Let $H(C) = \{A_1, A_2, ..., A_k, ...\}$. A semantic tree $\mathcal{T}_C$ is *complete* iff, for every tip node (i.e. leaf) N, $I(N)$ contains either A_i or $\neg A_i$ for $i = 1, 2, ...$ N is a *failure node* if $I(N)$ falsifies some ground instance of some $\mathcal{C}$ in C, but $I(N')$ does not falsify any ground instance of some $\mathcal{C}$ in C for every ancestor node N' of N. A branch of a semantic tree $\mathcal{T}_C$ *is closed* iff it terminates at a failure node; otherwise, it is said to be *open*. A semantic tree $\mathcal{T}_C$ is closed iff each of its branches is closed.

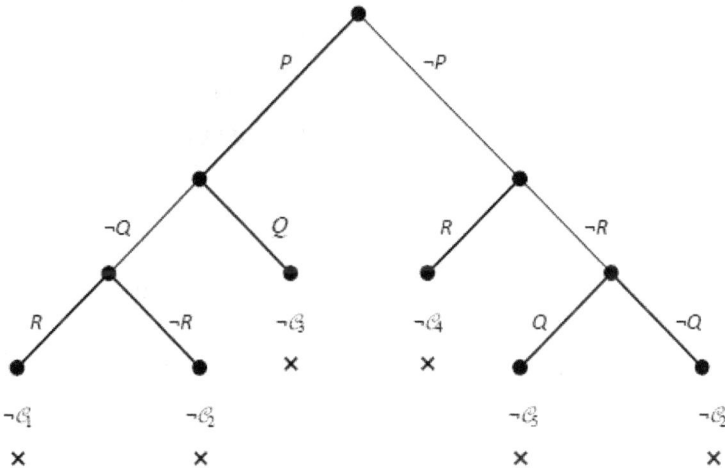

Figure 7.3.2.: Closed semantic tree of $C = \{\mathcal{C}_1, \mathcal{C}_2, \mathcal{C}_3, \mathcal{C}_4, \mathcal{C}_5\}$ in Example 7.3.3.

If $I(N)$, which is in fact a partial interpretation for C (i.e., $I(N)$ can be seen as an assignment of truth values to ground atoms of $H(C)$), falsifies C, then we can stop expanding nodes from N. This means that if C is unsatisfiable, its semantic tree $\mathcal{T}_C$ cannot fail to be finite.

Example 7.3.3. Let $C_1 = Q \vee \neg R$, $C_2 = Q \vee R$, $C_3 = \neg P \vee \neg Q$, $C_4 = P \vee \neg R$, and $C_5 = P \vee \neg Q \vee R$. The atom set of $C = \{C_1, C_2, C_3, C_4, C_5\}$ is $A(C) = \{P, Q, R\}$. The above conditions 7.3.9.1-2 are satisfied for C. Figure 7.3.2 shows the closed semantic tree of C. A closed branch is marked with $\times$.

Example 7.3.4. Let $C = \{\neg P(x) \vee Q(x), P(f(a)), \neg Q(z)\}$. Then,
$$H_C = \{a, f(a), f(f((a))), ...\}$$
$$H(C) = \{P(a), Q(a), P(f(a)), Q(f(a)), ...\}.$$
C is unsatisfiable. Figure 7.3.3 shows the semantic tree for C. A failure node is marked with $\times$.

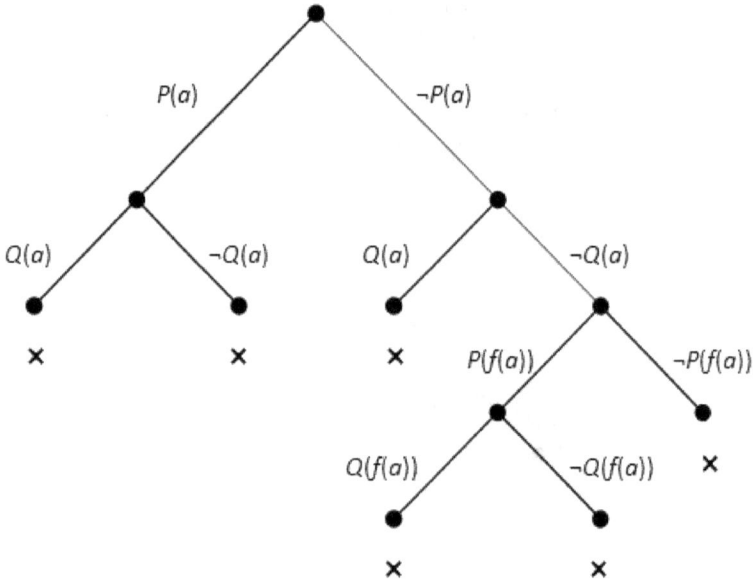

Figure 7.3.3.: A closed semantic tree.

Theorem 7.3.6. *(Herbrand, 1930 - version 1) A set C of clauses is unsatisfiable iff corresponding to every complete semantic tree of C, there is a finite closed semantic tree.*

Proof: The proof follows immediately from the above:

($\Rightarrow$) Assume C is unsatisfiable and $\mathcal{T}_C$ a complete semantic tree for C. For every branch there is the set of labels $I(N)$, which is an interpretation, because the tree is complete. Hence, $I(N)$ falsifies some ground instance $\mathcal{C}'$ of a clause $\mathcal{C} \in C$. Since there are only finitely many literals in $\mathcal{C}'$, there must be a failure node N in a finite distance from the root. Since every branch terminates at a failure node, there is a corresponding closed semantic tree $\mathcal{T}'_C$ that is finite.

($\Leftarrow$) Assume that for every complete semantic tree $\mathcal{T}_C$ there is a corresponding finite closed tree $\mathcal{T}'_C$. Then, every branch terminates at a failure node and hence every interpretation $I(N)$ falsifies C. Hence, C is unsatisfiable. **QED**

The reason why we chose last to introduce version 1 of Herbrand's theorem is that it is more directly connected to *resolution*, while version 2 has more to do with *refutation*. Nevertheless, the two versions make it clear that resolution is a refutation procedure, and a proof of version 1 can be greatly simplified if we already can apply the important definition of semantic tree:

Proof: (Herbrand's theorem, version 2). ($\Rightarrow$) Assume C is unsatisfiable and $\mathcal{T}_C$ is a complete semantic tree for C. Then, by Herbrand's theorem, version 1, there is a finite closed semantic tree $\mathcal{T}'_C$ of C. Let C' be the set of ground instances of clauses that are falsified at all failure nodes of $\mathcal{T}'_C$: C' is finite and is falsified by every interpretation. It follows that C' is unsatisfiable.

($\Leftarrow$) The proof is by contraposition. **QED**

Remark. **7.3.11.** Version 2 of Herbrand's theorem can be turned into a proof procedure by successively generating the sets $C'_0, C'_1, \dots$ in which C'_i is the set of all the ground instances of clauses of C, and by testing them for unsatisfiability by means available to the propositional calculus: the theorem tells us that for some finite N there is a set C'_N that is unsatisfiable if C is unsatisfiable.

Gilmore (1960) was the first to implement this idea with the *multiplication method*: as each C'_i is generated, it is multiplied out into a DNF; any conjunction in the DNF containing a complementary pair is then removed, and if some C'_i is found to be empty, a proof for its unsatisfiability has been found. However, as said above, this method is highly inefficient, as for, say, a set of ten two-literal ground clauses, there are 2^{10} conjunctions.

Exercises

Exercise 7.3.3.1. Show that finite satisfiability is semi-decidable for FOL formulae.

Exercise 7.3.3.2. Show that the set of unsatisfiable FO formulae is recursively enumerable under Herbrand semantics.

Exercise 7.3.3.3. Show that the sets of clauses of Exercise 7.3.2.1 are unsatisfiable by constructing their semantic trees.

Exercise 7.3.3.4. Prove (Complete the proof of) the theorems above left as exercises (left with a sketchy proof, respectively).

---

8. Automated theorem proving

Together with logic programming, automated theorem proving (ATP) is a major field of application of computational logic, with top applications in software verification and knowledge representation. Born in the late 1950s out of the desire to automate the ways humans produce mathematical proofs, ATP soon became an indispensable component of computational logic.[1] Although mathematical proof remains a central concern, nowadays automated deduction has many more applications, namely in technology and industry. Indeed, this field has extended itself to automating deduction with non-classical logics, an extension that was motivated by new key technologies for which classical logic is not appropriate.[2] Nevertheless, automated deduction with classical logic remains a fruitful area of research, as many technological applications still call for classical deductive computation with CFOL.

In this Chapter, we elaborate on two classical refutation calculi that are particularly well suited to automation, one–resolution–fully automated by now, and the other–analytic tableaux–still in progress in terms of automation. Although this elaboration gives the essentials of producing proofs "by hand," this is mostly for pedagogical reasons; reliable proofs, especially in the case of complex theories, can only be guaranteed by means of automated provers.

With respect to the resolution calculus, we implement it with Prover9-Mace4, which implements *binary resolution* and *unit deletion*, a first-order generalization of the Davis-Putnam one-literal rule (it also implements proof procedures for equational logic, i.e., *paramodulation*). This is a two-component prover, with Prover9 producing a proof by refutation and Mace4 generating a counter-model on the same input; one can use only one of these, or both, especially when Prover9 appears not to stop on a given input. Prover9-Mace4 is a free software available at https://www.cs.unm.edu/~mccune/prover9/download/.

As for the analytic tableau calculus, though some proof *assistants* are available, which require interaction between the human computer

---

[1]See Davis (2001) for a detailed early history of ATP. See MacKenzie (1995) for a broader survey, both in time and in subject.

[2]See Augusto (2020a) for automated deduction with many-valued logics, as well as for bibliography on other logics.

and the software, there is as yet no full-blown *prover* for this calculus for CFOL. Nevertheless, there are already a few web-based interactive provers that can be used for low-complexity FO theories (e.g., Tableaux Package, available at http://hackage.haskell.org/package/tableaux, and Tree Proof Generator, available at https://www.umsu.de/logik/trees/).

Fitting (1996) also concentrates on resolution and analytic tableaux from the perspective of ATP and the reader looking for further contents on these calculi can benefit from this text.

8.1. Resolution

In our elaboration on the resolution calculus, we follow closely Chang & Lee (1973) and Leitsch (1997), borrowing many examples from these sources. The reader may benefit from further sources such as, for example, Bachmair & Ganzinger (2001). We introduce this Section by giving the general algorithm for binary resolution (Algorithm 8.1); the specifications for the specific sub-calculi will be given in the corresponding subsections.

8.1.1. The resolution principle for propositional logic

The resolution principle is an extension of the one-literal rule first devised by Davis & Putnam (1960). Hence, a brief discussion of this inference rule is called for.

Definition. 8.1.1. The *one-literal rule* is an inference rule stating that if there is a unit ground clause $\mathcal{C} = \|L\|$ in a set of clauses C, we can obtain C' from C by deleting all the ground clauses in C that contain L. There are then two possible cases:

1. If C' is empty, then C is satisfiable.

2. If C' is not empty, we obtain a set C'' from C' by deleting $\neg L$ from C'. C'' is unsatisfiable iff C is.

Example 8.1.1. Let $C = \{\neg P \vee Q \vee \neg R, \neg P \vee \neg Q, P, R, U\}$. We have the set $C' = \{\neg P \vee Q \vee \neg R, \neg P \vee \neg Q, R, U\}$ by applying the one-literal rule for P, and then $C'' = \{Q \vee \neg R, \neg Q, R, U\}$ by Def. 8.1.1.2. We have $C \equiv_{sat} C''$. We repeat the rule for R in C'' and obtain the

---

Algorithm 8.1 Binary resolution

Input: A set of clauses C

Output: An inverted binary tree $\mathcal{T}_C$ for C that constitutes a proof of validity of C

---

1. Input the (factorized) clauses $\mathcal{C}_1, ..., \mathcal{C}_n \in C$, for finite $n \geq 2$, as the *leaves* of an inverted binary tree.

2. Resolve a pair of clauses $\mathcal{C}_i, \mathcal{C}_j$ for $1 \leq i \leq n, 1 \leq j \leq n$, and $i \neq j$, for some l, m such that $L_l \in \mathcal{C}_i$ and $L_m \in \mathcal{C}_j$ are complementary literals.

3. The derived clause $\mathcal{C}_{n+1}$–called resolvent–is a *node* in the tree and is added to the search space of Step 2, so that now we have $\mathcal{C}_1, ..., \mathcal{C}_{n+1}$.

4. Repeat Steps 2 and 3 until (i) the empty clause is derived or (ii) there are no more pairs of clauses to be resolved, in which cases the *root* of the tree is respectively (i) $\square$ or (ii) some non-empty resolvent $\mathcal{C}_{n+r}$ for some r.

   a) If (i), then C is unsatisfiable.

   b) Otherwise, C is satisfiable.

---

set $C''' = \{Q, \neg Q, U\}$. We have $C \equiv_{sat} C'''$. As C''' contains the empty clause $(Q \wedge \neg Q = \Box$; cf. Def. 4.3.3), C is unsatisfiable because C''' is unsatisfiable.

By extending the one-literal rule to *any* pair of clauses, we obtain the *resolution principle*:

Definition. 8.1.2. For any two clauses C_1 and C_2 and two complementary literals $L_1 \in C_1$ and $L_2 \in C_2$, we delete L_1 and L_2 from C_1 and C_2, respectively, and construct the disjunction of the remaining clauses, i.e., $C_1' \vee C_2'$.

Definition. 8.1.3. The constructed clause, $C_1' \vee C_2'$, is called a *resolvent* of C_1 and C_2.

Theorem 8.1.1. *A resolvent* $C = C_1' \vee C_2'$ *of two clauses* $C_1 = C_1' \vee L$ *and* $C_2 = C_2' \vee \neg L$ *is a logical consequence of* $C_1 \wedge C_2$*, i.e.:*

$$\frac{C_1' \vee L \quad C_2' \vee \neg L}{C_1' \vee C_2'}$$

Proof: Let $C_1 = L \vee C_1'$, $C_2 = \neg L \vee C_2'$, and $C = C_1' \vee C_2'$, C_1' and C_2' are disjunctions of literals. Supposing that C_1 and C_2 are both true in an interpretation $\mathcal{I}$, their resolvent C must also be true in $\mathcal{I}$. Obviously, either L or $\neg L$ is false in $\mathcal{I}$. Assume L is false in $\mathcal{I}$; then C_1 must not be a unit clause, otherwise it would be false in $\mathcal{I}$. Hence, C_1' must be true in $\mathcal{I}$, and the resolvent $C_1' \vee C_2'$ is true in $\mathcal{I}$. Assume $\neg L$ is false in $\mathcal{I}$ and proceed in the same way. Hence, $C_1' \vee C_2'$ is true in $\mathcal{I}$. **QED**

Recall the definition of contradiction (Def. 3.4.30.2). Recall also that the empty clause is always false (Prop. 7.2.3), thus equating with a contradiction (cf. $\Box_{df}$). Given this, we have from Theorem 3.4.10 that a set of clauses is unsatisfiable iff we can derive from it the empty clause.

Definition. 8.1.4. A *resolution deduction* of C from a set of clauses C, denoted by $C \vdash_{res} C$, is a finite sequence $C_1, C_2, ..., C_k$ of clauses such that each C_i is either a clause in C or a resolvent of clauses preceding C_i, and $C_k = C$.

1. We call the deduction of the empty set $\Box$ from C a *refutation*, or *proof* of C.[3]

---

[3]Compare with Defs. 3.4.16-7

2. A *deduction*, or *refutation*, *tree* is a labeled tree whose leaves, placed at the top, are the clauses in C, the nodes are the resolvents C_j for $j < k$, and the root is either $C_k = \square$ or some non-empty clause C_k. In the latter case, we speak of a *refutation-failure tree*.

Proposition. 8.1.5. *We can obtain a resolution deduction by means of a* deduction, *or* refutation, *tree.*

Proof: Left as an exercise. (Hint: Example 8.1.2.)

Example 8.1.2. Let $C = \{P \vee Q, \neg P \vee Q, P \vee \neg Q, \neg P \vee \neg Q\}$. Figure 8.1.1 shows the refutation tree for C.

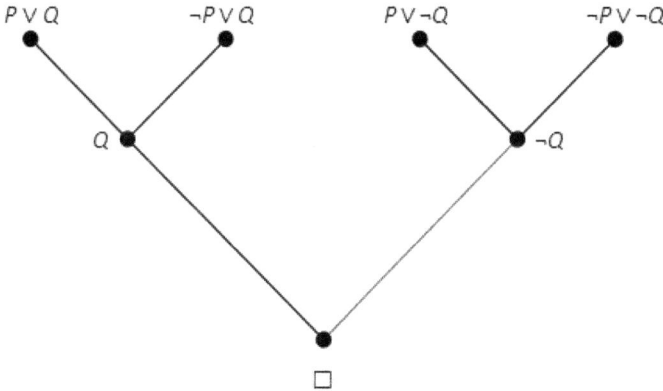

Figure 8.1.1.: A refutation tree.

Example 8.1.3. Let there be given the formalized propositional argument of Example 5.2.2. We implement resolution in the prover Prover9-Mace4.[4] The input is entered as follows: on the Prover9 "Formulas" window, write the premises in the "Assumptions" section, and the conclusion in the "Goals" section. Alternatively, because Prover9 is a refutation-based prover, simply enter the negated conclusion as another premise in the "Assumptions" section. Note in Figure 8.1.2 the notation used in Prover9-Mace4, and the fact that every formula must end with a full stop. In Figure 8.1.3, we give the output by Prover9. An analysis of the output of Prover9, which is easily readable, shows the automatic implementation of the resolution principle: the prover deduces the empty clause (denoted by $F) from the set of propositions

---

[4]See above for availability of this free software.

(assumptions and conclusion) that constitute the theory and which it clausifies.

Assumptions:
_____
```
(P|R)->-(Q&T).
-S->Q.
P&T.
```

Goals:
_____
```
S.
```

Figure 8.1.2.: A propositional argument as input in Prover9-Mace4.

Example 8.1.4. We now wish to test for the validity of a single formula. We can either enter the formula in the "Goals" section, or negate it in the "Assumptions" section of Prover9-Mace4. Figure 8.1.4 shows the proof by Prover 9 of the formula obtained from the argument of Example 8.1.3 by one application of the deduction theorem, i.e.

$$[((P \vee R) \to \neg (Q \wedge T)) \wedge (\neg S \to Q) \wedge (P \wedge T)] \to S.$$

Example 8.1.5. If we input the argument in Example 3.2.2 in Prover9-Mace4, Prover9 does not stop. However, we obtain the counter-model from Mace4 in Figure 8.1.5.

Exercises

Exercise 8.1.1.1. Prove the soundness of the propositional resolution calculus, i.e.

$$\text{if } \vdash_{res} \phi, \text{ then } \models \phi$$

for a formula $\phi \in \text{L0}$.

Exercise 8.1.1.2. Prove the completeness of the propositional resolution calculus, i.e.

$$\text{if } \models \phi, \text{ then } \vdash_{res} \phi$$

for a formula $\phi \in \text{L0}$.

```
====================== PROOF =========================
% -------- Comments from original proof --------
% Proof 1 at 0.06 (+ 0.05) seconds.
% Length of proof is 11.
% Level of proof is 3.
% Maximum clause weight is 3.
% Given clauses 0.

1 P | R -> -(Q & T) # label(non_clause).  [assumption].
2 -S -> Q # label(non_clause).  [assumption].
3 P & T # label(non_clause).  [assumption].
4 S # label(non_clause) # label(goal).  [goal].
5 -P | -Q | -T. [clausify(1)].
7 S | Q. [clausify(2)].
8 P. [clausify(3)].
9 T. [clausify(3)].
10 -S. [deny(4)].
11A -Q | -T. [resolve(8,a,5,a)].
11 -Q. [resolve(9,a,11A,b)].
12A Q. [resolve(10,a,7,a)].
12 $F. [resolve(11,a,12A,a)].
====================== end of proof ==================
```

Figure 8.1.3.: Output by Prover9: A valid propositional argument.

8. Automated theorem proving

```
======================= PROOF ==========================
% -------- Comments from original proof --------
% Proof 1 at 0.01 (+ 0.05) seconds.
% Length of proof is 8.
% Level of proof is 3.
% Maximum clause weight is 3.
% Given clauses 0.

1 (P | R -> -(Q & T)) & (-S -> Q) & P & T -> S
    # label(non_clause) # label(goal).   [goal].
2 -P | -Q | -T. [deny(1)].
4 S | Q. [deny(1)].
5 P. [deny(1)].
6 T. [deny(1)].
7 -S. [deny(1)].
8A -Q | -T. [resolve(5,a,2,a)].
8 -Q. [resolve(6,a,8A,b)].
9A Q. [resolve(7,a,4,a)].
9 $F. [resolve(8,a,9A,a)].
===================== end of proof ==================
```

Figure 8.1.4.: Output by Prover 9: A valid formula.

378

```
% number = 1 % seconds = 0
% Interpretation of size 2

A : 0
B : 1
C : 1
D : 0
E : 0
F : 0
K : 0
P : 0
R : 1
```

Figure 8.1.5.: Output by Mace4: A counter-model.

Exercise 8.1.1.3. Is the propositional resolution calculus strongly complete? Account for your answer.

Exercise 8.1.1.4. Test the validity of the following arguments by applying resolution. First prove "by hand" and then in Prover9-Mace4. Given the first output by Prover9, click the button "Reformat" on the top left corner of the proof window; next, in "Options" select "Expand proof." Compare both proofs, yours and the expanded proof by Prover9.

1. $\{Q \to R, R \to (P \wedge Q), P \to (Q \vee R)\} \vdash P \leftrightarrow Q$.

2. $\{(\neg S \wedge V) \to \neg P, P, V\} \vdash S$.

3. $\{\neg R \leftrightarrow (\neg P \vee \neg Q), P \wedge Q\} \vdash R$.

4. $(P \vee Q) \wedge (P \vee R) \vdash \neg P \to (Q \wedge R)$.

5. $\{(P \vee Q) \wedge (P \vee R), P \to S, Q \to S, P \to T, R \to T\} \vdash S \wedge T$.

8.1.2. The resolution principle for FOL

We now consider the resolution principle for FOL. In order to do so we require the definitions of substitution and unification for FOL (cf. Section 4.1.2), as well as a few new definitions.

Definition. 8.1.6. A *factor* of a clause C is a clause $C\sigma$, where σ is a MGU of some $C' \subseteq C$. If $C\sigma$ is a unit clause, then it is called a *unit factor* of C.

Definition. 8.1.7. Let C_1 and C_2 be two clauses (called *parent clauses*) with no variables in common. Let L_1 and L_2 be two complementary literals in C_1 and C_2, respectively. Then the clause

$$(C_1 - L_1)\sigma \cup (C_2 - L_2)\sigma$$

where σ is a MGU of L_1 and L_2, is called a *binary resolvent* of C_1 and C_2, and the literals L_1 and L_2 are the *literals resolved upon*.

Definition. 8.1.8. A *resolvent* of (parent) clauses C_1 and C_2 is one of the following binary resolvents:

1. a binary resolvent of C_1 and C_2;

2. a binary resolvent of C_1 and a factor of C_2;

3. a binary resolvent of a factor of C_1 and C_2;

4. a binary resolvent of a factor of C_1 and a factor of C_2.

Theorem 8.1.2. *(Binary resolution). A resolvent $C = C_1' \vee C_2'$ of two (parent) clauses $C_1 = C_1' \vee L$ and $C_2 = C_2' \vee \neg L$ of first-order logic is a logical consequence of C_1 and C_2, if there is a substitution σ such that σ unifies the pair of complementary literals L and $\neg L$, i.e.:*

$$\frac{C_1' \vee L \quad C_2' \vee \neg L}{(C_1' \vee C_2')\sigma}$$

Proof: The proof is left as an exercise.

Example 8.1.6. Let $C_1 = P(x) \vee Q(x)$ and $C_2 = \neg P(z) \vee R(x) \vee \neg P(a)$. In order to apply binary resolution to this pair of clauses we first must rename the variable x in C_2; renaming x as y will do. We next factor $\neg P(z)$ and $\neg P(a)$ in C_2, namely by applying the substitution set $\sigma = \{z \mapsto a\}$. We now have $C_1 = P(x) \vee Q(x)$ and $C_2 = \neg P(a) \vee R(y)$. Given the MGU $\theta = \{x \mapsto a\}$, we then have the binary resolvent $Q(a) \vee R(y)$. Obviously, the set $C = (C_1, C_2)$ is satisfiable. Figure 8.1.6 shows the refutation failure in a deduction tree.

$$\lambda = \{x \mapsto y\}$$

$$\theta = \{x \mapsto a\}$$

$$\sigma = \{z \mapsto a\}$$

$$\mathscr{C_1}\theta = P(a) \vee Q(a)$$

$$(\mathscr{C_2}\lambda)\sigma = \neg P(a) \vee R(y)$$

$$Q(a) \vee R(y)$$

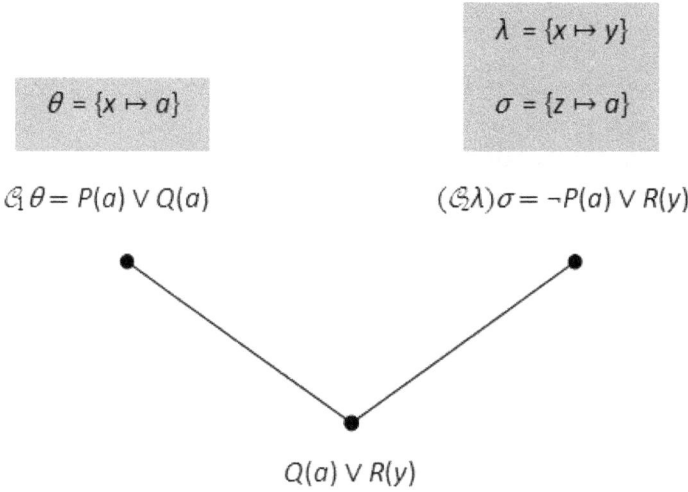

Figure 8.1.6.: A refutation-failure tree.

Example 8.1.7. We show how Prover9-Mace4 implements resolution in theorem proving in FOL by beginning with a basic geometric property (cf. Example 3.2.3). We introduce the input in Prover9-Mace4 in Figure 8.1.7. The proof is given in Figure 8.1.8.

Example 8.1.8. Let the following theory be given:

1. $\forall x \, (F\,(x) \rightarrow \exists y \, (G\,(y) \wedge H\,(x,y)) \wedge \exists y \, (G\,(y) \wedge \neg H\,(x,y)))$
2. $\exists x \, (J\,(x) \wedge \forall y \, (G\,(y) \rightarrow H\,(x,y)))$
3. $\exists x \, (J\,(x) \wedge \neg F\,(x))$

We want to know whether the conclusion is valid. The repetition of atoms in premise 1 makes this a somehow complex theory to prove "by hand," reason why Prover9-Mace4 comes in handy. In merely 14 steps, and by applying binary resolution alone, we have a proof of validity (Fig. 8.1.9).

Example 8.1.9. In 1978, L. Schubert proposed a challenge to automated theorem proving that would become known in the literature as *Schubert's steamroller*. This is indeed a complex theory, namely in combinatorial terms. Despite this complexity, the challenge was solved by a resolution calculus in Walther (1985). We provide the text in English (Fig. 8.1.10) and our translation into FOL (Fig. 8.1.11).

Figure 8.1.12 shows the output by Prover9. This proof is particularly interesting as, despite the complexity of the theory, it is massively carried

---

Assumptions:

% *Definition of a trapezoid*
```
all x all y all u all v (T(x,y,u,v)->P(x,y,u,v)).
```
% *Alternate interior angles of parallel lines are*
equal
```
all x all y all u all v (P(x,y,u,v)->E(x,y,v,u,v,y)).
```
% *Trapezoid in consideration*
```
T(a,b,c,d).
```

Goals:

% *Alternate interior angles formed by a diagonal of a*
trapezoid are equal
```
E(a,b,d,c,d,b).
```

Figure 8.1.7.: Input in Prover9-Mace4: A FO theory.

out by means of resolution, with only a few applications of *unit deletion*, a FO generalization of the Davis-Putnam one-literal rule, and of *unit-resulting resolution*, which, just like hyper-resolution (see below), allows for the resolution in a single step of more than one nucleus clause with other satellite clauses.

Exercises

Exercise 8.1.2.1. Determine whether the following sets of clauses are satisfiable or unsatisfiable by means of the resolution calculus:

1.

$$
C = \left\{
\begin{array}{c}
\|\neg P\left(y\right), Q\left(z, h\left(z\right)\right)\|, \\
\|S\left(z, f\left(x\right), x\right), \neg P\left(x\right)\|, \\
\|P\left(a\right)\|, \\
\|\neg R\left(x, h\left(y\right)\right), \neg S\left(g\left(z\right), z, a\right)\|, \\
\|\neg T\left(g\left(x\right)\right), \neg R\left(x, h\left(x\right)\right)\|, \\
\|R\left(x, y\right), \neg Q\left(z, y\right)\|
\end{array}
\right\}
$$

```
===================== PROOF =========================
% -------- Comments from original proof --------
% Proof 1 at 0.00 (+ 0.06) seconds.
% Length of proof is 10.
% Level of proof is 4.
% Maximum clause weight is 0.
% Given clauses 0.


1 (all x all y all z all u (T(x,y,z,u) -> P(x,y,z,u)))
    # label(non_clause).  [assumption].
2 (all x all y all z all u (P(x,y,z,u) ->
E(x,y,u,z,u,y)))
    # label(non_clause).  [assumption].
3 E(a,b,d,c,d,b) # label(non_clause) # label(goal).
[goal].
4 T(a,b,c,d).  [assumption].
5 -T(x,y,z,u) | P(x,y,z,u).  [clausify(1)].
6 P(a,b,c,d).  [resolve(4,a,5,a)].
7 -P(x,y,z,u) | E(x,y,u,z,u,y).  [clausify(2)].
8 E(a,b,d,c,d,b).  [resolve(6,a,7,a)].
9 -E(a,b,d,c,d,b).  [deny(3)].
10 $F.  [resolve(8,a,9,a)].
===================== end of proof ==================
```

Figure 8.1.8.: Output by Prover9.

```
============================ PROOF ============================
%
% -------- Comments from original proof --------
% Proof 1 at 0.00 (+ 0.06) seconds.
% Length of proof is 14.
% Level of proof is 5.
% Maximum clause weight is 0.
% Given clauses 0.

1 (all x (F(x) -> (exists y (G(y) & H(x,y))) & (exists y (G(y) & -H(x,y)))))
    # label(non_clause).   [assumption].
2 (exists x (J(x) & (all y (G(y) -> H(x,y)))))  # label(non_clause).   [assumption].
3 (exists x (J(x) & -F(x)))  # label(non_clause) # label(goal).
  [goal].
4 -J(x) | F(x).    [deny(3)].
7 -F(x) | G(f2(x)).    [clausify(1)].
8 -F(x) | -H(x,f2(x)).    [clausify(1)].
10 J(c1).    [clausify(2)].
12 -J(x) | G(f2(x)).    [resolve(4,b,7,a)].
13 -J(x) | -H(x,f2(x)).    [resolve(4,b,8,a)].
15 -G(x) | H(c1,x).    [clausify(2)].
16 G(f2(c1)).    [resolve(12,a,10,a)].
17 -H(c1,f2(c1)).    [resolve(13,a,10,a)].
19 H(c1,f2(c1)).    [resolve(16,a,15,a)].
20 $F.  [resolve(19,a,17,a)].
============================ end of proof ============================
```

Figure 8.1.9.: Output by Prover9.

Wolves, foxes, birds, caterpillars, and snails are animals, and there are some of each of them. Also there are some grains, and grains are plants. Every animal either likes to eat all plants or all animals much smaller than itself that like to eat some plants. Caterpillars and snails are much smaller than birds, which are much smaller than foxes, which are in turn much smaller than wolves. Wolves do not like to eat foxes or grains, while birds like to eat caterpillars but not snails. Caterpillars and snails like to eat some plants. Prove there is an animal that likes to eat a grain-eating animal.

Figure 8.1.10.: Schubert's steamroller in natural language.

2.
$$
C = \left\{ \begin{array}{c}
\|P\left(x\right),\neg Q\left(x,f\left(y\right)\right),\neg R\left(a\right)\|, \\
\|R\left(x\right),\neg Q\left(x,y\right)\|, \\
\|\neg P\left(x\right),\neg Q\left(y,z\right)\|, \\
\|P\left(x\right),\neg R\left(x\right)\|, \\
\|Q\left(x,y\right),\neg P\left(y\right)\|, \\
\|R\left(f\left(b\right)\right)\|
\end{array} \right\}
$$

Exercise 8.1.2.2. Prove the following arguments by applying resolution. First prove "by hand" and then in Prover9-Mace4. Given the first output by Prover9, click the button "Reformat" on the top left corner of the proof window; next, in "Options" select "Expand proof." Compare both proofs, yours and the expanded proof by Prover9.

1. $\forall x\left(P\left(x\right)\rightarrow Q\left(x\right)\right)\vdash\forall x\left(P\left(x\right)\right)\rightarrow\forall x\left(Q\left(x\right)\right)$

2. $\{\forall x\left(F\left(x\right)\right)\rightarrow\forall x\left(G\left(x\right)\right),\neg\forall x\left(G\left(x\right)\right)\}\vdash\neg\forall x\left(F\left(x\right)\right)$

3. $\{\forall x\left(F\left(x\right)\rightarrow G\left(x\right)\right),\forall x\left(Fx\right)\}\vdash G\left(a\right)$

4. $\{\exists x\left(P\rightarrow F\left(x\right)\right),\exists x\left(F\left(x\right)\rightarrow P\right)\}\vdash\exists x\left(P\leftrightarrow F\left(x\right)\right)$

5.
$$
\frac{\begin{array}{c}
Q\left(x\right)\vee R\left(x\right)\\
R\left(x\right)\vee P\left(x\right)\vee\neg Q\left(f\left(x\right)\right)\\
R\left(x\right)\vee P\left(x\right)\vee T\left(x,y\right)\\
\neg R\left(x\right)\\
Q\left(a\right)
\end{array}}{\exists x\forall y\left(P\left(x\right)\vee T\left(x,y\right)\right)}
$$

$$
\begin{array}{ll}
(P_1) & \forall x\,(W(x) \rightarrow A(x)) \wedge \exists x\,(W(x)) \\
(P_2) & \forall x\,(F(x) \rightarrow A(x)) \wedge \exists x\,(F(x)) \\
(P_3) & \forall x\,(B(x) \rightarrow A(x)) \wedge \exists x\,(B(x)) \\
(P_4) & \forall x\,(C(x) \rightarrow A(x)) \wedge \exists x\,(C(x)) \\
(P_5) & \forall x\,(S(x) \rightarrow A(x)) \wedge \exists x\,(S(x)) \\
(P_6) & \exists x\,(G(x)) \wedge \forall x\,(G(x) \rightarrow P(x)) \\
(P_7) & \forall x\,(A(x) \rightarrow (\forall y\,(P(y) \rightarrow Eats(x,y)) \vee \forall y\,((A(y) \wedge Smaller(y,x) \wedge \exists z\,(P(z) \wedge Eats(y,z))) \rightarrow Eats(x,y)))) \\
(P_8) & \forall x \forall y\,(C(x) \wedge B(y) \rightarrow Smaller(x,y)) \\
(P_9) & \forall x \forall y\,(S(x) \wedge B(y) \rightarrow Smaller(x,y)) \\
(P_{10}) & \forall x \forall y\,(B(x) \wedge F(y) \rightarrow Smaller(x,y)) \\
(P_{11}) & \forall x \forall y\,(F(x) \wedge W(y) \rightarrow Smaller(x,y)) \\
(P_{12}) & \forall x \forall y\,(W(x) \wedge (F(y) \vee G(y)) \rightarrow \neg Eats(x,y)) \\
(P_{13}) & \forall x \forall y\,(B(x) \wedge C(y) \rightarrow Eats(x,y)) \\
(P_{14}) & \forall x \forall y\,(B(x) \wedge S(y) \rightarrow \neg Eats(x,y)) \\
(P_{15}) & \forall x\,(C(x) \rightarrow \exists y\,(P(y) \wedge Eats(x,y))) \\
(P_{16}) & \underline{\forall x\,(S(x) \rightarrow \exists y\,(P(y) \wedge Eats(x,y)))} \\
(P_{17}) & \exists x \exists y\,(A(x) \wedge (A(y) \wedge \exists z\,(G(z) \wedge Eats(y,z))) \wedge Eats(x,y))
\end{array}
$$

Figure 8.1.11.: Schubert's steamroller in FOL.

```
% -------- Comments from original proof --------
% Proof 1 at 0.01 (+ 0.06) seconds.
% Length of proof is 65.
% Level of proof is 9.
% Maximum clause weight is 13.
% Given clauses 31.


1 (all x (W(x) -> A(x))) & (exists x W(x)) # label(non_clause).  [assumption].
2 (all x (F(x) -> A(x))) & (exists x F(x)) # label(non_clause).  [assumption].
3 (all x (B(x) -> A(x))) & (exists x B(x)) # label(non_clause).  [assumption].
5 (all x (S(x) -> A(x))) & (exists x S(x)) # label(non_clause).  [assumption].
6 (exists x G(x)) & (all x (G(x) -> P(x))) # label(non_clause).  [assumption].
7 (all x (A(x) -> (all y (P(y) -> Eats(x,y))) | (all y (A(y) & Smaller(y,x) & (exists z (P(z) &
Eats(y,z))) -> Eats(x,y)))))
# label(non_clause).  [assumption].
9 (all x all y (S(x) & B(y) -> Smaller(x,y))) # label(non_clause).  [assumption].
10 (all x all y (B(x) & F(y) -> Smaller(x,y))) # label(non_clause).  [assumption].
11 (all x all y (F(x) & W(y) -> Smaller(x,y))) # label(non_clause).  [assumption].
12 (all x all y (W(x) & (F(y) | G(y)) -> -Eats(x,y))) # label(non_clause).  [assumption].
14 (all x all y (B(x) & S(y) -> -Eats(x,y))) # label(non_clause).  [assumption].
16 (all x (S(x) -> (exists y (P(y) & Eats(x,y))))) # label(non_clause).  [assumption].
17 (exists x exists y (A(x) & A(y) & (exists z (G(z) & Eats(y,z))) & Eats(x,y))) # label(non_clause) #
label(goal).  [goal].
18 W(c1).  [clausify(1)].
19 -W(x) | A(x).  [clausify(1)].
20 -F(x) | -W(y) | Smaller(x,y).  [clausify(11)].
21 -W(x) | -F(y) | -Eats(x,y).  [clausify(12)].
22 -W(x) | -G(y) | -Eats(x,y).  [clausify(12)].
23 F(c2).  [clausify(2)].
24 -F(x) | A(x).  [clausify(2)].
25 -B(x) | -F(y) | Smaller(x,y).  [clausify(10)].
26 -F(x) | Smaller(x,c1).  [resolve(20,b,18,a)].
27 -F(x) | -Eats(c1,x).  [resolve(21,a,18,a)].
28 B(c3).  [clausify(3)].
29 -B(x) | A(x).  [clausify(3)].
31 -S(x) | -B(y) | Smaller(x,y).  [clausify(9)].
33 -B(x) | -S(y) | -Eats(x,y).  [clausify(14)].
34 -B(x) | Smaller(x,c2).  [resolve(25,b,23,a)].
41 S(c5).  [clausify(5)].
42 -S(x) | A(x).  [clausify(5)].
43 -S(x) | P(f2(x)).  [clausify(16)].
44 -S(x) | Eats(x,f2(x)).  [clausify(16)].
45 -S(x) | Smaller(x,c3).  [resolve(31,b,28,a)].
46 -S(x) | -Eats(c3,x).  [resolve(33,a,28,a)].
47 -G(x) | P(x).  [clausify(6)].
48 G(c6).  [clausify(6)].
49 -A(x) | -A(y) | -G(z) | -Eats(y,z) | -Eats(x,y).  [deny(17)].
50 -G(x) | -Eats(c1,x).  [resolve(22,a,18,a)].
51 Smaller(c2,c1).  [resolve(26,a,23,a)].
52 -A(x) | -P(y) | Eats(x,y) | -A(z) | -Smaller(z,x) |-P(u) | -Eats(z,u) | Eats(x,z).  [clausify(7)].
53 Smaller(c3,c2).  [resolve(34,a,28,a)].
54 Smaller(c5,c3).  [resolve(45,a,41,a)].
55 A(c1).  [resolve(18,a,19,a)].
56 A(c2).  [resolve(23,a,24,a)].
57 -Eats(c1,c2).  [resolve(27,a,23,a)].
58 A(c3).  [resolve(28,a,29,a)].
63 A(c5).  [resolve(41,a,42,a)].
64 P(f2(c5)).  [resolve(43,a,41,a)].
65 Eats(c5,f2(c5)).  [resolve(44,a,41,a)].
66 -Eats(c3,c5).  [resolve(46,a,41,a)].
67 P(c6).  [resolve(47,a,48,a)].
68 -A(x) | -A(y) | -Eats(y,c6) | -Eats(x,y).  [resolve(49,c,48,a)].
69 -Eats(c1,c6).  [resolve(50,a,48,a)].
70 -A(c1) | -P(x) | Eats(c1,x) | -A(c2) | -P(y) | -Eats(c2,y) | Eats(c1,c2).  [resolve(51,a,52,e)].
71 -P(x) | Eats(c1,x) | -P(y) | -Eats(c2,y).[copy(70),unit_del(a,55),unit_del(d,56),unit_del(g,57)].
72 -A(c2) | -P(x) | Eats(c2,x) | -A(c3) | -P(y) | -Eats(c3,y) | Eats(c2,c3).  [resolve(53,a,52,e)].
73 -P(x) | Eats(c2,x) | -P(y) | -Eats(c3,y) | Eats(c2,c3).  [copy(72),unit_del(a,56),unit_del(d,58)].
74 -A(c3) | -P(x) | Eats(c3,x) | -A(c5) | -P(y) | -Eats(c5,y) | Eats(c3,c5).  [resolve(54,a,52,e)].
75 -P(x) | Eats(c3,x) | -P(y) | -Eats(c5,y).  [copy(74),unit_del(a,58),unit_del(d,63),unit_del(g,66)].
79 -P(x) | Eats(c2,x) | -Eats(c3,x) | Eats(c2,c3).  [factor(73,a,c)].
86 -Eats(c2,c6).  [ur(71,a,67,a,b,69,a,c,67,a)].
92 -P(x) | Eats(c3,x).  [resolve(75,d,65,a),unit_del(c,64)].
94 Eats(c3,c6).  [resolve(92,a,67,a)].
98 Eats(c2,c3).  [resolve(94,a,79,c),unit_del(a,67),unit_del(b,86)].
103 $F.  [ur(68,a,56,a,b,58,a,c,94,a),unit_del(a,98)].
```

Figure 8.1.12.: Proof of Schubert's steamroller by Prover9.

6.
$$
\begin{array}{c}
P\left(f\left(x\right)\right) \rightarrow S\left(g\left(y\right),y\right) \\
S\left(y,g\left(y\right)\right) \rightarrow \left(P\left(y\right) \vee R\left(y\right)\right) \\
Q\left(x\right) \rightarrow \left(P\left(f\left(x\right)\right) \vee R\left(x\right)\right) \\
\neg\left(\neg Q\left(x\right) \wedge \neg R\left(x\right)\right) \\
\neg S\left(x,y\right) \\
\hline
R\left(x\right)
\end{array}
$$

7.
$$
\begin{array}{c}
P\left(x\right) \rightarrow P\left(g\left(x\right)\right) \\
\neg\left(P\left(x\right) \wedge Q\left(a,g\left(x\right)\right)\right) \\
Q\left(g\left(x\right),g\left(y\right)\right) \rightarrow \left(\neg\left(P\left(x\right) \wedge P\left(y\right)\right) \vee Q\left(x,y\right)\right) \\
Q\left(g\left(f\left(x,g\left(y\right)\right)\right),f\left(g\left(x\right),g\left(y\right)\right)\right) \\
P\left(a\right) \\
Q\left(f\left(x,a\right),x\right) \\
\hline
\neg Q\left(g\left(a\right),g\left(g\left(a\right)\right)\right)
\end{array}
$$

Exercise 8.1.2.3. Consider the theory of addition of natural numbers as formulated on the Prolog program *Addition* in Exercise 9.1.2.6. Enter the rules as axioms and prove in Prover9 that $0 + 2 = 2$. (Note: Invert $\leftarrow$.)

Exercise 8.1.2.4. Prove in the FO resolution calculus the following reasoning instances:

1. Exercise 3.2.2.7.

2. Exercise 3.2.2.8.

3. Exercise 3.2.4.

Exercise 8.1.2.5. Prove Theorem 8.1.2.

Exercise 8.1.2.6. Prove the soundness and the completeness of the FO resolution calculus.

8.1.3. Completeness of the resolution principle

The resolution principle is both sound and complete. We give the proof of completeness and leave the proof of soundness as an exercise. Two rules of inference describe in an essential way the resolution calculus: binary resolution (Theorem 8.1.3) and (positive) factoring (Def. 8.1.6). In order to prove the completeness of the resolution principle for FOL we require one further theorem known as the lifting lemma:

Theorem 8.1.3. *(Lifting lemma) Let* C_1', C_2' *be instances of* C_1 *and* C_2, *respectively. If* C' *is a resolvent of* C_1' *and* C_2', *then there exists a resolvent* C *of* C_1 *and* C_2 *such that* C' *is an instance of* C.

Proof: (Sketch) Let

$$C' = \left(C_1'\gamma - L_1'\gamma\right) \cup \left(C_2'\gamma - L_2'\gamma\right)$$

where γ is a MGU of the complementary literals L_1' and L_2'. We need to show that C' is an instance of C,

$$C = ((C_1\lambda)\,\sigma - L_1\sigma) \cup ((C_2\lambda)\,\sigma - L_2\sigma)$$

where σ is a MGU of L_1 and L_2 and λ_i is a MGU for $\{L_i^1, ..., L_i^{r_i}\}$ with $L_i = L_i^1\lambda_i, i = 1, 2$, if $r_i > 1$; if $r_i = 1$, then let $\lambda_i = \epsilon$ and $L_i = L_i^1\lambda_i$. In order to do so it suffices to show that there is a substitution θ such that $C_1' = C_1\theta$, $C_2' = C_2\theta$ and $(\lambda \circ \sigma) \leq_s (\theta \circ \gamma)$. **QED**

Theorem 8.1.4. *(Completeness of the resolution principle) A set C of clauses is unsatisfiable iff there is a deduction of the empty clause $\square$ from C, i.e.*

$$C \vdash \square.$$

Proof: We provide a sketch of the proof.

($\Rightarrow$) Given that a contradiction (i.e., the empty clause) can be deduced from any unsatisfiable set of clauses, the search for one proceeds by saturating the clause set, i.e. by systematically and exhaustively applying the inference rules until the empty clause is derived. In terms of a semantic tree, this means that the tree for a set C consisting only of the root node is generated, after, by Theorem 8.1.3, a process of obtaining subsequently smaller trees $\mathcal{T}_C', \mathcal{T}_C'', ...$ for $C \cup \{\mathcal{C}\}$, where $\mathcal{C}$ is a resolvent of clauses C_1 and C_2. The root node is generated only when $\square$ is derived. Therefore, there is a deduction of $\square$ from C.

($\Leftarrow$) Suppose there is a deduction of $\square$ from C. Let $R_1, ..., R_k$ be the resultants in the deduction. Assume C is satisfiable. Then there is a model $\mathcal{M}$ of C. By Theorem 8.1.1, $\mathcal{M}$ satisfies $R_1, ..., R_k$. But this is impossible, because one of these resolvents is $\square$. Therefore, C must be unsatisfiable. **QED**

Definition. 8.1.9. Let now Ψ be a complete resolution prover. Applying Ψ to a set of clauses C will produce *one* of the following results:

1. Derivation of $\square$, i.e. $\square \in \Psi(C)$. Obviously, C is unsatisfiable.

2. $\Psi(C)$ is finite (i.e. given C, Ψ is verified to terminate) and does not contain a refutation of C. Given that Ψ is complete, we know that C is satisfiable.

3. $\Psi(C)$ is infinite and does not contain a refutation of C. Given C, it is verified that Ψ does not terminate. Given that Ψ is complete, C must be satisfiable.

Exercises

Exercise 8.1.3.1. With respect to Theorem 8.1.3:

1. Summarize the idea of the lifting lemma.

2. Complete the given proof.

Exercise 8.1.3.2. Complete the proof of Theorem 8.1.4.

Exercise 8.1.3.3. Is it enough to prove the completeness of the FO resolution calculus, or do we have to prove that it is:

1. Actually refutation-complete?

2. Only refutation-complete?

Exercise 8.1.3.4. Prove the soundness of the resolution principle for FOL.

8.1.4. Resolution refinements

Although binary resolution is indeed a very efficient proof procedure, it often produces redundant resolvents. In order to minimize this problem we apply resolution refinements. We next give some of these important refinements; see Chang & Lee (1973) and Leitsch (1997), which we follow closely here, for many other important resolution refinements.

Definition. 8.1.10. Let Res_{rf} be a mapping from the set $\mathscr{C}$ of all finite sets of clauses to the set Res of all resolution deductions (i.e. *Res-deductions*). Let further ϱ denote a ground resolution. We say that Res_{rf} is a *resolution refinement* iff for every set of clauses $C \in \mathscr{C}$,

1. we have $Res_{rf}(C) \subseteq Res(C)$;

2. the set $\{\varrho | \varrho \in Res_{rf}(C)\}$ is decidable;

3. there is an algorithm Ψ that constructs $Res_{rf}(C)$;

4. if $C_1 \subseteq C_2$, then we have $Res_{rf}(C_1) \subseteq Res_{rf}(C_2)$.

Definition. 8.1.11. A resolution refinement Res_{rf} is said to be *complete* if, given an arbitrary unsatisfiable set of clauses C, $Res_{rf}(C)$ contains a refutation of C.

Recall Definition 8.1.9.

Definition. 8.1.12. Let Res_{rf} be a complete resolution refinement. Given a set of clauses C as input, there are three possible outputs:

1. $Res_{rf}(C)$ terminates with $\square$, i.e. $\square \in Res_{rf}(C)$. Obviously, C is unsatisfiable.

2. $Res_{rf}(C)$ terminates but does not produce $\square$. Given that Res_{rf} is complete, we know that C is satisfiable.

3. $Res_{rf}(C)$ does not terminate. $Res_{rf}(C)$ is infinite and $\square \notin Res_{rf}(C)$. Then C must be satisfiable, but $Res_{rf}(C)$ does not allow us to detect this property.

8.1.4.1. A-ordering

Definition. 8.1.13. Let F_0 be the set of all atomic formulae and let σ be some substitution. An *A-ordering* (i.e. *atom ordering*) $<_A$ is a binary relation on F_0 such that

1. $<_A$ is

    a) irreflexive
    b) transitive

2. and for every $A, B \in F_0$, $A < B$ entails $\sigma A <_A \sigma B$.

Given condition 8.1.13.1(a), it is obvious that A and B are *not* unifiable, i.e. $\sigma A \neq \sigma B$. The most important point with respect to $<_A$ is the specification of the ordering; this falls typically on the complexity of the terms and of the atom variables.

Definition. 8.1.14. We denote by $\vartheta \left(\cdot \right)$ and define *term depth*

1. of a term as
$$\vartheta \left(t \right) = 0$$
   for $t \in \left(Vi \cup Cons \right)$; for $f \in Fun$, $t_1, ..., t_n$ are terms, we have
$$\vartheta \left(f \left(t_1, ..., t_n \right) \right) = 1 + max \left\{ \vartheta \left(t_i \right) | i = 1, ..., n \right\}.$$

2. of a literal L as
$$\vartheta \left(L \right) = max \left\{ \vartheta \left(t \right) | t \in arg \left(L \right) \right\}.$$

3. of a clause $\mathcal{C}$ as
$$\vartheta \left(\mathcal{C} \right) = max \left\{ \vartheta \left(L \right) | L \in \mathcal{C} \right\}.$$

4. of a set of clauses C as
$$\vartheta \left(C \right) = max \left\{ \vartheta \left(\mathcal{C} \right) | \mathcal{C} \in C \right\}.$$

Definition. 8.1.15. We denote by $\vartheta_{max} \left(x, E \right)$ and define *maximal depth of a variable x in an expression E* as:

1. for E a term t,
$$\vartheta_{max} \left(x, t \right) = \begin{cases} 0 & \text{if } * \\ 1 + max \left\{ \vartheta_{max} \left(x, t_i \right) | i = 1, ..., n \right\} & \text{if } ** \end{cases},$$
$$* = x \neq Vi \left(t \right) \text{ or } x = t$$
$$** = x \in Vi \left(t \right) \text{ and } t = f \left(t_1, ..., t_n \right), f \in Fun.$$

2. for E an atom $P(t_1, ..., t_n)$,

$$\vartheta_{max}(x, P(t_1, ..., t_n)) = max\{(x, t_i)\,|i = 1, ..., n\}.$$

3. for E a literal L,

$$\vartheta_{max}(x, L) = \vartheta_{max}(x, arg(L)).$$

4. for E a clause $C = L_1 \vee ... \vee L_n$,

$$\vartheta_{max}(x, L_1 \vee ... \vee L_n) = max\{\vartheta_{max}(x, L_i)\,|i = 1, ..., n\}.$$

Example 8.1.10. Let $A = P(x, f(f(y)))$, $B = Q(f(x))$, and $C = \|A, \neg B\|$. Then, $\vartheta(A) = 2$, $\vartheta(B) = 1$, $\vartheta_{max}(x, C) = 1$, and $\vartheta_{max}(y, C) = 2$.

Given these definitions, we can now specify an *ordering* $<_\alpha$:

Definition. 8.1.16. For every atoms A and B we have $A <_\alpha B$ iff:

1. $\vartheta(A) < \vartheta(B)$, and

2. for every $x \in Vi(A)$ we have $\vartheta_{max}(x, A) < \vartheta_{max}(x, B)$ (entailing that $Vi(A) \subseteq Vi(B)$).

Remark. **8.1.17.** The above conditions (8.1.16.1-2) guarantee the irreflexivity and the transitivity of $<_\alpha$. If both conditions are satisfied, then for all substitutions σ and for all $y \in Vi(\sigma A)$ we have

1. $\vartheta(\sigma A) < \vartheta(\sigma B)$ and

2. $\vartheta_{max}(y, \sigma A) < \vartheta_{max}(y, \sigma B)$.

Example 8.1.11. $P(x, x) <_\alpha Q(f(x), y)$ because $\vartheta(P(x, x)) = 0$, $\vartheta(Q(f(x), y)) = 1$, $\vartheta_{max}(P(x, x)) = 0$, and $\vartheta_{max}(Q(f(x), y)) = 1$; also, for any substitution $\sigma = \{x \mapsto t\}$ for any term t, $\vartheta(\sigma[P(x, x)]) < \vartheta(\sigma[Q(f(x), y)])$ and $\vartheta_{max}(\sigma[P(x, x)]) < \vartheta_{max}(\sigma[Q(f(x), y)])$.

Example 8.1.12. $P(x, f(a)) \not<_\alpha Q(x, f(x))$ because condition 8.1.16.1 is violated. $P(x, a) \not<_\alpha P(f(a), x)$, since condition 8.1.16.2 is violated.

8. Automated theorem proving

Definition. 8.1.18. We say that a clause C is *condensed* if there is no factor of C that is a proper subclass of C, and we say that C' is *the condensation* of C if C' is a condensation of C such that $C' \subseteq C$.

Condensations are unique up to renaming.

Definition. 8.1.19. The *condensation rule* stipulates that in a proof procedure a clause be immediately replaced by its condensation, if it exists.

Example 8.1.13. The clause $\|P(x,y), P(y,x)\|$ is condensed. On the contrary, $\|P(x,y), P(x,a)\|$ is not condensed; its condensation is $\|P(x,a)\|$.

Definition. 8.1.20. Let C be a set of condensed clauses and $<_A$ an A-ordering. Let C be a resolvent of clauses $C_1, C_2 \in C$. Then, (the condensation of) C is a $<_A$-*resolvent* of C_1 and C_2 if there is no literal L in C such that $B <_A L$ for B the resolved atom in the resolution of C_1 and C_2.

We denote the fact that C is a $<_A$-resolvent of a set of clauses C by $C \in \rho <_A (C)$.

Example 8.1.14. Let $C_1 = \|Q(f(x_1), x_1), \neg R(f(x_1))\|$ and $C_2 = \|R(f(x_1)), \neg Q(x_1, x_2)\|$. Clearly, C_1 and C_2 are condensed. Let now $C = \{C_1, C_2\}$; we want to obtain a resolvent $C \in \rho <_A (C)$.

We begin by renaming the variables, obtaining, for instance, $C_1' = \|Q(f(x), x), \neg R(f(x))\|$ and $C_2' = \|R(f(y)), \neg Q(y, z)\|$. There are now two possible ways to obtain a resolvent $C \in \rho <_\alpha (C)$: given $\sigma = \{y \mapsto x\}$, $\sigma = mgu\left(C_1', C_2'\right)$, we obtain $C^* = Q(f(x), x) \vee \neg Q(x, z)$; with $\theta = \{y \mapsto f(x), z \mapsto x\}$, $\theta = mgu\left(C_1', C_2'\right)$, we obtain $C_* = \neg R(f(x)) \vee R(f(f(x)))$. However, only for C^* do we have that $C^* \in \rho <_\alpha (C)$, as $R(f(x)) \not<_\alpha L$ for $L = Q(f(x), x)$ or $L = \neg Q(x, z)$; on the other hand, $C_* \notin \rho <_\alpha (C)$ because $Q(f(x), x) <_\alpha R(f(f(x)))$.

This last example shows that we have to use *a-posteriori* criteria and that, among these, we have to choose the strongest one. An *a-priori* criterion for $<_\alpha$ would not have barred the resolution of C_*, even because *a priori* there is no $<_\alpha$-ordering relation between the atoms of C_1 and C_2.

Recall from above the definition of a proof or derivation (Def. 3.4.17). The following should be read with this definition in mind.

Definition. 8.1.21. Let C be a set of condensed clauses and $<_A$ an A-ordering. A *Res $<_A$-deduction* of a condensed clause $\mathcal{C}$ is a sequence $\mathcal{C}_1, ..., \mathcal{C}_n$ such that

1. $\mathcal{C}_n = \mathcal{C}$,

2. and for every $i = 1, ..., n$, either $\mathcal{C}_i \in C$ or $\mathcal{C}_i \in \rho <_A(\{\mathcal{C}_j, \mathcal{C}_k\})$ for any $j, k < i$.

Theorem 8.1.5. *(Completeness of Res $<_A$-deduction) Let C be a finite set of condensed clauses and $<_A$ an A-ordering. If C is unsatisfiable, then there is a Res $<_A$-refutation of C.*

Proof: ($\Rightarrow$) We build a closed semantic tree (cf. Def. 7.3.10): Given the set of atoms $At(C) = \{A_1, ..., A_n\}$ and an ordering such that $A_i < A_j$ for $i < j \leq n$, we can build a semantic tree with the labels A_1 and $\neg A_1$ in the two edges starting immediately at the root node and with the labels A_n and $\neg A_n$ in the edges of failure nodes, it being the case that every branch has a failure node. Recall Theorem 7.3.6. By this theorem, every failure node falsifies a clause of C. A semantic tree $\mathcal{T}_C$ is closed if all its branches end in a failure node. If we take a closed semantic tree for C, by reverting its building process up to its collapse in the empty clause (its root), i.e. $\square \in Res <_A$, we show that there cannot fail to exist a *Res $<_A$-deduction* from C (cf. Theorem 7.3.6).

($\Leftarrow$) The proof runs as for Theorem 7.3.6, namely by Definition 8.1.20 and by the fact that $\square \in Res <_A$ for any $\mathcal{C}_i = \square$. **QED**

8.1.4.2. Hyper-resolution and semantic resolution

Hyper-resolution is a kind of *macro-resolution*, i.e. the contraction of a sequence of resolution steps into a single inference.

Definition. 8.1.22. Let $\mathcal{C}$ be any non-positive clause and let $\mathcal{D}_1, ..., \mathcal{D}_n$ be positive clauses, $\mathcal{C}, \mathcal{D}_1, ..., \mathcal{D}_n \in C$. Then, $\Omega = (\mathcal{C}; \mathcal{D}_1, ..., \mathcal{D}_n)$ is a *clash sequence* in which $\mathcal{C}$ is the *nucleus* and $\mathcal{D}_1, ..., \mathcal{D}_n$ are the *satellites*. Let $\mathcal{C}_0 = \mathcal{C}$ and $\mathcal{C}_{i+1} \in Res(\{\mathcal{C}_i, \mathcal{D}_{i+1}\})$ for $i = 1, ..., n - 1$. If $\mathcal{C}_n$ is defined and positive, then we say that $\mathcal{C}_n$ is a *hyper-resolvent of* Ω. We denote by $R_H(C)$ the set of all the hyper-resolvents of a set of clauses C and we say that the corresponding operator R_H^* is a *hyper-resolution operator*.

Example 8.1.15. Let $C = \{C_1, C_2, C_3, C_4\}$, $C_1 = P(a, b)$, $C_2 = P(b, a)$, $C_3 = \neg P(x, y) \vee \neg P(y, z) \vee P(x, z)$, and $C_4 = \neg P(a, a)$. In the following resolution refutation (Fig. 8.1.13), one of the resolving clauses is always positive. In effect, it is a resolution refutation of $\Omega = (C_3; C_1, C_2)$, in which C_1 and C_2 are the satellites (or electrons).

We say that $C_6 = P(a, a)$ is a hyper-resolvent of $\Omega = (C_3; C_1, C_2)$ insofar as we can say that $C_5 = \neg P(x, b) \vee P(x, a)$ is an intermediate result with respect to C_6. Note that in C_5 the negative literal belongs to the nucleus and the positive literal is in fact the satellite $(C_1) \lambda$ (or $(C_2) \lambda$) for $\lambda = \{b \mapsto a\}$. The *macro-resolvents* of C are thus C_6 and C_7. We have then $R_H^*(C) = C \cup \{P(a, a), \square\}$, it being the case that all the produced clauses have at most one atom.

$$
\begin{array}{lll}
C_1 & P(a, b) & \\
C_2 & P(b, a) & \\
C_3 & \neg P(x, y) \vee \neg P(y, z) \vee P(x, z) & \\
C_4 & \neg P(a, a) & \\
C_5 & \neg P(x, b) \vee P(x, a) & \text{Resolvent of } (C_2, C_3)\,\sigma, \\
& & \sigma = \{y \mapsto b, z \mapsto a\} \\
C_6 & P(a, a) & \text{Resolvent of } (C_1, C_5)\,\theta, \\
& & \theta = \{x \mapsto a\} \\
C_7 & \square & \text{Resolvent of } C_4 \text{ and } C_6
\end{array}
$$

Figure 8.1.13.: Hyper-resolution of $\Omega = (C_3; C_1, C_2)$.

This is an example of positive hyper-resolution.

Definition. 8.1.23. Hyper-resolution is called *positive hyper-resolution* when all the electrons and all the hyper-resolvents are positive, and *negative hyper-resolution* when the hyper-resolvents and the electrons are negative.

In either case, in positive or negative hyper-resolution, it is all about imposing an interpretation when resolving a set of clauses, reason why hyper-resolution is in fact a kind of *semantic resolution*.

Definition. 8.1.24. Given a set of clauses C and the set $Pred(C)$ of predicates of C, let $\mathcal{I} = (\mathcal{D}, \Theta, \varpi)$ be an interpretation such that for every m-place predicate $P \in Pred(C)$ and for every $d_1, ..., d_m \in \mathcal{D}$ we have $\Theta(P)(val_\mathcal{I}(d_1), ..., val_\mathcal{I}(d_m)) = \mathtt{f}$. Then all the positive clauses are false in $\mathcal{I}$ and the remaining clauses are true in $\mathcal{I}$.

Definition. 8.1.25. Let C be a set of clauses and $\mathcal{I}$ an interpretation for C. Let $\mathcal{C}_1$ and $\mathcal{C}_2$ be clauses in the signature of C such that at least one of $\{\mathcal{C}_1, \mathcal{C}_2\}$ is false in $\mathcal{I}$. Then, we say that all the resolvents of $\mathcal{C}_1$ and $\mathcal{C}_2$ are $\mathcal{I}$-*semantic resolvents*. Consider now the set $C \cup D$ where C is the set of true clauses in $\mathcal{I}$ and D the set of false clauses in $\mathcal{I}$. Let

$$\mathscr{R}_{\mathcal{I}}(D) = \{E | E \text{ is a } \mathcal{I}\text{-semantic resolvent of } D\}.$$

We define the *semantic resolution operator* $\mathscr{R}_{\mathcal{I}}^*$ as

$$\mathscr{R}_{\mathcal{I}}^*(D) = D \cup \mathscr{R}_{\mathcal{I}}(D).$$

Definition. 8.1.26. Given a set of clauses C, let $\mathcal{I}$ be an interpretation and $\mathcal{O}$ an ordering of $P, Q, \ldots \in Pred(C)$. A finite set of clauses of C constitutes a *semantic clash* (or a *PI-clash*) $\Omega_{\mathcal{OI}} = (N; E_1, \ldots, E_n)$ in which N denotes the nucleus and $E_1, \ldots, E_n$ the electrons or satellites of $\Omega(C)$ iff

1. $E_1, \ldots, E_n$ are false in $\mathcal{I}$;

2. Let $\mathcal{C}_1 = N$. For every $i = 1, \ldots, n$ there is a resolvent $\mathcal{C}_{i+1}$ of $\mathcal{C}_i$ and E_i;

3. The resolved literal of E_i contains the highest predicate symbol in E_i, $i = 1, \ldots, n$;

4. $\mathcal{C}_{n+1}$ is false in $\mathcal{I}$.[5]

$\mathcal{C}_{n+1}$ is called the *PI-resolvent* of $\Omega_{\mathcal{OI}}(C) = (N; E_1, \ldots, E_n)$.

Example 8.1.16. Let $E_1 = P \vee R$, $E_2 = Q \vee R$, and $N = \neg P \vee \neg Q \vee R$. Given the interpretation $\mathcal{I} = \{\neg P, \neg Q, \neg R\}$ and the ordering $P > Q > R$, $(N; E_1, E_2)$ is a PI-clash and the resolvent of this clash is R. Note that R is false in $\mathcal{I}$.

Besides the restriction concerning the signs of the clash sequence, hyper-resolution imposes other restrictions on the space of the selection of the clauses to resolve, hindering in this way the production of many redundant formulae and their addition to the search space.

We prove solely the completeness of semantic resolution, of which hyper-resolution is a kind. Because by applying any interpretation $\mathcal{I}$ and

---

[5] Recall that the empty clause is always false in any interpretation.

any ordering $\mathcal{O}$ we can always obtain a PI-deduction of the empty clause from an unsatisfiable set of ground clauses, we prove the completeness of semantic resolution via a proof of PI-resolution.

Theorem 8.1.6. *(Completeness of PI-resolution) If $\mathcal{O}$ is an ordering of predicate symbols in an unsatisfiable finite set C of clauses and if $\mathcal{I}$ is an interpretation on C, then there is a PI-deduction of $\Box$ from C.*

Proof: The proof is by induction on the number of atoms of C. Let C be an unsatisfiable set of ground clauses. Let $At(C) = \{P\}$. Then, $C = \{P, \neg P\} = C'$. Clearly, the resolvent from C' is $\Box$, regardless of the interpretation $\mathcal{I}$ (i.e. $\mathcal{I} = \{P\}$ or $\mathcal{I} = \{\neg P\}$). Therefore, either P or $\neg P$ is false in $\mathcal{I}$ and $\Box$ is equally false in $\mathcal{I}$, and we have it that $\Box$ is a PI-resolvent.

We thus showed that the theorem holds for $n = 1$. Assume now that the theorem holds for $|At(C)| = i$, $1 \leq i \leq n$. In order to complete the induction, we shall consider that $|At(C)| = n+1$. We start by searching for a unit clause $C = L$ that is false in $\mathcal{I}$.

(i) C contains a unit clause $\|L\|$ that is false in $\mathcal{I}$. Then, by deleting in C the clauses containing L and by deleting $\neg L$ from the remaining clauses of C, we obtain C'. Clearly, C' is unsatisfiable (cf. one-literal rule: Def. 8.1.1). C' contains n or fewer than n atoms, so by the induction hypothesis there is a PI-deduction of $\Box$ from C'. Let us denote this PI-deduction by D'. We can obtain from C a PI-deduction of $\Box$ from D': it suffices to replace every clash sequence $\left(N'; E'_1, ..., E'_q\right)$, in which $N', E'_1, ..., E'_q$ are clauses connected to the initial nodes of D' and N' was obtained from N by deleting $\neg L$, by the PI-clash sequence $\left(N, L; E'_1, ...E'_q\right)$. If E'_i was obtained from E_i by deleting $\neg L$ in it, we add the PI-clash $(L; E_i)$ above the node of E'_i. In this way, we obtain a PI-deduction of $\Box$ from C.

(ii) C does not contain a unit clause $\|L\|$ that is false in $\mathcal{I}$. Then, we can obtain a PI-deduction D' of $\Box$ from C', in which C' is obtained by applying the one-literal rule to a literal L that is the symbol of the lowest predicate in some set $\{B, \neg B\} \subseteq At(C)$ and is false in $\mathcal{I}$. Clearly, C' is unsatisfiable. C' contains n or fewer than n atoms, so by the induction hypothesis there is a PI-deduction D' of $\Box$ from C'. Replace now again literal L in the clauses from which it was firstly removed and denote by D_1 the deduction obtained from D' by means of this operation: D_1 remains a PI-deduction, given that L contains the lowest predicate symbol and is false in $\mathcal{I}$. It is evident that either $D_1 = \Box$ or $D_1 = L$. In the first case, the proof is finished. In the second case, by (i) we obtain a PI-deduction $D_2 = \Box$ from $C \cup \{L\}$ in which L is a unit clause and false in $\mathcal{I}$. By the combination of D_1 and D_2 we obtain a PI-deduction of $\square$ from C. **QED**

Herbrand's theorem (version 2; Theorem 7.3.4) and the lifting lemma (Theorem 8.1.3) assure us that this result (i.e. the completeness of PI-resolution) holds for any unsatisfiable set of clauses, i.e. a set of non-ground clauses (cf. Chang & Lee, 1973, p. 107).

Example 8.1.17. Let there be given the theory of distributive lattices. We wish to know whether meet is a commutative operation in such a lattice. The input in Prover9-Mace4 is shown in Figure 8.1.14. As it can be easily seen in the output by Prover 9 (Fig. 8.1.15), hyper-resolution accounts for most of the proof.

Assumptions:

```
% axioms for a partial order
x <= x.
x <= y & y <= x -> x=y.
x <= y & y <= z -> x <= z.
% lattice axioms
x <= 1.
z <= x ^ y <-> z <= x & z <= y.
0 <= x.
x v y <= z <-> x <= z & y <= z.
% distributivity
(x v y) ^z = (x^z) v (y^z).
```

Goals:

```
% commutativity of meet
x ^ y = y ^ x.
```

Figure 8.1.14.: Theory of distributive lattices and commutativity of meet: Input in Prover9-Mace4.

Exercises

Exercise 8.1.4.1. Solve Exercises 8.1.2.1.1-2 and 8.1.2.2.6-7 by applying A-ordering.

```
===================== PROOF =========================
% -------- Comments from original proof --------
% Proof 1 at 0.05 (+ 0.09) seconds.
% Length of proof is 14.
% Level of proof is 5.
% Maximum clause weight is 11.
% Given clauses 46.


1 x <= y & y <= x -> x = y # label(non_clause).   [assump-
tion].
3 x <= y ^ z <-> x <= y & x <= z # label(non_clause).
[assumption].
5 x ^ y = y ^ x # label(non_clause) # label(goal).
[goal].
6 x <= x.   [assumption].
7 -(x <= y) | -(y <= x) | y = x.   [clausify(1)].
10 -(x <= y ^ z) | x <= y.   [clausify(3)].
11 -(x <= y ^ z) | x <= z.   [clausify(3)].
12 x <= y ^ z | -(x <= y) | -(x <= z).   [clausify(3)].
19 c2 ^ c1 != c1 ^ c2.   [deny(5)].
20 x ^ y <= x.   [hyper(10,a,6,a)].
21 x ^ y <= y.   [hyper(11,a,6,a)].
57 x ^ y <= y ^ x.   [hyper(12,b,21,a,c,20,a)].
484 x ^ y = y ^ x.   [hyper(7,a,57,a,b,57,a)].
485 $F.  [resolve(484,a,19,a)].
===================== end of proof ==================
```

Figure 8.1.15.: Proof by Prover9 of the commutativity of meet in a distributive lattice.

Exercise 8.1.4.2. Let $C = \{P, \neg P \lor Q, R \lor \neg P, \neg P \lor \neg Q \lor \neg R\}$. Prove that the set C is unsatisfiable by means of semantic resolution. The following cases should be considered:

1. $\mathcal{I} = \{\neg P, \neg Q, \neg R\}$; $R < Q < P$.

2. $\mathcal{I} = \{P, Q, R\}$; $R < P < Q$.

3. $\mathcal{I} = \{\neg P, \neg Q, R\}$; $P < Q < R$.

Exercise 8.1.4.3. Prove the reasoning in Example 8.1.1 in Prover9-Mace4. Analyze the unreformatted output by Prover9 in detail, paying particular attention to possible new rules or refinements. (You might need to consult the following webpage: https://www.cs.unm.edu/ ~mccune/prover9/manual/ 2009-11A/inf-rules.html.)

Exercise 8.1.4.4. Prove that the following are refinements of resolution:

1. A-ordering.

2. Hyper-resolution.

3. Semantic resolution.

Exercise 8.1.4.5. Given a set of clauses C, we say that a clause $\mathcal{C}$ is a *linear resolution* deduction from C, and write $C \vdash_{lres} \mathcal{C}$, if there is a sequence of pairs $(\mathcal{C}_0, \mathcal{D}_0), ..., (\mathcal{C}_n, \mathcal{D}_n)$ such that $\mathcal{C} = \mathcal{C}_{n+1}$ and (i) $\mathcal{C}_0$, called the *starting clause*, and the $\mathcal{D}_i$ are elements of C or some $\mathcal{C}_j, j < i$, (ii) each $\mathcal{C}_{i+1}, i \leq n$, is a resolvent of $\mathcal{C}_i$ and $\mathcal{D}_i$. The elements of C are called the *input clauses*, the $\mathcal{C}_i$ are the *center clauses* and the $\mathcal{D}_i$ the *side clauses*. If $\mathcal{C} = \square$, we say that there is a linear-resolution refutation of C, and write $C \vdash_{lres} \square$. Figure 8.1.16 shows the linear resolution refutation of the obviously unsatisfiable set of clauses $C = \{\|p, q\|, \|p, \neg q\|, \|\neg p, q\|, \|\neg p, \neg q\|\}$.

1. Show that linear resolution is a resolution refinement.

2. Prove the completeness of linear resolution.

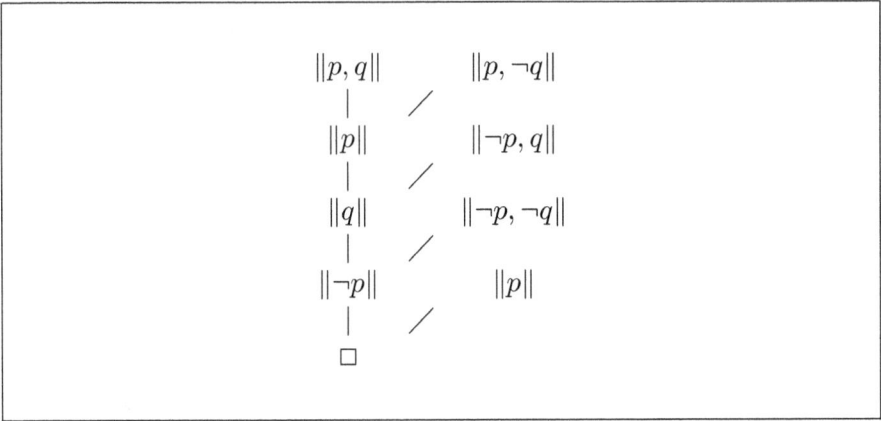

Figure 8.1.16.: A linear-resolution refutation.

3. Show that the following set C of clauses is unsatisfiable by applying linear resolution:

$$C = \left\{ \begin{array}{c} \|P(x), \neg Q(x, f(y)), \neg R(a)\|, \\ \|R(x), \neg Q(x, y)\|, \\ \|\neg P(x), \neg Q(y, z)\|, \\ \|P(x), \neg R(x)\|, \\ \|R(f(b))\|, \|Q(x, y), \neg P(y)\| \end{array} \right\}.$$

Exercise 8.1.4.6. Linear resolution (cf. Exercise 8.1.4.5) is called *linear input resolution* (abbr.: *LI resolution*) if, given a sequence of pairs $(\mathcal{C}_0, \mathcal{D}_0), ..., (\mathcal{C}_n, \mathcal{D}_n)$, then all the $\mathcal{D}_i$ are variants of clauses in C.

1. Show that LI resolution is not generally complete.

2. Show that LI resolution is complete for Horn clauses.

3. Show that LI resolution is a resolution refinement.

4. Show that the set of clauses of Exercise 8.1.4.5.3 is unsatisfiable by applying LI resolution.

Exercise 8.1.4.7. Let $\mathcal{C} = \|\neg L_1, ..., \neg L_n\|$, $\mathcal{D} = \|M, \neg M_1, ..., \neg M_m\|$ be ordered clauses. The following rule of inference, where we have $\sigma = mgu(M, L_i)$ and the resolvent is an ordered (or definite) clause, is called *linear definite resolution* (abbr.: *LD resolution*):

$$\frac{\neg L_1 \vee ... \vee \neg L_n \qquad M \vee \neg M_1 \vee ... \vee \neg M_m}{(\neg L_1 \vee ... \vee \neg L_{i-1} \vee \neg M_1 \vee ... \vee \neg M_m \vee \neg L_{i+1} \vee ... \vee \neg L_n)\sigma}$$

1. Show that LD resolution is complete for Horn clauses.

2. Show that LD resolution is a resolution refinement.

3. Prove the unsatisfiability of the set of clauses of Exercise 8.1.4.5.3 by applying LD resolution.

Exercise 8.1.4.8. The following inference rule, where $\sigma = mgu\,(M, L_i)$ and **r** is a selection rule or function, is called *selective linear definite resolution* (abbr.: *SLD resolution*):

$$\frac{\mathbf{r}\,(\neg L_1 \vee ... \vee \neg L_n) \qquad M \vee \neg M_1 \vee ... \vee \neg M_m}{(\neg L_1 \vee ... \vee \neg L_{i-1} \vee \neg M_1 \vee ... \vee \neg M_m \vee \neg L_{i+1} \vee ... \vee \neg L_n)\,\sigma}$$

1. Elaborate on the SLD resolution rule.

2. Prove that SLD resolution is complete for Horn clauses.

3. Prove that SLD resolution is a LD-resolution refinement.

4. Prove the unsatisfiability of the set of clauses of Exercise 8.1.4.5.3 by applying SLD resolution.

Exercise 8.1.4.9. Recall from the Introduction that computational logic is often summarized as the equation *Algorithm = Logic + Control*. Explain in which way this might be true of resolution refinements.

8.1.5. Paramodulation

As seen in Section 4.5, equality is a fundamental relation in many theories, namely in mathematical theories, and we often need to extend CL to CL$^=$. Thus, one can say that a proof calculus is not complete, and an automated prover is not full-blown, if it is not capable of tackling equality. *Paramodulation* is the way conceived to introduce equality in the resolution calculus. We give here the essentials of this proof technique; a deeper treatment of paramodulation can be found in, for example, Nieuwenhuis & Rubio (2001).

Suppose you are given the clauses $\mathcal{C}_1 = P\,(a)$ and $\mathcal{C}_2 = (a = b)$. Then, we should be able to derive $\mathcal{C} = P\,(b)$, i.e. we should be able to have $\{\mathcal{C}_1, \mathcal{C}_2\} \vdash_{res} \mathcal{C}$, by applying a simple substitution of b for a. Obvious as this might be, the resolution calculus above cannot handle this without being augmented with a *generalization* of substitution that, in turn, requires a clear definition of equality substitution. We next give these definitions.

Definition. 8.1.27. *Equality substitution* – Given $C(t)$, a clause C in which a term t occurs, and a unit clause $(t = s)$, we can derive a new clause $C(s)$ by substituting s for a single occurrence of t.

Definition. 8.1.28. Consider the clauses $C_1 = L(t) \vee C_1'$, where $L(t)$ denotes that term t occurs in literal L, and $C_2 = (r = s) \vee C_2'$ where r and s are any terms. Then a *paramodulant* of C_1 and C_2 is the clause

$$L(s) \vee C_1' \vee C_2'.$$

Example 8.1.18. Given the clauses $C_1 = P(a) \vee Q(b)$ and $C_2 = (a = b) \vee R(b)$, we can derive the paramodulant $P(b) \vee Q(b) \vee R(b)$.

The above is a definition of *ground paramodulation*. This is now generalized to formulae with individual variables in the following way:

Definition. 8.1.29. Let the clauses $C_1 = L(t) \vee C_1'$ be given, where $L(t)$ denotes that term t occurs in literal L, and $C_2 = (r = s) \vee C_2'$ where r and s are any terms. Let further $\sigma = mgu(t, r)$, C_1 and C_2 have no variables in common. Then a *binary paramodulant* of C_1 and C_2 is the clause

$$\left(L(s\sigma) \vee C_1' \vee C_2' \right) \sigma.$$

We say that L and $r = s$ are the literals *paramodulated upon* in the paramodulation from C_2 to C_1.

Example 8.1.19. Given the clauses $C_1 = P(x) \vee Q(x)$ and $C_2 = (a = b)$, by applying the substitution $\sigma = \{x \mapsto a\}$ we can obtain the paramodulants $P(b) \vee Q(a)$ or $P(a) \vee Q(b)$.

Example 8.1.19 shows an important feature of paramodulation, to wit, substitutions are applied in one argument position at a time. In this example, we have

$$(P(b) \vee Q(x))\sigma = \underbrace{P(b)}_{L\sigma(r\sigma)} \vee \underbrace{Q(a)}_{C_1'\sigma} \vee \underbrace{\square}_{C_2'\sigma}$$

or

$$(P(x) \vee Q(b))\sigma = \underbrace{P(a)}_{C_1'\sigma} \vee \underbrace{Q(b)}_{L\sigma(r\sigma)} \vee \underbrace{\square}_{C_2'\sigma}.$$

Example 8.1.20. A slightly more complex example can be given by the paramodulation from $C_2 = (f(g(b)) = a) \vee R(g(c))$ to $C_1 = P(g(f(x))) \vee Q(x)$. Following Definition 8.1.29, we make t be $f(x)$, $L(t)$ is $P(g(f(x)))$, r is $f(g(b))$; the MGU of r and t is $\sigma = \{x \mapsto g(b)\}$. We apply the paramodulation inference rule and obtain the paramodulant $P(g(a)) \vee Q(g(b)) \vee R(g(c))$ in the following two steps:

$$\underbrace{P(g(f(g(b))))}_{L(r\sigma)} \vee Q(x) \vee R(g(c))$$

$$\underbrace{P(g(a))}_{L\sigma(r\sigma)=(L(s\sigma))\sigma} \vee \underbrace{Q(g(b))}_{C_1'\sigma} \vee \underbrace{R(g(c))}_{C_2'\sigma}$$

We now further specify the definition of paramodulant:

Definition. 8.1.30. A paramodulant of (parent) clauses C_1 and C_2 is one of the following binary paramodulants:

1. A binary paramodulant of C_1 and C_2;

2. A binary paramodulant of a factor of C_1 and C_2;

3. A binary paramodulant of C_1 and a factor of C_2;

4. A binary paramodulant of a factor of C_1 and a factor of C_2.

However, it is not often the case that one is confronted with the simplicity above; more often than not, there are more than one equality literals and/or more than one literal to be paramodulated upon, so that strategies are required such as ordering of the equality literals and simultaneous paramodulation.

Definition. 8.1.31. The rule of inference

$$\frac{L(t) \vee C_1' \qquad (r = s) \vee C_2'}{\left(L(s\sigma) \vee C_1' \vee C_2'\right)\sigma}$$

where $\sigma = mgu(r,t)$, $L\sigma \not< L\sigma(s\sigma)$, $(r\sigma = s\sigma) \not< M$ for all the literals $M \in \left(C_2'\right)\sigma$ and $L\sigma \not< M$ for all the literals $M \in \left(C_1'\right)\sigma$, is called *ordered paramodulation*.

Definition. 8.1.32. The rule of inference

$$\frac{L\,(t) \vee S\,(t) \qquad (r = s) \vee C}{(L\,(s\sigma) \vee S\,(s\sigma) \vee C)\,\sigma}$$

is called *simultaneous paramodulation*.

Example 8.1.21. Let there be given the clauses $C_1 = P(f(x, g(x))) \vee Q(x)$ and $C_2 = (a = b) \vee (g(a) = a) \vee (f(a, g(a)) = b)$. It will be useful to indicate the constituents of these clauses as follows:

$$\underbrace{C_1}_{1} = \underbrace{P(f(x, g(x)))}_{a(1,2)} \vee \underbrace{Q\,(x)}_{b}$$

and

$$\underbrace{C_2}_{2} = \underbrace{(a = b)}_{a} \vee \underbrace{(g\,(a) = a)}_{b} \vee \underbrace{(f\,(a, g\,(a)) = b)}_{c}$$

We may obtain the following sample of all the paramodulants of the paramodulation from C_2 into C_1:
By an application of paramodulation with the MGU $\sigma_1 = \{x \mapsto a\}$ we obtain the paramodulants

para[2(a),1(a,1)]: $P(f(b, g(a))) \vee Q(a) \vee (g(a) = a) \vee (f(a, g(a)) = b)$
para[2(a),1(a,2)]: $P(f(a, g(b))) \vee Q(a) \vee (g(a) = a) \vee (f(a, g(a)) = b)$
para[2(a),1(a,1-2)]: $P(f(b, g(b))) \vee Q(a) \vee (g(a) = a) \vee (f(a, g(a)) = b)$
para[2(a),1(b)]: $P(f(a, g(a))) \vee Q(b) \vee (g(a) = a) \vee (f(a, g(a)) = b)$
para[2(a),1(a,b)]: $P(f(b, g(b))) \vee Q(b) \vee (g(a) = a) \vee (f(a, g(a)) = b)$

We can obtain a factor of $C_2' = (g\,(a) = b) \vee f\,(a, g\,(a)) = b$ by the transitivity of the equality relation on $g\,(a) = a$ and $a = b$, which gives us $g\,(a) = b$. Applying paramodulation with the MGU $\sigma_2 = \{x \mapsto g(a)\}$, we obtain the paramodulants

para[2'(a),1(a,1)]: $P(f(b, g(g(a)))) \vee Q(g(a)) \vee (f(a, g(a)) = b)$
para[2'(a),1(a,2)]: $P(f(g(a), g(b))) \vee Q(g(a)) \vee (f(a, g(a)) = b)$
para[2'(a),1(a,1-2)]: $P(f(b, g(b))) \vee Q(g(a)) \vee (f(a, g(a)) = b)$
para[2'(a),1(b)]: $P(f(g(a), g(g(a)))) \vee Q(b) \vee (f(a, g(a)) = b)$
para[2'(a),1(a,b)]: $P(f(b, g(b))) \vee Q(b) \vee (f(a, g(a)) = b)$

C_2' can further be factorized into $C_2'' = (g\,(a) = f\,(a, g\,(a)))$, which in turn gives us simpler paramodulants.

Example 8.1.22. Given the theory $\Theta_{\mathcal{G}}$, we prove in Prover9-Mace4 that if $x \star x = e$ for all x in a group $\mathcal{G} = (G, \star)$, then $\mathcal{G}$ is a commutative group. Figure 8.1.17 shows the input, and Figure 8.1.18 shows the proof ($! =$ stands for $\neq$).

Assumptions:

```
f(e,x)=x.
f(x,e)=x.
f(x,f(y,z))=f(f(x,y),z).
f(x,x)=e.
f(a,b)=c.
```

Goals:

```
c=f(b,a).
```

Figure 8.1.17.: Theory of commutative groups: Input in Prover9-Mace4.

Example 8.1.23. For further, very complex, proofs in group theory by applying paramodulation alone and resolution, see Robinson & Wos (1969).

Exercises

Exercise 8.1.5.1. Find the binary paramodulants of the following pairs of clauses:

1. $P\left(f\left(x,a\right),y\right) \vee R\left(y\right)$ and $\left(f\left(c,a\right) = g\left(b\right)\right) \vee R\left(g\left(b\right)\right)$

2. $\left(f\left(x\right) = b\right) \vee P\left(x\right)$ and $Q\left(f\left(a\right)\right) \vee R$

3. $\left(f\left(g, g\left(x\right)\right) = a\right) \vee P\left(x\right)$ and $R\left(y, f\left(g\left(y\right), z\right)\right) \vee S\left(z\right)$

4. $\left(f\left(g\left(b\right)\right) = a\right) \vee R\left(g\left(c\right)\right)$ and $P\left(g\left(f\left(x\right)\right)\right) \vee Q\left(x\right)$

5. $Q\left(g\left(a\right)\right) \vee \left(f\left(h\left(z\right), z\right) = g\left(z\right)\right), \neg P\left(f\left(x, g\left(y\right)\right)\right) \vee \left(k\left(x, g\left(y\right)\right) = h\left(y\right)\right)$

Exercise 8.1.5.2. Resolution-based paramodulation makes implicit use of the following equality axioms (cf. Section 4.5):

```
===================== PROOF =========================
% --------- Comments from original proof ---------
% Proof 1 at 0.01 (+ 0.05) seconds.
% Length of proof is 15.
% Level of proof is 5.
% Maximum clause weight is 11.
% Given clauses 9.


1 c = f(b,a) # label(non_clause) # label(goal).  [goal].
2 f(e,x) = x.  [assumption].
3 f(x,e) = x.  [assumption].
4 f(x,f(y,z)) = f(f(x,y),z).  [assumption].
5 f(f(x,y),z) = f(x,f(y,z)).  [copy(4),flip(a)].
6 f(x,x) = e.  [assumption].
7 f(a,b) = c.  [assumption].
8 c = f(a,b).  [copy(7),flip(a)].
9 f(b,a) != c.  [deny(1)].
10 f(b,a) != f(a,b).  [copy(9),rewrite([8(4)])].
11 f(x,f(x,y)) = y.  [para(6(a,1),5(a,1,1)),
    rewrite([2(2)]),flip(a)].
12 f(x,f(y,f(x,y))) = e.  [para(6(a,1),5(a,1)),flip(a)].
15 f(x,f(y,x)) = y.  [para(12(a,1),11(a,1,2)),
    rewrite([3(2)]),flip(a)].
19 f(x,y) = f(y,x).  [para(15(a,1),11(a,1,2))].
20 $F.  [resolve(19,a,10,a)].
===================== end of proof ==================
```

Figure 8.1.18.: Output by Prover9.

$\mathcal{E}1$ $\forall x\,(x = x)$

$\mathcal{E}2$ $\forall x_i \forall y_i\,[\neg\,(x_1 = y_1) \lor ... \lor \neg\,(x_n = y_n) \lor (f\,(x_1, ..., x_n) = f\,(y_1, ..., y_n))]$

$\mathcal{E}3$ $\forall x_i \forall y_i\,[\neg\,(x_1 = y_1) \lor ... \lor \neg\,(x_n = y_n) \lor \neg P\,(x_1, ..., x_n) \lor P\,(y_1, ..., y_n)]$

Show that $\mathcal{E}2$ and $\mathcal{E}3$ above are equivalent to $\mathcal{E}4$ and $\mathcal{E}5$ of Proposition 4.5.1, respectively.

Exercise 8.1.5.3. Given the theory of distributive lattices (cf. Example 8.1.17), we wish to know whether the property in Definition 1.2.16.1 belongs to a distributive lattice. Prove that it does in Prover9-Mace4 and analyze the proof by Prover9.

Exercise 8.1.5.4. Apply ordered paramodulation to the clauses of

Example 8.1.21.

Exercise 8.1.5.5. Prove the refutation-completeness of:

1. Ordered paramodulation.

2. Simultaneous paramodulation.

Exercise 8.1.5.6. RUE resolution, denoting "resolution with unification and equality," is based on the rule

$$\frac{\begin{array}{c} L\left(t_1, ..., t_n\right) \vee \mathcal{C} \\ \neg L\left(s_1, ..., s_n\right) \vee \mathcal{D} \end{array}}{\sigma\mathcal{C} \vee \sigma\mathcal{D} \vee \left(t_1 \neq s_1\right) \vee ... \vee \left(t_n \neq s_n\right)}$$

where σ is a substitution. See, for instance, Digricoli & Harrison (1986).

1. Elaborate on the RUE resolution principle.

2. Prove the soundness of the RUE resolution rule.

3. Prove the completeness of RUE resolution.

8.2. Analytic tableaux

This proof system was firstly conceived as *semantic tableaux* by Beth (1955) and Hintikka (1955), whose concerns were mostly semantical; it was later greatly simplified by Smullyan (1968) into the variant known as *analytic tableaux*. We concentrate on the latter, whose general procedure is given as Algorithm 8.2. Further specifications will be given in the appropriate places.

Analytic tableaux is a remarkably efficient refutation-proof procedure for classical logic, and some non-classical logics as well (e.g., Augusto, 2020a; Priest, 2008), based on labeled binary trees. Briefly, analytic tableaux are binary trees whose nodes are formulae that are (sub)goals in the proofs; the tree structure concretizes the logical dependence among the (sub)goals. A tree constitutes a proof iff it is a closed tableau, i.e. a tree whose every branch has a contradictory pair of literals.

Recall the definitions above (Section 2.3) of (semi-)decidability. In fact, given a finite set of formulae Γ, the analytic tableaux proof is guaranteed to *terminate* in either a closed or an open tableau, being

---

Algorithm 8.2 Analytic tableaux

Input: A set of formulas $X \subseteq F_L$
Output: A closed/open tableau $\mathcal{T}_X$ of X

Steps:

1. **INITIALIZATION**: In the root of a downwards-growing binary tree, put:

   a) $\neg\chi$ if $X = \{\chi\}$;

   b) $\chi_1 \wedge ... \wedge \chi_{n-1} \wedge \neg\chi_n$ if $X = \{\chi_1, ..., \chi_{n-1}, \chi_n\}$, $\chi_1, ..., \chi_{n-1}$ are the premises and χ_n is the conclusion of some formalized argument.

2. Apply the **EXPANSION RULES** to the sub-formulas on the tree until they can no longer be applied (i.e. there are only literals).

3. **CLOSE** contradictory branches (i.e. branches containing both L and $\neg L$, where L is a literal).

4. **TERMINATE** successfully iff all branches are closed; unsuccessfully, otherwise.

---

thus indeed a decision procedure–albeit only for propositional logic. If we allow for Γ being infinite, then the tableau construction is guaranteed to terminate only if Γ is unsatisfiable, running forever if Γ is satisfiable. Thus, in the latter case we speak of semi-decidability.

8.2.1. Analytic tableaux as a propositional calculus

We now expand on and formalize the above in terms of a propositional calculus for classical logic.

Definition. 8.2.1. Let the language L0 be given. A *tableau* is a finite binary tree $\mathcal{T}$ whose nodes are formulae from F_{L0}.[6] A *tableau proof system* (or *calculus*) over F_{L0} is a set $RT = \{\mathbf{I}, \mathbf{A}, \mathbf{B}, \mathsf{X}\}$ of *initialization*

---

[6]More properly put, a tableau is *implemented* by a tree (cf. Fitting, 1999).

rule(s) **I**, *expansion rules* **A** and **B** each having a set $\Gamma = F'_{L0}$ of formulae from F_{L0}, and *closure rule(s)* **X**. For each Γ, the transitive closure of the rules in RT defines a set of tableaux constructed with RT for Γ.

Definition. 8.2.2. In a tableau $\mathcal{T}_\Gamma$ for a set of formulae $\Gamma = F'_{L0}$, a branch $\mathcal{B}$ of $\mathcal{T}_\Gamma$ is *closed* iff $\mathcal{B} \cup \Gamma$ contains either a pair $(L, \neg L) \in F_{L0}$ or $\bot$. Otherwise, $\mathcal{B}$ is *open*. A tableau is closed iff all its branches are closed.

Definition. 8.2.3. A *tableau proof* of a set $\Gamma = F'_{L0}$ is a closed tableau $\mathcal{T}_\Gamma$.

Proposition. 8.2.4. *Analytic tableaux constitute a refutation-proof procedure for propositional logic.*

Proof: Left as an exercise. (Hint: Algorithm 8.2, Step 1.)

Algorithm 8.2 can be considered a decision proof procedure for propositional logic, not really requiring further specifications besides the following ones:

Definition. 8.2.5. Given a set $\Gamma = F'_{L0}$ of formulae from F_{L0}, a *tableau for Γ* is defined as a tableau constructed according to the following rules:

1. *Initialization rule:* The tree $\mathcal{T}_\Gamma$ consisting of a single node t is a tableau for Γ.

2. *Expansion rule:* Let $\mathcal{T}_\Gamma$ be a tableau for Γ, and $\mathcal{B}$ a branch of $\mathcal{T}_\Gamma$. Let further ϕ be a formula in $\mathcal{B} \cup \Gamma$. Obtain the tree $\mathcal{T}'_\Gamma$ by expanding $\mathcal{B}$ with n new subtrees whose nodes are the formulae in the expansion of the rule instance. Then, $\mathcal{T}'_\Gamma$ is a tableau for Γ.[7]

3. *Closure rule:* Given a tableau $\mathcal{T}_\Gamma$ and a branch $\mathcal{B}$ thereof, if we have $(L, \neg L) \in \mathcal{B} \cup \Gamma$, then $\mathcal{B}$ is a closed branch of $\mathcal{T}_\Gamma$. If all branches of $\mathcal{T}_\Gamma$ are closed, then $\mathcal{T}_\Gamma$ is a closed tableau for Γ.

The closing of a branch is typically indicated by the symbol X. The notions of unicity of decomposition and immediate sub-formula (cf. Prop. 3.1.4 and Def. 3.1.5) allow us to act in such a way as to have a final

---

[7]The order of application of the expansion rules is non-deterministic. Moreover, one can repeat formulae, though this may increase the complexity of the tree. We say that a tableaux implementation is *fair* if we apply every rule that is applicable in the case at hand.

set of (negated) atoms by proceeding to a step-by-step decomposition of complex formulae. Underlying the construction of the tree is the rewriting of all the formulae into equivalent (negations of) conjunctions and disjunctions in the usual ways, being a tree thus in fact a disjunction of conjunctions.[8] Now, recall that for a literal L we have $(L \wedge \neg L) = \bot$, and that in turn $(\bot \vee \bot)$ is also a contradiction; this explains why a disjunction of finitely many $\bot$-containing branches constitutes a closed tree, i.e. a contradiction of the negated formula. Recall now the classicality conditions for the $\heartsuit$-consequences in Section 4.2.1.

Definition. 8.2.6. The expansion rules and the closure rule, expressed in the language of set theory and accounted for in terms of the classical logical consequence relation, are defined as follows:

1.

$$(\wedge_{RE}) \qquad \frac{X \cup \{\phi \wedge \psi\}}{X \cup \{\phi, \psi\}}$$

given that $\wedge$ is classical in terms of $\Vdash$ iff we have $X, \phi \wedge \psi \Vdash \chi$ iff $X, \phi, \psi \Vdash \chi$.

2.

$$(\vee_{RE}) \qquad \frac{X \cup \{\phi \vee \psi\}}{X \cup \{\phi\} \mid X \cup \{\psi\}}$$

given that $\vee$ is classical in terms of $\Vdash$ iff we have $X, \phi \vee \psi \Vdash \chi$ iff $X, \phi \Vdash \chi$ and $X, \psi \Vdash \chi$.

3.

$$(\mathsf{X}) \qquad \frac{X \cup \{L, \neg L\}}{\mathsf{X}}$$

because classically we have $X, \phi, \neg \phi \Vdash \chi$, i.e. *ex contradictione quodlibet*, or "explosion."

Remark. **8.2.7.** The expansion rules can be conveniently reduced to two according to what is known as the $\alpha\beta$-classification for the formulae (Fig. 8.2.1).[9]

---

[8] Indeed, a tableau can be a conjunction of disjunctions, but this typically increases the size of the tableau, and it is thus a less efficient proof procedure.

[9] By convention, doubly negated formulae are treated as formulae of type α, rather than as an elimination rule. One can further treat $\neg\top$ and $\neg\bot$ as type-α formulae, resulting in the nodes labeled with $\bot$ and $\top$, respectively.

α	α_1	α_2
$A \wedge B$	A	B
$\neg(A \vee B)$	$\neg A$	$\neg B$
$\neg(A \to B)$	A	$\neg B$
$\neg\neg A$	A	A

β	β_1	β_2
$\neg(A \wedge B)$	$\neg A$	$\neg B$
$A \vee B$	A	B
$A \to B$	$\neg A$	B

Figure 8.2.1.: Analytic tableaux expansion rules: $\alpha\beta$-classification.

Definition. 8.2.8. Let $\mathcal{T}_\Gamma$ be an analytic tableau for $\Gamma = \{\alpha, \beta\}$, Γ is a set of propositional formulae. The expansion of a formula α in a single branch $\mathcal{B}$ of $\mathcal{T}_\Gamma$ as

$$\textbf{(A)} \qquad \frac{\alpha}{\begin{array}{c}\alpha_1 \\ \alpha_2\end{array}}$$

is called an A-*rule*. The expansion of a formula β into two branches $\mathcal{B}'$ and $\mathcal{B}''$ of $\mathcal{T}_\Gamma$ as

$$\textbf{(B)} \qquad \frac{\beta}{\beta_1 \mid \beta_2}$$

is called a B-*rule*.

Informally, these two rules mean that if α is a conjunction of α_1 and α_2, then both these two sub-formulae are logical consequences of α and thus are nodes in the one and same branch of the tree (rule **A**); if β is a disjunction of β_1 and β_2, then these sub-formulae originate two different branches as a logical consequence of β (rule **B**). As said above, a tree implementing an analytic tableau is a disjunction of conjunctions; this is the result of giving priority to α-type formulae over β-type formulae in terms of decomposing (Smullyan, 1968).

Theorem 8.2.1. *For any propositional tableau, after a finite number of steps no more expansion rules will be applicable.*

Proof: (Sketch) The theorem holds assuming that we analyze each formula at most once. Let us begin by assuming that we have a formula with $n = 0$ connectives. Then, this is a propositional atom, and no expansion rules apply. Let us now assume that the theorem holds for any formula with at most n connectives. Then we can prove it for a formula ϕ with $n + 1$ connectives. This can be done in two ways, depending on the type of the formula:

1. ϕ is a formula of type α. We apply rule **A**, and we mark ϕ as analyzed once. Clearly, α_1 and α_2 contain each fewer connectives with

relation to ϕ; we apply the inductive hypothesis, and say that a tableau
can be built such that each formula is analyzed at most once. After
a finite number of steps, no more expansion rules can be applied. We
combine the two proofs, and the theorem is proved for a type-α formula.

2. ϕ is a formula of type β. We apply rule **B**, and we mark ϕ as analyzed once. Clearly, β_1 and β_2 contain each fewer connectives with relation to ϕ; we apply the inductive hypothesis and say that two tableaux
can be built for β_1 and β_2 such that each formula is analyzed at most
once. After a finite number of steps, no more expansion rules can be
applied. We combine the two proofs, and the theorem is proved for a
type-β formula. **QED**

Recall the sub-formula property of a sequent calculus proof (cf. Prop.
5.3.5). It is evident that this property is at play in the proof above.
In effect, we say that an expansion rule is *analytic* whenever its every
application yields the sub-formula property. Formalizing this:

Definition. 8.2.9. We say that an expansion rule is *analytic* if every
formula that occurs as a conclusion of the rule–i.e. α_i, β_i, for $i = 1, 2$–is
a sub-formula of the formula that occurs as the premise of the rule (i.e.
α, β).

This shows the interesting relation between the sequent calculus and
analytic tableaux: a downward growing tree in the latter is an upward
growing tree in the former. This has to do with the *invertibility* of the
rules of the sequent calculi (e.g., D'Agostino, 1999).

Example 8.2.1. Figure 8.2.2 shows an analytic tableaux proof of
$((A \to B) \land ((A \land B) \to C)) \to (A \to C)$. Note on the left the numbered steps and on the right the application of the corresponding rules
of expansion.

Although analytic tableaux is a proof system, it actually is a hybrid
system in the sense that a tableau proof is a *counter-model* with respect
to some semantics. In effect, a tableaux proof procedure is so with respect to the negation of the formula one wishes to prove. Informally, a
counter-model corresponds to a tree whose branches are partial descriptions of the model, where the $\alpha\beta$-classification is to be understood as
follows:

Proposition. 8.2.10. *Under any interpretation $\mathcal{I}$, the following facts
clearly hold:*

1. $\neg(((A \to B) \land ((A \land B) \to C)) \to (A \to C))$

2.	$(A \to B) \land ((A \land B) \to C)$	A (1)
3.	$\neg(A \to C)$	A (1)
4.	$A \to B$	A (2)
5.	$(A \land B) \to C$	A (2)
6.	A	A (3)
7.	$\neg C$	A (3)

8. $\quad \neg A \quad$ B (4) $\qquad\qquad$ 9. $\quad B \quad$ B (4)

X

10. $\quad \neg(A \land B) \quad$ B (5) $\qquad$ 11. $\quad C \quad$ B (5)

X

12. $\quad \neg A \quad$ B (10) $\qquad$ 13. $\quad \neg B \quad$ B (10)

X $\qquad\qquad\qquad$ X

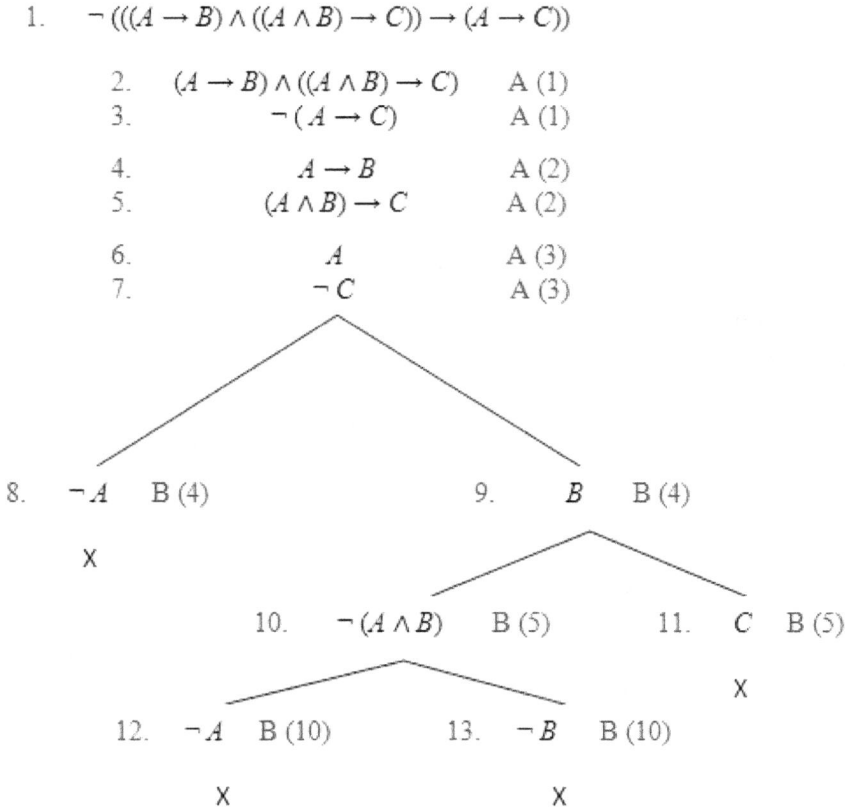

Figure 8.2.2.: A propositional tableau proof.

1. $(\alpha_\top)$ $\quad \alpha$ *is true iff* α_1, α_2 *are both true;*

2. $(\beta_\top)$ $\quad \beta$ *is true iff at least one of* β_1, β_2 *is true.*

Proof: Left as an exercise.

As we already know, a branch in a tree is said to close if both literals L and $\neg L$ are in it (i.e. if they are nodes of the same branch), and the tree itself is said to close if all its branches close. We next formalize this, leaving all the proofs as exercises.

Proposition. 8.2.11. *A tableau $\mathcal{T}_\Gamma$ for a set of formulae $\Gamma \subset F_{L0}$ is satisfiable iff there is a model $\mathcal{M}_\Gamma$ of Γ such that for every valuation* val *of the formulae of Γ there is a branch $\mathcal{B}$ of $\mathcal{T}_\Gamma$ with $\models_{val_\mathcal{M}} \mathcal{B}$. We then say that $\mathcal{M}$ is a model of $\mathcal{T}_\Gamma$, and we write $\models_\mathcal{M} \mathcal{T}_\Gamma$.*

8. Automated theorem proving

Proof: Left as an exercise.

Proposition. 8.2.12. *No model or valuation can satisfy a closed branch, so a closed tree $\mathcal{T}_\Gamma$ is a proof of the unsatisfiability of Γ.*

Proof: Left as an exercise.

In terms of satisfiability, we have the following facts with respect to the expansion rules for formulae of types α and β:

Proposition. 8.2.13. *Let $X \subseteq F_{L0}$ be a set of formulae. Then the following facts hold:*

1. (α_{sat}) *If X is satisfiable and $\alpha \in X$, then $\{X, \alpha_1, \alpha_2\}$ is satisfiable.*

2. (β_{sat}) *If X is satisfiable and $\beta \in X$, then at least one of $\{X, \beta_1\}$, $\{X, \beta_2\}$ is satisfiable.*

Proof: Left as an exercise.

Propositions 8.2.10 and 8.2.13 are fundamental in that they state that the decomposition of formulae in the propositional tableaux calculus preserves truth and satisfiability.

We now require a few further definitions in order to address the soundness and completeness of the propositional analytic tableaux calculus.

Definition. Let $\mathcal{T}$ be a tableau. We say that a branch $\mathcal{B} \in \mathcal{T}$ is *complete* if for every α occurring in $\mathcal{B}$, both α_1 and α_2 occur in $\mathcal{B}$, and for every β occurring in $\mathcal{B}$, at least one of β_1, β_2 occurs in $\mathcal{B}$. We call a tableau $\mathcal{T}$ *complete(d)* if every branch of $\mathcal{T}$ is either closed or complete.

Theorem 8.2.2. *(Soundness of propositional analytic tableaux) Every formula provable by the analytic tableaux calculus is a tautology.*

Proof: (Sketch) Let Γ be a set of formulae, possibly a singleton, and let $\mathcal{I}$ be an interpretation. Let $\models_{\mathcal{I}} \mathcal{T}_\Gamma^{(0)}$, $\mathcal{T}_\Gamma^{(0)}$ is the completed tableau for Γ. Then, for any subtree $\mathcal{T}_\Gamma^i$, $0 < i \leq n$, we have $\models_{\mathcal{I}} \mathcal{T}_\Gamma^i$ by Propositions 8.2.10 and 8.2.13, i.e. any interpretation satisfying all the formulae in Γ must also satisfy all the formulae contained in at least one complete branch of any completed tableau for Γ. Thus, for any completed tableau $\mathcal{T}$, and applying induction and the analyticity property of the tableaux calculus (Def. 8.2.9), if the root is true under an interpretation $\mathcal{I}$, then

416

$\mathcal{T}$ must be true under $\mathcal{I}$. It is evident (by Proposition 8.2.12) that a closed tableau cannot be true under any interpretation, and therefore its root cannot be true under any interpretation. Hence, every formula provable by the analytic tableaux method must be a tautology. **QED**

Corollary 8.2.3. *(Consistency of propositional analytic tableaux) The tableau method for classical propositional logic is consistent.*

Proof: From the proof of the soundness theorem, we conclude that no formula and its negation are both provable in the analytic tableaux calculus. **QED**

We turn now to the completeness theorem of analytic tableaux. In order to prove this theorem a few more fundamental notions are required.

Definition. 8.2.15. Let $X \subseteq F_{L0}$. We say that X is a *downward saturated set* iff

1. if $\alpha \in X$, then $\alpha_1 \in X$ and $\alpha_2 \in X$;

2. if $\beta \in X$, then $\beta_1 \in X$ or $\beta_2 \in X$.

Note that this entails that a tableau $\mathcal{T}_X$ for a downward saturated set X is completed in the sense that every branch $\mathcal{B}$ of $\mathcal{T}_X$ is complete.

Definition. 8.2.16. A downward saturated set that does not contain an (atomic) formula and its negation is called an *(atomic) Hintikka set.*

Lemma 8.2.4. *(Hintikka's lemma) Every downward saturated set X (whether finite or infinite) is satisfiable.*

Proof: Let X be a Hintikka set. We need to show that some Herbrand interpretation HI_X of X is a model of X.[10] Let H_X be the set of ground terms in X. Because $H(X)$ by definition (of X as a Hintikka set) does not contain an atomic formula and its negation, it immediately defines a Herbrand interpretation HI_X. Assume that $HI_\phi = \mathbf{t}$ for all formulae $\phi \in X$ with complexity less than n, where by complexity of a formula χ it is meant the number n of occurrences of formulae in χ. Now consider any complex formula $\psi \in X$ with complexity n. It is evident that the complexity of any tableau sub-formula of ψ is less than n. Clearly, ψ must be a α- or a β-type formula, and we have two cases.

---

[10]Recall the material from Section 6.2.

Case 1: If ψ is a α-type formula, then by definition of downward saturation every α_i is in X. By induction, and by $\alpha_\top$ (Prop. 8.2.10.1), we have $H\mathcal{I}_{\alpha_i} = \mathbf{t}$, $H\mathcal{I}_{\phi,\psi} = \mathbf{t}$, and of course $H\mathcal{I}_X = \mathbf{t}$.

Case 2: If ψ is a β-type formula, then by definition of downward saturation some β_i is in X. By induction, and by $\beta_\top$ (Prop. 8.2.10.2), we have $H\mathcal{I}_{\beta_i} = \mathbf{t}$, $H\mathcal{I}_{\phi,\psi} = \mathbf{t}$, and of course $H\mathcal{I}_X = \mathbf{t}$. **QED**

The proof of Hintikka's lemma, together with α_{sat} and β_{sat} (Prop.s 8.2.13.1-2), actually proves the following important theorem:

Theorem 8.2.5. *(Smullyan, 1968) Any complete open branch of any tableau is (simultaneously) satisfiable.*

In turn, this theorem implies the completeness of the propositional analytic tableaux calculus.

Theorem 8.2.6. *(Completeness of propositional analytic tableaux) (a) If ψ is a tautology, then every complete tableau starting with $\neg\psi$ must close. (b) Every tautology is provable in the tableaux calculus.*

The derivation of (a) from Theorem 8.2.5 runs as follows: Assume that $\mathcal{T}_\psi$ is a complete tableau starting with $\neg\psi$. By Theorem 8.2.5, if $\mathcal{T}_\psi$ is open, then $\neg\psi$ is satisfiable. Therefore, ψ cannot be a tautology. Hence, if ψ is a tautology, then $\mathcal{T}_\psi$ must be closed.

The proofs of Theorems 8.2.5-6 are left as exercises.

Exercises

Exercise 8.2.1.1. Prove items 1-10 of Exercise 5.1.1 in the analytic tableau calculus.

Exercise 8.2.1.2. Test the validity of the argument in Example 5.2.2 in the analytic tableaux calculus.

Exercise 8.2.1.3. Test the validity of the arguments of Exercise 8.1.1.4 in the analytic tableaux calculus.

Exercise 8.2.1.4. Prove Propositions 8.2.10-13.

Exercise 8.2.1.5. Complete the proof of Theorems 8.2.1-2 and prove Theorems 8.2.5-6.

8.2.2. Analytic tableaux as a FO predicate calculus

The introduction of analytic tableaux rules for the quantifiers motivates some problems, in particular because the set of constants is infinite, a problem that actually accounts for the non-terminating character of tableaux proofs in FOL. In effect, Theorem 8.2.1 does *not* hold for FO tableaux. In any case, these problems can be overcome in several ways, so that the tableaux calculus is sound and complete for FOL. Our elaboration of this FO calculus falls only on the essential points thereof; in particular, we leave the proofs of the soundness and completeness as exercises, and we refer the reader interested in applying this calculus to CL$^=$ to Fitting (1996).

Definition. 8.2.17. Let Υ be a signature for the language L1. A *tableau* (over Υ) is a finite tree whose nodes are formulae from F_{L1}. A *tableau proof system* (or *calculus*) over F_{L1} is a set $RT = \{I, A, B, C, D, X\}$ where I, A, B, X are essentially as in Definition 8.2.1 and C, D are additional *rules of extension for FO formulae*.

Definitions 8.2.1-3 are thus generalizable to classical FOL; the same holds for formulae of types α and β (with the proviso that now by "formula" it is understood "closed formula of quantification theory"; cf. Smullyan, 1968). This sets the formal scenario for approaching tableaux for FOL.

Remark. **8.2.18.** Just as in the case of propositional formulae, there are only two expansion rules when quantifiers are involved: Given the FO formulae γ and δ, we now have the $\gamma\delta$-classification (Fig. 8.2.3). It is evident that Definition 8.2.9 generalizes to FOL.

γ	$\gamma\,(a)$
$\forall x A\,(x)$	$A\,(a)$
$\neg\exists x A\,(x)$	$\neg A\,(a)$

δ	$\delta\,(a)$
$\exists x A\,(x)$	$A\,(a)$
$\neg\forall x A\,(x)$	$\neg A\,(a)$

Figure 8.2.3.: Analytic tableaux expansion rules: $\gamma\delta$-classification.

Definition. 8.2.19. Let $\phi \in F_{L1}$. Formula ϕ is said to be of type γ if either $\phi = \forall x A\,(x)$ or $\phi = \neg\exists x A\,(x)$, in which case we apply the expansion rule

$$\textbf{(C)} \qquad \frac{\gamma}{\gamma\,(a)}$$

known as C-*rule*, where a is a ground term or a Skolem constant (more strictly: a parameter). Formula ϕ is said to be of type δ if either $\phi = \exists x A\,(x)$ or $\phi = \neg\forall x A\,(x)$, in which case we apply the expansion rule

$$(\mathbf{D}) \qquad \frac{\delta}{\delta\,(a)}$$

where a is new to the branch, known as D-*rule*.

By $\gamma\,(a)$ or $\delta\,(a)$ it is meant, for a formula ϕ, ϕ_a^x or $\neg\phi_a^x$, i.e. the formula ϕ or its negation with x substituted by a.

Just as in the case of the $\alpha\beta$-classification, there is a semantical account for the expansion rules for formulae of types γ and δ that assures us that the decomposition of these latter types of formulae preserves truth and satisfiability. We leave the proofs of the following propositions as exercises.

Proposition. 8.2.20. *Under any interpretation $\mathcal{I}$ for a domain $\mathcal{D}$, the following facts hold:*

1. (γ_T) γ *is true iff $\gamma\,(a)$ is true for every $a \in \mathcal{D}$;*

2. (δ_T) δ *is true iff $\delta\,(a)$ is true for at least one $a \in \mathcal{D}$.*

Proof: Left as an exercise.

In terms of satisfiability, we have the following facts:

Proposition. 8.2.21. *Let $X \in F_{L1}$ be a set of formulae. Then,*

1. (γ_{sat})   *If X is satisfiable and $\gamma \in X$, then for every constant (more strictly: parameter) a the set $\{X, \gamma\,(a)\}$ is satisfiable.*

2. (δ_{sat})   *If X is satisfiable and $\delta \in X$, and if a is a constant (more strictly: parameter) that does not occur in any element of X, then $\{X, \delta\,(a)\}$ is satisfiable.*

Proof: Left as an exercise.

However, the above classifications are all too general. Importantly, we need to discuss separately tableaux for FOL with and without unification, as the expansion rules for the quantifiers are different for each.

8.2.2.1. FOL tableaux without unification

To begin with, some useful heuristics: whenever possible, apply propositional rules before quantifier rules; in the latter case, apply the tableau rules on δ-formulae before γ-formulae. "Beyond this, you are on your own" (Fitting, 1996), a remark that works as a reminder that analytic tableaux is *not* a decision procedure for FOL, for the simple reason that no such procedure exists (see above). In particular, misapplication of the tableau rule for γ-formulae may cause the tableau not to close. The analytic tableaux rules for the quantifiers without unification are defined as in Definition 8.2.19, but with the following specifications:

Definition. 8.2.22. The expansion C-*rule without unification* for a formula γ is

$$\textbf{(C)} \qquad \frac{\forall x \gamma (x)}{\gamma (t)}$$

where t is any ground term. The expansion D-*rule without unification* for a formula δ is

$$\textbf{(D)} \qquad \frac{\exists x \delta (x)}{\delta (c)}$$

where the constant symbol c is new to the branch.

Note that in $\exists x \delta (x)$, the variable x does not occur within the scope of a universal quantifier. Therefore, the Skolemization of $\exists x \delta (x)$ generates solely a constant, i.e. a 0-ary function.

Example 8.2.2. To prove by means of analytic tableaux the validity of the formula $\forall x \, (F (x) \wedge G (x)) \leftrightarrow \forall x \, (F (x)) \wedge \forall x \, (G (x))$ we apply rules **C** and **D**, i.e. the expansion rules for the universal quantifier without unification. See Fig. 8.2.4 for the respective tableau proof.

Definition. 8.2.23. Given a tree $\mathcal{T}_\Gamma$, for $\Gamma \subset F_{L1}$, and a branch $\mathcal{B}$ thereof, let $(L, \neg L) \in \mathcal{B} \cup \Gamma$, L is an atom. Then $\mathcal{B}$ is a closed branch of $\mathcal{T}_\Gamma$. If all branches of $\mathcal{T}_\Gamma$ are closed, then $\mathcal{T}_\Gamma$ is a closed tableau for Γ.

The proof of soundness of the predicate analytic tableaux calculus without unification is as for the propositional case, but appealing now also to Propositions 8.2.20-1. In order to prove the completeness of tableaux for FOL we need to extend our definition of a downward saturated set (Def. 8.2.15) to sets containing γ- and δ-type formulae:

Definition. 8.2.24. Let $X \subseteq F_{L1}$. We say that X is a downward saturated set iff conditions 1 and 2 of Def. 8.2.15 hold, and additionally

1. $\neg[\forall x(F(x) \wedge G(x)) \leftrightarrow (\forall x(F(x) \wedge \forall x(G(x)))]$

2. $\forall x(F(x) \wedge G(x))$ B, A (1)
3. $\neg[\forall x(F(x)) \wedge \forall x(G(x))]$ B, A (1)

4. $\neg[\forall x(F(x) \wedge G(x))]$ B, A (1)
5. $\forall x(F(x)) \wedge \forall x(G(x))$ B, A (1)

6. $\neg\forall x(F(x))$ B (3)
7. $\neg\forall x(G(x))$ B (3)

8. $\neg F(a)$ D (6)
9. $F(a) \wedge G(a)$ C (2)

10. $F(a)$ A (9)
11. $G(a)$ A (9)
  x

12. $\neg G(a)$ D (7)
13. $F(a) \wedge G(a)$ C (2)

14. $F(a)$ A (13)
15. $G(a)$ A (13)
  x

16. $\forall x(F(x))$ A (5)
17. $\forall x(G(x))$ A (5)

18. $\neg(F(a) \wedge G(a))$ D (4)

19. $\neg F(a)$ B (18)
20. $\neg G(a)$ B (18)

21. $F(a)$ C (16)
22. $G(a)$ C (17)
  x   x

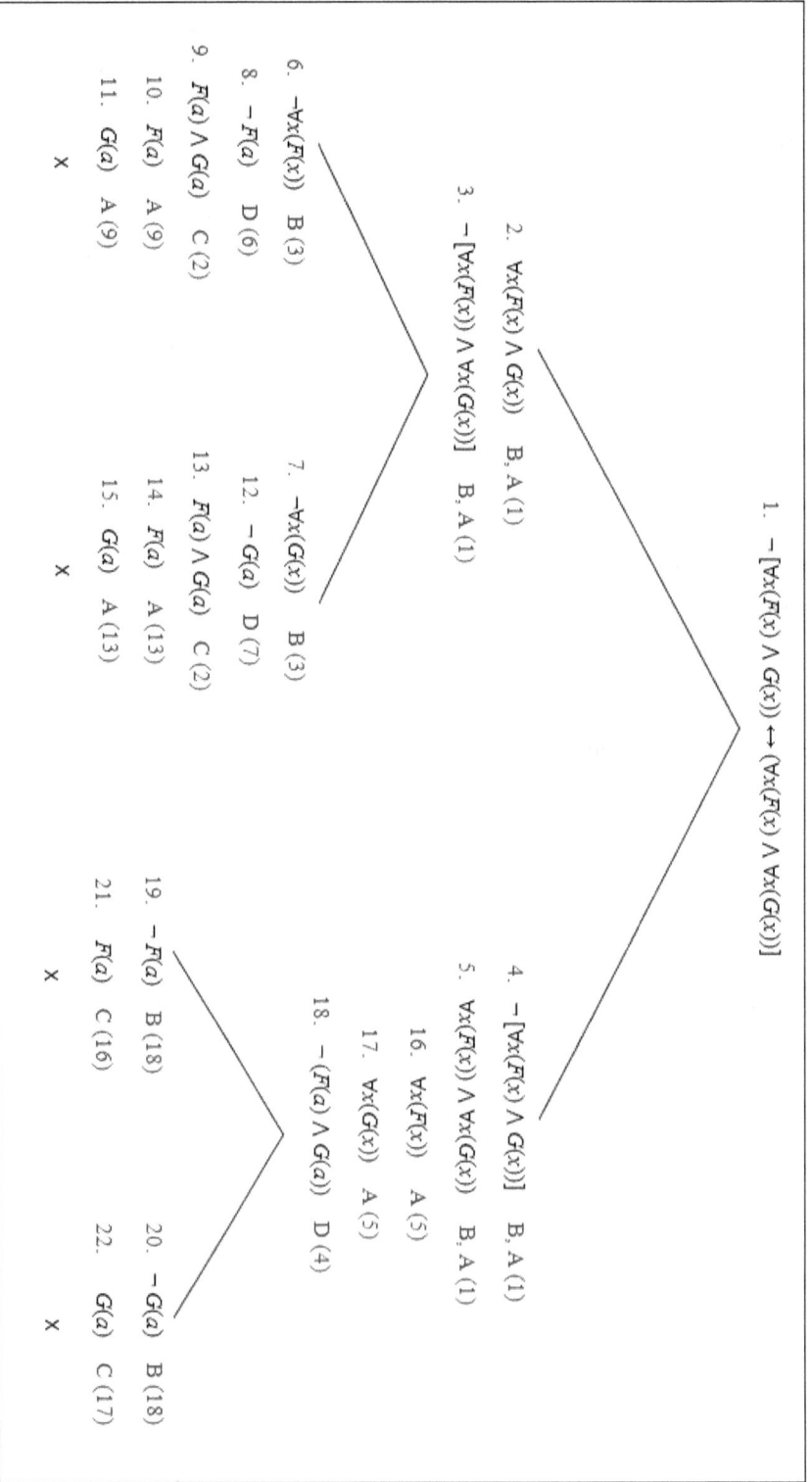

Figure 8.2.4.: A FO tableau proof without unification.

1. if $\gamma \in X$, then $\gamma(t) \in X$ for all $\gamma(t)$ with $t \in H_X$;

2. if $\delta \in X$, then $\delta(c) \in X$ for at least one $c \in Cons \subseteq \Upsilon_X$.

The definition of Hintikka set (Def. 8.2.16) generalizes immediately to the sets above, and so does Lemma 8.2.4. Thus, the proof of completeness of the predicate analytic tableaux calculus without unification runs as for the propositional case, with the additional assumptions that $H\mathcal{I}_{\gamma(t)} = \mathtt{t}$ for any $t \in H_X$ and $H\mathcal{I}_{\delta(c)} = \mathtt{t}$ for some constant $c \in Cons \subseteq \Upsilon_X$. In the former case, let $\phi \in X$, ϕ is a formula of type $\gamma = \forall x \phi'$, where ϕ' is an *immediate tableau sub-formula of* ϕ (in this case, $\phi' = \phi[x/t]$); by the property of being a downward saturated set and by the induction assumption, we have $H\mathcal{I}_{\gamma(t)} = \mathtt{t}$ for any term $t \in H_X$. Because H_X is the universe of $H\mathcal{I}_X$, and $H\mathcal{I}_X$ maps any term t to itself (cf. Def. 6.2.3), for all variable assignments ϖ to H_X we have $H\mathcal{I}_{\phi'}^{\varpi} = H\mathcal{I}_{\phi'[x/\varpi(x)]} = \mathtt{t}$. In the latter case, $\phi \in X$ is a δ-formula; by downward saturation and the induction hypothesis we have it that $H\mathcal{I}_{\delta(c)} = \mathtt{t}$ for some constant $c \in T$ and therefore $H\mathcal{I}_\phi = \mathtt{t}$.

8.2.2.2. FOL tableaux with unification

Note that when applying rule **C** without unification we have the problem of choosing the ground term t. In effect, any possible ground term can be chosen, or even every one, but this would certainly not be relevant for the tableau proof. The problem is particularly acute in terms of automation, as this is basically impossible without guidance on what term(s) to choose. The strategy then is to delay the choice of the ground term until $\gamma(t)$ allows the closure of *at least* one of the branches of the tree, though evidently a substitution σ that will *simultaneously* close *all* branches is what is required. To do this, we simply use a variable instead of a ground term, so that the C-rule now is as follows:

Definition. 8.2.25. The expansion C-*rule with unification* for a formula γ is

$$(\mathbf{C'}) \qquad \frac{\forall x \gamma(x)}{\gamma(x')}$$

where x' is a variable not occurring anywhere else in the tableau.

Just as for rule **D**, we apply Skolemization again, but now Skolem terms must not be constants, because the application of unification may create free variables that are implicitly universally quantified. This means that a formula $\exists x \delta(x)$ can now be within the scope of one or more universal quantifiers. According to Def. 7.2.9, we now have:

Definition. 8.2.26. The expansion D-*rule with unification* for a formula δ is

$$(\mathbf{D'}) \qquad \frac{\exists x \delta(x)}{\delta(f(x_1, ..., x_n))}$$

where f is a new function symbol and $x_1, ..., x_n$ are the n free variables of δ.

The Skolemization procedure can be made technically more rigorous by making it explicit that f is a function assigning to each $\delta \in \breve{\Upsilon}$ a symbol $\bigcirc \in Fun_{Sko}$, the set of Skolem functions (cf. Def. 7.2.10), $Fun_{Sko} \cap \Upsilon = \emptyset$, $\breve{\Upsilon} = \Upsilon \cup Fun_{Sko}$, such that $\bigcirc > f$ for all $f \in Fun_{Sko}$ occurring in δ for $>$ an arbitrary but fixed ordering on Fun_{Sko}, and for all $\delta, \delta' \in \Upsilon$ the symbols $\bigcirc$ and $\bigcirc'$ are identical iff δ and δ' are identical up to variable renaming, so that we have the rule

$$(\mathbf{D'}) \qquad \frac{\exists x \delta(x)}{\delta(\bigcirc(x_1, ..., x_n))}$$

where $x_1, ..., x_n$ are the free variables in δ. This δ-rule is formulated in Beckert, Hähnle, & Schmitt (1993).

Definition. 8.2.27. The above entail a new rule for FO tableaux, known as *substitution rule for FO tableaux*:

$$(\mathcal{T}\mathbf{Sub}) \qquad \text{Modify } \mathcal{T} \text{ to } \mathcal{T}_\sigma$$

where σ is free for all formulae in $\mathcal{T}$.

$\mathcal{T}\mathbf{Sub}$ is specific to FO tableaux with unification. As a matter of fact, it is a "destructive" rule, whereas all previous rules are conservative in the sense that the initial tableau is not modified but merely expanded.[11]

In the case of free-variable tableaux, a reformulation of satisfiability is required (Hähnle & Schmitt, 1994):

Definition. 8.2.28. $\forall$-*satisfiability* – A collection $\mathscr{C}$ of sets of FO formulae is said to be $\forall$-satisfiable if there is an interpretation $\mathcal{I}$ such that, for every variable assignment ϖ, $\mathcal{I}$ is a $\{\varpi\}$-model for some element of $\mathscr{C}$.

Then, for closed formulae, for some element of $\mathscr{C}$ we have it that $\forall$-satisfiability coincides with satisfiability. Propositions 8.2.13 and 8.2.21 can easily be reformulated for $\forall$-satisfiability.

We finally rephrase the closure rule above (Def. 8.2.23) for our FO language L1 with formulae with free variables as follows:

---

[11]See Letz (1999) for an elaboration.

Definition. 8.2.29. Given a tableau $\mathcal{T}_\Gamma$ for $\Gamma \subset F_{L1}$ and a branch $\mathcal{B}$ thereof, let $L, \neg L \in (\mathcal{B} \cup \Gamma)$, L is an atom. Let further L and $\neg L$ be unifiable by means of the MGU σ. Then $\mathcal{B}$ is a closed branch of $\mathcal{T}_{\sigma\Gamma}$, the tree for $\sigma\Gamma$ constructed by applying σ to all formulae of Γ. If all branches of $\mathcal{T}_{\sigma\Gamma}$ are closed, then $\mathcal{T}_{\sigma\Gamma}$ is a closed tableau for $\sigma\Gamma$.

Example 8.2.3. Fig. 8.2.5 shows a proof by means of a tableau with unification that the set $A = \{\forall x\, (P\,(x))\,, \exists x\, (\neg P\,(x) \vee \neg P\,(f\,(x)))\}$ is unsatisfiable. Rule $\mathbf{D'}$ was not applied, as $\exists x \delta\,(x)$ does not fall within the scope of a universal quantifier. For further examples, see Letz (1999) and Fitting (1996).

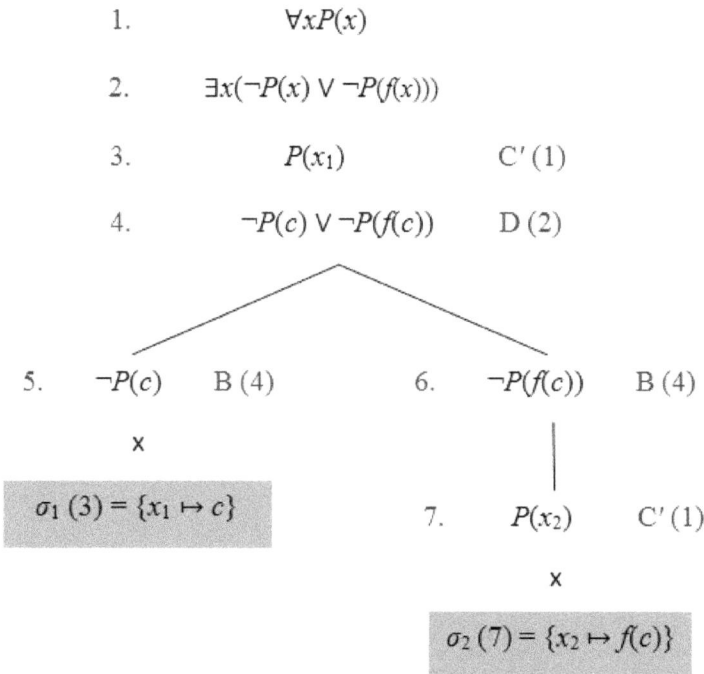

1. $\forall x P(x)$

2. $\exists x(\neg P(x) \vee \neg P(f(x)))$

3. $P(x_1)$ C' (1)

4. $\neg P(c) \vee \neg P(f(c))$ D (2)

5. $\neg P(c)$ B (4)   6. $\neg P(f(c))$ B (4)

         ×

$\sigma_1\,(3) = \{x_1 \mapsto c\}$ 7. $P(x_2)$ C' (1)

         ×

       $\sigma_2\,(7) = \{x_2 \mapsto f(c)\}$

Figure 8.2.5.: A FO tableau with unification.

Exercises

Exercise 8.2.2.1. Prove in the FO analytic tableaux calculus the items 11-15 of Exercise 5.1.1.

Exercise 8.2.2.2. Prove in the FO analytic tableaux calculus the arguments of Exercise 8.1.2.2.

Exercise 8.2.2.3. Prove Propositions 8.2.20-1.

Exercise 8.2.2.4. Reformulate Proposition 8.2.21 in terms of ∀-satisfiability.

Exercise 8.2.2.5. Prove the soundness and completeness of the FO analytic tableaux calculus.

9. Programming

Logic programming (LP) is a programming paradigm based on the notion of a program as defining a set of consequences, so that a computation is actually a deduction of consequences of the program. We call this *deductive computation* in general (cf. Section 3.5); more narrowly, we speak of *deductive programming*.

In effect, a *logic program* is a set of rules or axioms. In particular, these rules and axioms are so of relations between objects, reason why it should be obvious that logic programming languages rely heavily on first- or even second-order logic.[1] Importantly, the semantics of LP is essentially Herbrand semantics (cf. Section 6.2); as it will become evident below, an interpretation for a logic program just is a subset of the Herbrand base.

In this introduction to LP, we provide a general theory and some fundamental implementations. We first approach LP at the metalanguage level, and then discuss it at the object-language level. This latter discussion focuses on more strictly logical aspects of LP; readers seeking a more comprehensive elaboration on this topic can benefit from Gabbay, Hogger, & Robinson (1998) and Doets (1994). As for the implementations, we give the essentials of the main LP language, to wit, Prolog. The objective is to enable the reader to understand both the declarative and the procedural components of this programming language by focusing on the equation *Algorithm = Logic + Control*, i.e. we are interested in both the logical aspects of Prolog and the impact that control has on this fundamental component of this programming language. This understanding having been gained, the reader can easily complement the knowledge of this topic with material provided by the diverse environments available.

Importantly, mastering Prolog allows one to grasp and use easily other related (sub-)languages such as Datalog and Answer Set Programming. Of these we elaborate on Datalog, namely as a solely declarative-interpretatable language. Our interest in Datalog is mainly the integration of LP and databases into *deductive databases*, which we see as an essential component of classical deductive computing with classical

---

[1]We restrict our discussion to FOL.

logic, namely CFOL, or, more briefly in our approach, *classical deductive programming*. In effect, Prolog and Datalog are both subsets of **L1**. This commonality notwithstanding, Datalog does not allow function symbols, reason why we may consider it a "dialect" of Prolog. The case is that deductive databases *extend* relational databases by means of LP. Deductive databases are particularly important from the viewpoint of applications, as, in fact, Prolog is–still–of limited application for classical deductive programming over large collections of data.

For constraints to do with how voluminous this book already is, we do not discuss here LP from the viewpoint of formal grammars; the reader is referred to Deransart & Małuszyński (1993) for this topic that is highly relevant for classical deductive programming.

9.1. Logic programming as deductive programming

9.1.1. Query systems and programming systems

The notion of deductive programming finds theoretical support on a few rather intuitive analogies, or equivalences, between the logical and the computing domains: besides the already mentioned notions of a *program* as a *set of formulae* and a *computation* as a *proof*, further equivalences can be established between *inputs* and *queries*, as well as between *outputs* and *replies* to queries, where both queries and replies are formulae.

These intuitive analogies or equivalences are also clearly formalizable. We leave unchanged the symbols for the logical objects *(set of) formulae*, *proof*, *syntactical consequence*, and *semantical consequence*. We begin by reviewing the notion of a proof system.

Definition. 9.1.1. A *proof system* is a triple $\mathcal{P} = (\Theta, P, \vdash)$ where Θ is a set of formulae, P is a set of proofs, and $\vdash \subseteq 2^{\Theta} \times P \times \Theta$ is a ternary relation of logical consequence. For some $\blacksquare \in P$, the relation $\Theta \vdash_{\blacksquare} \phi$ means that $\blacksquare$ is a proof of ϕ from premises in Θ. $\mathcal{P}$ is a *monotonic* proof system if it is verified that, for $F \subseteq \Theta$ and $\blacksquare \in P$,

$$\text{if } F \vdash_{\blacksquare} \phi \text{ and } F \subseteq G, \text{ then } G \vdash_{\blacksquare} \phi.$$

From the logical viewpoint, the novelty in the above summary is the existence of the logical objects *queries* and *replies*, as well as the relation between both. We formalize this novel aspect.

Definition. 9.1.2. Let the symbol ¿ denote a *query*. A *query system* is a triple $\mathcal{Q} = (\Theta_{¿}, Q, \succ)$, where $\Theta_{¿}$ is a set of formulae that are replies to queries, Q is a set of queries, and $> \subseteq Q \times \Theta_{\dot{\iota}}$ is a binary relation.

Definition. 9.1.3. A proof system for a query system $\mathcal{Q} = (\Theta_{\dot{\iota}}, Q, >)$ is a triple $\mathcal{P}_\mathcal{Q} = (\Theta, P, \vdash)$ where $\Theta \supseteq \Theta_{\dot{\iota}}$ such that $\phi \in \Theta_{\dot{\iota}}$ is a *provably correct* reply to a query $\dot{\iota} \in Q$ if, for some $F \subset 2^\Theta$, we have

$$\dot{\iota} > F \vdash_\blacksquare \phi$$

formalizing the fact that, given the premises (of Θ) in F, we may have a proof $\blacksquare \in P$ that ϕ *as a reply* (denoted by $>$) to the query $\dot{\iota}$. This can be specified according to the condition

$$\dot{\iota} > F \vdash_\blacksquare \phi \quad \text{iff} \quad \dot{\iota} > \phi \text{ and } F \vdash_\blacksquare \phi.$$

We now introduce the fundamental notion of a programming system.

Definition. 9.1.4. A *programming system* is a 6-tuple

$$\mathfrak{P} = (\underline{\varPi}, I, O, \varXi, \triangleright, \vdash)$$

where $\underline{\varPi} = \{\varPi_1, ..., \varPi_n\}$ is a set of programs, $I = \{\iota_1, ..., \iota_n\}$ is a set of inputs, $O = \{o_1, ..., o_n\}$ is a set of outputs, $\varXi = \{\xi_1, ..., \xi_n\}$ is a set of computations, $\vdash$ is a ternary relation on $\underline{\varPi} \times \varXi \times O$ such that for each $\varPi \in \underline{\varPi}$ there is a $\xi \in \varXi$ and an $o \in O$ with

$$(\mathfrak{P}_\vdash) \qquad \varPi \vdash_\xi o$$

and $\triangleright$ is a relation on $I \times \mathfrak{P}_\vdash$ such that for each $\iota \in I$ we have

$$\iota \triangleright \varPi \vdash_\xi o$$

denoting that given input ι one possible computation ξ carried out by program $\varPi$ yields output o. If $\iota \triangleright \varPi \vdash o$ iff there is a computation $\xi \in \varXi$ such that $\iota \triangleright \varPi \vdash_\xi o$, then this defines the *computed-output* relation on $I \times \underline{\varPi} \times O$.

The above definition shows clearly that deductive programming is a specific instance of the more general concept of deductive computation.

Definition. 9.1.5. A programming system $\mathfrak{P} = (\underline{\varPi}, I, O, \varXi, \triangleright, \vdash)$ is said to be *deterministic* iff, for each input $\iota \in I$ and program $\varPi \in \underline{\varPi}$, there is a *unique* computation $\xi \in \varXi$ such that $\iota \triangleright \varPi \vdash_\xi (o)$, and it is said to be *determinate* iff for each input $\iota \in I$ and each program $\varPi \in \underline{\varPi}$ there is *at most one* output $o \in O$ such that $\iota \triangleright \varPi \vdash_{(\xi)} o$. These are the properties of *determinism* and *determinacy*, respectively, of a programming system.

Proposition. 9.1.6. *For a query system* $Q = (\Theta_{\dot{c}}, Q, \gg)$ *with a mono-tonic proof system* $P_Q = (\Theta, P, \vdash)$, $\Theta \supseteq \Theta_{\dot{c}}$, *and a programming system* $\mathfrak{P} = (\underline{\Pi}, I, O, \Xi, \triangleright, \vdash)$ *the following equivalence holds:*

$$\left(\overset{LP}{\underset{\vdash}{\Longleftrightarrow}}\right) \qquad \dot{c} \gg F \vdash_{\blacksquare} \phi \Longleftrightarrow \iota \triangleright \Pi \vdash_{\xi} o.$$

Proof: Left as an exercise.

This equivalence expresses the correspondence between the objects of logic and the objects of computation that account for the notion of deductive computation, and, in this particular case, deductive program-ming, or LP. This equivalence is particularly well established by the fact that we identify the symbol for syntactical *consequence* with the symbol for *computational yield*, an identity that we actually first employed al-ready in Chapter 2 above and formalize more clearly now, namely from the semantical viewpoint.

As seen above, defined by means of a proof system the relation $\gg$ merely determines the *formal acceptability* of a reply to a query. In order to determine the *semantical correctness* of a reply to a query we need to specify a semantics adequate for Q.

Definition. 9.1.7. A *semantical system* for a query system $Q = (\Theta_{\dot{c}}, Q, \gg)$ is a triple $\mathfrak{S} = (\Theta, M, \models)$ where $\Theta \supseteq \Theta_{\dot{c}}$ and M is a set of models such that $\phi \in \Theta_{\dot{c}}$ is a *semantically correct* reply to a query $\dot{c}$, denoted by $\dot{c} \gg K \models_{\mathcal{M}} \phi$ for $K \subseteq \Theta$ and a model $\mathcal{M} \in M$, according to the condition

$$\dot{c} \gg K \models_{\mathcal{M}} \phi \quad \text{iff} \quad \dot{c} \gg \phi \text{ and } K \models_{\mathcal{M}} \phi.$$

If in the above definition we let K denote "explicit knowledge," we begin to see that a program of LP is in fact a *knowledge base*, and the role of the *user* of the program is merely that of asking questions concerning the knowledge base.

Proposition. *9.1.8.* *For a query system* $Q = (\Theta_{\dot{c}}, Q, \gg)$ *with a se-mantical system* $\mathfrak{S} = (\Theta, M, \models)$, $\Theta \supseteq \Theta_{\dot{c}}$, *and a programming system* $\mathfrak{P} = (\underline{\Pi}, I, O, \Xi, \triangleright, \vdash)$ *the following equivalence holds:*

$$\left(\overset{LP}{\underset{\models}{\Longleftrightarrow}}\right) \qquad \dot{c} \gg F \models_{\mathcal{M}} \phi \Longleftrightarrow \iota \triangleright \Pi \vdash_{\xi} o.$$

Proof: Left as an exercise.

The adequateness of a query system $Q = (\Theta_{\dot{\iota}}, Q, >)$ is then established if, if $\dot{\iota} > K \vdash_\blacksquare \phi$ then $\dot{\iota} > K \models_\mathcal{M} \phi$ (soundness), and if $\dot{\iota} > K \models_\mathcal{M} \phi$ then $\dot{\iota} > K \vdash_\blacksquare \phi$ (completeness). Note, however, that by *adequateness* of a query system $Q = (\Theta_{\dot{\iota}}, Q, >)$ it is now meant that a formula $\phi \in \Theta_{\dot{\iota}}$ is both formally acceptable and semantically correct as a reply to a query $\dot{\iota} \in Q$. Generalizing the adequateness of $Q = (\Theta_{\dot{\iota}}, Q, >)$ to LP is then a straightforward matter.

However, we may start with a sound and complete proof system but end up with an incomplete programming system if we do not pay attention to the proof strategies to be of use in the latter. Typically, these proof strategies are actually restricted forms of proofs (e.g., cut elimination), and thus this is a problem that is rather easy to avoid. We concentrate on semantical aspects as these are rather more complex and can lead to serious specification errors.

Definition. 9.1.9. Let $Q = (\Theta_{\dot{\iota}}, Q, >)$ be a query system and $\mathfrak{S} = (\Theta, M, \models)$ be a semantical system with $\Theta_{\dot{\iota}} \subseteq \Theta$. We say that a formula ϕ is a *consequentially strongest correct reply* to the query $\dot{\iota}$ iff for $K \subseteq \Theta$

1. $\dot{\iota} > K \models \phi$;

2. for all $\psi \in \Theta_{\dot{\iota}}$, whenever $\dot{\iota} > K \models \psi$, then $\phi \models \psi$.

The above gives us the semantical equivalence of all consequentially strongest correct replies to a query $\dot{\iota}$. In LP we are particularly interested in questions of the form

$$\dot{\iota} x_1, ..., x_k$$

where $x_1, ..., x_k$ is a list of non-repeating variables.

Definition. 9.1.10. Let now

$$Q = \{(\dot{\iota} x_1, ..., x_k : \phi) \,|\, \phi \in \Theta_{\dot{\iota}}\}.$$

We define the relation $> \subseteq Q \times \Theta_{\dot{\iota}}$ by

$$(\dot{\iota} x_1, ..., x_k : \phi) > \psi$$

iff

$$\psi = \phi\,[x_1/t_1, ..., x_k/t_k]$$

for some terms $t_1, ..., t_k \in K \subseteq \Theta_{\dot{\iota}}$.

9. Programming

In a query as specified above, we are interested in knowing for what terms $t_1, ..., t_k$ does $\phi\,[x_1/t_1, ..., x_k/t_k]$ hold.

Proposition. 9.1.11. *The set*

$$\{(t_1, ..., t_k)\,|\,\dot{\xi} > K \models \phi\,[x_1/t_1, ..., x_k/t_k]\}$$

is semi-computable.

Proof: Left as an exercise.

Proposition. 9.1.12. *Let ψ' be an instance of ψ. Then, $\psi \models \psi'$, and it follows immediately that if $\dot{\xi} > K \models \psi$, then $\dot{\xi} > K \models \psi'$. If ψ' is a variable renaming of ψ, then $\psi' \equiv \psi$.*

Proof: Trivial. **QED**

Definition. 9.1.13. Let $K \subseteq \Theta$ be a set of formulae and let $q = (\dot{\xi}x_1, .., x_k : \phi) \in Q$ be a query. We say that a reply ψ is a *most general reply* to q iff

1. $q > K \models \psi$;

2. for all $\psi' \in \Theta_{\dot{\xi}}$, if ψ is an instance of ψ' and $q > K \models \psi'$, then ψ' is a variable renaming of ψ.

Definition. 9.1.14. A set $R \subseteq \Theta_{\dot{\xi}} \subseteq \Theta$ is a *most general set of correct replies* to q for K iff

1. each $\phi \in R$ is a most general reply to q for K;

2. for all $\psi \in \Theta$, if $q > K \models \psi$, then ψ is an instance of some $\phi \in R$;

3. for all $\phi_1, \phi_2 \in R$, if ϕ_2 is an instance of ϕ_1, then $\phi_2 = \phi_1$.

Definition. 9.1.15. Let now

$$Q_\wedge = \left\{ \left(\overset{\dot{\xi}}{\wedge} x_1, ..., x_k : \phi \right) \,|\, \phi \in \Theta_{\dot{\xi}} \right\}$$

and define the relation $> \,\subseteq Q_\wedge \times \Theta_{\dot{\xi}}$

$$\left(\overset{i}{\wedge} x_1, ..., x_k : \phi \right) > \psi$$

iff

$$\psi = \phi \left[x_1/t_1^1, ..., x_k/t_k^1 \right] \wedge ... \wedge \phi \left[x_1/t_1^n, ..., x_k/t_k^n \right]$$

for some terms $t_1^1, ..., t_k^n \in K \subseteq \Theta_i$. Then, the triple $\mathcal{Q}_\wedge = \left(\Theta_\wedge^i, Q_\wedge, > \right)$ is a query system for *conjunctive replies* to queries of the form "for what terms $t_1, ..., t_k$ does $\phi \left[x_1/t_1, ..., x_k/t_k \right]$ hold?"

In $\mathcal{Q}_\wedge$, replies to $\overset{i}{\wedge} x_1, ..., x_k : \phi$ are conjunctions of replies to $\underset{\iota}{x_1}, ..., x_k : \phi$. The former may have a consequentially strongest reply even if the latter does not. In fact, if the most general set of replies to $\underset{\iota}{x_1}, ..., x_k : \phi$ is finite, their conjunction is a consequentially strongest reply to $\overset{i}{\wedge} x_1, ..., x_k : \phi$.

Exercises

Exercise 9.1.1.1. With respect to Proposition 9.1.6:

1. Account for the monotonicity of the proof system.

2. Prove this proposition.

Exercise 9.1.1.2. Formalize in more complete detail the adequateness of a programming system.

Exercise 9.1.1.3. Given a programming system $\mathfrak{P}$, determinism implies determinacy, but the reverse does not hold. Justify.

9.1.2. LP programs and their meaning

In the above discussion, we elaborated on the semantical systems that allow us to *specify* a LP language and on the proof systems that allow us to *implement* a LP language. Importantly, there is no question of precedence of one over the other, implementation or specification, in the design of a LP language.

We now move from the metalanguage of LP to the object-language level. LP as a *programming language* can be considered as a synonym for *pure Prolog*, a proper subset of *real Prolog*, which we approach in Section 9.2 below. Unless otherwise stated, the points that we next discuss, as well as the examples given, are so with pure Prolog in mind. However, the distinction between pure and real Prolog, though an important one, is of no import in this Section, and the examples below can be implemented in any Prolog environment. We suggest the latest version of the LP implementation SWI-Prolog.[2]

It will be easy to see how the above metalanguage definitions apply to the object-language constructs of LP. The definitions above of expressions, substitution, and unification for FO classical logic (cf. Section 4.1) hold generally, with the following specifications:[3]

Definition. 9.1.16. A *LP formula* is a logical expression of the form

$$(F_{LP}) \qquad \forall x_1...\forall x_l \, (A_1, ..., A_m \leftarrow B_1, ..., B_n)$$

where $l, m, n \geq 0$, $A_1, ..., A_m, B_1, ..., B_n$ are atoms, $A_1, ..., A_m = A_1 \vee ... \vee A_m$, and $B_1, ..., B_n = B_1 \wedge ... \wedge B_n$. A LP formula is *closed* (*ground*) in the same circumstances as a classical logic formula is closed (ground, respectively).

More specifically, a LP formula is always a clause.

Definition. 9.1.17. A *LP clause* is a LP formula of the form

$$(\mathcal{C}_{LP}) \qquad \forall x_1...\forall x_l \, (A \leftarrow B_1, ..., B_n) \, .$$

A LP clause is typically simplified as

$$(\mathcal{C}_{LP}) \qquad A \leftarrow B_1, ..., B_n.$$

In $\mathcal{C}_{LP}$, A is called the *head*, and $B_1, ..., B_n$ is the *body*. This is in fact a *Horn clause*, as inverting the symbol $\leftarrow$ we have the formula $B_1 \wedge ... \wedge B_n \rightarrow A$; applying now $\rightarrow_{def}$ and DeM$_\wedge$, we have

$$B_1 \wedge ... \wedge B_n \rightarrow A \equiv \neg (B_1 \wedge ... \wedge B_n) \vee A \equiv \neg B_1 \vee ... \vee \neg B_n \vee A.$$

Obviously, $\mathcal{C}_{LP}$ is a *definite clause*.

Intuitively, $\mathcal{C}_{LP}$ can be interpreted as, for $0 \leq i \leq n$, if every B_i is true, then A is true; or, what is the same, A can be proved by proving all the B_i. In other terms, we say that A is implied by $\bigwedge_{i=1}^{n} B_i$ (the *declarative* reading), or that in order to answer query A we have to answer the query $\bigwedge_{i=1}^{n} B_i$ (the *procedural* reading).[4]

---

[2]Freely available at www.swi-prolog.org.
[3]A complete specification is given for real Prolog in Section 9.2.1 below.
[4]See Apt (1996) for an elaboration on these two readings or interpretations.

Definition. 9.1.18. The basic constructs of LP are *terms* and *statements*.

1. A LP *term* can be simple or compound. A *simple term* is a variable or a *constant*. A *compound term* comprises a *functor* and a sequence of one or more terms called *arguments*. A compound term of arity n has the form $p(t_1, ..., t_n)$, where p is the *name* of the functor and $t_1, ..., t_n$ are the arguments of p. A functor p with arity n is denoted by p/n. A functor can be a *relation symbol* or a *function symbol*. The name of a functor is an *atom*. A constant just is a functor of arity 0, so that it is also an atom.[5]

2. LP *statements* can be *facts*, *goals*, *rules*, and *queries*. Let $\mathcal{C}_{LP}$ be given; then, the unit clause A is a fact, the B_i are goals, and $\mathcal{C}_{LP}$, for $n \geq 0$, is a rule. Thus, a fact is the special case of a rule when $n = 0$. When $n = 1$, we have an *iterative clause*. The empty clause $\square$ is considered a goal. A *query* is a clause with a question mark.

It should be obvious that *relation symbol* just is another name for *predicate (symbol)*, and we shall favor the latter over the former for consistency reasons. Contrary to the language **L1**, the same functor name can be used for functions or predicates of different arity, a feature that is responsible for what is spoken of as *ambivalent syntax*. This is a useful feature when there is a natural relation between predicate and function symbols.

Example 9.1.1. X, Y, John, john, sara, father, father (X, Y), and male(john) are terms of LP.[6]

---

[5]Note that variables, too, can be atoms in the language Prolog that we shall use for real Prolog (see Section 9.2.1 below).

[6]This example illustrates clearly that it is very useful to use **this font** when writing LP terms and statements: It helps to distinguish the natural language English from the formal language(s) of LP, a distinction that is crucial given the "denotational" character of the latter. We shall carry this practice over to both Prolog and Datalog. Precisely because of these, we also adopt the common practice–actually required by most software–of always ending a fact or a rule with an end mark. This said, we shall often relax these practices, writing simply, say, $p(X,Y) \leftarrow q(Y,X), r(X)$ instead of p (X, Y) ← q (Y, X), r (X)., namely when LP languages are considered more immediately as logical languages. Further variations are $p(X,Y) \leftarrow q(Y,X) \wedge r(X)$ and p (X, Y) : −q (Y, X), r (X)..

- **X, Y** and **John** are individual variables, **john** and **sara** are atoms (constants, or names of individuals), and **father** (X, Y), as well as **male(john)**, are functors whose arity is denoted by **father/2** and **male/1**, respectively. It is easy to see that individual variables are written with initial uppercase letters and atoms are written with initial lowercase letters; variables can also start with an underscore "_".[7]

- **father** (X, Y) is a non-ground predicate and **male(john)** is a ground predicate.

- **father** (X, sara). and **male(john)**. are facts built from the predicates **father** (X, sara) and **male** (john).

- **father(john, sara)**? is a query asking whether the relation "*X* is the father of *Y*" holds between John and Sara, i.e. whether John is the father of Sara.

- **daughter** (X, Y) ← **father** (Y, X) , **female** (X). is a rule for the relation *daughter-of*. In this rule, **father** (Y, X) and **female** (X) are the goals, but the head, **daughter** (X, Y), can also be a goal.

- **father** (john, **father** (rita)) is a legal atom of LP.

Definition. 9.1.19. A *logic program* in its simplest form is a finite set of facts. More typically, a logic program is a finite set of rules formulated as definite clauses, reason why we call this a *definite program*.

We shall often write *Prolog program* as a synonym for *logic*, or *LP*, *program*. More properly, though, a LP program is a finite *sequence* of rules. In effect, conjunction, as well as disjunction, is not a commutative operation, with $\phi_1 \overset{\wedge}{\underset{\vee}{}} \phi_2 \neq \phi_2 \overset{\wedge}{\underset{\vee}{}} \phi_1$ in terms of processing for ϕ a rule or a literal. These aspects, and their import to deductive programming, will become clearer in Sections 9.1.3 and 9.2.2 below.

Example 9.1.2. These facts can constitute the program *Fatherhood*:

---

[7] Variables with "_" are called *anonymous variables* and each such occurrence in a clause or query denotes a different variable.

father (john, sara).	male (john).
father (john, peter).	male (rick).
father (john, rick).	male (peter).
father (rick, carl).	male (carl).
father (harry, louis).	male (harry).
father (harry, mary).	male (louis).
father (harry, jane).	female (sara).
	female (mary).
	female (jane).

Example 9.1.3. The following program, called *Addition*, has only two rules:

plus (0, X, X).
plus (s (X), Y, s (Z)) ← plus (X, Y, Z).

The basic operation on a LP program is *unification*. It is essentially the same as for CFOL (cf. Section 4.1.2) and we provide now only a few specifications.

Definition. 9.1.20. A LP *substitution* is a set of pairs of the form $X_i = t_i$, $0 \le i \le n$, where X_i is a variable and t_i is a term, $X_i \ne X_j$ for every $i \ne j$, and X_i does not occur in t_j for any i and j. Let σ be a substitution and A a term; then the result of applying substitution σ to term A, denoted by $A\sigma$, is the term obtained by substituting t for every occurrence of X in A for every pair $(X = t) \in \sigma$.

Definition. 9.1.21. We say that B is an *instance* of A if there is a substitution σ such that $A\sigma = B$. C is a *common instance* of A and B if it is an instance of A and an instance of B, i.e. if there are substitutions σ_1, σ_2 such that $A\sigma_1 \equiv B\sigma_2$.

Example 9.1.4. Consider the program *Fatherhood*. Let there be given the substitution $\sigma = \{X = \text{john}, Y = \text{sara}\}$. Then, the result of applying σ to the term father (X, Y), denoted by (father (X, Y)) σ, is the term father(john, sara). The goal father(john, sara) is an instance of the goal father (X, Y) (under substitution σ). Equally,

(daughter (Y, X) ← father (X, Y), female (Y).) σ

gives us

daughter (sara, john) ← father (john, sara), female (sara).

This unification for the goal ? − daughter (sara, john). can be represented in a tree as shown in Figure 9.1.1.

Consider now the program *Addition*. Let there be given the substitutions $\sigma_1 = \{X = 1\}$ and $\sigma_2 = \{Y = 1\}$. Then the goals plus $(0, 1, X)$ and plus $(0, Y, Y)$ have the common instance plus $(0, 1, 1)$ by applying to them the substitutions σ_1 and σ_2, respectively.

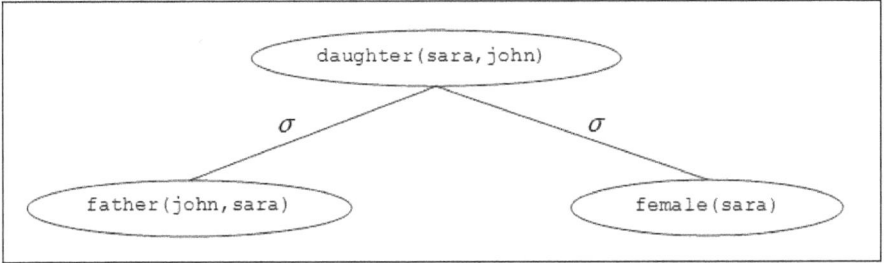

Figure 9.1.1.: Unification via a substitution σ.

Common instances are important because in fact we can reply to a query by finding a common instance of both a query and a fact.

Variables in facts are implicitly universally quantified, whereas variables in queries are implicitly existentially quantified.

Definition. 9.1.22. A fact $p(t_1, ..., t_n)$ reads "for all $X_1, ..., X_n$, where the X_i are variables in the fact, $p(t_1, ..., t_n)$ holds or is true," i.e.

$$\forall X_1, ..., \forall X_n \, (p(t_1, ..., t_n)).$$

This definition holds for rules: the variables appearing in the head are universally quantified and their scope is the whole rule. However, variables occurring in the body of a rule but not in its head are considered to be existentially quantified.

Example 9.1.5. grandfather $(X, Y) \leftarrow$ father (X, Z), father (Z, Y). is read "for all X and Y, X is the grandfather of Y if there exists a Z such that X is the father of Z and Z is the father of Y."

Definition. 9.1.23. A query $p(t_1, ..., t_n)$? reads "are there variables $X_1, ..., X_n$ such that $p(t_1, ..., t_n)$ holds or is true?", i.e.

$$\exists X_1, ..., \exists X_n \, (p(t_1, ..., t_n))?$$

For convenience reasons, the universal quantifiers are omitted in facts and the existential quantifiers are so in queries; both quantifiers are omitted in rules.

Definition. 9.1.24. For any substitution σ,

1. from a universal fact P deduce an instance $P\sigma$ of it. We call this *instantiation* and denote it $\vdash_{inst}$.

2. an existential query P is a logical consequence of an instance $P\sigma$ of it. We call this *generalization* and denote it by $\vdash_{gen}$.

By combining 1 and 2 above, we have a reply to a query by means of a common instance, i.e. we have

$$P. \vdash_{inst} P\sigma \vdash_{gen} (P?)$$

Definition. 9.1.25. A *solution* to a query is a fact that is a (common) instance of the query.

Example 9.1.6. The goal father(john, sara) implies that there exists an X such that father(X, sara) is *true*, in this case X = john. Then, father(john, sara) is a solution to the query father(X, sara)?, and the solution is represented by the substitution X = john.

An existential query may have several solutions.

Example 9.1.7. Consider the program *Addition*. plus $(X, Y, 4)$?, the query asking for numbers X and Y that add up to four, has as possible solutions $\{X = 0, Y = 4\}$ and $\{X = 3, Y = 1\}$.[8]

Definition. 9.1.26. A query that is a goal is actually a special case of a *conjunctive query*, i.e. a conjunction of goals of the form

$$q_1, ..., q_n? \equiv q_1 \wedge ... \wedge q_n?$$

We denote a conjunctive query by $Q_\wedge$.

---

[8]Note that in fact we have the query plus $(X, Y, s (s (s (s (0)))))$? and the possible solutions $\{X = 0, Y = s (s (s (s (0))))\}$ and $\{X = s (s (s (0))), Y = s (0)\}$, as the natural numbers are defined by means of the function $s (0)$ such that $s (0) = 1$, $s (s (0)) = 2$, etc.

9. Programming

Definition 9.1.23 above determines that the scope of the existential quantifier in a conjunctive query is the whole conjunction, so that a query $p(X), q(X)$? actually asks whether there is an X such that both $p(X)$ and $q(X)$ hold.

Definition. 9.1.27. In a conjunctive query of the form $p(X), q(X)$? we say that X is a *shared variable*.

Definition. 9.1.28. A conjunctive query $Q_\wedge = q_1, ..., q_n$? is a logical consequence of a program Π, denoted by

$$\Pi \vdash q_1, ..., q_n?$$

if for every goal $q_i \in Q_\wedge$, $0 < i \leq n$, we have

$$\Pi \vdash q_i$$

where shared variables in the q_i are instantiated to the same values.

In order to allow a logic program to compute deductively, we need to define one further rule:

Definition. 9.1.29. *Universal modus ponens (UMP)* – From the rule $r = A \leftarrow B_1, ..., B_n$ and the facts $B'_1., ..., B'_n$. we can deduce A' if $A' \leftarrow B'_1, ..., B'_n$ is an instance of r.

For convenience reasons, let us denote a quantified goal by G.

Definition. 9.1.30. An existentially quantified goal G is a logical consequence of a program Π, denoted by $\Pi \vdash G$, if there is a clause in Π with a ground instance $A \leftarrow B_1, ..., B_n$, $n \geq 0$, such that $\Pi \vdash B_i$ for all $0 < i \leq n$ and A is an instance of G. In other words, G is a logical consequence of Π iff G can be deduced from Π by a finite number of applications of UMP.

Definition. 9.1.31. An *abstract interpreter* Ψ for a logic program is an algorithm that takes as input a program Π and a goal G, answering *true* (or *yes*) if G is a logical consequence of Π and *false* (or *no*) otherwise. In the first case we say that Ψ performs a *true-computation*, and in the second case we say that Ψ performs a *false-computation*.

Definition. 9.1.32. The *meaning of a program* Π, denoted by $M(\Pi)$, is the set of unit ground goals $G = \{B_1, ..., B_n\}$ such that for all $0 < i \leq n$ we have

$$\Pi \vdash B_i.$$

A program Π is said to be *correct* with respect to some intended meaning $IM(\Pi)$ iff $M(\Pi) \subseteq IM(\Pi)$, and it is said to be *complete* with respect to some intended meaning $IM(\Pi)$ iff $IM(\Pi) \subseteq M(\Pi)$. A program Π is *adequate*, i.e. both correct and complete, with respect to some intended meaning M iff $IM(\Pi) = M(\Pi)$.

Informally, a program Π is correct iff it does not "say" unintended "things," and it is complete if every "thing" that is intended can be "said." The meaning of a basic program built up solely of ground facts is the program itself. Put differently, the program "means" just what it "says." The meaning of a regular logic program (i.e. a logic program comprising rules) contains explicitly whatever the program states implicitly.

Example 9.1.8. The meaning of the program *Fatherhood* (Example 9.1.2), $M(Fatherhood)$, just is the program itself. If we add to this program the rule `parent (X, Y) ← father (X, Y) .`, then $M(Fatherhood)$ additionally contains all goals of the form `parent (X, Y)` for every pair (X, Y) such that `father (X, Y) .` is in the program.

This said, it should be obvious that the *intended meaning* of a program Π–a set $IM(\Pi)$ of unit ground goals–is intuitively given by the choice of names in the program. This allows a semantics of quasi-truth values in the following way:

Definition. 9.1.33. Given a program Π, we say that a ground goal G is *true* with respect to $IM(\Pi)$ if $G \in IM(\Pi)$; otherwise, we say that G is *false*.

Exercises

Exercise 9.1.2.1. With respect to the program *Fatherhood* (Example 9.1.2), ask the following queries in the correct format:

1. Who is Sara's father?
2. Is Rick Carl's father?
3. Who are John's children?
4. Who are John's sons?
5. Who are John's daughters?
6. Who is Mary's father?

7. Has Rick got any children?

8. Are Louis, Mary, and Jane all Harry's children?

Exercise 9.1.2.2. With respect to the program *Fatherhood*:

1. Create a rule that determines that being a father is a predicate of a male.

2. Create a rule that allows us to find if one of the males is a grandfather. By means of this rule, find whether Harry, John, or Rick are grandfathers.

3. Create a rule that allows us to determine who a grandfather's grandchildren are.

Exercise 9.1.2.3. Add rules to the program *Fatherhood* so that this becomes the program *Parenthood*.

Exercise 9.1.2.4. Consider the predicates male/1, sibling/2, and parent/2, where "/integer" indicates the arity of the predicates.

1. From these, define a predicate uncle/2.

2. Change the program obtained in 1 so that it is

   a) incomplete.
   b) incorrect.

Exercise 9.1.2.5. Consider the following two rules:
nearby $(X, Y) \leftarrow$ connected (X, Y, W).
nearby $(X, Y) \leftarrow$ connected (X, Z, W), connected (Z, Y, W).

1. Write a program *European_borders* for the countries Spain, Portugal, France, Italy, Germany, Switzerland.

2. Instantiate the variable W to other values to produce different programs.

3. Add one or more rules to allow the program *European_borders* to give "true" replies to queries such as nearby (France, Italy)? and nearby (Italy, France)?.

Exercise 9.1.2.6. Add a predicate `natural_number` (X) to the program *Addition* in Example 9.1.3, so that the program is as follows:

plus $(0, X, X) \leftarrow$ natural_number(X).
plus $(s(X), Y, s(Z)) \leftarrow$ plus (X, Y, Z).
natural_number (0).
natural_number $(s(X)) \leftarrow$ natural_number (X).

In terms of deductive computation, what would be the gains of this addition?

Exercise 9.1.2.7. With respect to the argument in Exercise 3.2.4:

1. Write the premises of the argument as a LP program.

2. Formulate the following as a query:

  a) What country is Daffy an enemy of?

  b) Who is native of country C?

  c) What does the West sell to Daffy?

  d) Who is hostile?

  e) Is M1 a weapon?

  f) What does Daffy own?

  g) Does the West sell a missile to Daffy?

  h) Does Daffy own a M1 missile?

9.1.3. Resolution and LP computations

We now expand on the abstract interpreter Ψ of Definition 9.1.31, namely as a search algorithm of LP when computing a goal. Interestingly enough, the resolution calculus of Section 8.1 above, and more specifically so some of its refinements, will prove to have a crucial role in this topic.

Definition. 9.1.34. We call *resolvent* the current (usually conjunctive) stage of a LP computation. The *empty resolvent* (or *empty clause*), denoted by $\Box$, is the clause with empty head and empty body. The sequence of resolvents produced during a computation is called the *trace* of the interpreter.

Algorithm 9.1 Ground reduction

Input: (Π, G)
Output: True or False

Initialize the resolvent to G

while resolvent $\neq \square$ **do**
    choose a goal A from the resolvent
    choose a ground instance of a clause $A' \leftarrow B_1, ..., B_n \in \Pi$ s.t. $A = A'$
      if no such goal and clause exist, leave the while loop
    replace A by $B_1, ..., B_n$ in the resolvent
if resolvent $= \square$, **then** output *true*, **else** output *false*

Definition. 9.1.35. Given a LP program Π and a goal G, the replacement of G by the body of an instance of a clause $C \in \Pi$ whose head is identical to G is called a *reduction*. A reduction is *ground* if both the goal G and the instance of the clause C are ground. The goal replaced in a reduction is said to be *reduced* and we say that the new goals are *derived*.

The algorithm for this procedure is given as Algorithm 9.1. If goal G is not deducible from program Π, then Ψ may fail to terminate. Note that each iteration of the "while loop" is a single application of UMP, i.e. a reduction. It is easy to see that reduction is the basic computational step in LP. The selection of the goal to be reduced and the order of the reductions thereof is arbitrary, as all the goals in a given resolvent must be reduced. The selection of a clause and a suitable instance thereof is non-deterministic but critical.

Recall from the above discussion that given a query system $Q = (\Theta_{\dot{c}}, Q, >)$ and a proof system for it $\mathcal{P}_Q = (\Theta \supseteq \Theta_{\dot{c}}, P, >)$, for some query $\dot{c}$ and a set of formulae $F \subseteq 2^\Theta$ we have a provably correct reply $\phi \in \Theta_{\dot{c}}$ to $\dot{c}$, i.e. $\dot{c} > F \vdash_\blacksquare \phi$, iff we have

$$\dot{c} > \phi \text{ and } F \vdash_\blacksquare \phi.$$

This, by Proposition 9.1.6, is equivalent to $\iota \rhd \Pi \vdash_\xi o$ in terms of LP. This entails that given input ι, a computation ξ producing output o from a program Π actually is a proof that the query follows from the program. Such a proof is implicitly represented in the trace of a query, but we can represent it explicitly in the form of a tree.

Definition. 9.1.36. A *(reduction) proof tree* is a (directed) tree whose nodes represent the goals that are reduced during a computation, there being a (directed) edge from a node to each node that corresponds to a derived goal of a reduced goal. In a proof tree, the number of nodes corresponds to the number of reduction steps in a computation. The root of a proof tree for a simple query is the query itself. The proof tree for a conjunctive query is the collection of all the proof trees for its individual goals.[9]

Figure 9.1.1 above shows in fact the (reduction) proof tree for the goal ? − daughter (sara, john) . given the program *Fatherhood* augmented as in Example 9.1.4.

However, knowledge of the resolution calculus (cf. Section 8.1) and a little thought will reveal that reduction, a generalization of UMP, can be re-expressed in the terms of this calculus. Indeed, it is easy to see that we can apply resolution to find a contradiction from the combination of a goal clause, with solely negative literals, and a fact (in a rule), a positive literal. It is important to remark that a goal clause $\mathcal{G}$ is not a *program clause*, which can be only either a rule or a fact. This should be born in mind when considering $\Pi \cup \{\mathcal{G}\}$, i.e. when we add a goal clause $\mathcal{G}$ to a LP program Π. The goal clause $\mathcal{G} = \|\neg q_1, ..., \neg q_n\|$ is added to the program, in order to test if $\mathcal{Q}_\wedge = q_1 \wedge ... \wedge q_n$ follows from it, and this is the case iff $\Pi \cup \{\mathcal{G}\}$ is unsatisfiable. This is so iff we can deduce the *empty goal clause* $\square$ from $\Pi \cup \{\mathcal{G}\}$ by an application of resolution. More specifically, we refer here to *linear input resolution* (abbr.: LI resolution), as this has been proven complete for Horn clauses. LI resolution, in turn, is a refinement of linear resolution (cf. Exercise 8.1.4.6), and we begin by giving important results for LP with relation to it.

Example 9.1.9. Consider the program *Addition*. Figure 9.1.2 shows a LI-resolution proof that $0 + 2 = 2$, i.e. plus (0, s (s (0)) , s (s (0))). Notice that each resolvent is a reduced goal. Figure 9.1.3 is the corresponding resolution-proof tree, in which for convenience we label the tree only with the numbers in the proof of Figure 9.1.2.

Lemma 9.1.1. *Let Π be a LP program and $\mathcal{G} = \|\neg q_1, ..., \neg q_n\|$ a goal clause. Then, all the q_i are consequences of Π iff $\Pi \cup \{\mathcal{G}\}$ is unsatisfiable.*

Proof: Left as an exercise.

---

[9]Basically, a (reduction) proof tree shows the instantiation of goals up to the queried goal. See Fig, 9.1.1.

1. ← plus (X, Y, s (s (0))) .   Goal
2. plus (s (X1) , Y1, s (Z1)) ← plus (X1, Y1, Z1) .   Variable renaming
3. ← plus (X1, Y1, s (0)) .   Resolution (1, 2)
                                                                    σ = {X = s (X1), Y = Y1, Z1 = s (0)}
4. plus (s (X2) , Y2, s (Z2)) ← plus (X2, Y2, Z2) .   Variable renaming (2)
5. ← plus (X2, Y2, 0) .   Resolution (3, 4)
                                                                    τ = {X1 = s (X2), Y1 = Y2, Z2 = 0}
                                                                    Variable renaming
6. plus (0, X3, X3) .   Resolution (5, 6)
7. □   λ = {X2 = 0, X3 = 0, Y2 = 0}

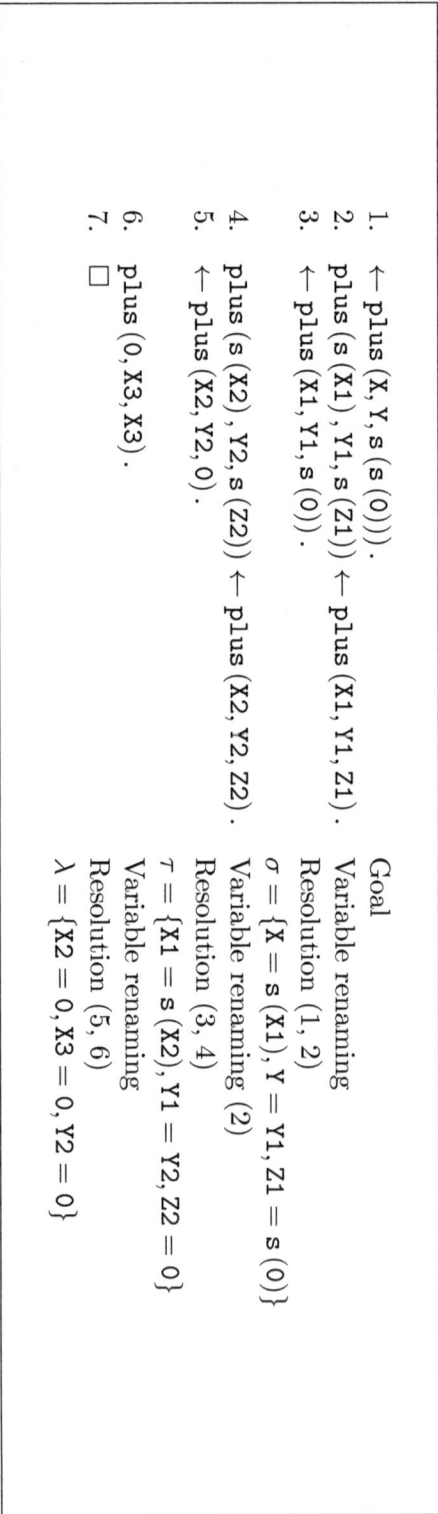

Figure 9.1.2.: A Ll-resolution proof on a LP program.

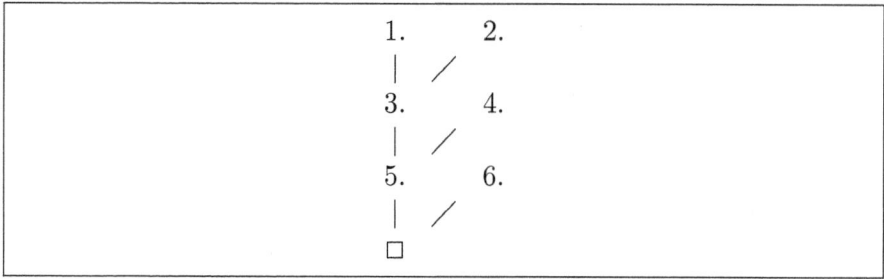

Figure 9.1.3.: A LI-resolution proof tree.

Theorem 9.1.2. *(Refutation completeness of linear resolution for Horn clauses) If C is an unsatisfiable set of Horn clauses, then there is a linear resolution proof that is a refutation of C, i.e. $C \vdash_{lres} \square$.*

Proof: (Idea) Assume that C is finite and proceed by induction on the elements of C. **QED**

Although LI resolution is the general resolution rule for LP, when given as input the query $q_1, ..., q_n$? a LP interpreter actually searches for a *SLD-resolution proof of* $\square$ (cf. Exercise 8.1.4.8), in turn a case of LD resolution (cf. Exercise 8.1.4.7). We now define these two resolution refinements with respect to LP.

Definition. 9.1.37. Let $\Pi \cup \{\mathcal{G}\}$ be given as a set of ordered clauses. Then, a LD-resolution refutation of $\Pi \cup \{\mathcal{G}\}$, denoted by $\Pi \cup \{\mathcal{G}\} \vdash_{ldres} \square$, is a sequence

$$(\mathcal{G}_0, \mathcal{C}_0), ..., (\mathcal{G}_n, \mathcal{C}_n)$$

where the $\mathcal{G}_i, \mathcal{C}_i$, $0 \leq i \leq n$, are ordered clauses, such that $\mathcal{G}_0 = \mathcal{G} = \|\neg p_1, ..., \neg p_n\|$ and $\mathcal{G}_{n+1} = \square$. More specifically, we have

$$\mathcal{G}_i = \left\|\neg p_{i,1}, ..., \neg p_{i,n(i)}\right\|, \, |\mathcal{G}_i| = n(i)$$

are the goal clauses, and for the $\mathcal{C}_i \in \Pi$ we have

$$\mathcal{C}_i = \left\|q, \neg q_{i,1}, ..., \neg q_{i,m(i)}\right\|, \, |\mathcal{C}_i| = m(i) + 1 \text{ or } 1 \text{ if } \mathcal{C}_i = \|q\|.$$

Then, for each $i < n$ there is a resolution rule

$$\frac{\mathcal{G}_i \quad \mathcal{C}_i}{\mathcal{G}_{i+1}}$$

where $\mathcal{G}_{i+1} = \left\|\neg p_{i,1}, ..., \neg p_{i,k-1}, \neg q_{i,1}, ..., \neg q_{i,m(i)}, \neg p_{i,k+1}, ..., \neg p_{i,n(i)}\right\|$, an ordered clause with $|\mathcal{G}_{i+1}| = (n(i) - 1) + m(i)$, is the resolvent.

Lemma 9.1.3. *For a LP program Π and a goal clause $\mathcal{G} = \|\neg q_1, ..., \neg q_n\|$, if $\Pi \cup \{\mathcal{G}\}$ is an unsatisfiable set of ordered clauses, then there is a LD-resolution refutation of $\Pi \cup \{\mathcal{G}\}$ beginning with $\mathcal{G}$.*

Proof: Left as an exercise.

We obtain SLD resolution by introducing a *selection rule* r to choose the literal $p_i \in \mathcal{G}_i$ to be resolved upon with LD resolution, so that $r(\mathcal{G}_i)$ is the literal resolved upon in the $(i+1)$-th step of the proof.

Definition. 9.1.38. The LD-resolution rule

$$\frac{r(\mathcal{G}_i) \qquad C_i}{\mathcal{G}_{i+1}}$$

where r is a selection rule, is called a *SLD-resolution rule*.

In LP, the selection rule simply chooses the leftmost literal in the goal clause to be resolved upon.

Theorem 9.1.4. *(Completeness of SLD resolution for LP) Given a LP program Π and a goal clause $\mathcal{G} = \|\neg q_1, ..., \neg q_n\|$, if $\Pi \cup \{\mathcal{G}\}$ is an unsatisfiable set of ordered clauses, then there is a selection rule r such that there is a SLD-resolution refutation of $\Pi \cup \{\mathcal{G}\}$ via r beginning with $r(\mathcal{G})$.*

Proof: (Sketch). Lemma 9.1.3 assures us that there is a LD-resolution refutation of $\Pi \cup \{\mathcal{G}\} \notin SAT$ beginning with $\mathcal{G}$. We only need to prove that there is a SLD-resolution refutation of $\Pi \cup \{\mathcal{G}\} \notin SAT$ via $r(\mathcal{G})$. The proof is by induction on the length of $\mathcal{G}$. If $|\mathcal{G}| = 1$, then $\mathcal{G}_0$ is a unit clause and $r(\mathcal{G}_0)$ is irrelevant. We now let $(\mathcal{G}_0, C_0), ..., (\mathcal{G}_n, C_n)$ be a LD-resolution refutation of $\Pi \cup \{\mathcal{G}_0\} \notin SAT$ and we suppose that the selection rule r chooses the literal $\neg p_{0,k} \in \mathcal{G}_0$. Because $\Pi \cup \{\mathcal{G}_0\} \notin SAT$, we must have $\mathcal{G}_{n+1} = \square$, and hence there must be some $j < n$ at which we resolve on $\neg p_{0,k}$. If $j = 0$, we are done. If $j \geq 1$, then there must be some C that is a resolvent of $\mathcal{G}_0$ and C_j. Then, there must be a LD-resolution refutation of length $n - 1$ of $\Pi \cup \{C\}$ beginning with C. By induction, this refutation can be replaced by a SLD-resolution refutation via r. We add this refutation onto the single step resolution of $\mathcal{G}_0$ and C_j obtaining the SLD-resolution of $\Pi \cup \{\mathcal{G}\} \notin SAT$ via $r(\mathcal{G})$ beginning with $\mathcal{G} = \mathcal{G}_0$. **QED**

We now elaborate on how SLD resolution corresponds to the search process in LP when a query is entered as input. A LP-proof tree corresponds to the search process known as *depth-first search with backtracking*: by "depth-first search" it is meant that, given a finitely branching tree, all the descendants of a node are checked before their siblings on the right of the tree and no edge is traversed more than once; if a *fail* leaf is encountered, then the search "backtracks" to the immediate ancestor of this leaf and the depth-search process is resumed. If a success leaf (denoted by □) is found, the search stops until we prompt the search to proceed further by means of an *expand* "command" that makes the search retake.[10] The search is considered successful if at least one □-resolvent is found on the tree; otherwise, the search fails and "false" is the output to the query.

Put briefly, given a LP program Π and a goal clause $\mathcal{G}$, every branch of a complete LP-proof tree is either a successful SLD-resolution proof or a failed SLD-resolution proof. For this reason, we refer to this tree as a *SLD-resolution tree*.

Example 9.1.10. Let there be given the following LP program Π_1:

1. $p\,(X,Y) \leftarrow q\,(X,Z)\,,r\,(Z,Y)\,.$
2. $p\,(X,X) \leftarrow s\,(X)\,.$
3. $q\,(X,b)\,.$
4. $q\,(b,a)\,.$
5. $q\,(X,a)\,.$
6. $r\,(b,a)\,.$
7. $s\,(X) \leftarrow t\,(X,a)\,.$
8. $s\,(X) \leftarrow t\,(X,b)\,.$
9. $s\,(X) \leftarrow t\,(X,X)\,.$
10. $t\,(a,b)\,.$
11. $t\,(b,a)\,.$

Our query is $? - p\,(X,X)$, i.e. $\|\neg p\,(X,X)\|$, where we use the symbol $\neg$ for convenience (see next Section). The depth-first algorithm starts by checking premise 1 and then moves to premise 2. Beginning with premise 1, we have it that there is a successful SLD-resolution proof when we apply SLD resolution to the premises 1, 3, and 6, in this exact order, with, after renaming of variables, substitutions $\sigma = \{X_1 \rightarrow X, Y \rightarrow X\}$, $\theta = \{X_2 \rightarrow X, Z \rightarrow b\}$, and $\lambda = \{X \rightarrow a\}$ (cf. Figure 9.1.4). Applying SLD resolution to the sequences 1 and 4 or 1 and 5 will not produce successful proofs. The search starting by checking premise 2 produces two successful proofs and a failure. Figure 9.1.5 shows the complete SLD-proof tree for $\Pi_1 \cup \{\|\neg p\,(X,X)\|\}$ with the further substitutions $\omega = \{X \rightarrow b, Z \rightarrow a\}$, $\varsigma = \{Z \rightarrow a\}$, and $\mu = \{X \rightarrow b\}$; renaming of variables was omitted and ε denotes the empty substitution.

---

[10]In the SWI-Prolog interpreter, we simply enter ";".

$$\|\neg p\,(X,X)\| \qquad\qquad 1.\,\|p\,(X_1,Y)\,,\neg q\,(X_1,Z)\,,\neg r\,(Z,Y)\|$$
$$|\ \diagup\sigma$$
$$\|\neg q\,(X,Z)\,,\neg r\,(Z,X)\| \qquad\qquad 3.\,\|q\,(X_2,b)\|$$
$$|\ \diagup\theta$$
$$\|\neg r\,(b,X)\| \qquad\qquad 6.\,\|r\,(b,a)\|$$
$$|\ \diagup\lambda$$
$$\square$$

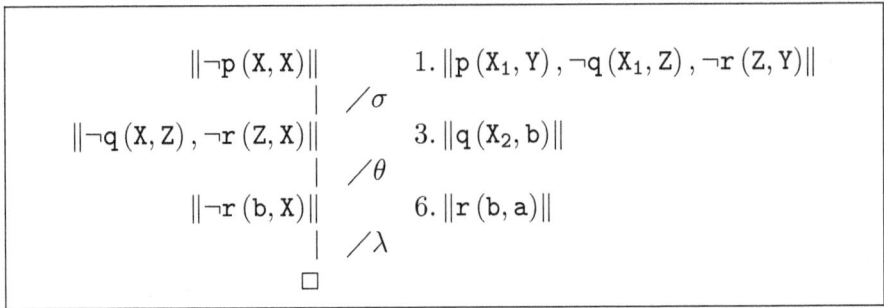

Figure 9.1.4.: A SLD-resolution proof.

Example 9.1.11. Figure 9.1.6 shows how SWI-Prolog answers the query $? - p(X,X).$ when given program Π_1 and how it answers the request to produce a trace of some instantiations. In the first case, given the input (query) $? - p(X,X).$, SWI-Prolog gives the first answer, to wit, $X = a$. Asked to provide more answers by means of the prompt ";", SWI-Prolog gives the replies $X = b$ and $X = a$, finally replying that there are no more instantiations (denoted by false). Asked to output traces of $? - p(X,X).$, $? - p(a,X).$, and $? - p(b,X).$, SWI-Prolog does so, in each case adding that $X = a$. Compare these with the SLD-proof tree in Figure 9.1.5. Figure 9.1.7 shows both the case of a successful instantiation $? - p(b,b).$ and a failed instantiation $? - p(c,d)..$ In these last traces, "redo" indicates backtracking.

Exercises

Exercise 9.1.3.1. Show that a set of Horn clauses is unsatisfiable iff it contains at least one fact and one goal clause.

Exercise 9.1.3.2. With respect to the argument in Exercise 3.2.4, write it as a Prolog program in SWI-Prolog.

1. Verify if the conclusion holds and construct the corresponding SLD-resolution refutation tree.

2. Input the questions in Exercise 9.1.2.7.2.a-h as queries and obtain the replies thereto.

Exercise 9.1.3.3. Prove lemmata 9.1.1 and 9.1.3.

Exercise 9.1.3.4. What is the result in terms of computation of adding a rule of the form $p \leftarrow p$ to a LP program?

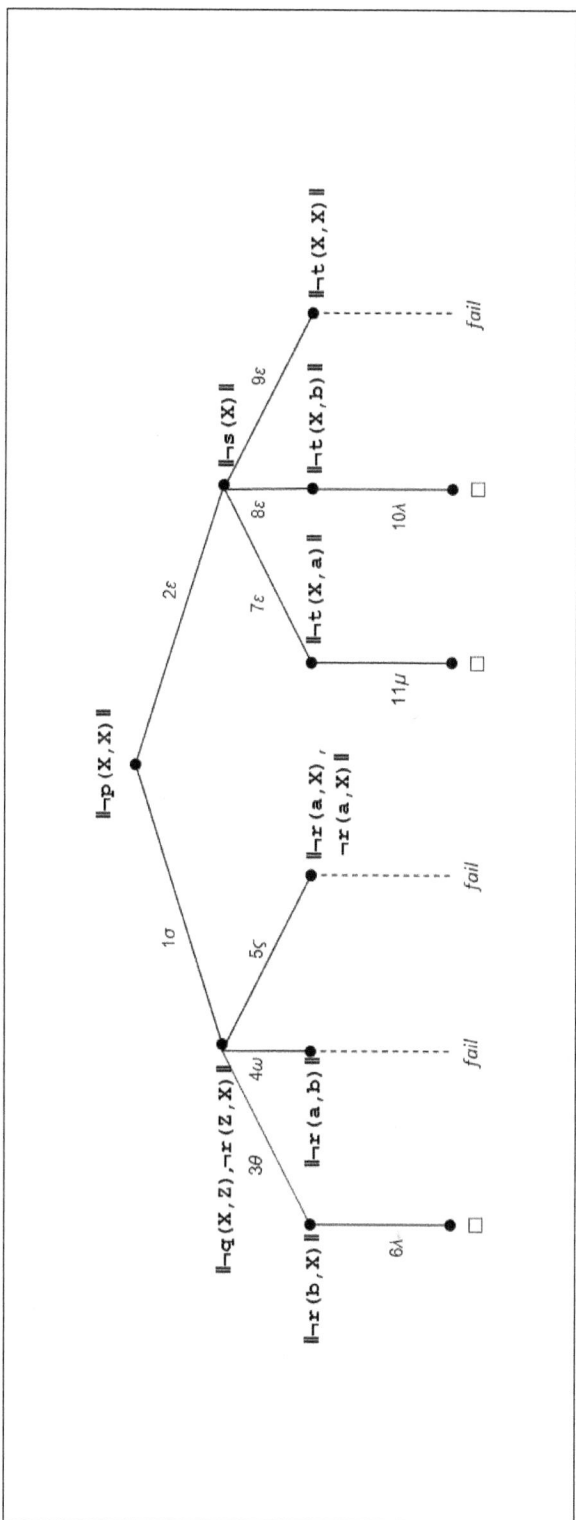

Figure 9.1.5.: A complete SLD-proof tree for a Prolog program.

```
?- p(X,X).
X = a ;
X = b ;
X = a ;
false.

?- trace.
true.

[trace] ?- p(X,X).
   Call:  (8) p(_2768, _2768) ?  creep
   Call:  (9) q(_2768, _2986) ?  creep
   Exit:  (9) q(_2768, b) ?  creep
   Call:  (9) r(b, _2768) ?  creep
   Exit:  (9) r(b, a) ?  creep
   Exit:  (8) p(a, a) ?  creep
X = a .

[trace] ?- p(a,X).
   Call:  (8) p(a, _2770) ?  creep
   Call:  (9) q(a, _2986) ?  creep
   Exit:  (9) q(a, b) ?  creep
   Call:  (9) r(b, _2770) ?  creep
   Exit:  (9) r(b, a) ?  creep
   Exit:  (8) p(a, a) ?  creep
X = a .

[trace] ?- p(b,X).
   Call:  (8) p(b, _2770) ?  creep
   Call:  (9) q(b, _2986) ?  creep
   Exit:  (9) q(b, b) ?  creep
   Call:  (9) r(b, _2770) ?  creep
   Exit:  (9) r(b, a) ?  creep
   Exit:  (8) p(b, a) ?  creep
X = a .
```

Figure 9.1.6.: SWI-Prolog answering a query and outputting traces for some "true" instantiations.

```
[trace] ?- p(b,b).
   Call:  (8) p(b, b) ?  creep
   Call:  (9) q(b, _2950) ?  creep
   Exit:  (9) q(b, b) ?  creep
   Call:  (9) r(b, b) ?  creep
   Fail:  (9) r(b, b) ?  creep
   Redo:  (9) q(b, _2950) ?  creep
   Exit:  (9) q(b, a) ?  creep
   Call:  (9) r(a, b) ?  creep
   Fail:  (9) r(a, b) ?  creep
   Redo:  (9) q(b, _2950) ?  creep
   Exit:  (9) q(b, a) ?  creep
   Call:  (9) r(a, b) ?  creep
   Fail:  (9) r(a, b) ?  creep
   Redo:  (8) p(b, b) ?  creep
   Call:  (9) s(b) ?  creep
   Call:  (10) t(b, a) ?  creep
   Exit:  (10) t(b, a) ?  creep
   Exit:  (9) s(b) ?  creep
   Exit:  (8) p(b, b) ?  creep
true.

[trace] ?- p(c,d).
   Call:  (8) p(c, d) ?  creep
   Call:  (9) q(c, _2950) ?  creep
   Exit:  (9) q(c, b) ?  creep
   Call:  (9) r(b, d) ?  creep
   Fail:  (9) r(b, d) ?  creep
   Redo:  (9) q(c, _2950) ?  creep
   Exit:  (9) q(c, a) ?  creep
   Call:  (9) r(a, d) ?  creep
   Fail:  (9) r(a, d) ?  creep
   Redo:  (8) p(c, d) ?  creep
   Fail:  (8) p(c, d) ?  creep
false.
```

Figure 9.1.7.: SWI-Prolog traces of a "true" and a "false" instantiation.

Exercise 9.1.3.5. Given a LP program Π and a query clause $\mathcal{G}$, exactly one of the five following SLD trees $\mathcal{T}_{\Pi,\mathcal{G}}$ for $\Pi \cup \{\mathcal{G}\}$ exists. Explain when each SLD tree is the case.

1. $\mathcal{T}_{\Pi,\mathcal{G}}$ terminates.

2. $\mathcal{T}_{\Pi,\mathcal{G}}$ diverges.

3. $\mathcal{T}_{\Pi,\mathcal{G}}$ potentially diverges.

4. $\mathcal{T}_{\Pi,\mathcal{G}}$ gives infinitely many answers.

5. $\mathcal{T}_{\Pi,\mathcal{G}}$ fails.

Exercise 9.1.3.6. A *breadth-first search* traverses a SLD-resolution tree level by level. What is the computational advantage, in terms of complexity, of a depth-first over a breadth-first search?

Exercise 9.1.3.7. Prove the soundness of SLD resolution for LP.

Exercise 9.1.3.8. Complete the proofs of theorems 9.1.2 and 9.1.4.

Exercise 9.1.3.9. Another way to prove the completeness of SLD resolution for LP is via the *least* or *minimal Herbrand model* for a LP program Π, denoted by $\underline{H}\mathcal{M}_\Pi$. (Typically, the least H-model is a minimal H-model that is unique.)

1. Research into this proof and give its details, not forgetting to define $\underline{H}\mathcal{M}_\Pi$.

2. Explain why it can be stated

$$\underline{H}\mathcal{M}_\Pi = M(\Pi)$$

where $M(\Pi)$ denotes the meaning of a LP program Π (cf. Def. 9.1.32).

Exercise 9.1.3.10. LP programs are often *recursive* (cf. Exercise 1.1.1.10).

1. Define this property in the framework of LP and explain what it entails for LP.

2. A simple way to check for recursion in a LP program Π is by means of a *(simple) dependency graph* for Π, a directed graph $\vec{\mathfrak{G}}_\Pi$ in which the arcs are of the form $p \longrightarrow q$ whenever there is a rule $r \in \Pi$ with $q \in Head_r$ and $p \in Body_r$. A program Π is recursive iff there is at least one loop in $\vec{\mathfrak{G}}_\Pi$. Check for recursion in the following programs:

   a) *Addition* (cf. Exercise 9.1.2.6).

   b) Π_1 (cf. Example 9.1.10).

Exercise 9.1.3.11. Draw the search tree for the query `?-p(b,b).` in Figure 9.1.7.

Exercise 9.1.3.12. Consider the program *Fatherhood* in Example 9.1.2 with the following additional rule and fact:

$$\text{son}(X, Y) \leftarrow \text{father}(Y, X), \text{male}(X).$$

$$\text{son}(\text{louis}, \text{harry}).$$

Given the input `son(louis, harry)?`, provide the trace of an interpreter Ψ and the final output by Ψ.

9.1.4. Negation as failure

We begin this Section by retaking the LP program Π_1 of Example 9.1.10. Consulting this program, SWI-Prolog will reply `true` to the queries $? - p(a, a)., ? - p(b, a).,$ and $? - p(b, b).$; but the queries $? - p(a, b).$ and $? - p(c, c).,$ for example, are given the reply `false`. In CFOL, this would mean that, given some interpretation $\mathcal{I}$, there is a valuation $val_{\mathcal{I}}$ such that $val_{\mathcal{I}}(p(a, a)) = \text{t}$ iff $val_{\mathcal{I}}(\neg p(a, a)) = \text{f}$, and $val_{\mathcal{I}}(p(c, c)) = \text{f}$ iff $val_{\mathcal{I}}(\neg p(c, c)) = \text{t}$. In this resides the bivalent character of CFOL. However, in LP this simply means that

$$\Pi_1, ? - p(a, a). \vdash_{sldres} \Box \quad \Rightarrow \quad p(a, a) = \text{true}$$

indicating *success*, and

$$\Pi_1, ? - p(c, c). \nvdash_{sldres} \Box \quad \Rightarrow p(c, c) = \text{false}$$

indicating *failure*. In other words, given a LP program Π and a goal $\mathcal{G} = \|\neg q\|$, SWI-Prolog replies to a query with `true` if $\Pi \cup \{\mathcal{G}\} \notin SAT$ and with `false` if $\Pi \cup \{\mathcal{G}\} \in SAT$. This latter case just means that

SWI-Prolog failed to prove goal $\mathcal{G}$ by means of SLD resolution; it means neither that $val\,(\mathcal{G}) = \mathbf{f}$ nor that $val\,(\neg\mathcal{G}) = \mathbf{t}$.

However, in the last paragraph of Section 0.4, we wrote that LP is deductive computation *also* because it is truth-preserving computation, and above in this Section 9.1 we elaborated at length on the completeness of LP, so that we can say that there is a truth-based semantics for LP. Indeed, this is the case; what there is *not* is a semantics of falsity, namely with respect to the classical negation denoted by $\neg$. In LP, falsity with respect to a query q is the case when, as seen, q cannot be deduced from the program, or q does not match any of the data in the program. This is well expressed in terms of the meaning of a program Π, denoted by $M\,(\Pi)$, in Definition 9.1.33. But this does not mean that we have $\neg q$, instead. It just means that we failed to prove that q is deducible from the program or matches some data in it. This should clarify why above we spoke of a "semantics of quasi-truth values" for LP.

This is perfectly evident if we bear in mind that SLD resolution is a convenient proof calculus to "express" the proof mechanism proper of LP, which is *reduction*, in turn based on UMP (cf. Defs. 9.1.29 and 9.1.34-5). In this proof mechanism, what is sought is a repeated replacement of a goal A by the head of a rule $A' = A$ until no more replacements can take place: if we obtain the empty resolvent, then the reduction is considered successful, and A is deducible from the database via UMP; otherwise, this algorithm fails. In a sense, then, and a rather implicit one, we can say that if a LP interpreter fails to prove that q, then $\neg q$ must be the case.

This *implicitness* of negation can be stated as follows:

Definition. 9.1.39. Given a LP program Π and a query ? which is a ground atom q, we have either

$$(?_\square) \qquad \Pi, \neg q \vdash_{sldres} \square \quad \Rightarrow \quad q$$

or

$$(?_\boxtimes) \qquad \Pi, \neg q \nvdash_{sldres} \square \quad \Rightarrow \quad \neg q$$

Then, $?_\boxtimes$ can be interpreted in terms of $\neg$ as "$\neg q$ succeeds if q finitely fails" and $?_\square$ as "$\neg q$ fails if q finitely succeeds", where by "finitely succeeds" ("finitely fails") it is meant that there is a SLD-resolution tree $\mathcal{T}_{\Pi,\neg q}$ with at least one $\square$-leaf (with no $\square$-leaf, respectively). We accordingly call this interpretation of the symbol $\neg$ *negation by failure (NBF)*.

Definition. 9.1.40. Let the ground atom A be a goal and let $?A$ denote a query. Then, the following rule of inference is called negation by failure:

$$(\text{NF}) \qquad \frac{?(\neg A) \quad ?A \text{ fails}}{\square} \quad .$$

Rather than a rule of inference, NF is a *meta-rule*. There are two cases in which NBF may be an interpretation for $\neg$, depending on the way we define *completeness* with respect to a program. Firstly, the program, or database, is *complete* in the sense that it contains all the information that is "true" about some domain. We can assume that what is "true" is also known to be "true," and what is not known to be "true" is "false." We call this *closed-world assumption* (CWA).

Example 9.1.12. CWA is actually a very frequent kind of reasoning. Suppose you want a direct flight from Madrid, Spain to Sydney, Australia. You consult the flights available in a specific airline. If there is no direct flight Madrid-Sydney listed, you conclude that there is no such flight in this airline, or that the contrary statement is false. In either case, you can be considered to employ NBF.

CWA was originally proposed by R. Reiter (1978), and the CWA interpretation of $\neg$ in LP was defended in Shepherdson (1984). This interpretation is highly relevant, as it makes of LP a kind of *non-monotonic* deduction.[11] In fact, CWA is, too, an inference meta-rule.

Definition. 9.1.41. Let Π be a definite LP program and let the ground atom A be a query. The following meta-rule of inference is called CWA:

$$(\text{CWA}) \qquad \frac{\Pi \nvdash_{sldres} A}{\neg A}$$

Secondly, a program, or database, is said to be *complete* in the sense that the statements of the definite-clause program or database are assumed to *axiomatize completely* all the possible reasons that make atomic ground formulae true. This assumption, known as *complete-database* (CDB) (Clark, 1978), is concretized in the replacement of all $\leftarrow$ by $\leftrightarrow$.

Example 9.1.13. Let there be given the following LP program Π_2:

1. $p(X) \leftarrow q(X)$.
2. $p(X) \leftarrow r(X)$.
3. $p(X) \leftarrow t(X)$.

---

[11]Cf. Exercise 9.1.4.11 below. See, for example, Section 3.3 of Augusto (2020b) for a development.

4. q (a) .
5. q (c) .
6. r (b) .

Then, we can simply replace statements 1-3 by

$$7.\ p\,(X) \leftrightarrow (q\,(X) \lor r\,(X) \lor t\,(X)) .$$

called the *Clark formula for the predicate* p (X). The remaining Clark formulae are

$$8.\ q\,(X) \leftrightarrow (X = a \lor X = c) .$$

and

$$9.\ r\,(X) \leftrightarrow X = b.$$

However, the Clark formula for the predicate t (X) is

$$10.\ \neg t\,(X) .$$

expressing the fact that there is no X such that X is instantiated by a ground term in t (X). In effect, recall from above that a fact in LP is implicitly universally quantified, so that we have

$$\forall X\,(\neg t\,(X)).$$

As in Example 9.1.12, we employ here NBF. In order to have a Clark-complete database Δ corresponding to program Π_2, we have

$$\Delta = \{7, 8, 9, 10, 11, 12, 13\}$$

where 7-10 are as above but universally quantified, and the remaining statements are

$$11.\ \neg\,(a = b) .$$

$$12.\ \neg\,(a = c) .$$

$$13.\ \neg\,(b = c) .$$

Definition. 9.1.42. Given a definite-clause theory or database Θ, the *Clark completion of* Θ, denoted by $Compl\,(\Theta)$, is the theory consisting of (i) all the Clark formulae for every predicate $P \in \Theta$, and (ii) statements of the form $\neg\,(t_1 = t_2)$ for every non-unifiable pair of terms $(t_1, t_2) \in T\,(\Theta)$.

Given a definite-clause theory or database Θ, Algorithm 9.2 provides an efficient procedure for constructing the Clark completion $Compl\,(\Theta) =$

Δ.

We have the following important results:

Proposition. 9.1.43. *For a theory Θ to be Clark-complete, all functions $f, g \in Fun(\Theta)$ must obey the following three non-logical axioms:*

$$(\mathcal{E}f1) \qquad \forall f \forall g \left[\neg (f = g) \Rightarrow \neg (f(x_1, ..., x_n) = g(y_1, ..., y_n)) \right]$$

$$(\mathcal{E}f2) \ \forall f x y \left[(f(x_1, ..., x_n) = f(y_1, ..., y_n)) \Rightarrow (x_1 = y_1) \wedge ... \wedge (x_n = y_n) \right]$$

$$(\mathcal{E}f3) \qquad \forall f \forall x \neg (f(x) = x)$$

Proof: Left as an exercise.

Theorem 9.1.5. *Let Π be a definite program and q a ground atom. Then,*

$$\Pi \models q \quad \Rightarrow \quad Compl(\Pi) \models q.$$

Proof: Left as an exercise.

The above notwithstanding, for most practical applications we actually have to extend $\mathcal{C}_{LP}$ to the form

$$\left(\mathcal{C}_{LP}^{-} \right) \qquad A \leftarrow B_1, ..., B_n$$

where the B_i are either positive literals or negative literals.

Definition. 9.1.44. $\mathcal{C}_{LP}^{-}$ is called a *general clause*. A LP program in which negation $\neg$ is allowed to occur in the body of a rule, i.e. a LP program with $\mathcal{C}_{LP}^{-}$, is called a *general program*.

We leave the essentials of negation in general programs as exercises (cf. Exercises 9.1.4.6, 9.1.4.9, and 9.1.4.14 below). We can remark, however, that just as in the case of definite programs, it is not possible for a negative literal to be a logical consequence of a general program.

We introduce now a LP notion–*stratification*–that will also be relevant for Section 9.3:

Definition. 9.1.45. We say that a general LP program Π is *stratified* if the set $Pred(\Pi)$ of the predicates in Π can be partitioned into $Pred_0(\Pi), ..., Pred_n(\Pi)$ such that if $A \leftarrow B_1, ..., B_m$ is a rule of Π and $A \in Pred_k(\Pi)$, $0 \leq k \leq n$, then:

1. If there occurs no negation in B_i for $1 \leq i \leq m$, then

$$B_i \in \bigcup_{j=0}^{k} Pred_j(\Pi).$$

Algorithm 9.2 Clark completion

Input: A definite-clause theory or database Θ
Output: $Compl\left(\Theta\right) = \Delta$

---

1. Rewrite every individual definite clause

$$(\mathcal{C}_{LP}) \quad \forall \vec{X}\left(p\left(\vec{t}\right) \leftarrow B_1 \wedge \ldots \wedge B_n\right)$$

   as

$$(CC_{LP}) \quad \forall \vec{Y} \left[\exists \vec{X}\left(B_1 \wedge \ldots \wedge B_n \wedge \left(\vec{Y} = \vec{t}\right)\right) \rightarrow p\left(\vec{Y}\right)\right]$$

   where $\vec{Y}$ is a sequence of new variables.

   a) If $n = 0$, then we have

$$(CC_{LP}) \quad \forall \vec{Y}\left[\exists \vec{X}\left(\vec{Y} = \vec{t}\right) \rightarrow p\left(\vec{Y}\right)\right].$$

   b) When $\mathcal{C}_{LP}$ is a ground formula, then we have

$$(CC_{LP}) \quad \forall \vec{Y}\left[\left(B_1 \wedge \ldots \wedge B_n \wedge \left(\vec{Y} = \vec{t}\right)\right) \rightarrow p\left(\vec{Y}\right)\right].$$

2. Let us simplify CC_{LP} as

$$\forall \vec{Y}\left(E \rightarrow p\left(\vec{Y}\right)\right).$$

   Suppose that there are k such clauses, i.e. there are $E_1, \ldots, E_k$. Then, we have the single formula

$$\forall \vec{Y}\left[p\left(\vec{Y}\right) \leftrightarrow (E_1 \vee \ldots \vee E_k)\right].$$

   a) If $k = 0$, then we have the Clark formula

$$\forall \vec{Y}\left(\neg p\left(\vec{Y}\right)\right).$$

3. For every non-unifiable pair of terms $(t_1, t_2) \in \Theta$, construct the statement

$$\neg\left(t_1 = t_2\right).$$

---

2. If, however, $B_i = \neg C_i$, then

$$C_i \in \bigcup_{j=0}^{k-1} Pred_j\left(\Pi\right).$$

We leave further theoretical aspects on stratification as an exercise (cf. Exercise 9.1.4.9). A simple way to check whether a LP program is stratifiable is by means of a dependency graph, which, because now negation may be present, will no longer be called *simple* (cf. Exercise 9.1.3.10.2). Because a general LP program can be seen as a generalization of a Datalog program with negation, we leave the discussion on stratification with relation to dependency graphs for Section 9.3.5; the application of this discussion to general LP programs is direct.

Exercises

Exercise 9.1.4.1. Are the following tautologies/laws of CL (cf. Prop. 4.3.1) valid in PL?

1. DN.

2. MT.

3. Law of contraposition.

4. RA.

Exercise 9.1.4.2. With respect to Algorithm 9.2, show that $\mathcal{C}_{PL} \equiv CC_{LP}$.

Exercise 9.1.4.3. With respect to a definite-clause theory or database Θ, show that:

1. $\Theta \subseteq Compl\left(\Theta\right)$, i.e. $Compl\left(\Theta\right)$ is an extension of Θ.

2. $Compl\left(\Theta\right)$ is consistent.

3. If the PL query $? - q$ succeeds, for q an atomic ground formula, then $Compl\left(\Theta\right) \vdash \neg q$.

Exercise 9.1.4.4. Construct the Clark completion of the theory in Exercise 3.2.4.

Exercise 9.1.4.5. State intuitively the meaning of the axioms in Proposition 9.1.43.

Exercise 9.1.4.6. Show that, if Π is a general program, *Compl* (Π) can be inconsistent.

Exercise 9.1.4.7. Consider SLDNF, i.e. SLD resolution with rule NF, for definite programs.

1. Formulate and prove the soundness of SLDNF.

2. Is SLDNF complete? Give a constructive proof, i.e. a proof in which you actually provide an example.

Exercise 9.1.4.8. Prove Proposition 9.1.43 and Theorem 9.1.5.

Exercise 9.1.4.9. Let Π be a general program.

1. Show that if Π is stratified, then *Compl* (Π) is consistent.

2. What are the advantages, from the viewpoint of negation, of a stratified program, which allows negated atoms in the body of a rule, compared to applying NF?

Exercise 9.1.4.10. Consider the following definite Program *On*:

above $(A, B) \leftarrow$ on (A, B).
above $(A, B) \leftarrow$ on (A, C), above (C, B).
on (c, b).
on (b, a).

1. Apply NBF to it by means of:

   a) *Compl* (On).
   b) CWA.

2. Are the results of these applications identical?

3. Is program *On* recursive? Use a dependency graph to account for your answer.

Exercise 9.1.4.11. CL is said to be *monotonic*, because given a theory Θ and an extension thereof Θ^*, for a formula ϕ we have:

$$(M) \qquad \text{if } \Theta \Vdash \phi, \text{ then } \Theta \cup \Theta^* \Vdash \phi$$

A logical system is said to be *non-monotonic* if M is not necessarily the case.

1. Elaborate on how non-monotonicity may be a better reflection of human reasoning.

2. Explain in which way CWA gives to LP the property of non-monotonicity.

Exercise 9.1.4.12. Recall the notion of a least Herbrand model in Exercise 9.1.3.9. What relation can you establish between this model, as well as between a Herbrand model in general, and CWA in LP?

Exercise 9.1.4.13. Show that LP with CWA is unsound for definite programs.

Exercise 9.1.4.14. Prove the soundness of SLDNF for general programs.

Exercise 9.1.3.15. In the context of SLDNF, *floundering* is an issue one must beware of. Define it and elaborate on its impact on LP.

Exercise 9.1.3.16. CWA can be contrasted with what is known as *open-world assumption* (OWA).

1. Give the essential aspects of OWA.

2. How (dis)advantageous is OWA in terms of the (intended) meaning of a program Π when compared to CWA?

9.2. Declarative + procedural interpretation: Prolog

Prolog, abbreviating the French expression *programmation en logique*, is the main language (family) of LP. Although not the most commonly used

programming language for commercial or industrial applications, Prolog
has the advantage of being a Turing-complete language, which partly
explains its considerable success within the (European) artificial intelligence community since its original development in the early 1970s by
A. Colmerauer and P. Roussel. In effect, any computable function–and
we know that this is any function that can be computed by a Turing
machine–can be represented by means of Prolog; in other words, Prolog
can simulate any Turing machine.

As elsewhere in this book, we concentrate on what in Prolog relates
more directly to logic–or impacts on its deductive properties. Because
this is basically the material above on LP, or *pure* Prolog, we now have
only a few remarks on *real* Prolog. Readers seeking a more hands-
on approach comprising aspects not discussed here but fundamental to
Prolog (e.g., lists) can benefit from Sterling & Shapiro (1994).

9.2.1. Prolog and Prolog

By "real Prolog" we intend to capture pure Prolog extended with the
predicates $not/1$, $!/0$, and $fail/0$. The distinction between real and
pure Prolog is actually an important one, as only the latter is Turing-
complete; the addition of *cut*, denoted by the symbol "!", divests Prolog
of this desired property.[12] But most of all, real Prolog, though of a
more procedural type than the more declarative pure Prolog, shows how
deductive programming can be carried out by procedures, too.[13]

In what follows, when we write simply "Prolog" we mean "real Prolog,"
unless otherwise stated.

Definition. 9.2.1. A *Prolog rule* has the form $\mathcal{C}_{LP}$ and is written

$$A : -B_1, ..., B_n.$$

for $n \geq 1$; if $n = 0$, then we have a *Prolog fact* and we write simply $A.$.
Similarly, *Prolog goals* and *queries* are as in LP, and the same holds for
Prolog terms.

Definition. 9.2.2. A *Prolog program* Π is a sequence of Prolog facts
and rules.

Henceforth, we shall only write "Prolog rule," "Prolog fact," etc. if we
need to disambiguate; otherwise, we write simply "rule," "fact," etc. This
similarity notwithstanding, Prolog is more procedural than LP, which is

---

[12]We remark that this is not an established distinction in the field, with the label
"pure Prolog" capturing many different versions of Prolog or LP.
[13]See the introductory discussion in the last paragraph of Section 0.4.

more declarative, and this is expressed in a few features present solely in Prolog. In turn, these features are associated to a specific language that we can call Prolog.[14] We define some of these features:

Definition. 9.2.3. A *meta-variable* is a variable that can occur as an atom.

Definition. 9.2.4. A *built-in predicate* is a predicate that is defined by internal rules of Prolog, i.e. it need not be defined explicitly when writing a program.

Built-in predicates are particularly relevant for (autonomous) management of the data in a program, but some were conceived with interaction with the user in mind. Most built-in predicates have a procedural interpretation, but this can have an impact on the declarative interpretation of a program. This is the case for the predicates !/0 and `fail/0`. We discuss them below, given their importance, but anticipate that they cannot occur in the head of a rule, i.e. they are always goals. Another built-in predicate of particular interest from the viewpoint of the declarative interpretation is `not/1`, and we discuss it at length in Section 9.2.3.

Example 9.2.1. Further built-in predicates relevant for deductive programming (a sample) are:

- The order predicates $<, >, <=, >=$ corresponding to the common symbols $<, >, \leq, \geq$.

- The arithmetical predicates $==$ and $= \backslash =$ for arithmetical equality and difference, and $+$, $*$, etc.

- The infix predicate $=$ for unification or matching.

- The predicate read(X) allows the user to unify the variable X with a specific constant, thus allowing the user to manipulate domains and instantiations at will.

---

[14]To be formally correct, we should write "Prolog" instead of "Prolog" and also often "Logic programming" instead of "Logic programming." However, this would require a fastidious distinction between the languages and the programs, and then between these and the corresponding programming systems–a distinction that actually goes against the common practice in the field of programming. As a matter of fact, we often write "CFOL" instead of "L1" for similar reasons. Thus, we use this font only when the discussion is strictly restricted to the language at hand.

- The infix predicate `is` expresses equality in Prolog; in `X is Y`, X must be instantiated to numbers or other arithmetical expressions (e.g., `X is 5 * 2`).

- The predicate `true/0` that, like `!/0` and `fail/0`, is a goal that always succeeds, can be used to force the attempt to satisfy subsequent subgoals regardless of the failure of an earlier goal.

- The predicate `call (X)` allows us to instantiate a variable X to a term that can be interpreted as a goal.

We can now define the language **Prolog** in the Backus-Naur notation as follows:

Definition. 9.2.5. Let there be given the set

$$O_{Prolog} = \left\{ \begin{array}{c} :- \\ , \\ ; \end{array} \right\}$$

of operators and the punctuation marks "," (between arguments), ".", and left and right parentheses. If we define inductively terms, atoms, and statements over a signature $\Upsilon = (Pred, Fun, ar)$ with $ar \geq 0$ denoted by $\cdot/\mathbf{n}$ for $t_1, ..., t_n$ terms, as

			Variables	X	::=	e.g., X | John | _9
Terms	t	::=	*Constants*	c	::=	e.g., a | john | 9
			Functors	$p; f$	::=	p/n; f/n
Atoms				$A\ (B, ...)$	::=	$p \| f \| X$
			Facts	A	::=	A.
Statements			*Rules*	r	::=	A : −B$_1$, ..., B$_n$.
			Goals	$G\ (A, ...)$	::=	B$_1$, ..., B$_n$

then we have the language (basic FO) **Prolog** and its syntax.[15]

We can extend O with the operator ";" denoting disjunction. We have already used it in queries to prompt all the possible replies, but it can

---

[15] In Prolog, a predicate $p\,(t_1, ..., t_n)$ with n terms, denoted by p/n, may itself be a term in the negative literal *not* $(p\,(t_1, ..., t_n))$. This is a feature that contributes to the ambivalent syntax of Prolog, a feature already mentioned above for LP and which we further discuss below.

also be employed in the body of rules. However, it is interpreted as different clauses. Say you have a rule $A : -B_1 ; B_2$. Then, the interpreter will consider this as two rules, to wit $A : -B_1$. and $A : -B_2$.. Example 9.2.5 below shows the utility of this operator in the body of rules.

Strictly, Prolog statements are solely facts and rules; goals are parts of rules, and only as such can they be considered statements. Queries are goals with a question mark, i.e. "statements" of the form $? - q$ where $q = G$; in other words, queries are not statements of the programs. However, we can extend O with the operator ?. As a matter of fact, full **Prolog** contains more operators, some of which are actually built-in predicates.

The above allows for a redefinition of a Prolog program:

Definition. 9.2.6. A Prolog program is a sequence of facts and rules in which there can occur (i) meta-variables and (ii) built-in predicates.

Comparison of Definition 9.2.5 with Definition 4.1.1 shows that, though a FO logical language, **Prolog** differs from L1 in important ways. The occurrence of meta-variables and of functions as atoms, as well as of predicates as arguments (see below), accounts for what we above referred to as the ambivalent syntax of LP. It should be noted, however, that this ambivalent syntax does not entail that **Prolog** is an orderless language. We omitted quantification in Definition 9.2.5, because this is a rather implicit business in LP, as seen above. In effect, we can use **Prolog** at a purely propositional level, but then Prolog is a rather uninteresting programming system; and we can use it at second or higher orders, but this requires further specifications that we do not discuss in this text.[16]

Exercises

Exercise 9.2.1.1 Consider the theory of Exercise 3.2.4 as a Prolog program. How do you account for a negative answer to the following queries?

1. ?- sells(west,missile,daffy).

2. ?- sells(west,weapon,daffy).

3. ?- owns(daffy,weapon).

---
[16]See, for instance, Sterling & Shapiro (1994), Chapter 16, for 2nd-order Prolog programming.

4. ?- owns(daffy,missile).

Exercise 9.2.1.2. Create Prolog programs for:

1. Multiplication of two natural numbers.

2. Exponentiation.

3. Sum of three natural numbers.

4. The ordering $<$ for the natural numbers.

Exercise 9.2.1.3. Write Prolog programs for the following theories (given in the indicated examples) and check the respective conclusions.

1. Example 3.2.3.

2. Example 8.1.17

Exercise 9.2.1.4. Consider the simple graph in Figure 1.2.5. The following is an incomplete program for this graph:

```
path(V,V).
path(U,V):-path(U,W),edge(W,V).
edge(1,2).
...
```

1. Complete the program.

2. Obtain the list of all the connected edges in this graph.

3. Further complete the program so that one can find out whether this is a connected graph.

Exercise 9.2.1.5. Let the following Prolog program be called *Height*.

taller $(X, Y) : -$height $(X, \_, Z)$, height $(Y, \_, W)$, $Z > W$.
tall $(X) : -$height $(X, male, Y)$, $Y => 180$cm.
tall $(X) : -$height $(X, female, Y)$, $Y => 175$cm.
normal $(X) : -$height $(X, male, Y)$, $Y > 170$cm, $Y < 180$cm.
normal $(X) : -$height $(X, female, Y)$, $Y > 160$cm, $Y < 175$cm.
short $(X) : -$height $(X, male, Y)$, $Y =< 170$cm.
short $(X) : -$height $(X, female, Y)$, $Y =< 160$cm.

1. Which relevant queries may it answer?

2. What facts need to be added to it? Indicate the functor(s) and the respective arguments.

Exercise 9.2.1.6. Show that pure Prolog is Turing-complete.

Exercise 9.2.1.7. Determine the consequence(s) from the computational viewpoint of introducing in a Prolog program a rule of the form

$$A \leftarrow B_1, ..., A, ..., B_n.$$

Exercise 9.2.1.8. What is the procedural interpretation of the operator ";" in Prolog?

Exercise 9.2.1.9. Consider the FA in Figure 2.2.3. Write a program for this FA having in mind Σ^+. Hints:

- Use the constant symbols q0, q1, q2 for the states;

- Use facts of the form trans (Qi, X, Qj) and final (Qr);

- A list in Prolog can be defined recursively as (i) the empty list $[]$ is a list, and (ii) $[H|T]$, where H denotes the head and T does so for the tail, is a list. Define the rules

$$\text{accept} (Qr, []) : -\text{final} (Qr).$$

and

$$\text{accept} (Qr, [H|T]) : -\text{trans} (Qi, H, Qj), \text{accept} (Qj, T).$$

1. Determine whether the FA accepts the following strings by means of the query $? - \text{accept} \left(qi, \underbrace{[X, Y, ...]}_{n} \right)$.:

   a) 001.

   b) 1000001.

   c) 1.

   d) 0000.

   e) 00011.

2. Include the acceptance of the empty string in the program for the FA.

Exercise 9.2.1.10. Write Prolog programs for the following automata:

1. The NDFA in Figure 2.2.4.

2. The FA in Figure 2.2.5.2.

3. The Mealy and Moore machines in Figure 2.2.8.1-2.

4. The PDA of Example 2.2.8.

5. The Turing machine of Example 2.2.12.

6. The Turing machine of Example 2.2.16.

7. The Turing machine in Figure 2.2.18.

Exercise 9.2.1.11. A truly challenging exercise would be to write a Prolog program for:

1. The Turing machine in Figure 2.2.15.

2. Schubert's steamroller (cf. Example 8.1.9).

9.2.2. Logic + control: ! and fail

In the Introduction, we touched upon the aspect of computational logic summarized as the equation *Algorithm = Logic + Control*. As seen above, SLD resolution is sound and complete for LP; but this does not mean that it is efficient. In fact, it is not, as it entails the construction and complete traversal of often complex SLD-resolution trees. This is precisely the point where we add the factor "Control" to "Logic," making real Prolog more efficient than pure Prolog as far as computation is concerned. The cost of controlling deductive computation may be high (see below), but the trade-off may be favorable in the end, at least in some cases.

Control is added to Logic in Prolog by means of special atoms that have a procedural impact on the deductive computation without for that changing the meaning of a program. In other words, the programmer controls how the deductive computation takes place. This is done by means of predicates that, inside the body of a rule, have an interpretation of the kind "If *X*, then do *Y*," rather than the more logical interpretation "If *X*, then *Y* follows." We speak here of the predicates

!/0, read "cut," and `fail/0`. As 0-ary predicates, they are interpreted as operators rather than predicates.

We begin with the operator !: Given a SLD-resolution tree, its use is intended to cut off failing branches and to prune succeeding branches.

Definition. 9.2.7. Given a LP program $\Pi = \{\mathcal{C}_1, ..., \mathcal{C}_n\}$, where $\mathcal{C}$ abbreviates $\mathcal{C}_{LP}$, let $\mathcal{C}_i \subseteq \Pi$ be the clause

$$A \leftarrow B_1, ..., B_j, !, B_{j+2}, ..., B_n$$

and let G be a goal. Then, if G unifies with $A \in \mathcal{C}_i$, and $B_1, ..., B_j \in \mathcal{C}$ succeed, $! \in \mathcal{C}$ has the following effects:

1. The program is so to say committed to $\mathcal{C}_i$ to reduce G, no alternative to $A \in \mathcal{C}_k$, $k > i$, being considered.

2. In case B_i, $i > j+1$, fails, then backtracking goes no further back than !, the $B_1, ..., B_j$ being pruned from the search tree.

3. If the backtracking search goes as far back as to !, then ! fails, and the search goes back to the last $\mathcal{C}_j$ prior to the choice of $\mathcal{C}_i$.

Although real Prolog is often seen as a good example of a programming system that allies declarative and procedural paradigms, this alliance is not without issues. As a matter of fact, this alliance entails the loss of Turing-completeness, a much desired property of pure Prolog.

Example 9.2.2. Fathers of graduate children are typically proud. This can be deduced from the following program called *Proud_fathers*:

1. `proud (X) : −father (X, Y) , graduate (Y) .`
2. `father (X, Y) : −parent (X, Y) , male (X) .`
3. `parent (tom, sheila) .`
4. `parent (tom, lucy) .`
5. `male (tom) .`
6. `graduate (lucy) .`

Given the query ? − `proud (tom)`. as input, SWI-Prolog will answer "true." Edit the program by adding ! at the end of rule 2, so that you have

2′. `father (X, Y) : −parent (X, Y) , male (X) , !.`

Given the same input, SWI-Prolog now replies "false." This is so because Sheila, who is not graduate (by NBF), is the first of Tom's children in the program sequence, and after ! there are no more attempts to find any other children. Figure 9.2.1 shows–with a broken line–the pruned successful branch of the SLD-proof tree with the substitution $\lambda = \{Y = \text{lucy}\}$ and the first, failed, branch with substitution $\theta = \{Y = \text{sheila}\}$ ($\sigma = \{X = \text{tom}\}$). If we now interchange the positions of facts 3 and 4, so that Lucy appears as the first child of Tom, the same query will be answered "true," even with 2'; additionally, the failure branch will be pruned from the tree, thus making the computation more efficient.

Example 9.2.2 illustrates clearly the fact that ! is a problematic operator in deductive programming. Indeed, the addition of ! to the program *Proud_fathers* does impact significantly on the meaning of the program (cf. Def. 9.1.32). Let us denote this program by Π and the query proud(tom) by G. Then we have

$$\Pi \vdash G$$

accordingly to its meaning, to wit, $M(\Pi) = G = \{\text{proud}(\text{tom})\}$. Let us now consider the same program but with ! added, i.e. $\Pi \cup \{!\}$. Then we have

$$\Pi \cup \{!\} \nvdash G$$

as well as[17]

$$\Pi \cup \{!\} \vdash \neg G.$$

Clearly, $M(\Pi \cup \{!\}) \neq G$. This shows how a procedural interpretation of a program may in fact impact on the declarative meaning thereof. The fact that interchanging facts 3 and 4 corrects the problem can be used to "blame" the programmer for the undesirable results above, which is partly true, but it should also highlight the problems that may arise with the addition of ! to a program, particularly so in the case of complex programs. When the operator ! does change the meaning of a program, we call it a *red cut*–the opposite being a *green cut*.

Theorem 9.2.1. *Prolog with the operator ! is not Turing-complete.*

Proof: (Idea) In pure Prolog we are assured that, given a program Π and a query q, Prolog either proves or refutes $\Pi \vdash q$. In effect, Prolog carries out the unification algorithm and is guaranteed to stop when all

---

[17] That is to say, if given the goal ? − not (G) . as input, SWI-Prolog, for instance, will reply "true." See next Section for the predicate not/1.

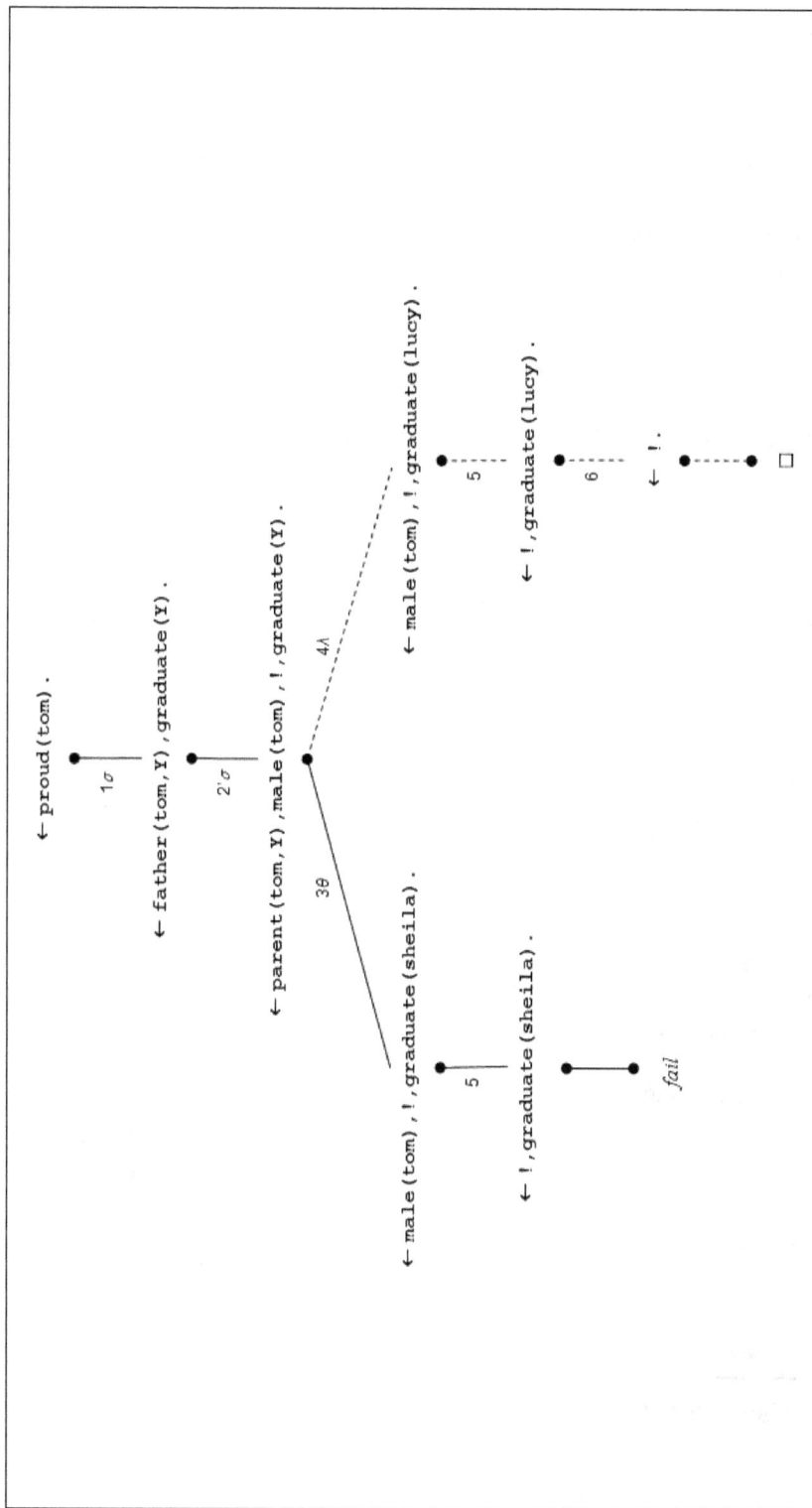

Figure 9.2.1.: A SLD-proof tree for a Prolog program with !.

the variables have been instantiated or when $\Pi \vdash q$ has been disproved. The operator ! "cuts off" branches of the search tree, possibly precisely those that contain the proof of $\Pi \vdash q$. **QED**

Yet another way to control deductive computation in Prolog is by means of the predicate `fail/0`. We already know *fail* from the context of SLD-resolution proofs for given goals: Given a LP program Π, if a goal G succeeds, then the goal $\neg G$ fails at all the possible attempts to satisfy it by backtracking, and we have

$$\Pi \vdash G$$

otherwise, i.e. if a goal G fails (for the reasons above), $\neg G$ succeeds and we have

$$\Pi \nvdash G.$$

In the latter case, to the query ? − G. SWI-Prolog replies "false." The predicate `fail/0`, which just like ! is rather an operator, is a means to control this deductive effect: the programmer elects the goals that are to fail. But contrarily to !, `fail` has the procedural effect that no alternative solution whatsoever is considered to satisfy a goal followed by it: the goal immediately fails.

Definition. 9.2.8. The 0-ary predicate `fail` is a built-in predicate with the empty definition whose procedural interpretation is as follows: given this predicate in the body of a rule, the head of the rule fails.

Example 9.2.3. Let it be given the Prolog program with the single rule `crazy (daffy) : −fail`. Then, given the query ? − `crazy (daffy)`, SWI-Prolog will reply "false."

According to Definition 9.2.5, both ! and `fail` are goals; as built-in undefined goals, they have the property that they always succeed, so they should be cautiously employed when writing a Prolog program. Their combination, in particular, is a powerful means to control deductive computation (see next Section).

Exercises

Exercise 9.2.2.1. Describe the SLD-resolution tree of a program with the single rule $A : -!, B_2, ..., B_n$.

Exercise 9.2.2.2. Consider the following Prolog program.

```
move (0, _, _, _) : −!.
move (N, X, Y, Z) : −N1 is N − 1, move (N1, X, Y, Z),
      print_move (X, Y), move (N1, Y, Z, X).
print_move (X, Y) : −write ("move from X to Y").
tower (N) : −move (N, left, middle, right).
```

1. Identify the built-in predicates and their operations.

2. What is the program expected to compute?

3. Determine whether the nature of the program is declarative or procedural.

Exercise 9.2.2.3. Consider the following program:

```
min (X, Y, X) : −X < Y, !.
min (X, Y, Y).
```

1. Identify the undesirable consequences of this program.

2. Correct the program.

Exercise 9.2.2.4. Consider the following Prolog program and determine the answers to the queries below by drawing the respective SLD-proof trees:

```
q(a).
q(b).
q(c).
r(b,boo).
r(c,coo).
r(a,awa).
r(a,owo).
r(a,waw).
p(X,Y):-q(X),r(X,Y).
p(d,dee).
paa(X,Y):-q(X),r(X,Y),!.
pho(X,Y):-q(X),!,r(X,Y).
pi(X,Y):-!,q(X),r(X,Y).
pi(d,dee).
```

1. `?-p(X,Y).`

2. `?-paa(X,Y).`

3. `?-pho(X,Y).`

4. `?-pi(X,Y).`

Exercise 9.2.2.5. Determine whether ! is green or red in the following (parts of) programs.

1.
```
if_then_else(X,Y,_):-X,!,Y.
if_then_else(_,_,Z):-Z.
```

2.
```
p(a).
p(b).
p(c).
q(X):-p(X),!.
q(d).
```

3.
```
abs_value(X,X):-X>=0,!.
abs_value(X,Y):-X<0,!,Y is -X.
```

4.
```
mammal(X):-warm_blooded(X),!,vertebrate(X).
warm_blooded(dolphins).
warm_blooded(humans).
vertebrate(dolphins).
vertebrate(humans).
```

5.
```
elect_fault(X):-capacitor(X),!,impedance(X,R),R<5000.
elect_fault(X):-fuse(X),!,impedance(X,R),R>100.
```

Exercise 9.2.2.6. Let there be given the following step function:

$$f(x) = \begin{cases} 0 & \text{if } x < 5 \\ 1 & \text{if } 5 \le x < 10 \\ 2 & \text{if } x \ge 10 \end{cases}.$$

1. Write a Prolog program for this function with green cuts.

2. Draw the SLD-resolution tree for the queries:

   a) $?-f(2,y), 4 < y.$

   b) $?-f(12,y), 1 < y.$

Exercise 9.2.2.7. Put the operator ! into relation with deterministic computation (cf. Def. 9.1.5).

Exercise 9.2.2.8. Give a complete proof of Theorem 9.2.1.

Exercise 9.2.2.9. By Theorem 9.2.1, real Prolog is not Turing-complete. Elaborate on the trade-off between introducing ! in pure Prolog and the loss of Turing-completeness.

Exercise 9.2.2.10. Show that the Prolog predicate `fail/0` has as a logical support the classical law *ex contradictione quodlibet* (ECQ or EFQ; cf. Fig. 3.2.1).

9.2.3. Negation in Prolog: The predicate `not`

As seen, the operator ! may–and often does–lead to incorrect programs, and `fail` requires cautious employment as well. An alternative to each of these, but actually employing them in combination, is *negation*. In effect, a form of NBF can be implemented in Prolog by means of ! and `fail` in this exact order. We express it by means of the predicate `not/1` for convenience, but if using SWI-Prolog, the unary predicate \+ must be used instead.

Definition. 9.2.9. The Prolog predicate `not/1` is defined as:

not (X) : −X, !, fail.
not (X).

We call this combination of ! and `fail` *cut-failure negation*.

This is a built-in predicate, so the programmer does not need to write this definition as part of a program.

Example 9.2.4. Given the program *Proud_fathers* (cf. Example 9.2.2), SWI-Prolog gives the following replies to the indicated queries:

? − graduate (sheila).
false

? − not (graduate (sheila)).
true

? − not (graduate (lucy)).
false

With respect to the first query, SWI-Prolog simply applies NBF: the goal graduate (sheila) does not match any data in the program. As

for the second and third queries, SWI-Prolog applies the definition of
not/1. In these two queries, the goal G is $\leftarrow$ not (graduate (Y)). and
the subgoal G' is $\leftarrow$ graduate (Y).. If G' succeeds, then the operator
! cuts off the possibility that G may be the case and G fails. In this
case (the third query above), SWI-Prolog replies "false." If, however, G'
finitely fails, the search algorithm tries G by means of backtracking and
G succeeds. In this case (the second query above), the answer is "true."

It should be noted that cut-failure negation is indeed a form of NBF:
Given the program *Proud-fathers*, the query ? $-$ not (graduate (a)). will
be answered "true" for any ground term a such that $a \neq lucy$. This leaves
us with an infinite domain in which the predicate not (graduate (X)) $\in$
Proud_fathers is "true."

Example 9.2.5. As said above, for practical reasons the predicate
not/1 is often essential in a program. For instance, suppose an avian
center requires a database from which it can be easily known which
bird(s) might have escaped the closed precincts by flying. This center
has all sorts of birds, a few of which do not normally fly, but most of
which do fly. By adding a rule to the database like

$$\text{fly}(X) : -\text{bird}(X), \text{not}(\text{abnormal}(X)).$$

we can eliminate all non-flying birds, i.e. the "abnormal" birds, such as
penguins, ostriches, etc. This possibly still leaves a lot of flying birds,
but we can further reduce the possibilities by employing another negated
atom, say, not (quarantined (X)). The program *Avian_center* can be
sketched as follows:[18]

```
fly(X):-bird(X),\+(abnormal(X)),\+(quarantined(X)).
bird(X):-canary(X);nightingale(X);penguin(X);ostrich(X);
    crow(X);emu(X);woodpecker(X);turkey(X);duck(X);hen(X).
abnormal(X):-penguin(X);ostrich(X);emu(X);turkey(X);hen(X).
penguin(toto).
ostrich(sheila).
emu(tom).
turkey(sam).
turkey(sandra).
hen(lolita).
```

---

[18]In, for example, the predicate penguin (toto), *toto* is just a constant; it can stand
equally for a single penguin (say, in a shelter for birds) or for all the penguins (in
a large avian center, for instance). Note also the useful operator ";" in the body
of the second and third rules of this program.

```
canary(roberto).
nightingale(sarita).
crow(bob).
woodpecker(lola).
duck(cassandra).
duck(samantha).
quarantined(roberto).
quarantined(bob).
```

The query $? - \mathtt{fly}\,(\mathtt{X})$. and the operator ";" will output all the reduced possibilities:

```
X = sarita;
X = lola;
X = cassandra;
X = samantha;
false
```

More negated goals will help us to obtain a further reduced sample of the birds that could have escaped by flying, from which sample it might be much easier to determine the fleeing bird(s).

In Definition 9.2.9, it should be noted that X is a meta-variable, a syntactical feature that is absent in CFOL, as already commented upon. This entails yet another feature that violates CFOL: given the n-ary function symbol $\mathtt{f}\,(\mathtt{t_1}, ..., \mathtt{t_n})$, then the query $? - \mathtt{not}\,(\mathtt{f}\,(\mathtt{t_1}, ..., \mathtt{t_n}))$. in Definition 9.2.9 requires a computation in which $\mathtt{f/n}$ is both a function and a predicate symbol. The same is true of a rule such as

$$\mathtt{p}\,(\mathtt{t_1}, ..., \mathtt{t_n}) : -\mathtt{not}\,(\mathtt{p}\,(\mathtt{t_1}, ..., \mathtt{t_n})).$$

in which $\mathtt{p/n}$ is a predicate symbol in the head and a function symbol in the body of the rule. This requires that we accept that, given a signature for a Prolog program Π, we may have

$$Pred\,(\Pi) \cap Fun\,(\Pi) \neq \emptyset.$$

In other words, we take ambivalent syntax to be a feature of Prolog.

One of the main goals when writing a program should be to make the deductive computation on it run error-free. Useful as negated atoms may be in Prolog, negation is unfortunately a source of errors. Let us say that a program is *safe* if no errors are generated, namely with respect to the SLD-resolution process.

Definition. 9.2.10. We say of an atom A that it is *unsafe* if

1. A is a meta-variable.

2. $A = \text{not}\,(\text{X})$, for X a variable.

3. $A = \text{not}\,(\text{not}\,(\text{t}))$, for a term t.

Definition. 9.2.11. A Prolog query is *meta-safe* if none of its atoms is unsafe.

If a meta-variable is selected in the SLD-resolution process, then errors are bound to occur. Hence, we have a problem, as Definition 9.2.9 does indeed contain a meta-variable.[19] The possible solution runs as follows, from definitions to theorem:

Definition. 9.2.12. A Prolog program Π containing the two rules of Definition 9.2.9 as a prefix and in which for every LP atom $\neg A$ we write $\text{not}\,(\text{A})$ instead is called a *restricted Prolog program* if, when omitting the mentioned prefix and writing $\neg A$ instead of $\text{not}\,(\text{A})$, we still have a syntactically correct general LP program. In the same way, we say that a Prolog query in which we write $\text{not}\,(\text{A})$ instead of $\neg A$ is a *restricted Prolog query* if it is a syntactically correct LP query when writing $\neg A$ instead of $\text{not}\,(\text{A})$.

Lemma 9.2.2. *Let q be a meta-safe Prolog query and Π a restricted Prolog program. Then all resolvents of q are meta-safe.*

Proof: Left as an exercise.

Theorem 9.2.3. *The computation of a restricted Prolog query and a restricted Prolog program generates no errors.*

Proof: Left as an exercise.

---

[19]It should be noted that even if we choose to define the first rule without a meta-variable, namely as

$$\text{not}\,(\text{p}\,(\text{t}_1, ..., \text{t}_n)) : -\text{p}\,(\text{t}_1, ..., \text{t}_n)\,, !, \text{fail}.$$

we have the ambivalent-syntax problem. We leave it as a research task (cf. Exercise 9.2.3.6) for the reader to find out what other issues the ambivalent syntax of Prolog can cause.

Exercises

Exercise 9.2.3.1. In which sense can it be said that `not/1` ∈ Prolog is a stronger form of negation than ¬ ∈ L1? (Hint: Consider resolution without NF.)

Exercise 9.2.3.2. Comment on how negative goals contribute to non-monotonic reasoning on the program *Avian_center*.

Exercise 9.2.3.3. Write the programs with the given rules:

1. A person who is big is assumed to be strong, unless there is reason to believe they are weak. A person who is small is assumed to be weak, unless there is reason to believe they are strong. A person who is weak but muscular is assumed to be strong, except if there is some reason to believe the opposite.

2. Cats are typically finicky. Nevertheless, finicky cats can eat what they would otherwise reject if they are truly hungry. But super-finicky cats would rather starve to death than eat what they typically reject.

3. Being the focus of attention can be delightful, unless you have no privacy. But some stars who have no privacy like being the focus of attention anyway, except when they are grieving.

Exercise 9.2.3.4. Write the premises of the argument in Example 3.2.1 as a Prolog program and test it for the query

?-have(sheila,problem).

(Hint: You may need to make changes to the set of premises. For example, you might add the premise "Susan likes all animals but dogs.")

Exercise 9.2.3.5. Above we mentioned the generation of errors when meta-variables are used in Prolog programs. Specify these errors.

Exercise 9.2.3.6. Research further into the ambivalent syntax of Prolog and its implications from the viewpoint of deductive computation.

Exercise 9.2.3.7. Prove Lemma 9.2.2 and Theorem 9.2.3.

9.3. Purely declarative interpretation: Datalog

Elsewhere in this book, we have been mostly concerned with *logical theories* and *logical programs*, which we occasionally spoke of rather loosely as *databases*.[20] Now, we shall be focusing on databases proper, namely on *relational databases*. More precisely, we shall concentrate on these as *deductive databases*. Our discussion is, so to say, biased by our main subject–classical deductive computing with classical logic–, and the reader is referred to the following works for a strictly database approach, with or without LP: For the large topic of *databases*, the reader may benefit from Date (2004); for *databases and LP* we refer the reader to Ceri, Gottlob, & Tanca (1989, 1990), Gallaire, Minker, & Nicolas (1984), Minker (1997)–which also provides historical aspects on the topic of logic and databases–, and Abiteboul, Hull, & Vianu (1995), a comprehensive textbook. Finally, Greco & Molinaro (2016) provides an extensive elaboration on Datalog.

In terms of deductive programming, Datalog has a purely declarative interpretation. This means that as far as the equation *Algorithm = Logic + Control* is concerned, we have now the identity *Algorithm = Logic*. This identity is supported by a FO language–a subset of Prolog–that, besides being capable of representing knowledge in terms of relations, also allows deduction to be carried out over sets of formulae expressing relations. This means that Datalog both is a *relational language* over which we can build *relational databases* and allows deduction over those databases. Although our interest falls mainly on Datalog as a deductive database, these relational notions need to be briefly elaborated on, as in fact Datalog is the addition of logical features pertaining to CFOL to relational databases. In other words, Datalog is the formalization of relational databases by means of a subset of L1, a formalization that entails the deductive properties inherent in this subset when considered in the framework of a calculus under (mostly) Herbrand semantics. Anticipating the due formal definitions, this subset is function-free–which, as seen above (cf. Section 6.2), guarantees decidability under Herbrand semantics–and includes solely the connectives for conjunction and (inverse) implication, so that, just as in the case of Prolog, we shall be employing SLD resolution over sets of Horn clauses.

For our purposes, the contents of Section 9.1.1 hold *mutatis mutandis*, as we include deductive databases in deductive programming, for which the notions *query* and *query system* are central. Exceptions and specifications will be clearly stated. In particular, we go on denoting arbitrary

---

[20]For the former, see especially Section 3.4.5; for the latter, see Sections 9.1-2.

variables by the uppercase letters $X, Y, ...$ and arbitrary predicate symbols by lowercase letters (e.g., p).

9.3.1. Relational languages and databases

Definition. 9.3.1. Let L be a formal language over a signature $\Upsilon = (Pred, \emptyset, Cons)$.[21] Then, if the assertions of L are of the form $R(t_1, ..., t_n)$, where R is a *n-ary relation symbol* and the t_i are variables and/or constants, L is called a *relational language*.

Although, as it will be seen, we may consider a relational language as a function-free fragment of $\mathsf{L1}$, in the more narrowly defined context of relational databases the following specifications need to be made:

Definition. 9.3.2. Let a domain $\mathscr{D}_i$, for $1 \leq i \leq n$, be a set of *values*.

1. We say that the (commutative) Cartesian product $\mathscr{D}_1 \times ... \times \mathscr{D}_n$ of $i = 1, ..., n$ domains is a *(finite) relation* $R \subseteq \mathscr{D}_1 \times ... \times \mathscr{D}_n$ of arity $i = 1, ..., n$.

2. The values of domain $\mathscr{D}_i$ are called *attributes*. Let $A_1, ..., A_n$ be attributes. Then, a relation R with n attributes defines a (non-ordered) n-tuple $(A_1, ..., A_n)$ such that $R(A_1, ..., A_n)$ is called a *relation scheme* and $R = \{(c_1^1, ..., c_n^1), ..., (c_1^k, ..., c_n^k)\}$, where the c_i^j, $1 \leq j \leq k$, denote specific elements (items or individuals), is said to be an *instance* (or *extension*) thereof.

3. A finite set of instances of relations $R_l \subseteq \mathscr{D}_1^l \times ... \times \mathscr{D}_n^l$, $l = 1, ..., m$, is a *(finite) relational database*.

We can envisage a relation R as a table of which the attributes A_i, $i = 1, ..., n$, are the columns and the instances $u_j = \left(c_1^j, ..., c_n^j\right)$, $1 \leq j \leq k$, are the rows.

Example 9.3.1. Let us revisit the avian center of Example 9.2.5 above. A relational database for this center can be constructed by means of a relation $R(A_1, A_2)$ where $R = BIRD$, $A_1 = SPECIES$, and $A_2 = NAME$. Attributes A_1, A_2 are associated with the corresponding domains $\mathscr{D}_1, \mathscr{D}_2$ of, respectively, *bird species* and *proper nouns*. The table for this relation is shown in Figure 9.3.1.
Then, the relational database *Avian_center* is constituted by the finite instances of the relation scheme $BIRD(SPECIES, NAME)$. (Note

---

[21]Or $\Upsilon = (Pred, \emptyset, ar \geq 0)$.

| BIRD ||
SPECIES	NAME
Penguin	Toto
Ostrich	Sheila
Emu	Tom
Turkey	Sam
Turkey	Sandra
Hen	Lolita
Canary	Roberto
Nightingale	Sarita
Crow	Bob
Woodpecker	Lola
Duck	Cassandra
Duck	Samantha

Figure 9.3.1.: Table for $BIRD\,(SPECIES, NAME)$.

how for the values *Turkey* and *Duck* there are two rows on the table for this relation scheme.) The extension of this relation scheme is the set:

$$BIRD = \left\{ \begin{array}{c} (penguin, toto)\,, (ostrich, sheila)\,, (emu, tom)\,, \\ (turkey, sam)\,, (turkey, sandra)\,, (hen, lolita)\,, \\ (canary, roberto)\,, (nightingale, sarita)\,, \\ (crow, bob)\,, (woodpecker, lola)\,, \\ (duck, cassandra)\,, (duck, samantha) \end{array} \right\}$$

More frequently than not, a relational database has more than a single relation scheme. As a matter of fact, a relational database can be further specified as a set of ground assertions over a relational language L *together* with the following axioms:

Proposition. 9.3.3. *If* $c_1, ..., c_p$ *are all the constant symbols of a relational language* L *and* $R\left(c_1^1, ..., c_n^1\right), ..., R\left(c_1^m, ..., c_n^m\right)$ *denote all the facts under* R *for each relational symbol* $R \in \Sigma_L$*, then the following,* together *with the axioms for equality* $\mathcal{E}1, \mathcal{E}2, \mathcal{E}3, \mathcal{E}5$ *(cf. Prop. 4.5.1),*[22] *are the axioms of a relational database over* L*:*

(UN) $(c_1 \neq c_2)\,, ..., (c_1 \neq c_p)\,, (c_2 \neq c_3)\,, ..., (c_{p-1} \neq c_p)$

---

[22]To simplify, consider $\mathcal{E}5$ as formalized in FOL (cf. Exercise 4.5.4).

$$(\text{DC}) \qquad \forall X \left[(X = c_1) \vee ... \vee (X = c_p) \right]$$

$$(\text{CO}) \qquad \forall X_1 ... \forall X_n [R(X_1, ..., X_n) \rightarrow$$
$$\left((X_1 = c_1^1) \wedge ... \wedge (X_n = c_n^1) \right) \vee ... \vee \left((X_1 = c_1^m) \wedge ... \wedge (X_n = c_n^m) \right)]$$

Proof: Left as an exercise.

We denote the set of all the axioms in Proposition 9.3.3, often called the *particularization axioms*, by AX_Δ, where Δ denotes an arbitrary relational database.[23] This axiomatization, firstly formulated in Reiter (1984), formalizes three assumptions of relational databases: the *unique-name assumption* (axiom UN), which states that every distinct individual in the database has a different name or, which is the same, individuals with different names are distinct; the *domain-closure assumption* (axiom DC), according to which there are no other individuals than those in the database; the *completion assumption* (axiom CO), which states that the only tuples that a relation R can have are those specified in the relational database. Importantly, the completion assumption "translates" the *closed-world assumption* (CWA) already known from our discussion on Prolog (cf. Section 9.1.4): That there are no other instances of some relation R than those implied by the database entails that $\neg R(c_1, ..., c_n)$ is assumed to be true if the tuple $(c_1, ..., c_n)$ does not constitute an instance of R in the database.

Definition. 9.3.4. Given a finite set of ground assertions R, a relational database Δ is defined as:

$$\Delta = R \cup AX_\Delta$$

Example 9.3.2. Given the relational database *Avian_center*, by axiom UN we can assume that Sam and Sandra are two distinct turkeys, i.e. *sam* $\neq$ *sandra*; by DC, we know that these are the only two turkeys in the database, as for all the birds X in the database and the value *Turkey* we have $(X = sam) \vee (X = sandra)$; and by CO we can safely assume that Sheila is not a turkey, as $(turkey, sheila) \notin BIRD$, and hence we can conclude $\neg BIRD(turkey, sheila)$.

Definition. 9.3.5. Let there be given a relational database Δ. Then:

---

[23]Not to be confused with a general database Δ, as used above in the context of Prolog programs.

9. Programming

1. The set of ground assertions of the form $R(c_1, ..., c_n)$, is called the *extensional database* (EDB).

2. The set of axioms of the form $R_1(\vec{x}_1) \leftarrow R_2(\vec{x}_2), ..., R_m(\vec{x}_m)$, where the $\vec{x}_i$ are tuples of appropriate arities, is called the *intensional database* (IDB).

Exercises

Exercise 9.3.1.1. Consider the countries of the European Union as a domain $\mathscr{D}$.

1. Create the EDB of a relational database *European_Union_Borders* with the relations $COUNTRY(a)$ and $BORDER(a, b)$, where $a, b \in \mathscr{D}$.

2. Design rules for the corresponding IDB such that

   a) $BORDER(X, Y)$ is a symmetrical relation.

   b) the relation $REACHABLE(X, Y)$ depends on the relation $BORDER(X, Y)$.

   c) the relation $REACHABLE(X, Y)$ can be defined recursively.

Exercise 9.3.1.2. Consider the typical underground train (a.k.a. metro) network.

1. List the relations (with respective attributes and domains) required to build a complete EDB.

2. Which rules will allow a traveler to use the network with ease?

Exercise 9.3.1.3. Consider the book archive of a library as a relational database.

1. List some of the relations (with respective attributes and domains) for the corresponding EDB.

2. Which rules will allow a reader to use the archive with ease?

9.3.2. Deductive databases and Datalog

Given a relational language, for some relation $R \subseteq \mathscr{D}_1 \times ... \times \mathscr{D}_n$, the corresponding attributes $A_1, ..., A_n$ can be expressed as variables $X_1, ..., X_n$ (as constants $c_1, ..., c_n$), so that we actually have $p(X_1, ..., X_n)$ (respectively, $p(c_1, ..., c_n)$), where p is a *predicate symbol* of arity n. In other words, we can express n-ary relations of a relational database as atoms of the form $p(t_1, ..., t_n)$ where $t_i \in [(Vi \cup Cons) \subseteq \mathsf{L1}]$. The language Datalog–that, for the same reasons stated for Prolog, we shall write simply Datalog–allows us to do this.

Definition. 9.3.6. Let there be given the set

$$O_{Dlog} = \left\{ \begin{array}{c} :- \\ , \end{array} \right\}$$

of operators and the punctuation marks "," (between arguments), ".", and left and right parentheses. If we define inductively terms, functors (predicate, or relation, symbols of arity n), atoms, and statements over a signature $\Upsilon = (Pred, \emptyset, Cons)$ as

Terms	t	::=	*Variables*	X ::=	*e.g.*, X $\mid$ John
			Constants	c ::=	*e.g.*, a $\mid$ john $\mid$ 8
Functors				p ::=	$p(t_1, ..., t_n)$
Atoms			$A(B, ...)$	::=	p
Statements			*Facts* A	::=	A.
			Rules r	::=	A : $-B_1, ..., B_n$.
			Goals $G(A, ...)$	::=	$B_1, ..., B_n$

then we have the language Datalog and its syntax if the following *safety conditions* are satisfied:

1. Each fact A is a ground atom.

2. Given a rule $\mathbf{r} = $ A : $-B_1, ..., B_n$., if a variable X occurs in A, the *head* of the rule (i.e. $\{A\} \subseteq Head_{\mathbf{r}}$), then it must occur at least in one of the (non-arithmetical) $B_i \in Body_{\mathbf{r}}$ constituting the *body* of $\mathbf{r}$ (for $\{B_1, ..., B_n\} \subseteq Body_{\mathbf{r}}$).

From the above definitions, it is obvious that Datalog is a *function-free* subset of $\mathsf{L1}$. As a matter of fact, if $Cons$ is a finite set, then we can consider Datalog to be a subset of the FO fragment L1$_{ff}$ (cf. Section 6.2) and provide it with the desirable results formulated above in Section 6.2.

Proposition. 9.3.7. Datalog $\subseteq$ L1$_{ff}$ *is a relational language.*

Proof: Trivial. **QED**

Trivial as the proof above might be, some remarks need to be made. Let $\mathscr{RTL}$ be the class of relational languages; then, Datalog is a relational language in the sense that we have Datalog $\supseteq \mathscr{RTL}$. We shall consider this containment relation as formalizing the fact that Datalog is more expressive than a relational language. This is particularly so with respect to recursion (cf. Exercise 1.1.1.10): While this property is expressible in Datalog, it is not a feature of the class $\mathscr{RTL}$. But the impact of applying Datalog on a relational database has further interesting and important consequences. We next elaborate on this topic.

The main objective of "logicizing" relational databases, namely by means of Datalog, is that of allowing deduction to be carried out over them. We call these "logicized" relational databases *deductive databases* (abbrev.: DDBs). In effect, given a query q and a relational database Δ, we may wish to know whether $q \in \Delta$. This is clearly a decision problem, and, as seen above, we can formulate it as a logical problem: Given the pair (Δ, q), we wish to find out a "Yes/No" answer to the question

$$\Delta \overset{?}{\vdash} q.$$

This logical formulation, in turn, requires that our relational database be capable of having deduction carried out over it, so that $q = R\left(c_1^i, ..., c_n^i\right)$ may be a *new* relation instance–a *view*, in DDB jargon–deducible from Δ, but not an explicit assertion in Δ. Furthermore, we aim at adequate deduction over our database, so that we want the proof system $\mathcal{P}_\Delta$ for Δ to correspond to some semantics $\mathfrak{S}_\Delta$ in the sense that we have $\Delta \vdash q$ iff we have $\Delta \models q$.

Interestingly enough, all we have to do is to see Δ as a FO logical theory. In effect, just compare Definition 9.3.4 with Definition 3.4.39, armed with the knowledge that $AX_\Delta \subseteq (EDB \cup IDB)$. We shall consider this FO formalization of Δ to be a deductive database.

Definition. 9.3.8. A *deductive database* (DDB), denoted by Δ, is defined as:
$$\Delta = EDB \cup IDB$$

Proposition. 9.3.9. *A deductive database Δ is a logical theory. In particular, a deductive database Δ is a logical theory over $\mathsf{L1}_{ff}$.*

Proof: Trivial. **QED**

For reasons to do with implementation, we shall consider that every formula in Δ satisfies the particularization axioms, but shall not consider explicitly $AX_\Delta \subseteq \Delta$. One of the reasons for this omission is that we do not wish to employ $\mathsf{L1}^=$ or any of its reducts containing the symbol "=" (see Section 4.5); in particular, AX_Δ would contribute to combinatorial complexity leading to inefficient implementations. This restriction can be satisfied by considering a DDB as

$$\Delta = EDB \cup IDB \cup \{\text{NF}\}$$

(cf. Def. 9.1.40).[24] It is evident that we can construct such a DDB by means of the logical-relational language Datalog.[25] As a matter of fact, we can speak of a *Datalog database* as a synonym for DDB.

Definition. 9.3.10. Let there be given a Datalog database Δ. Then, a *Datalog program* $\Pi_\Delta \subseteq \Delta$ is defined as:

$$\Pi_\Delta = IDB \backslash EDB$$

In other words, a Datalog program is a finite set of Datalog rules. This may appear at first as a very narrow definition, accepted on the basis that the EDB be physically stored in a relational database, a feature that is explained by the need to store a large number of assertions. Indeed, it may appear that an *assertion* in a relational language just is a *fact* in LP, in which a fact just is a special rule, namely a rule without a body (cf. Def. 9.1.18.2). But this is actually not supported by the semantics for Datalog. Thus, we have both pragmatic and formal reasons to define a Datalog program as above.

We next elaborate on this, for which end we shall require a more precise definition of a Datalog program:

Definition. 9.3.11. Let there be given a Datalog DB $\triangle$. Then, the finite set

$$\Pi_\Delta = \underbrace{Datalog\,rules}_{IDB \subseteq \Delta} \backslash \underbrace{Assertions}_{EDB \subseteq \Delta}$$

---

[24] We shall have more to say about negation in Datalog below.

[25] A DDB also typically comprises a set of *integrity constraints*, FO formulae expressing facts such as "an individual cannot be both mother and father of a child." However, these constraints are not essential in DDBs. If a constraint is indeed essential, then it can be formulated as a rule in the DDB.

is a Datalog program if

1. the head predicate of every rule in Π_Δ does not occur in *EDB*, or, in other words, is an *intensional predicate* (or *relation*), and

2. predicates in *EDB* occur only in the bodies of rules in Π_Δ, being thus called *extensional predicates* (or *relations*).

Example 9.3.3. Let us retake the relational database of Examples 9.3.1-2, based on the Prolog program *Avian_center* (cf. Example 9.2.5). We extend this relational database with the supplementary relation schemas $ABNORMAL(c)$, $QUARANTINED(c)$, and $EATS(c_1, c_2)$. The corresponding attributes and domains should be evident from the values given. The complete relational database *Avian_Center_DB* is as follows:

$$BIRD = \left\{ \begin{array}{c} (penguin, toto), (ostrich, sheila), (emu, tom), \\ \dots \\ (duck, cassandra), (duck, samantha) \end{array} \right\}$$

$$ABNORMAL = \{(penguin), (ostrich), (emu), (turkey), (hen)\}$$

$$QUARANTINED = \{(roberto), (bob)\}$$

$$EATS = \left\{ \begin{array}{c} (penguin, fish), (ostrich, all), (emu, all), \\ (turkey, seeds), (hen, all), (canary, seeds), \\ (nightingale, seeds), (crow, all), \\ (woodpecker, bugs), (duck, all) \end{array} \right\}$$

The corresponding complete EDB *Avian_Center_EDB* is shown in Figure 9.3.2. We can now build a Datalog DB *Avian_Center_DDB* constituted by the EDB *Avian_Center_EDB* and the following Datalog rules, which, in turn, constitute the program *Avian_Sick_Prog*:

r_1 :   sick (Y) : −quarantined (Y).
r_2 :   on_diet (Y, Z) : −bird (X, Y), sick(Y), eats (X, Z).

Here, the predicate on_diet (Y, Z) denotes the fact that the sick bird Y is on an extra portion of Z, the food adequate for its species, for quick convalescence. After querying the DB with respect to the currently sick birds (query: ? − sick (Y).), staff responsible for quarantined birds can check on their dietary needs by means of the query ? − on_diet (Y, Z). For instance, knowing that Bob is quarantined, one can query the DB

$$\left\{ \begin{array}{c} bird\,(penguin,toto)\,,bird\,(ostrich,sheila)\,,bird\,(emu,tom)\,,\\ bird\,(turkey,sam)\,,bird\,(turkey,sandra)\,,bird\,(hen,lolita)\,,\\ bird\,(canary,roberto)\,,bird\,(nightingale,sarita)\,,\\ bird\,(crow,bob)\,,bird\,(woodpecker,lola)\,,\\ bird\,(duck,cassandra)\,,bird\,(duck,samantha)\,,\\ abnormal\,(penguin)\,,abnormal\,(ostrich)\,,\\ abnormal\,(emu)\,,abnormal\,(turkey)\,,abnormal\,(hen)\,,\\ quarantined\,(roberto)\,,quarantined\,(bob)\,,\\ eats\,(penguin,fish)\,,eats\,(ostrich,all)\,,eats\,(emu,all)\,,\\ eats\,(turkey,seeds)\,,eats\,(hen,all)\,,eats\,(canary,seeds)\,,\\ eats\,(nightingale,seeds)\,,eats\,(crow,all)\\ eats\,(woodpecker,bugs)\,,eats\,(duck,all) \end{array} \right\}$$

Figure 9.3.2.: The EDB *Avian_Center_EDB*.

with the query ? − on_diet (bob, Z).. (Actually, one can simply query the DB by means of ? − on_diet (Y, Z)., getting at once the information on the quarantined birds and their corresponding diets.)

This example illustrates the following facts: Firstly, it shows why the separation between the relational DB and the program proper is of advantage. Indeed, the DB requires regular, perhaps daily, updating, as the number of birds in the center changes frequently and quarantines need regular updates, too. An industrial-scale relational DB may be so complex as to actually require some sort of automatic updating. On the contrary, the rules of the program do not change. Secondly, a large institution typically has different departments. In this particular example, the reader can easily imagine a department in the avian center for the quarantined birds. Staff working in this department might very likely be the only ones who need to use this program, but the DB must be available to the whole center. This accounts for the head predicates in the program rules not occurring in the DB, as well as for the fact that the DB, in turn, has predicates that do not occur in the program. For instance, in the program of Example 9.3.3, the predicate name *ABNORMAL* does not occur.

The practical account for Definitions 9.3.10-11 being given, we focus now on the formal account. This requires yet a further specification for a Datalog program:

Definition. 9.3.12. Given a Datalog program Π_Δ, let us call the finite set of all extensional predicate names or symbols the *extensional schema*, denoted by $EDB\,(\Pi_\Delta)$, and the finite set of all intensional predicate names or symbols the *intensional schema*, denoted by $IDB\,(\Pi_\Delta)$. We set

$$EDB\,(\Pi_\Delta) \cap IDB\,(\Pi_\Delta) = \emptyset.$$

Then, the *schema* of a Datalog program is defined as the set

$$Sch\,(\Pi_\Delta) = [EDB\,(\Pi_\Delta) \subseteq Pred\,(EDB)] \cup IDB\,(\Pi_\Delta)\,.$$

Intuitively, $Sch\,(\Pi_\Delta)$ gives us the *structure* of the corresponding Datalog program.

Example 9.3.4. We abbreviate the name of the program in Example 9.3.3 as A. With respect to this program, we have the sets

$$EDB\,(A) = \{\texttt{bird}, \texttt{quarantined}, \texttt{eats}\}$$

and

$$IDB\,(A) = \{\texttt{sick}, \texttt{on\_diet}\}\,.$$

The schema for this program is

$$Sch\,(A) = \{\texttt{bird}, \texttt{quarantined}, \texttt{eats}, \texttt{sick}, \texttt{on\_diet}\}\,.$$

Note that the predicate $\texttt{abnormal} \in Pred\,(EDB)$ does not belong to the schema of this program, as we have $\texttt{abnormal} \notin EDB\,(A)$. Also, the intensional predicate name $\texttt{sick} \in IDB\,(A)$ occurs also in the body of rule r_2, which illustrates the fact that the body of a Datalog program rule can have both extensional and intensional predicate names or symbols. It is this fact that accounts for recursion as a property of Datalog programs: In effect, recursion is the case when some intensional predicate occurs both in the head and in the body of rules.

Exercises

Exercise 9.3.2.1. Consider the safety conditions in Definition 9.3.6.

1. What do they guarantee with respect to a Datalog program?

2. Determine which of the following Datalog facts and rules are safe:

   a) $\texttt{r}\,(\texttt{X}) : -\texttt{p}\,(\texttt{Y}), \texttt{Y} < \texttt{X}.$

   b) $\texttt{buy}\,(\texttt{john}, \texttt{Z}).$

   c) $\texttt{meet}\,(\texttt{X}, \texttt{Y}) : -\texttt{see}\,(\texttt{X}, \texttt{Y}).$

d) love (John, pet).

e) ancestor $(X, Y) : -\text{parent}(X, Z), \text{ancestor}(Z, Y)$.

Exercise 9.3.2.2. Given the rules in the following Datalog programs Π_Δ, determine for each program both $EDB(\Pi_\Delta)$ and $IDB(\Pi_\Delta)$:

1. $\Pi_{\Delta_1} = \left\{ \begin{array}{l} \text{r}(X, Y) : -\text{t}(X, Y). \\ \text{r}(X, Y) : -\text{t}(X, Z), \text{r}(Z, Y). \\ \text{r}(X, \text{peter}) : -\text{n}(X). \\ \text{n}(X) : -\text{e}(X, Y). \end{array} \right\}$

2. $\Pi_{\Delta_2} = \left\{ \begin{array}{l} \text{p}(X, Y) : -\text{p}(Y, X). \\ \text{q}(X, Y) : -\text{p}(X, Y). \\ \text{q}(X, Y) : -\text{q}(X, Z), \text{p}(Z, Y). \end{array} \right\}$

3. $\Pi_{\Delta_3} = \left\{ \begin{array}{l} \neg a \vee b \\ \neg a \vee c \\ \neg c \vee \neg b \vee d \end{array} \right\}$

4. $\Pi_{\Delta_4} = \left\{ \begin{array}{l} \text{reach}(X, X) : -\text{links}(L, X, Y). \\ \text{reach}(X, X) : -\text{links}(L, Y, X). \\ \text{reach}(X, Y) : -\text{links}(L, X, Z), \text{reach}(Z, Y). \\ \text{stations}(X) : -\text{reach}(\text{odeon}, X). \end{array} \right\}$

Exercise 9.3.2.3. Show in which way a Datalog program can be seen as a CFG. (Hint: Parse trees.)

Exercise 9.3.2.4. One of the advantages of Datalog over a purely relational calculus is recursion. Comment on in which way(s) this is an advantage. (Hint: Graph transitive closure.)

Exercise 9.3.2.5. Why is Datalog not Turing-complete?

Exercise 9.3.2.6. List all the differences between Prolog and Datalog.

Exercise 9.3.2.7. Research into:

1. The Null-Value Problem.

2. Extended disjunctive deductive databases (EDDDBs).

3. Indefinite deductive databases (IDDDBS).

4. Temporal deductive databases.

9.3.3. Semantics for Datalog DDBs

We are now ready to introduce semantics for a Datalog DB, namely for a Datalog program. There are at least two semantics for a Datalog program, to wit, Herbrand semantics and fixed-point semantics. Importantly, these two semantics are equivalent. Although with different weights, we discuss here both semantics.

9.3.3.1. Herbrand semantics

For the exposition that follows, recall the contents of Sections 3.5.2.1 and 6.2. Recall also the general definitions and properties of the consequence operation Cn (Def. 3.4.4) and of the consequence relation $\Vdash$ (Def. 3.4.7), in particular the semantical consequence relation $\models$ (see Section 3.4.3).

Definition. 9.3.13. Given a Datalog DB Δ, a semantics $\mathfrak{S}_\Delta$ for Δ is the set of models based on the function

$$f_\Delta : EDB\,(\Pi_\Delta) \longrightarrow IDB\,(\Pi_\Delta)$$

mapping *EDB facts* to *IDB facts*.

    This definition, which introduces the notion of an IDB fact, entails a semantics $\mathfrak{S}_\Delta$ for a DDB Δ such that $\mathfrak{S}_\Delta$ is the set of models based on a mapping from inputs over the *EDB* to outputs over the *IDB*. Hence, a semantics $\mathfrak{S}_\Delta$ for Δ just is the semantics for the program Π_Δ associated with it. This motivates the following definition that gives a new meaning to the concept *query*:

Definition. 9.3.14. Given a Datalog DB Δ, a program Π_Δ is a *query* against $E_\Delta \subseteq EDB$, where E_Δ is the set of Datalog formulae expressing the known facts of Δ that can be queried via Π_Δ.[26]

    In other words, a Datalog program Π_Δ provides a means of querying a subset of an associated relational DB Δ. It should not be hard to see

---

[26]We emphasize the fact that E_Δ is not necessarily identical to *EDB*. Precisely because of this inequality, and because the *IDB* of a Datalog DB may also contain further rules, namely integrity constraints and particularization axioms, $EDB\,(\Pi_\Delta)$ and $IDB\,(\Pi_\Delta)$ are relevant notions. In particular, if a predicate symbol $p \in [EDB\,(\Pi_\Delta) \subset Sch\,(\Pi_\Delta)]$, then p occurs in a formula of E_Δ. As for the set $IDB\,(\Pi_\Delta)$, we may assume that $IDB\,(\Pi_\Delta) = Pred\,(IDB\,(\Delta))$. If $IDB\,(\Delta)$ contains more rules than those whose predicate names are to be found in $IDB\,(\Pi_\Delta)$, then we may specify the facts formed by means of $IDB\,(\Pi_\Delta) \subsetneq Pred\,(IDB\,(\Delta))$ as *IDB facts proper*. In effect, the *IDB* of any DDB is the output of querying the associated *EDB* by means of a Datalog program.

that both the EDB facts in E_Δ and the IDB facts proper are constituted by the ground instances of the predicates over $Sch\,(\Pi_\Delta)$, denoted by $Sch\,(\Pi_\Delta)_g$, which, in turn, constitute the set of ground instances of the goals of a Datalog program. This just is the Herbrand base obtained from the Herbrand universe of Δ. We can now further specify a semantics $\mathfrak{S}_{\Pi_\Delta}$ for a Datalog program as follows:

Definition. 9.3.15. Given the Herbrand universe H_Δ for a Datalog DB Δ with a program Π_Δ, we denote by $H\,(\Delta)$ the Herbrand base of Δ. Let us denote by $EDB\,(\Pi_\Delta)_g$ the ground atoms of $H\,(\Delta)$ corresponding to the predicates in $EDB\,(\Pi_\Delta)$, and by $IDB\,(\Pi_\Delta)_g$ the ground atoms of $H\,(\Delta)$ corresponding to the predicates in $IDB\,(\Pi_\Delta)$, so that we have:

$$Sch\,(\Pi_\Delta)_g = EDB\,(\Pi_\Delta)_g \cup IDB\,(\Pi_\Delta)_g$$

$Sch\,(\Pi_\Delta)_g$ is called an *instance* of Δ *over* $Sch\,(\Pi_\Delta)$.[27] Then, a semantics $\mathfrak{S}_{\Pi_\Delta}$ for a Datalog program Π_Δ

1. is a function
$$g_\Delta : EDB\,(\Pi_\Delta) \longrightarrow IDB\,(\Pi_\Delta)$$

   such that, given an extensional schema $EDB\,(\Pi_\Delta) \supseteq E_\Delta$, we have

$$g_\Delta\,(EDB\,(\Pi_\Delta)) = Cn\,(\Pi_\Delta \cup E_\Delta) = g_\Delta\,(E_\Delta)$$

   where $p\left(c_1^i, ..., c_n^i\right) \in Cn\,(\Pi_\Delta \cup E_\Delta)$ if either (i) $p \in EDB\,(\Pi_\Delta)$ and $p\left(c_1^i, ..., c_n^i\right) \in (E_\Delta)_g$, or (ii) there is some rule $\mathbf{r}$ in Π_Δ such that $p\,(X_1, .., X_n) = Head_{\mathbf{r}}$ and $Body_{\mathbf{r}} \subseteq Sch\,(\Pi_\Delta)_g$, and

2. given some goal $G =\, ? - \mathrm{p}\,(\mathbf{X_1}, ..., \mathbf{X_n})\,.$, we have

$$g_\Delta\,(G) = \bigcup \left\{ \left(c_1^i, ..., c_n^i\right) \mid \Pi_\Delta \cup E_\Delta \models p\left(c_1^i, ..., c_n^i\right) \right\}$$

   for $i = 1, 2, ..., k$ and where $p \in IDB\,(\Pi_\Delta)$ and $p\left(c_1^i, ..., c_n^i\right) \in g_\Delta\,(EDB\,(\Pi_\Delta))$ is a ground instance of G.

Example 9.3.5. Let us retake the Datalog DB *Avian_Center_DDB* and the respective Datalog program *Avian_Sick_Prog*, that we shall abbreviate as A. As seen in Example 9.3.4, the schema of this program

---

[27] Equivalently, given a domain $\mathscr{D}_\Delta \supseteq \mathscr{D}_1, ..., \mathscr{D}_n$ for a DDB Δ, a *database instance* $Sch\,(\Pi_\Delta)_g$ is a finite Herbrand interpretation over $\mathscr{D}_\Delta$. Note that whereas $Sch\,(\Pi_\Delta)$ specifies the structure of the deductive database Δ, $Sch\,(\Pi_\Delta)_g$ specifies its *content*.

```
bird (penguin, toto) .
bird (ostrich, sheila) .
bird (emu, tom) .
bird (turkey, sam) .
bird (turkey, sandra) .
bird (hen, lolita) .
bird (canary, roberto) .
bird (nightingale, sarita) .
bird (crow, bob) .
bird (woodpecker, lola) .
bird (duck, cassandra) .
bird (duck, samantha) .

quarantined (roberto) .
quarantined (bob) .

eats (penguin, fish) .
eats (ostrich, all) .
eats (emu, all) .
eats (turkey, seeds) .
eats (hen, all) .
eats (canary, seeds) .
eats (nightingale, seeds) .
eats (crow, all) .
eats (woodpecker, bugs) .
eats (duck, all) .

sick (roberto) .
sick (bob) .

on_diet (roberto, seeds) .
on_diet (bob, all) .
```

Figure 9.3.3.: An instance of the Datalog database *Avian_Center_DDB* with respect to the program *Avian_Sick_Prog*.

is

$$Sch\,(A) = \{\texttt{bird}, \texttt{quarantined}, \texttt{eats}, \texttt{sick}, \texttt{on\_diet}\}.$$

An instance over this schema, denoted by $Sch\,(A)_g$, is shown in Figure 9.3.3. Then, given goals $G_1 = ? - \texttt{sick(Y)}.$ and $G_2 = ? - \texttt{on\_diet}\,(\texttt{Y}, \texttt{Z}).$, we have

$$g_\Delta\,(G_1) = \{(\texttt{roberto})\} \cup \{(\texttt{bob})\}$$

and

$$g_\Delta\,(G_2) = \{(\texttt{roberto}, \texttt{seeds})\} \cup \{(\texttt{bob}, \texttt{all})\}$$

Note how $\texttt{sick}\,(\texttt{Y}) = Head_{r_1}$ and $\texttt{on\_diet}\,(\texttt{Y}, \texttt{Z}) = Head_{r_2}$, as well as that $Body_{r_1} \subset Sch\,(A)_g$ and $Body_{r_2} \subset Sch\,(A)_g$, thus satisfying condition 1 in Definition 9.3.15. Furthermore, taking into consideration the associated $IDB\,(A)$, the domain $\mathscr{D}_A$ is reduced to

$$\mathscr{D}'_A = \{canary, crow, roberto, bob, seeds, all\}.$$

This gives us the reduced instance of Figure 9.3.4, which in fact corresponds to a minimal Herbrand model for this program. In effect, we have, for A abbreviating *Avian_Sick_Prog* and AC doing so for the associated EDB *Avian_Center_EDB*, and for the goals above G_1 and G_2,

$$A \cup AC \models \left\{ \begin{array}{c} \texttt{sick}\,(\texttt{roberto}) \\ \\ \texttt{sick}\,(\texttt{bob}) \end{array} \right.$$

and

$$A \cup AC \models \left\{ \begin{array}{c} \texttt{on\_diet}\,(\texttt{roberto}, \texttt{seeds}) \\ \\ \texttt{on\_diet}\,(\texttt{bob}, \texttt{all}) \end{array} \right. .$$

In particular, we have

$$(E_{AC})_g = \left\{ \begin{array}{c} bird\,(canary, roberto)\,, \\ bird\,(crow, bob)\,, \\ quarantined\,(roberto)\,, \\ quarantined\,(bob)\,, \\ eats\,(canary, seeds)\,, \\ eats\,(crow, all) \end{array} \right\}$$

and

```
bird (canary, roberto).
bird (crow, bob).

quarantined (roberto).
quarantined (bob).

eats (canary, seeds).
eats (crow, all).

sick (roberto).
sick (bob).

on_diet (roberto, seeds).
on_diet (bob, all).
```

Figure 9.3.4.: $Cn(Avian\_Sick\_Prog \cup E_{Avian\_Center\_DDB})$.

$$g_\Delta(E_{AC}) = \left\{ \begin{array}{c} bird\,(canary, roberto)\,, \\ bird\,(crow, bob)\,, \\ quarantined\,(roberto)\,, \\ quarantined\,(bob)\,, \\ eats\,(canary, seeds)\,, \\ eats\,(crow, all)\,, \\ sick\,(roberto)\,, \\ sick\,(bob)\,, \\ on\_diet\,(roberto, seeds)\,, \\ on\_diet\,(bob, all) \end{array} \right\}.$$

In fact, $g_\Delta(E_{AC})$ can be further reduced to a subset $g_\Delta\left(E'_{AC}\right)$ if we consider a sub-domain; for instance, let us focus on Bob, so that we have

$$\mathscr{D}''_A = \{crow, bob, all\}$$

and

$$g_\Delta\left(E'_{AC}\right) = \left\{ \begin{array}{c} bird\,(crow, bob)\,, \\ quarantined\,(bob)\,, \\ eats\,(crow, all)\,, \\ sick\,(bob)\,, \\ on\_diet\,(bob, all) \end{array} \right\}.$$

Consider now Definition 9.3.15: in it, $\Pi_\Delta \cup E_\Delta \models p\left(c_1^i, ..., c_n^i\right)$ can be abbreviated as $\Pi_\Delta \models p\left(c_1^i, ..., c_n^i\right)$ if we conceive $\Pi_\Delta = E_\Delta \cup IDB_\Delta$. This we do if we conceive a Datalog program as a definite program.

498

Then, we can further specify

$$\Pi_\Delta \models_{Sch(\Pi_\Delta)_g} p\left(c_1^i, ..., c_n^i\right)$$

so that we actually have

$$A \cup AC \models_{Sch(A)_g} \left\{ \begin{array}{c} \texttt{sick (roberto)} \\ \\ \texttt{sick (bob)} \end{array} \right.$$

and

$$A \cup AC \models_{Sch(A)_g} \left\{ \begin{array}{c} \texttt{on\_diet (roberto, seeds)} \\ \\ \texttt{on\_diet (bob, all)} \end{array} \right. .$$

In effect, logical consequence with respect to a Datalog program Π_Δ is defined here in terms of the Herbrand interpretations therefor. Because there are no function symbols in Datalog, the Herbrand universe H_{Π_Δ} is finite. Also, because the domains are specified in the associated relational DB, the constants in HI_{Π_Δ} are all the constants occurring in $EDB\left(\Pi_\Delta\right)_g$. Finally, as there are no (explicitly) negative literals in our Datalog DDB, there is actually only one H-interpretation HI_{Π_Δ} for a Datalog program if all the predicate symbols of $Sch\left(\Pi_\Delta\right)$ are included, i.e. if $HI_{\Pi_\Delta} = \bigcup_i HI_{\Pi_\Delta}^i$. Then, we have the following result:

Proposition. 9.3.16. *Given a Datalog program Π_Δ, there is a least H-model*

$$\underline{H}\mathcal{M}_{\Pi_\Delta} =$$

$$\left\{ p\left(c_1^i, ..., c_n^i\right) \mid p\left(c_1^i, ..., c_n^i\right) \in Sch\left(\Pi_\Delta\right)_g \text{ and } \left(\Pi_\Delta \cup E_\Delta\right) \models p\left(c_1^i, ..., c_n^i\right) \right\}$$

such that

$$\underline{H}\mathcal{M}_{\Pi_\Delta} = g_\Delta\left(E_\Delta\right).$$

Proof: Left as an exercise. (Hint: Note how, in Example 9.3.5, $g_\Delta\left(E_{AC}'\right)$ is a minimal H-model, but *not* the least H-model for AC; $g_\Delta\left(E_{AC}\right)$ is.)

In other words–and informally–, $g_\Delta\left(E_\Delta\right)$ provides *all* the information, and *only* the information, expressed by a Datalog program Π_Δ. In terms of Herbrand semantics, this just is the least H-model $\underline{H}\mathcal{M}_{\Pi_\Delta}$.

9.3.3.2. Fixed-point semantics

We obtain an equivalent result in the following way: If working with fixed-point semantics for a DDB, the immediate consequence operator takes over the role of Cn and $\models$. We denote this operator by $\mathbf{T}$, and define it as follows:

Definition. 9.3.17. Let $I \subseteq Sch\left(\Pi_\Delta\right)_g$ be a ground instance of a Datalog program Π_Δ. Then, the *immediate consequence operator* is the mapping

$$\mathbf{T}_\Pi : 2^I \longrightarrow 2^I$$

where $\mathbf{T}_\Pi$ is a simplified notation for $\mathbf{T}_{\Pi_\Delta}$. For a Datalog program Π_Δ and $I \subseteq Sch\left(\Pi_\Delta\right)_g$, we define

$$\mathbf{T}_\Pi\left(I\right) = \{\sigma A \mid \left(A \leftarrow B_1, ..., B_n\right) \in \Pi_\Delta \text{ and } \sigma\left(B_1\right), ..., \sigma\left(B_n\right) \in I\}$$

for some ground substitution σ.[28]

1. We set
$$\mathbf{T}_\Pi^0\left(I\right) := I$$

   after which we iterate

   $$\mathbf{T}_\Pi^1\left(I\right) := \mathbf{T}_\Pi\left(I\right)$$

   $$\mathbf{T}_\Pi^2\left(I\right) := \mathbf{T}_\Pi\left(\mathbf{T}_\Pi^1\left(I\right)\right)$$

   $$\vdots$$

   $$\mathbf{T}_\Pi^{n+1}\left(I\right) := \mathbf{T}_\Pi\left(\mathbf{T}_\Pi^n\left(I\right)\right)$$

   until no more ground atoms can be output, i.e. until

   $$\mathbf{T}_\Pi\left(\mathbf{T}_\Pi^n\left(I\right)\right) = \mathbf{T}_\Pi^n\left(I\right)$$

   and $\mathbf{T}_\Pi^n\left(I\right)$ is a *fixed point* of Π_Δ.

2. Let now
$$\mathbf{T}_\Pi \uparrow^0\left(\emptyset\right) := \emptyset$$
$$\mathbf{T}_\Pi \uparrow^1\left(\emptyset\right) := \mathbf{T}_\Pi\left(\mathbf{T}_\Pi \uparrow^0\left(\emptyset\right)\right) = \mathbf{T}_\Pi^0\left(\emptyset\right)$$
$$\mathbf{T}_\Pi \uparrow^2\left(\emptyset\right) := \mathbf{T}_\Pi\left(\mathbf{T}_\Pi \uparrow^1\left(\emptyset\right)\right) = \mathbf{T}_\Pi^1\left(\emptyset\right)$$
$$\vdots$$

---

[28]More correctly, a *matching* σ (see the next Section).

$$\mathbf{T}_\Pi \uparrow^{n+1} (\emptyset) := \mathbf{T}_\Pi \left(\mathbf{T}_\Pi \uparrow^n (\emptyset)\right) = \mathbf{T}_\Pi^n (\emptyset)$$

such that

$$\mathbf{T}_\Pi \uparrow^\omega (\emptyset) := \bigcup_{n \in \mathbb{N}}^{\infty} \mathbf{T}_\Pi \uparrow^n (\emptyset) = \lim_{n \to \infty} \mathbf{T}_\Pi \uparrow^n (\emptyset)$$

so that for finite n we have $\mathbf{T}_\Pi \uparrow^\omega (\emptyset) = \mathbf{T}_\Pi \uparrow^n (\emptyset)$ and we say that $\mathbf{T}_\Pi \uparrow^\omega (\emptyset)$ is the *least fixed point* of Π_Δ, denoted by $lfp\,(\Pi_\Pi)$.

Example 9.3.6. Consider the following Datalog program as constituted by the following set of logical formulae:

$$\Pi_\Delta = \left\{ \begin{array}{c} p\,(a) \\ q\,(X) \leftarrow p\,(X) \\ r\,(X) \leftarrow q\,(X) \\ s\,(X) \leftarrow r\,(X)\,, q\,(X) \end{array} \right\}$$

We start with $\mathbf{T}_\Pi \uparrow^0 (\emptyset) = \emptyset$, and we then iterate until a least fixed-point is found:

$$\mathbf{T}_\Pi \uparrow^1 (\emptyset) = \mathbf{T}_\Pi (\emptyset) = \{p\,(a)\}$$

$$\mathbf{T}_\Pi \uparrow^2 (\emptyset) = \mathbf{T}_\Pi (\{p\,(a)\}) = \{p\,(a)\,, q\,(a)\}$$

$$\mathbf{T}_\Pi \uparrow^3 (\emptyset) = \mathbf{T}_\Pi (\{p\,(a)\,, q\,(a)\}) = \{p\,(a)\,, q\,(a)\,, r\,(a)\}$$

$$\mathbf{T}_\Pi \uparrow^4 (\emptyset) = \mathbf{T}_\Pi (\{p\,(a)\,, q\,(a)\,, r\,(a)\}) = \{p\,(a)\,, q\,(a)\,, r\,(a)\,, s\,(a)\}$$

$$\mathbf{T}_\Pi \uparrow^5 (\emptyset) = \mathbf{T}_\Pi (\{p\,(a)\,, q\,(a)\,, r\,(a)\,, s\,(a)\}) = \{p\,(a)\,, q\,(a)\,, r\,(a)\,, s\,(a)\}$$

and

$$\mathbf{T}_\Pi \uparrow^5 (\emptyset) = lfp\,(\Pi_\Delta)\,.$$

$\mathbf{T}_\Pi \uparrow^5 (\emptyset)$ is the least fixed-point of Π_Δ.

The equivalence of Herbrand semantics and fixed-point semantics with respect to a Datalog program is expressed in the following theorem:

Theorem 9.3.1. *(van Emden & Kowalski, 1976) Let Π_Δ be a set of definite clauses. Then:*

$$\underline{H}\mathcal{M}_{\Pi_\Delta} = lfp\,(\Pi_\Delta) = \mathbf{T}_\Pi \uparrow^\omega (\emptyset)$$

Proof: Left as an exercise.

Further aspects of fixed-point semantics are left as exercises.

Exercises

Exercise 9.3.3.1. With respect to a least H-model for a Datalog program:

1. Give an intuitive account of its meaning/role.

2. Put it into relation with CWA.

3. Establish the relation between a semantics for a relational theory Δ based on this model–called a *least-Herbrand-model semantics*–and the proof theory for the corresponding DDB Δ.

Exercise 9.3.3.2. With respect to the program *Avian_Sick_Prog* of Example 9.3.5, compute:

1. The Herbrand universe.

2. The Herbrand base.

3. All the H-models.

4. The least H-model.

Exercise 9.3.3.3. In which way does bottom-up evaluation coincide with fixed-point semantics for a Datalog DDB?

Exercise 9.3.3.4. For the following Datalog programs given as sets of formulae, find (i) the Herbrand universe, (ii) the Herbrand base, and (iii) the least Herbrand model (we abbreviate Π_{Δ_i} as Π_i):

1. $\Pi_1 = \{p, q \leftarrow p, s \leftarrow r \wedge q, r \leftarrow q, t \leftarrow s\}$

2. $\Pi_2 = \{r \leftarrow r\}$

3. $\Pi_3 = \{p \leftarrow q, q \leftarrow p\}$

4. $\Pi_4 = \{p \leftarrow q, q \leftarrow p, r \leftarrow p \wedge q\}$

5. $\Pi_5 = \{p, q, r, s \leftarrow p, t \leftarrow r \wedge q, v \leftarrow s \wedge r\}$

6. $\Pi_6 = \{p(d), q(Z) \leftarrow p(Z), r(Z) \leftarrow q(Z)\}$

7. $\Pi_7 = \{p(a), q(b), r(X) \leftarrow p(X), r(Y) \leftarrow q(Y)\}$

8. $\Pi_8 = \left\{ \begin{array}{c} p(a,b), p(a,f), p(b,c), p(c,d), p(c,e), \\ r(X,Y) \leftarrow p(X,Y), \\ r(X,Y) \leftarrow r(X,Z), p(Z,Y) \end{array} \right\}$

Exercise 9.3.3.5. Prove Proposition 9.3.16. (Hint: See Exercise 9.1.3.9.)

Exercise 9.3.3.6. Find the least fixed-point of the programs in Exercise 9.3.3.4.

Exercise 9.3.3.7. With respect to the immediate consequence operator for a Datalog program Π_Δ, prove that:

1. $\mathbf{T}_{\Pi_\Delta}$ is monotone, i.e. for $I_1 \subseteq I_2$ we have $\mathbf{T}_{\Pi_\Delta}(I_1) \subseteq \mathbf{T}_{\Pi_\Delta}(I_2)$.

2. $\mathbf{T}_{\Pi_\Delta}$ is continuous.

3. $\mathbf{T}_{\Pi_\Delta} \uparrow^\omega (\emptyset)$ is the least fixed-point of $\mathbf{T}_{\Pi_\Delta}$.

4. $\mathbf{T}_{\Pi_\Delta} \uparrow^\omega (\emptyset) \models \Pi_\Delta$.

5. For every H-model of Π_Δ, we have:

$$\mathbf{T}_\Pi \uparrow^\omega (\emptyset)\,(H\mathcal{M}_\Pi) \subseteq H\mathcal{M}_{\Pi_\Delta}$$

6. Every fixed-point of Π_Δ is a model of Π_Δ.

7. Not every model of Π_Δ is a fixed-point thereof.

Exercise 9.3.3.8. The following two statements are central theorems in fixed-point semantics. Give the main aspects of their proofs.

1. *(Knaster-Tarski theorem) Any monotone operator* $\mathbf{T}$ *on a complete lattice* $\mathcal{L} = (A, \leq)$ *has a least fixed-point* $lfp(\mathbf{T})$ *and*

$$lfp(\mathbf{T}) = inf\,(\{x \in A | \mathbf{T}(x) \leq x\}).$$

2. *(Kleene's least fixed-point theorem) Any continuous operator* $\mathbf{T}$ *on a complete lattice* $\mathcal{L} = (A, \leq)$ *has a least fixed-point* $lfp(\mathbf{T})$ *and*

$$lfp(\mathbf{T}) = sup\,(\{\mathbf{T}^i | i \geq 0\})$$

where $\mathbf{T}^0 = inf\,(A)$.

Exercise 9.3.3.9. Consider a Datalog database that includes rules of the form

$$A_1 \vee ... \vee A_m \leftarrow B_1, ..., B_n, \neg C_1, ..., \neg C_l$$

known as *disjunctive deductive database* (DDDB). The simplest example of a DDDB would be constituted by $\Delta = \{p \vee q\}$. What is the main difference between a DDB and a DDDB:

1. From the viewpoint of least-Herbrand-model semantics?

2. From the viewpoint of CWA?

Exercise 9.3.3.10. Research into the notion of *supported model* for a Datalog program and put it into relation with both Herbrand semantics and fixed-point semantics.

9.3.4. A proof system for Datalog definite programs: SLD resolution

It should be obvious that each set in $g_\Delta(G)$ for some goal G corresponds to a ground instantiation $G\sigma$ where σ is a substitution $\{X \mapsto c\}$.

Example 9.3.7. In Example 9.3.5, $g_\Delta(G_1)$ is obtained from the ground instantiations

$$\texttt{sick(roberto)} : -\texttt{quarantined(roberto)}.$$

given the substitution $\sigma_1 = \{Y \mapsto \texttt{roberto}\}$, and

$$\texttt{sick(bob)} : -\texttt{quarantined(bob)}.$$

given the substitution $\theta_1 = \{Y \mapsto \texttt{bob}\}$. As for $g_\Delta(G_2)$, we have the ground instantiations

$$\texttt{on\_diet(roberto, seeds)} : -\texttt{bird(canary, roberto)},$$

$$\texttt{sick(roberto), eats(canary, seeds)}.$$

given the substitution $\sigma_2 = \{Y \mapsto \texttt{roberto}, Z \mapsto \texttt{seeds}, X \mapsto \texttt{canary}\}$, and

$$\texttt{on\_diet(bob, all)} : -\texttt{bird(crow, bob)},$$

$$\texttt{sick(bob), eats(crow, all)}.$$

given the substitution $\theta_2 = \{Y \mapsto \texttt{bob}, Z \mapsto \texttt{all}, X \mapsto \texttt{crow}\}$.

The process of ground-instantiating a goal in a Datalog DB is by means of unification. However, we can make this process more precise with respect to a Datalog DB $\Delta = E_\Delta \cup \Pi_\Delta$ in the following way:

Definition. 9.3.18. Let $E_\Delta = \{A_1, ..., A_n\}$, where each A_i is a ground assertion or ground fact. Then, given some rule $(A \leftarrow B_1, ..., B_k) \in \Pi_\Delta$, we say that B_j *matches* some A_i if there is a substitution σ such that $B_j\sigma = A_i$ for $0 < j \le k, 0 < i \le n$.

Definition. 9.3.19. Given a Datalog DB $\Delta = E_\Delta \cup \Pi_\Delta$, we can *infer* a (new) fact A from $E_\Delta = \{A_1, ..., A_n\}$ and a rule $(A \leftarrow B_1, ..., B_k) \in \Pi_\Delta$ if there is a substitution σ such that $B_j\sigma = A_i$ is a matching for $0 < j \leq k, 0 < i \leq n$.

1. We call this inference rule *universal modus ponens* (UMP) and denote this inference by $\Delta \vdash_{UMP} A$.

2. The process of obtaining the inference $\Delta \vdash A$ by a finite number of applications of UMP is called *reduction*. We specify this inference by writing $\Delta \vdash_{red} A$.

Compare Definition 9.3.19 with Definitions 9.1.29-30 and 9.1.35 for Prolog: UMP and reduction are essentially the same rule and process as defined for (pure) Prolog, the single difference being in the fact that we now speak of matching as a special form of unification, namely a form thereof that does not involve function symbols and their terms. Basically, we say that given some substitution σ some IDB fact matches an EDB fact. This conceptual equivalence holds for the notion of a proof tree (cf. Def. 9.1.36), which, given the notion of matching, is more interesting in the context of Datalog, as every leaf thereof is an EDB fact. An example of a Datalog proof tree is given in Figure 9.3.5.

Example 9.3.8. Figure 9.3.5 shows the proof tree for the IDB fact on_diet(bob, all). obtained from the program *Avian_Sick_Prog* as applied over the EDB *Avian_Center_EDB*. The bold-line ellipses are facts in the EDB *Avian_Center_EDB* (cf. Fig. 9.3.2). $\sigma = \{Y \mapsto bob\}$ and $\theta = \{X \mapsto crow, Z \mapsto all\}$ are the substitutions used. Note in this proof tree that the goal–the IDB fact–is the root and each EDB fact is a leaf.

Proposition. 9.3.20. *The pair* $(\Delta, \vdash_{red})$ *constitutes a proof system for a Datalog DB* $\mathcal{P}_\Delta$.

Proof: Let Δ be a set of clauses and let A be a ground fact. Then, either $A \in \Delta$, in which case we have $\Delta \vdash A$ by the very definition of the logical consequence relation (namely by R1; cf. Def. 3.4.7), or A can be inferred by repeated applications of UMP to a rule $r \in \Delta$ and a set of ground facts $\{B_1, ..., B_k\} \subset \Delta$, so that we have $\Delta \vdash_{red} A$. **QED**

The reason why we accepted the LP notion of reduction with no reservations is because, just as in Prolog, this corresponds to a proof by *resolution*. In effect, note that in Proposition 9.3.20 there is no mention to a distinction between the ground assertions in EDB and the rules in IDB:

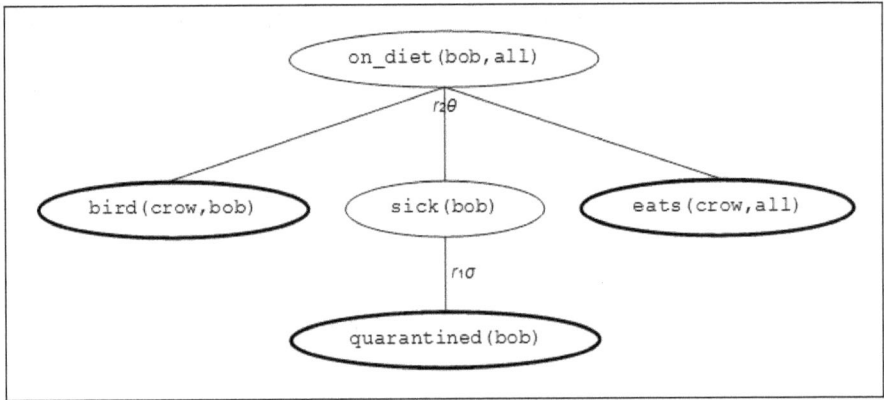

Figure 9.3.5.: A Datalog proof tree.

We now see Δ as a single set of clauses whose elements are both unit clauses (facts) and definite clauses (rules), it being the case that these clauses respect the safety conditions of Definition 9.3.6. In other words, Δ is a set of Datalog *formulae*.

Recall that

$$\text{Datalog} \subseteq \text{L1}_{ff} \subset \text{Prolog}$$

which means that Datalog can be seen as a function-free "dialect" of Prolog. Thus, besides the fact that all the facts in Δ are ground literals, a Datalog database Δ distinguishes itself from a Prolog database Δ in that no predicate has functions as arguments. Hence, just as in Prolog, in Datalog we have a database Δ as a basis for a *definite program* Π.

Example 9.3.9. Figure 9.3.6 shows the Datalog definite program *Avian_Center_Quarantine*.

With respect to the program of Example 9.3.9, it must be remarked that the order of the facts and rules is wholly irrelevant, as Datalog, contrarily to Prolog, satisfies the property of commutativity with respect to both $\wedge$ and $\vee$. This means that the Prolog operator ! for *cut* is not a Datalog operator, the same holding for the Prolog operator `fail`.

The importance of having a Datalog definite program is that, just as in the case of Prolog, SLD-resolution provides a complete proof system for it.[29]

Example 9.3.10. Figure 9.3.7 shows the SLD-resolution proof for

---

[29]Note, however, that for Datalog a breadth-first search is more adequate than the depth-first search characteristic of Prolog. Indeed, neither the order of the rules nor that of the goals affect the evaluation of a Datalog program.

```
bird (penguin, toto).
bird (ostrich, sheila).
bird (emu, tom).
bird (turkey, sam).
bird (turkey, sandra).
bird (hen, lolita).
bird (canary, roberto).
bird (nightingale, sarita).
bird (crow, bob).
bird (woodpecker, lola).
bird (duck, cassandra).
bird (duck, samantha).
abnormal (penguin).
abnormal (ostrich).
abnormal (emu).
abnormal (turkey).
abnormal (hen).
quarantined (roberto).
quarantined (bob).
eats (penguin, fish).
eats (ostrich, all).
eats (emu, all).
eats (turkey, seeds).
eats (hen, all).
eats (canary, seeds).
eats (nightingale, seeds).
eats (crow, all).
eats (woodpecker, bugs).
eats (duck, all).
sick (Y) : −quarantined (Y).
on_diet (Y, Z) : −bird (X, Y), sick (Y), eats (X, Z).
```

Figure 9.3.6.: Datalog definite program *Avian_center_Quarantine*.

the query $? - \text{on\_diet}(\text{bob}, \text{all}).$ given $\sigma = \{Y \mapsto \text{bob}, Z \mapsto \text{all}\},\ \theta = \{X \mapsto \text{crow}\}$, and $\lambda = \{Y_1 \mapsto \text{bob}\}$.

Because conjunction and disjunction are commutative operators in Datalog, we are assured that, if there is a resolution proof of a query, then there is a SLD-resolution proof of it. (We leave the proofs of soundness and completeness as exercises.) This means that a Datalog DDB can be seen as a theory, and we can check for goals by means of proving theorems (for instance, by using Prover9-Mace4).[30] In the proof above, we simply moved the goal quarantined(bob) to the leftmost position in the goal clause, but we could actually have placed the goal sick(Y) at the rightmost position in the body of rule r_2. However, contrarily to Prolog, to which resolution is so to say inherent, for Datalog resolution as a proof calculus is a matter of adopting a *top-down* vs. a *bottom-up* *evaluation technique*, where by "evaluation" we mean the process (the algorithm) and/or its implementation (e.g., a proof tree) of finding an answer to a query given a DDB.

Example 9.3.11. Figure 9.3.5 shows the top-down evaluation of the query $? - \text{on\_diet}(Y, Z).$ for the matching $\theta = \{Y = \text{bob}, Z = \text{all}\}$. We say that (given θ) the query $? - \text{on\_diet}(Y, Z).$ *evaluates to* the reply on_diet(bob, all).. In turn, in this evaluation we say that $r_1\sigma$ was evaluated before $r_2\theta$, so that the term "evaluation" applies also to every step of the algorithm at hand. Actually, given a rule $A \leftarrow B_1, ..., B_n$, we speak also of the evaluation of each of $A, B_1, ..., B_n$. Thus, in the case at hand, the sub-goal $? - \text{sick}(Y).$ in the body of r_2 was evaluated before the other sub-goals in this rule.

Exercises

Exercise 9.3.4.1. For each of the following Datalog programs one or more goals are given. Determine the replies to the goals by applying SLD resolution and construct the corresponding Datalog proof trees whenever appropriate.

1. $\Pi_\Delta = \left\{ \begin{array}{l} p(a,s), p(b,s), p(c,t), p(d,f) \\ q(e,s), q(g,t), q(h,i) \\ r(X,Y) \leftarrow p(X,Z), q(X,Z) \end{array} \right\}$; $\begin{array}{l} G_1 = r(X,e)? \\ G_2 = r(c,Y)? \\ G_3 = r(X,i)? \end{array}$.

---

[30] This is only possible for small EDBs. Note that a Datalog DDB may be implemented as a Prolog program–again, solely for small EDBs–only if one is aware of the non-commutativity of conjunction and disjunction in the latter.

← on_diet (bob, all) .

← bird (X, bob) , sick (bob) , eats (X, all) . $\diagdown \sigma$ on_diet (Y, Z) ← bird (X, Y) , sick (Y) , eats (X, Z) .

← sick (bob) , eats (crow, all) . $\diagdown \theta$ bird (crow, bob) ← .

← quarantined (bob) , eats (crow, all) . $\diagdown \lambda$ sick (Y₁) ← quarantined (Y₁) .

← eats (crow, all) . $\diagdown$ quarantined (bob) ← .

← := □ $\diagdown$ eats (crow, all) ← .

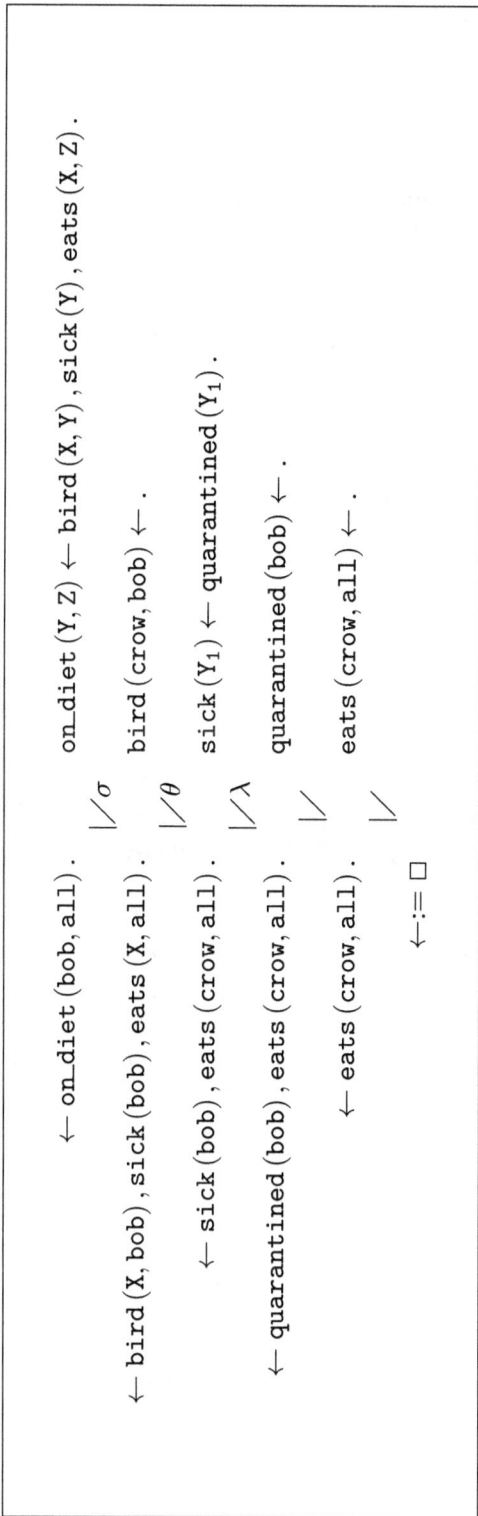

Figure 9.3.7.: A SLD-resolution proof of a Datalog query.

$$2.\ \Pi_\Delta = \left\{ \begin{array}{c} p\,(a,b)\,,p\,(a,f)\,,p\,(b,c) \\ p\,(c,d)\,,p\,(c,e) \\ r\,(X,Y) \leftarrow p\,(X,Y) \\ r\,(X,Y) \leftarrow r\,(X,Z)\,,p\,(Z,Y) \end{array} \right\} ;\ \begin{array}{l} G_1 = r\,(a,c)? \\ G_2 = (X,d)? \\ G_3 = (c,Y)? \end{array} .$$

Exercise 9.3.4.2. Consider the Datalog program of Exercise 9.3.2.2.4 above together with the EDB constituted by the single ternary relation *LINKS*:

$$LINKS = \left\{ \begin{array}{c} (4, stGermain, odeon)\,, (4, odeon, stMichel)\,, \\ (4, stMichel, chatelet)\,, (1, chatelet, louvre)\,, \\ (1, louvre, palaisRoyal)\,, (1, palaisRoyal, tuileries)\,, \\ (1, tuileries, concorde) \end{array} \right\}$$

Compute by applying SLD resolution the replies to the following queries:

1. $G = reach\,(stGermain, odeon)$?

2. $G = reach\,(tuileries, stGermain)$?

3. $G = stations\,(X)$?

Exercise 9.3.4.3. Consider an EDB constituted by the assertions for the relation symbols and respective attributes $PERSON\,(NAME)$ and $PARENT\,(CHILD, PARENT)$ as follows:

$$PERSON = \left\{ \begin{array}{c} (ann)\,, (bertrand)\,, (charles)\,, \\ (dorothy)\,, (evelyn)\,, (fred)\,, \\ (george)\,, (hilary) \end{array} \right\}$$

$$PARENT = \left\{ \begin{array}{c} (dorothy, george)\,, (evelyn, george)\,, \\ (bertrand, dorothy)\,, (ann, dorothy)\,, \\ (hilary, ann)\,, (charles, evelyn) \end{array} \right\}$$

Given the program

$$\Pi_\Delta = \left\{ \begin{array}{c} sgc\,(X,X) \leftarrow person\,(X) \\ sgc\,(X,Y) \leftarrow parent\,(X, X_1)\,, sgc\,(X_1, Y_1)\,, parent\,(Y, Y_1) \end{array} \right\}$$

where "sgc" abbreviates "same-generation cousins" reply to the following queries by applying SLD resolution and construct the Datalog proof trees whenever appropriate:

1. $G = sgc\,(charles, ann)$?

2. $G = sgc\,(ann, charles)$?

3. $G = sgc\,(bertrand, ann)$?

4. $G = sgc\,(evelyn, dorothy)$?

Exercise 9.3.4.4. With respect to a Datalog DDB, SLD resolution constitutes a proof system for it characterized by (i) soundness and (ii) completeness. Formulate the respective theorems and prove them.

Exercise 9.3.4.5. Show that, given a Datalog program Π_Δ and a goal G, the Datalog-decision problem

$$\Pi_\Delta \cup E_\Delta \overset{?}{\vdash} G$$

is **P**-complete.

9.3.5. Datalog with negation: Datalog$^\neg$

Just as in the case of Prolog, we may simply accept CWA as a meta-rule (Def. 9.1.41), and take NF to be an implicit rule of a Datalog DDB. As in Prolog, this entails non-monotonicity with respect to a Datalog DDB (cf. Exercise 9.1.4.11). However, there are circumstances that call for explicit negation in a Datalog DDB. For instance, given a program for some electronic device (e.g., a printer), a rule such as

$$ready\,(X) \leftarrow device\,(X)\,, \neg busy\,(X)$$

may be required for the correct functioning of the device. Yet another illustration: In the program for a game, the following rule expresses the fact that one wins when one forces the opponent to a situation in which they have no chance to move:

$$win\,(X) \leftarrow move\,(X, Y)\,, \neg win\,(Y)$$

From the viewpoint of programming, we say that negation increases the *expressiveness* of the language. This is especially relevant when computing over finite domains, which is typically the case of a Datalog DDB. Datalog$^\neg$ is a more expressive extension of Datalog; however, as we shall see, this increased expressiveness comes at the cost of syntactic restrictions.

Definition. 9.3.21. A Datalog rule **r** of the form

$$A \leftarrow B_1, ..., B_n$$

where each $B_i \in Body_\mathbf{r}$ is a positive or a negative literal, i.e. $B_i = p(c_1, ..., c_k)$ or $B_i = \neg p(c_1, ..., c_k)$, respectively, constitutes an extension of Datalog called *Datalog with negation*. We abbreviate it as, or denote it by, Datalog$^\neg$.

Just as in the case of Datalog, safety conditions apply:

Definition. 9.3.22. Let $\mathbf{r}$ be a rule in a Datalog$^\neg$ program, denoted by $\Pi_\Delta^\neg$. Then, $\mathbf{r}$ is said to be a *safe rule* if

1. negation does not occur in the head of $\mathbf{r}$, and

2. every variable occurring in a negative literal must also occur in a positive literal.

As seen above, one of the advantages of Datalog DDBs over relational DBs is the fact that the former allow for recursion; in the presence of negation, however, we might have *recursion through negation*, i.e. predicates being defined recursively in terms of their own negation. An example of this is the rule $p \leftarrow \neg p$, which can be read as "p is provable if *not-p* is provable (or p is not provable)," which is clearly a contradiction.[31]

It so happens that neither Herbrand semantics nor fixed-point semantics are adequate for Datalog$^\neg$. On the other hand, several other semantics can take over in Datalog$^\neg$, with more or less success. (We leave all these aspects as exercises.) We chose to discuss here the so-called *stratified semantics*, firstly elaborated on in van Gelder (1986) for general logic programs and in Apt, Blair, & Walker (1988). In effect, even if not all meaningful Datalog$^\neg$ programs are stratified, the handling of negation in this semantics is highly adequate in our view, namely from the viewpoint of deduction. Moreover, it is an extension of fixed-point semantics, already studied above. It is useful to consider stratified semantics as a natural extension of semi-positive Datalog$^\neg$:

Definition. 9.3.23. We say that a Datalog$^\neg$ program $\Pi_\Delta^\neg$ is *semi-positive* if, whenever $\neg p(\vec{x}) \in Body_\mathbf{r}$ for a rule $\mathbf{r} \in \Pi_\Delta^\neg$, then $p \in EDB(\Pi_\Delta^\neg)$.

This means that, given some relation $R(t_1, ..., t_n)$ in a Datalog DDB, $\neg R(t_1, ..., t_n)$ is *true* iff $t_i \in \mathscr{D}_i$ for all $i = 1, ..., n$ and $\{t_i\}_{i>0}^n \notin R$, so that we have $(\Pi_\Delta^\neg \cup EDB) \nvDash R(t_1, ..., t_n)$. But if $\{t_i\}_{i>0}^n \notin R$, then by transitive closure there is some relation R' such that $\{t_i\}_{i>0}^n \in R'$. This, in turn, means that for any negated relation $\neg R(t_1, ..., t_n)$ we

---

[31]Note how this collides with the classical equivalence $p \leftarrow \neg p \equiv \neg \neg p \vee p \equiv p$.

can compute its *complement* $\overline{R}(t_1, ..., t_n)$ such that $(\Pi_\Delta^- \cup EDB) \models$ $\overline{R}(t_1, ..., t_n)$ and thus obtain $\overline{EDB}$. In other words, given some predicate $p' \in EDB(\Pi_\Delta^-)$, we can replace all occurrences of $\neg p'$ by the *new* predicate $\overline{p'} \in EDB(\Pi_\Delta^-)$, thus eliminating negation and obtaining a program that is basically a (positive) Datalog program Π_Δ.

Example 9.3.12. Given $EDB = \{R(a,b), R(b,c), R(a,a), R(a,c)\}$, we consider the semi-positive Datalog$^-$ IDB

$$IDB = \left\{ \begin{array}{c} r'(x,y) \leftarrow v(x), \neg r'(x,y) \\ t(x,y) \leftarrow r'(x,y) \\ t(x,y) \leftarrow t(x,z), r'(z,y) \end{array} \right\}.$$

We replace $\neg r'$ by $\overline{r'}$, obtaining the "positive" IDB

$$IDB = \left\{ \begin{array}{c} r'(x,y) \leftarrow v(x), \overline{r'}(x,y) \\ t(x,y) \leftarrow r'(x,y) \\ t(x,y) \leftarrow t(x,z), r'(z,y) \end{array} \right\}.$$

The obtained program can be run as a (positive) Datalog program and we can compute

$$\overline{EDB} = \left\{ R'(b,a), R'(b,b), R'(c,c), R'(c,b), R'(c,a) \right\}.$$

Let now the notation $J|EDB(\Pi_\Delta^-)$ denote the *restriction* of instance J to $EDB(\Pi_\Delta^-)$. Then we have the following result:

Theorem 9.3.2. *Let Π_Δ^- be a semi-positive Datalog$^-$ program. Then, for every instance I over $EDB(\Pi_\Delta^-)$, there exists a least H-model / a least fixed-point satisfying $J|EDB(\Pi_\Delta^-)$ such that*

$$\underline{H\mathcal{M}}_{\Pi_\Delta^-} = lfp(\Pi_\Delta^-) = \lim_{i \to \infty} \left\{ \mathbf{T}_{\Pi_\Delta^-}^i(I) \right\}_{i>0}.$$

Proof: Left as an exercise.

Let there be given a Datalog$^-$ program Π_Δ^- such that there is some rule $\mathbf{r} \in \Pi_\Delta^-$ and $\neg R(\vec{t}) \in Body_\mathbf{r}$, where $\vec{t}$ abbreviates the sequence $\{t_i\}_{i=1}^n$ for some finite n. Intuitively, we are interested in knowing the value of $R(\vec{t})$ in order to evaluate $Body_\mathbf{r}$: If $R(\vec{t})$ is *false*, then $\neg R(\vec{t})$ is *true*; otherwise, if $R(\vec{t})$ is *true*, then $\neg R(\vec{t})$ is *false* and $\mathbf{r}$ is not applicable. Thus, the first evaluation falls on $R(\vec{t})$. If rule $\mathbf{r}$ is applicable, we can then move to the stratification of Π_Δ^-, for whose definition a few previous notions are required.

Definition. 9.3.24. Let $\vec{\mathfrak{G}} = \left(V, E, \vec{f}\right)$ be a directed graph such that, given a Datalog$^\neg$ program $\Pi_\Delta^\neg$, we have

$$V = Sch\left(\Pi_\Delta^\neg\right)$$

and for some rules $\mathbf{r}_i, \mathbf{r}_j \in \left(\Pi_\Delta^\neg \cup E_\Delta\right)$,

$$E = \left\{(R, S) \mid Head_{\mathbf{r}_i} = R \text{ and } S \in Body_{\mathbf{r}_j}\right\}$$

where for any pair of relations (R, S) there is at most one arc $R \longrightarrow S$. We call this the *dependency graph* of $\Pi_\Delta^\neg$ and denote it by $\vec{\mathfrak{G}}_{\Pi_\Delta^\neg}$ (abbreviated: $\vec{\mathfrak{G}}_\Pi$).

1. An arc (R, S) such that $Head_{\mathbf{r}_i} = R$ and $\neg S \in Body_{\mathbf{r}_j}$ is said to be a *negative* arc; otherwise it is called a *positive* arc.

Given a dependency graph $\vec{\mathfrak{G}}_\Pi$, for the sake of graphical convenience for each negative arc we draw an arc of the form $R \overset{\neg}{\longrightarrow} S$ in $\vec{\mathfrak{G}}_\Pi$.

Example 9.3.13. Let there be given the following Datalog$^\neg$ program $\Pi_{\Delta_1}^\neg$ (abbr.: $\Pi_1^\neg$):

$$\Pi_1^\neg = \left\{ \begin{array}{l} p\left(X, Y\right) \leftarrow q\left(X, Y\right) \\ p\left(X, Y\right) \leftarrow q\left(X, Z\right), p\left(Z, Y\right) \\ r\left(X, Y\right) \leftarrow s\left(X, Y\right), \neg p\left(X, Y\right) \\ t\left(X, Y\right) \leftarrow r\left(X, Y\right) \\ t\left(X, Y\right) \leftarrow r\left(X, Z\right), t\left(Z, Y\right) \end{array} \right\}$$

Figure 9.3.8 shows the dependency graph of $\Pi_1^\neg$.

Proposition. 9.3.25. *Given some dependency graph $\vec{\mathfrak{G}}_\Pi$, if*

$$((R = R_1) \longrightarrow R_2 \longrightarrow \dots \longrightarrow R_{k-1} \longrightarrow (R_k = S)) \in \vec{\mathfrak{G}}_\Pi$$

such that some $R_i \longrightarrow R_{i+1}$ for $1 \leq i \leq k - 1$ is a negative arc, then S must be evaluated prior to R.

Proof: Left as an exercise.

Proposition 9.3.25 is called the *stratification principle*. In order to elaborate on this principle, a few notions are required.

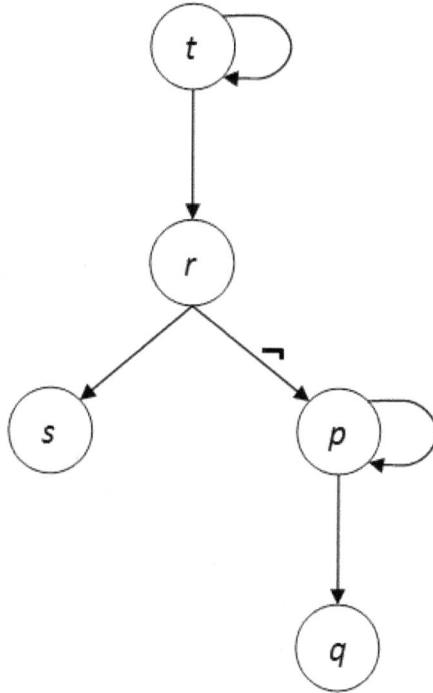

Figure 9.3.8.: Dependency graph $\vec{\mathfrak{G}}_{\Pi_1^{\neg}}$ of the Datalog$^{\neg}$ program $\Pi_1^{\neg}$.

Definition. 9.3.26. Consider the set $Sch\left(\Pi_\Delta^{\neg}\right)$ of predicate symbols of a Datalog$^{\neg}$ program $\Pi_\Delta^{\neg}$ and the function

$$\ell : Sch\left(\Pi_\Delta^{\neg}\right) \longrightarrow \mathbb{N}$$

such that $\ell\left(p\right) = 0$ if p is a leaf in $\vec{\mathfrak{G}}_{\Pi}$ and $\ell\left(p\right) = i$ for $1 \leq i \leq n$ if there is a path ε with $1, 2, ..., n$ arcs in $\vec{\mathfrak{G}}_{\Pi}$ originating in p and ending in some leaf q. We call $\ell\left(p\right)$ the *length* of predicate p and the corresponding set

$$\mathscr{R}\left(Sch\left(\Pi_\Delta^{\neg}\right)\right)_i = \{p | p \in Sch\left(\Pi_\Delta^{\neg}\right) \text{ and } \ell\left(p\right) = i \in [0, n]\}$$

is called the i-th *rank* of $Sch\left(\Pi_\Delta^{\neg}\right)$.

We abbreviate $\mathscr{R}\left(Sch\left(\Pi_\Delta^{\neg}\right)\right)_i$ simply as $\mathscr{R}_i$. It should be evident that rank $\mathscr{R}_0$ contains the predicates in $E_\Delta \subseteq EDB\left(\Pi_\Delta^{\neg}\right)$.

Example 9.3.14. The ranking of the predicates of $Sch\left(\Pi_1^{\neg}\right)$ (cf.

Example 9.3.13) is as follows:

$$\begin{aligned}
\mathscr{R}_0 &= \{q, s\}\\
\mathscr{R}_1 &= \{p\}\\
\mathscr{R}_2 &= \{r\}\\
\mathscr{R}_3 &= \{t\}
\end{aligned}$$

Definition. 9.3.27. A Datalog¬ program $\Pi_\Delta^\neg$ is said to be *stratified* if there is a partition

$$\mathscr{P}_{Sch(\Pi_\Delta^\neg)} = \bigcup_{i\geq 0}^n \Sigma_i$$

of Σ_i *strata* (singular: *stratum*) such that for two relations $R \in \mathscr{R}_i, S \in \mathscr{R}_j$ such that there is an arc $e = (R \longrightarrow S) \in \vec{\mathfrak{G}}_\Pi$ (i) if e is a positive arc, then $i \geq j$ and for every rule $\mathbf{r}$ such that $Head_\mathbf{r} \supseteq S$ we have $\mathbf{r} \in \bigcup_{i\geq j} \Sigma_i$; (ii) if e is a negative arc, then $i > j$ and for every rule $\mathbf{r}$ such that $Head_\mathbf{r} \supseteq S$ we have $\mathbf{r} \in \bigcup_{i>j} \Sigma_i$.

From this Definition, it is evident that in fact we have

$$\mathscr{P}_{Sch(\Pi_\Delta^\neg)} = \bigcup_{i\geq 1}^n \Sigma_i = \bigcup \{(\mathbf{r}, p) \,|\, Head_\mathbf{r} \supseteq p \text{ and } p \in \mathscr{R}_i\}$$

such that each stratum Σ_i corresponds to a rank $\mathscr{R}_i$.

Example 9.3.15. The computation of $\mathscr{P}_{Sch(\Pi_1^\neg)}$ yields

$$\Sigma_0 = \emptyset$$

$$\Sigma_1 = \{p\,(X,Y) \leftarrow q\,(X,Y)\} \cup \{p\,(X,Y) \leftarrow q\,(X,Z), p\,(Z,Y)\}$$
$$\Sigma_2 = \{r\,(X,Y) \leftarrow s\,(X,Y), \neg p\,(X,Y)\}$$

and

$$\Sigma_3 = \{t\,(X,Y) \leftarrow r\,(X,Y)\} \cup \{t\,(X,Y) \leftarrow r\,(X,Z), t\,(Z,Y)\}.$$

Intuitively, we have a finite sequence of subprograms $\Sigma_1, \Sigma_2, ..., \Sigma_n$ where each Σ_i for $1 \leq i \leq n$ defines at least one EDB relation. The evaluation order for the relations falls on the ranks in the following way:

Remark. **9.3.28.** *Evaluation order:* Let there be given the relations in $\mathscr{R}_0$ and $I = (E_\Delta)_g \subseteq EDB\,(\Pi_\Delta^\neg)_g$.

1. Evaluate the relations in $\mathscr{R}_1$: All relations $R \in Head_{\mathbf{r}}$ for $\mathbf{r} \in \Sigma_1$ are defined.[32] This evaluation yields

$$J_1 \subseteq Sch\left(\Sigma_1\right)_g.$$

2. Evaluate the relations in $\mathscr{R}_2$ considering the relations in $EDB\left(\Pi_{\Delta}^{\neg}\right)$ and $\mathscr{R}_1$ as $EDB\left(\Sigma_2\right)$, where $\neg R\left(\vec{t}\right)$ is true if $R\left(\vec{t}\right)$ is false in $I \cup J_1$: All relations $R \in Head_{\mathbf{r}}$ for $\mathbf{r} \in \Sigma_2$ are defined. This evaluation yields

$$J_2 \subseteq Sch\left(\Sigma_2\right)_g.$$

3. Repeat for $\mathscr{R}_i$, $i = 3, ..., n$ considering the relations in $EDB\left(\Pi_{\Delta}^{\neg}\right)$ and $\mathscr{R}_1, \mathscr{R}_2, ..., \mathscr{R}_{i-1}$ as $EDB\left(\Sigma_i\right)$, where $\neg R\left(\vec{t}\right)$ is true if $R\left(\vec{t}\right)$ is false in $I \cup J_1 \cup J_2 \cup ... \cup J_{i-1}$: All relations $R \in Head_{\mathbf{r}}$ for $\mathbf{r} \in \Sigma_i$ are defined. This evaluation yields

$$J_3 \subseteq Sch\left(\Sigma_3\right)_g$$
$$\vdots$$
$$J_n \subseteq Sch\left(\Sigma_n\right)_g$$

4.
$$\Pi_{\mathscr{P}}\left(I\right) = I \cup J_1 \cup J_2 \cup ... \cup J_n$$

where $\Pi_{\mathscr{P}}\left(I\right)$ denotes the evaluation of $\Pi_{\Delta}^{\neg}$ on I with respect to $\mathscr{P}$.

Example 9.3.16. With respect to $\Pi_1^{\neg}$, let us consider $EDB\left(\Pi_1^{\neg}\right) = \{q, s\}$ such that $I = \{q\left(a, b\right), s\left(a, b\right)\}$. Then, the evaluations of $\mathscr{R}_i$ for $i = 1, 2, 3$ give:

$$J_1 = \{p\left(a, b\right)\}$$
$$J_2 = \{\neg r\left(a, b\right)\}$$
$$J_3 = \{\neg t\left(a, b\right)\}$$
$$\Pi_{\mathscr{P}}\left(I\right) = \{q\left(a, b\right), s\left(a, b\right), p\left(a, b\right), \neg r\left(a, b\right), \neg t\left(a, b\right)\}$$

Note that $\neg r\left(a, b\right)$ and $\neg t\left(a, b\right)$ are here to be interpreted as "$r\left(a, b\right)$ is false in J_2" and "$t\left(a, b\right)$ is false in J_3", as facts, whether EDB or IDB ones, are never "negative." To play it one the safe side, one may in fact

---

[32]Note that Σ_1 does not have any negated relations for the reason that EDB facts cannot be negated relations.

consider that $J_2 = \emptyset$ and $J_3 = \emptyset$, so that

$$\Pi_{\mathscr{P}}(I) = \{q(a,b), s(a,b), p(a,b)\}.$$

From Definition 9.3.27, it is also obvious that no negative relation occurs in the body of a rule in Σ_0, which can actually be empty.

Theorem 9.3.3. *A Datalog$^\neg$ program $\Pi_\Delta^\neg$ is stratified iff in its dependency graph there are no cycles containing a negative arc.*

Proof: (Sketch) ($\Rightarrow$) Suppose that we have some program $\Pi_\Delta^\neg$ that is stratifiable. Let $(R_1 \longrightarrow \ldots \longrightarrow R_k \longrightarrow R_1) \in \vec{\mathfrak{G}}_{\Pi_\Delta^\neg}$ where some $R_i \longrightarrow R_j$ for $1 \leq i \leq k, 1 \leq j \leq k$ is a negative arc. Let this arc be $R_k \to R_1$. Then, $R_1 > R_1$ by Definition 9.3.27.ii, but this is a contradiction. ($\Leftarrow$) Left as an exercise.

Example 9.3.17. Let us add the rule $p(X, Y) \leftarrow t(X, Y)$ to program $\Pi_1^\neg$ of Example 9.3.13. The new program $\Pi_{1'}^\neg$ is not stratifiable. In effect, the dependency graph of this new program has a cycle containing a negative arc (see Fig. 9.3.9).

We conclude the study of Datalog$^\neg$ by giving two important statements whose proofs we leave as (research) exercises.

Proposition. 9.3.29. *For any given program $\Pi_\Delta^\neg$ that is stratifiable $\Pi_{\mathscr{P}}(I)$ is well defined.*

Proof: Left as an exercise.

Theorem 9.3.4. *For any given program $\Pi_\Delta^\neg$ that is stratifiable $\Pi_{\mathscr{P}}(I)$ is a minimal model K of $\Pi_\Delta^\neg$ such that $K|EDB(\Pi_\Delta^\neg) = I$.*

Proof: Left as an exercise.

Exercises

Exercise 9.3.5.1. Elaborate on the meaning of the following Datalog$^\neg$ programs from the viewpoint of expressiveness and give it formally:

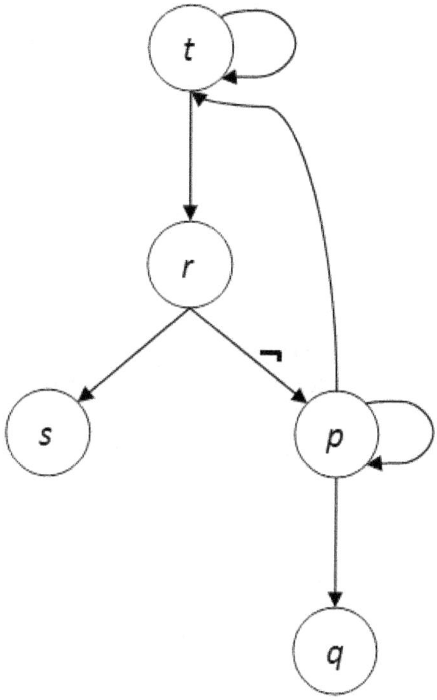

Figure 9.3.9.: Dependency graph of a non-stratifiable program.

1.

$$Monopoly =$$

$$\left\{ \begin{array}{c} redline\,(a,b) \\ redline\,(b,c) \\ greenline\,(a,b) \\ greenpath\,(X,Y) \leftarrow greenline\,(X,Y) \\ greenpath\,(X,Y) \leftarrow greenpath\,(X,Z)\,, greenpath\,(Z,Y) \\ monopoly\,(X,Y) \leftarrow \neg greenpath\,(X,Y)\,, redline\,(X,Y) \end{array} \right\}$$

2.

$$S = \left\{ \begin{array}{c} q \\ p \leftarrow q \\ q \leftarrow p \\ r \leftarrow p,q \\ s \leftarrow \neg p, \neg q \end{array} \right\}$$

Exercise 9.3.5.2. Consider the following Datalog¬ program:

$$\Pi_\Delta^\neg = \left\{ \begin{array}{l} p\,(a) \\ q\,(a) \\ p\,(b) \\ p\,(c) \\ s\,(X) \leftarrow p\,(X)\,,q\,(X) \\ r\,(X) \leftarrow \neg s\,(X)\,,p\,(X) \end{array} \right\}$$

1. Give the (formal) meaning of the program.

2. Replace the predicate symbols in the program by the following ones so that the intended meaning of the program is preserved:

$$Sch\,(\Pi_\Delta^\neg) = \{married, adult, single, is\_husband\}$$

Exercise 9.3.5.3. Consider the following Datalog¬ program where $X \neq Y$ is an inbuilt predicate:

$$\Pi_\Delta^\neg = \left\{ \begin{array}{l} r\,(X) \\ k\,(A, B) \\ l\,(A, B) \leftarrow k\,(A, B) \\ l\,(A, B) \leftarrow k\,(B, A) \\ m\,(A, B) \leftarrow l\,(A, B) \\ m\,(A, B) \leftarrow m\,(A, C)\,,l\,(C, B) \\ n\,(X, A, B) \leftarrow m\,(A, B)\,,r\,(X)\,,\neg o\,(X, A, B) \\ o\,(X, A, B) \leftarrow l\,(A, B)\,,X \neq A, r\,(X)\,,X \neq B \\ o\,(X, A, B) \leftarrow o\,(X, A, C)\,,o\,(X, C, B) \\ p\,(A, B) \leftarrow n\,(X, A, B)\,,X \neq A, X \neq B \\ q\,(A, B) \leftarrow m\,(A, B)\,,\neg p\,(A, B) \\ r\,(X) \leftarrow l\,(X, Y) \end{array} \right\}$$

1. Give the (formal) meaning of the program.

2. Replace the predicate symbols in the program by the following ones so that the intended meaning of the program is preserved:

$$Sch\,(\Pi_\Delta^\neg) = \left\{ \begin{array}{l} is\_connected, link, circumvents \\ station, is\_linked, has\_cutpoint \\ has\_0\_cutpoints, has\_ \geq 1\_cutpoints \end{array} \right\}$$

Exercise 9.3.5.4. Comment on why fixed-point semantics is problematic for Datalog¬ by applying the immediate consequence operator to the following programs (we abbreviate $\Pi_{\Delta_i}^\neg$ as $\Pi_i^\neg$):

1. $\Pi_1^{\neg} = \{p \leftarrow \neg p\}$

2. $\Pi_2^{\neg} = \{p \leftarrow \neg q, q \leftarrow \neg p\}$

3. $\Pi_3^{\neg} = \{r, p \leftarrow \neg r, r \leftarrow \neg p, p \leftarrow \neg p\}$

4. $\Pi_4^{\neg} = \{p \leftarrow p, q \leftarrow q, p \leftarrow \neg q, q \leftarrow \neg p\}$

5. $\Pi_5^{\neg} = \{r\,(a), p\,(X) \leftarrow \neg r\,(X) \wedge \neg q\,(X)\}$

6. $\Pi_6^{\neg} = \{r\,(b), p\,(X) \leftarrow \neg r\,(X) \wedge \neg q\,(X), q\,(X) \leftarrow \neg r\,(X) \wedge \neg p\,(X)\}$

7. $\Pi_7^{\neg} = \{q\,(a), p\,(X) \leftarrow q\,(X) \wedge \neg r\,(X), r\,(X) \leftarrow q\,(X) \wedge \neg p\,(X)\}$

8. $\Pi_8^{\neg} = \Pi_7^{\neg} \cup \{r\,(X) \leftarrow \neg r\,(X) \wedge p\,(X)\}$

Exercise 9.3.5.5. Consider the following Datalog program given as a set of formulae:

$$\Pi_\Delta = \left\{ \begin{array}{c} q\,(a) \\ q\,(b) \\ p\,(X) \leftarrow q\,(X), \neg p\,(X) \end{array} \right\}$$

1. Determine the problem(s) it raises from the viewpoint of the least H-model.

2. Comment on the reason(s) why Herbrand semantics, namely least H-models, is not adequate for Datalog$^{\neg}$.

Exercise 9.3.5.6. Let $Body_{\mathbf{r}} = Body_{\mathbf{r}}^{+} \cup Body_{\mathbf{r}}^{-}$, where $Body_{\mathbf{r}}^{+} \subseteq Body_{\mathbf{r}}$ contains solely positive literals and $Body_{\mathbf{r}}^{-} \subseteq Body_{\mathbf{r}}$ has only negative literals for members. Then, the following condition also applies in fixed-point semantics: Given a Datalog$^{\neg}$ program and some rule $\mathbf{r} \in \Pi_\Delta^{\neg}$, it is required that

$$Body_{\mathbf{r}}^{-} \cap (E_\Delta)_g = \emptyset.$$

Account for this condition taking the solutions to the previous exercise into account.

Exercise 9.3.5.7. Determine whether the following Datalog$^{\neg}$ programs are stratifiable and, if so, give their stratification:

1. $\Pi_1^{\neg} = \left\{ \begin{array}{c} p \leftarrow \neg r \\ q \leftarrow \neg r, p \\ s \leftarrow \neg t \\ t \leftarrow \neg q, s \\ u \leftarrow \neg t, p, s \end{array} \right\}$

9. *Programming*

2. $\Pi_2^{\neg} = \left\{ \begin{array}{l} p(X) \leftarrow \neg q(X) \\ q(X) \leftarrow \neg p(X) \\ r(a) \leftarrow q(a) \\ r(a) \leftarrow p(a) \end{array} \right\}$

3. $\Pi_3^{\neg} = \left\{ \begin{array}{c} p(X,Y) \leftarrow \neg q(X,Y), s(X,Y) \\ q(X,Y) \leftarrow q(X,Z), q(Z,Y) \\ q(X,Y) \leftarrow t(X,Y), \neg r(X,Y) \\ r(X,Y) \leftarrow t(Y,X) \\ s(X,Y) \leftarrow q(X,Z), q(Y,W), \neq (X,Y) \end{array} \right\}$

4. $\Pi_4^{\neg} = \left\{ \begin{array}{c} t(a) \\ s(X) \leftarrow p(X), q(X), \neg r(X) \\ p(X) \leftarrow t(X), \neg q(X) \\ q(X) \leftarrow t(X), \neg p(X) \\ r(X) \leftarrow t(X), t(b) \end{array} \right\}$

5. $\Pi_5^{\neg} = \left\{ \begin{array}{l} p(a) \\ r(X) \leftarrow \neg q(a) \\ p(X) \leftarrow \neg r(Y) \end{array} \right\}$

6. $\Pi_6^{\neg} = \left\{ \begin{array}{c} z(a) \\ r(X) \leftarrow p(X), \neg q(X) \\ q(X) \leftarrow \neg u(X), z(X) \\ p(X) \leftarrow \neg s(X), z(X) \\ u(X) \leftarrow t(X) \\ t(X) \leftarrow q(X) \end{array} \right\}$

7. $\Pi_7^{\neg} = \left\{ \begin{array}{l} husband(X) \leftarrow man(X), \neg bachelor(X) \\ bachelor(X) \leftarrow man(X), \neg husband(X) \end{array} \right\}$

8. $\Pi_8^{\neg} = \left\{ \begin{array}{c} v(X,Y) \leftarrow r(X,X), r(Y,Y) \\ u(X,Y) \leftarrow s(X,Y), s(Y,Z), \neg v(X,Y) \\ w(X,Y) \leftarrow \neg u(X,Y), v(Y,X) \end{array} \right\}$

Exercise 9.3.5.8. Determine whether the Datalog$^{\neg}$ programs of Exercises 9.3.5.1-3 above are stratifiable, and stratify them if so.

Exercise 9.3.5.9. Prove:

1. Theorem 9.3.2.

2. Proposition 9.3.25.

3. Proposition 9.3.29.

522

4. Theorem 9.3.4.

Exercise 9.3.5.10. Complete the proof of Theorem 9.3.3.

Exercise 9.3.5.11. Design an algorithm based on Theorem 9.3.3 such that the complexity class of testing for stratifiability of a Datalog¬ program is **DTIME**(n^k) for n the size of the program.

Exercise 9.3.5.12. The stratification $\mathscr{P}_{Sch(\Pi_\Delta^\neg)} = \bigcup_{i \geq 0}^n \Sigma_i$ for a stratifiable Datalog¬ program is not unique if we ignore the ranks $\mathscr{R}_i$.

1. Rephrase Definition 9.3.27 omitting the ranks $\mathscr{R}_i$.

2. Show that for a stratifiable Datalog¬ program all stratifications are equivalent.

3. Draw the consequences of the above statement from the viewpoint of a semantics for Datalog¬.

Exercise 9.3.5.13. Research into the following semantics for Datalog¬ (Hint: Consult the cited literature.):

1. Stable-model semantics (Gelfond & Lifschitz, 1988).

2. Perfect-model semantics (Przymusinski, 1989).

3. 3-valued semantics (van Gelder, Ross, & Schlipf, 1991).

4. Well-founded semantics (van Gelder, Ross, & Schlipf, 1991).

5. Inflationary vs. non-inflationary semantics (Abiteboul & Vianu, 1991).

Bibliography

References

Abiteboul, S., Hull, R., & Vianu, V. (1995). *Foundations of databases.* Reading, MA, etc.: Addison-Wesley.

Abiteboul, S. & Vianu, V. (1991). Datalog extensions for database queries and updates. *Journal of Computer and System Sciences,* *43*(1), 62-124.

Apt, K. R. (1996). *From logic programming to Prolog.* Upper Saddle River, NJ: Prentice Hall.

Apt, K. R., Blair, H. A., & Walker, A. (1988). Towards a theory of declarative knowledge. In J. Minker (ed.), *Foundations of deductive databases and logic programming* (pp. 89-148). Los Altos, CA: Morgan Kaufmann.

Aristotle (ca. 350 BC). *Metaphysics.* Trans. by W. D. Ross (1908). Available at http://classics.mit.edu//Aristotle/metaphysics.html.

Augusto, L. M. (2019). *Formal logic: Classical problems and proofs.* London: College Publications.

Augusto, L. M. (2020a). *Many-valued logics: A mathematical and computational introduction.* 2nd ed. London: College Publications.

Augusto, L. M. (2020b). *Logical consequences. Theory and applications: An introduction.* 2nd ed. London: College Publications.

Augusto, L. M. (2021). *Languages, machines, and classical computation.* 3rd ed. London: College Publications.

Baaz, M., Egly, U., & Leitsch, A. (2001). Normal form transformations. In A. Robinson & A. Voronkov (eds.), *Handbook of automated reasoning,* vol. 1 (pp. 273-333). Amsterdam: Elsevier / Cambridge, MA: MIT Press.

Bachmair, L. & Ganziger, H. (2001). Resolution theorem proving. In A. Robinson & A. Voronkov (eds.), *Handbook of automated reasoning,* vol. 1 (pp. 19-99). Amsterdam: Elsevier / Cambridge, MA: MIT Press.

Beckert, B., Hähnle, R., & Schmitt, P. H. (1993). The *even more* liberalized δ-rule in free variable semantic tableaux. In G. Gottlob, A. Leitsch, & D. Mundici (eds.), *Proceedings of the third Kurt Gödel Colloquium KGC'93, Brno* (pp. 108-119). Springer.

Beth, E. W. (1955). Semantic entailment and formal derivability. *Mededelingen der Koninklijke Nederlandse Akademie van Wetenschappen, 18*, 309-342.

Beth, E. W. (1960). Completeness results for formal systems. In J. A. Todd (ed.), *Proceedings of the International Congress of Mathematicians, 14-21 August 1958* (pp. 281-288). Cambridge: CUP.

Biere, A., Heule, M., van Maaren, H., & Walsh, T. (2009). *Handbook of satisfiability*. Amsterdam, etc.: IOS Press.

Bloch, E. D. (2011). *Proofs and fundamentals: A first course in abstract mathematics*. 2nd ed. New York, etc.: Springer.

Blum, M. (1967). A machine-independent theory of the complexity of recursive functions. *Journal of the Association for Computing Machinery, 14*, 322-336.

Boole, G. (1847). *The mathematical analysis of logic. Being an essay towards a calculus of deductive reasoning*. Cambridge: Macmillan, Barclay, and Macmillan.

Boole, G. (1854). *An investigation of the laws of thought, on which are founded the mathematical theories of logic and probabilities*. London: Walton and Maberly.

Börger, E., Grädel, E., & Gurevich, Y. (2001). *The classical decision problem*. Berlin, etc.: Springer.

Ceri, S., Gottlob, G., & Tanca, L. (1989). What you always wanted to know about Datalog (and never dared to ask). *IEEE Transactions on Knowledge and Data Engineering, 1*(1), 146-166.

Ceri, S., Gottlob, G., & Tanca, L. (1990). *Logic programming and databases*. Berlin & Heidelberg: Springer.

Chang, C.-L. & Lee, R. C.-T. (1973). *Symbolic logic and mechanical theorem proving*. New York & London: Academic Press.

Chomsky, N. (1956). Three models for the description of language. *IRE Transactions on Information Theory*, *2*(3), 113-124.

Chomsky, N. (1959). On certain formal properties of grammars. *Information and Control*, *2*(2), 113-124.

Church, A. (1936a). A note on the Entscheidungsproblem. *Journal of Symbolic Logic*, *1*(1), 40-41.

Church, A. (1936b). An unsolvable problem of elementary number theory. *American Journal of Mathematics*, *58*(2), 345-363.

Clark, K. L. (1978). Negation as failure. In H. Gallaire & J. Minker (eds.), *Logic and data bases* (pp. 293-322). New York: Plenum.

Cleave, J. P. (1991). *A study of logics.* Oxford: Clarendon Press.

Cormen, T. H., Leiserson, C. E., Rivest, R. L., & Stein, C. (2009). *Introduction to algorithms.* 3rd ed. Cambridge, MA & London, UK: MIT Press.

Curry, H. B. (1963). *Foundations of mathematical logic.* New York, etc.: McGraw-Hill.

D'Agostino, M. (1999). Tableau methods for classical propositional logic. In M. D'Agostino et al. (eds.), *Handbook of tableau methods* (pp. 45-123), Dordrecht: Kluwer.

Date, C. J. (2004). *Introduction to database systems.* 8th ed. Reading, MA: Addison-Wesley.

Davis, M. (2001). The early history of automated deduction. In A. Robinson & A. Voronkov (eds.), *Handbook of automated reasoning*, vol. 1 (pp. 1-15). Amsterdam: Elsevier / Cambridge, MA: MIT Press.

Davis, M. & Putnam, H. (1960). A computing procedure for quantification theory. *Journal of the ACM*, *7*(3), 201-215.

Davis, M. D. & Weyuker, E. J. (1983). *Computability, complexity, and languages. Fundamentals of theoretical computer science.* Orlando, etc.: Academic Press.

Davis, M., Logemann, G., & Loveland, D. (1962). A machine program for theorem-proving. *Communications of the ACM*, *5*(7), 394-397.

Deransart, P. & Małuszyński, J. (1993). *A grammatical view of logic programming*. Cambridge, MA: MIT Press.

Digricoli, V. J. & Harrison, M. C. (1986). Equality-based binary resolution. *Journal of the Association for Computing Machinery, 33*(2), 253-289.

Doets, K. (1994). *From logic to logic programming*. Cambridge, MA & London, England: The MIT Press.

Enderton, H. B. (2001). *A mathematical introduction to logic*. 2nd ed. San Diego, etc.: Harcourt Academic Press.

Etchemendy, J. (1999). *The concept of logical consequence*. Stanford: CSLI Publications.

Fitting, M. (1996). *First order logic and automated theorem proving*. 2nd ed. New York, etc.: Springer.

Fitting, M. (1999). Introduction. In M. D'Agostino et al. (eds.), *Handbook of tableau methods* (pp. 1-44). Dordrecht: Kluwer.

Frege, G. (1892). Über Sinn und Bedeutung. *Zeitschrift für Philosophie und philosophische Kritik, 100*(1), 25-50.

Gabbay, D. M. & Woods, J. (2003). *A practical logic of cognitive systems. Vol. 1: Agenda relevance. A study in formal pragmatics*. Amsterdam, etc.: Elsevier.

Gabbay, D. M., Hogger, C. J., & Robinson, J. A. (eds.) (1998). *Handbook of logic in artificial intelligence and logic programming. Vol. 5: Logic Programming*. Oxford: Clarendon Press.

Gallaire, H., Minker, J., & Nicolas, J.-M. (1984). Logic and databases: A deductive approach. *Computing Surveys, 16*(2), 153-185.

Gallier, J. (2011). *Discrete mathematics*. New York, etc.: Springer.

Garey, M. R. & Johnson, D. S. (1979). *Computers and intractability: A guide to the theory of NP-completeness*. New York: W. H. Freeman and Company.

Gelfond, M. & Lifschitz, V. (1988). The stable model semantics for logic programming. In R. Kowalski & K. Bowen (eds.), *Logic programming: Proceedings of the 5th international conference and symposium* (pp. 1070-1080). MIT Press.

Gentzen, G. (1934-5). Untersuchungen über das logische Schliessen. *Mathematische Zeitschrift*, *39*, 176-210, 405-431. (Engl. trans.: Investigations into logical deduction. In M. E. Szabo (ed.), *The Collected Papers of Gerhard Gentzen* (pp. 68-131). Amsterdam: North-Holland.)

Gilmore, P. (1960). A proof method for quantification theory: Its justification and realization. *IBM Journal of Research and Development*, *4*(1), 28-35.

Gödel, K. (1930). Die Vollständigkeit der Axiome des logischen Funktionkalküls. *Monatshefte für Mathematik*, *37*, 349-360. (Engl. trans.: The completeness of the axioms of the functional calculus of logic. In S. Feferman et al. (eds.), *Collected works. Vol. 1: Publications 1929-1936* (pp. 103-123). New York: OUP & Oxford: Clarendon Press, 1986.)

Gödel, K. (1931). Über formal unentscheidbare Sätze der *Principia Mathematica* und verwandter Systeme, I. *Monatshefte für Mathematik und Physik*, *38*, 173-198. (Engl. trans.: On formally undecidable propositions of *Principia Mathematica* and related systems, I. In S. Feferman et al. (eds.), *Collected works. Vol. 1: Publications 1929-1936* (pp. 144-195). New York: OUP & Oxford: Clarendon Press, 1986.)

Gödel, K. (1964). Postscriptum to Gödel (1934). In *Collected works I* (pp. 369-371), Oxford: OUP, 1986.

Greco, S. & Molinaro, C. (2016). *Datalog and logic databases*. Morgan & Claypool.

Grune, D. & Jacobs, C. J. H. (2010). *Parsing techniques: A practical guide*. 2nd ed. New York, NY: Springer.

Hähnle, R. & Schmitt, P. H. (1994). The liberalized δ-rule in free-variable semantic tableaux. *Journal of Automated Reasoning*, *13*(2), 211-221.

Henkin, L. (1949). The completeness of the first-order functional calculus. *Journal of Symbolic Logic*, *14*(3), 159-166.

Herbrand, J. (1930). *Recherches sur la théorie de la démonstration*. Thèses présentées à la Faculté des Sciences de Paris.

Hilbert, D. & Ackermann, W. (1928). *Grundzüge der theoretischen Logik*. Berlin: Springer.

Hintikka, J. (1955). Form and content in quantification theory. *Acta Philosophica Fennica, 8*, 7-55.

Hopcroft, J. E., Motwani, R., & Ullman, J. (2013). *Introduction to automata theory, languages, and computation.* 3rd ed. Boston, etc.: Pearson.

Hurley, P. J. (2012). *A concise introduction to logic.* 11th ed. Boston, MA: Wadsworth.

Jaśkowski, S. (1934). On the rules of suppositions in formal logic. *Studia Logica, 1*(1), 5-32.

Kleene, S. C. (1952). *Introduction to metamathematics.* Princeton, NJ: D. van Nostrand Co.

Kleene, S. C. (1956). Representation of events in nerve nets and finite automata. In C. E. Shannon & J. McCarthy (eds.), *Automata studies* (pp. 3-42). Princeton: Princeton University Press.

Leary, C. C. & Kristiansen, L. (2015). *A friendly introduction to mathematical logic.* 2nd ed. Geneseo, NY: Milne Library.

Leitsch, A. (1997). *The resolution calculus.* Berlin, etc.: Springer.

Letz, R. (1999). First-order tableau methods. In M. D'Agostino et al. (eds.), *Handbook of tableau methods* (pp. 125-196), Dordrecht: Kluwer.

Libkin, L. (2012). *Elements of finite model theory.* Berlin, etc.: Springer.

MacKenzie, D. (1995). The automation of proof: A historical and sociological exploration. *IEEE Annals of the History of Computing, 17*, 7-29.

Makinson, D. (2008). *Sets, logic, and maths for computing.* London: Springer.

Martin, N. M. & Pollard, S. (1996). *Closure spaces and logic.* Dordrecht: Kluwer.

Matiyasevich, Y. V. (1993). *Hilbert's Tenth Problem.* Cambridge, MA: MIT Press.

Mendelson, E. (2015). *Introduction to mathematical logic.* 6th ed. Boca Raton, FL: Taylor & Francis Group.

Minker, J. (1997). Logic and databases: Past, present, and future. *AI Magazine, 18*(3), 21-47.

Minsky, M. (1974). A framework for representing knowledge. Report AIM, 306, Artificial Intelligence Laboratory, MIT.

Newell, A. (1973). Production systems: Models of control structures. In W. G. Chase (ed.), *Visual information processing* (pp. 463-526), New York: Academic Press.

Newell, A. (1990). *Unified theories of cognition.* Cambridge, MA: Harvard University Press.

Nieuwenhuis, R. & Rubio, A. (2001). Paramodulation-based theorem proving. In A. Robinson & A. Voronkov (eds.), *Handbook of automated reasoning,* vol. 1 (pp. 371-443). Amsterdam: Elsevier / Cambridge, MA: MIT Press.

Ouyang, M. (1998). How good are branching rules in DPLL? *Discrete Applied Mathematics, 89*(1-3), 281-286.

Prawitz, D. (1965). *Natural deduction. A proof-theoretical study.* Stockholm: Almqvist & Wiksell.

Priest, G. (2008). *An introduction to non-classical logic.* Cambridge: Cambridge University Press.

Przymusinski, T. C. (1989). On the declarative and procedural semantics of logic programs. *Journal of Automated Reasoning, 5,* 167-205.

Quine, W. V. O. (1938). Completeness of the propositional calculus. *Journal of Symbolic Logic, 3*(1), 37-40.

Rahwan, I. & Simari, G. R. (eds.) (2009). *Argumentation in artificial intelligence.* Dordrecht, etc.: Springer.

Reiter, R. (1978). On closed world data bases. In H. Gallaire & J. Minker (eds.), *Logic and data bases* (pp. 55-76). New York: Plenum.

Reiter, R. (1984). Towards a logical reconstruction of relational database theory. In M. L. Brodie, J. Mylopolous, & J. W. Schmidt (eds.), *On conceptual modeling. Perspectives from artificial intelligence, databases, and programming languages* (pp. 191-238). New York: Springer.

Robinson, A. J. (1965). A machine-oriented logic based on the resolution principle. *Journal of ACM, 12*(1), 23-41.

Robinson, G. & Wos, L. (1969). Paramodulation and theorem-proving in first-order theories with equality. *Machine Intelligence, 4*, 135-150.

Scott, M. L. (2009). *Programming language pragmatics.* 3rd ed. Amsterdam, etc.: Morgan Kaufmann.

Sebesta, R. W. (2012). *Concepts of programming languages.* 10th ed. Boston, etc.: Pearson.

Shepherdson, J. C. (1984). Negation as failure: A comparison of Clark's completed data base and Reiter's closed world assumption. *Journal of Logic Programming, 1*(1), 1-48.

Siekmann, J. H. (ed.) (2014). *Handbook of the history of logic. Vol. 9: Computational logic.* Amsterdam, etc.: North-Holland, Elsevier.

Sippu, S. & Soisalon-Soininen, E. (1990). *Parsing theory. Vol. II: LR(k) and LL(k) parsing.* Berlin, Heidelberg: Springer.

Smullyan, R. M. (1968). *First-order logic.* Mineola, NY: Dover.

Stepney, S. et al. (2005). Journeys in non-classical computation I: A grand challenge for computing research. *International Journal of Parallel, Emergent and Distributed Systems, 20*(1), 5-19.

Sterling, L. & Shapiro, E. (1994). *The art of Prolog.* Cambridge, MA & London, England: The MIT Press.

Stone, M. H. (1936). The theory of representation for Boolean algebras. *Transactions of the American Mathematical Society, 40*(1), 37-111.

Tarski, A. (1930). Fundamentale Begriffe der Methodologie der deduktiven Wissenschaften. I. *Monatshefte für Mathematik und Physik, 37*, 361-404. (Engl. trans.: Fundamental concepts of the methodology of the deductive sciences. In A. Tarski, *Logic, semantics, metamathematics: Papers from 1923 to 1938* (pp. 60-109). Oxford: Clarendon Press, 1956.)

Tarski, A. (1935). Der Wahrheitsbegriff in formalisierten Sprachen. *Studia Philosophica, 1*, 261-405 (Engl. trans.: The concept of truth in formalized languages. In A. Tarski, *Logic, semantics, metamathematics: Papers from 1923 to 1938* (pp. 152-278). Trans. by J. H.

Woodger. Oxford: Clarendon Press, 1956) (Originally published in Polish in 1933.)

Tarski, A. (1994). *Introduction to logic and to the methodology of deductive sciences.* 4th ed. J. Tarski (ed.). New York & Oxford: Oxford University Press.

Troelstra, A. S. & Schwichtenberg, H. (2000). *Basic proof theory.* 2nd ed. Cambridge: Cambridge University Press.

Tseitin, G. S. (1968). On the complexity of derivations in the propositional calculus. In A. O. Slisenko (ed.), *Studies in constructive mathematics and mathematical logic. Part 2. Seminar in mathematics* (pp. 115-125). Steklov Mathematical Institute.

Turing, A. (1937). On computable numbers, with an application to the Entscheidungsproblem. *Proceedings of the London Mathematical Society, Series 2, 41*, 230-265.

van Emden, M. H. & Kowalski, R. A. (1976). The semantics of predicate logic as a programming language. *Journal of the Association for Computing Machinery, 23*(4), 733-742.

van Gelder, A. (1986). Negation as failure using tight derivations for general logic programs. In *Proceedings of the Third IEEE Symposium on Logic Programming* (pp. 137-146).

van Gelder, A., Ross, K. A., & Schlipf, J. S. (1991). The well-founded semantics for general logic programs. *Journal of the ACM, 38*(3), 620-650.

Walther, C. (1985). A mechanical solution of Schubert's Steamroller by many-sorted resolution. *Artificial Intelligence, 26*, 217-224.

Wójcicki, R. (1988). *Theory of logical calculi: Basic theory of consequence operations.* Dordrecht: Kluwer.

Younger, D. H. (1967). Recognition and parsing of context-free languages in time n^3. *Information and Control, 10*(2), 189-208.

Index

Index

www.ingramcontent.com/pod-product-compliance
Lightning Source LLC
Chambersburg PA
CBHW031345210326
41599CB00019B/2650